Student Solutions Manual

Elementary
Linear
Algebra

Applications Version   **Eighth Edition**

HOWARD ANTON
CHRIS RORRES
Drexel University

Prepared by
Elizabeth M. Grobe
Charles A. Grobe, Jr.
Bowdoin College

Chris Rorres
Drexel University

JOHN WILEY & SONS, INC.
New York Chichester Weinheim Brisbane Singapore Toronto

To order books or for customer service call 1-800-CALL-WILEY (225-5945).

ISBN 0-471-38248-5

Printed in the United States of America

10 9 8 7 6 5 4 3 2 1

Printed and bound by Quebecor Books- Fairfield

PREFACE

This manual contains detailed solutions to substantially more than half of the exercises in the Eighth Edition of *Elementary Linear Algebra, Applications Version* by Howard Anton and Chris Rorres.

Routine Exercises: Since the textbook contains answers to virtually all of the exercises, we include solutions to only a representative sample.

Less Routine Exercises: We give solutions to well over half of the less routine exercises, including a solution to at least one of every type of exercise in each problem set. Where there is more than one exercise of a given kind, we have omitted the solution to at least one problem so that it can be assigned for credit.

Solutions to most of the exercises which are new to the Eighth Edition have been included, and most of the solutions which also appeared in the *Solutions Manual* to earlier editions have been retained.

This book was typeset in LaTeX by Techsetters, Inc. We very much appreciate the excellent job that they have done.

We especially wish to thank Ms. Roseann Zappia of John Wiley & Sons for all her encouragement and assistance.

We would also like to acknowledge the excellent contributions of Professor Donald Passman of the University of Wisconsin and Professor Michael A. Carchidi of Drexel University to this Student Solutions Manual.

Elizabeth M. Grobe
Charles A. Grobe, Jr.
Chris Rorres

CONTENTS

Preface

CHAPTER THREE

CHAPTER FOUR

CHAPTER FIVE

CHAPTER SIX

CHAPTER ELEVEN

EXERCISE SET 1.1

1. **(b)** Not linear because of the term $x_1 x_3$.

   (d) Not linear because of the term x_1^{-2}.

   (e) Not linear because of the term $x_1^{3/5}$.

3. **(a)** Let $x = t$. Then $y = \dfrac{7t - 3}{5}$.

   Alternatively, let $y = s$. Then $x = \dfrac{5s + 3}{7}$.

   (c) One possibility is to let $x_1 = r$, $x_2 = s$, and $x_4 = t$. Then

   $$x_3 = \frac{-8r + 2s + 6t - 1}{5}$$

6. **(b)** Substituting the given expressions for x and y into the equation $x = 5 + 2y$ yields

   $$t = 5 + 2\left(\frac{1}{2}t - \frac{5}{2}\right) = 5 + t - 5 = t$$

   Since this equation is valid for all values of t, the proposed solution is, indeed, the general one.

   Alternatively, we let $x = t$ and solve for y, obtaining $y = \dfrac{1}{2}t - \dfrac{5}{2}$.

7. Since each of the three given points must satisfy the equation of the curve, we have the system of equations

$$ax_1^2 + bx_1 + c = y_1$$

$$ax_2^2 + bx_2 + c = y_2$$

$$ax_3^2 + bx_3 + c = y_3$$

If we consider this to be a system of equations in the three unknowns a, b, and c, the augmented matrix is clearly the one given in the exercise.

8. If the system is consistent, then we can, for instance, subtract the first two equations from the last. This yields $c - a - b = 0$ or $c = a + b$. Unless this equation holds, the system cannot have a solution, and hence cannot be consistent.

9. The solutions of $x_1 + kx_2 = c$ are $x_1 = c - kt$, $x_2 = t$ where t is any real number. If these satisfy $x_1 + \ell x_2 = d$, then $c - kt + \ell t = d$, or $(\ell - k)t = d - c$ for all real numbers t. In particular, if $t = 0$, then $d = c$, and if $t = 1$, then $\ell = k$.

10. If $x - y = 3$, then $2x - 2y = 6$. Therefore, the equations are consistent if and only if $k = 6$; that is, there are no solutions if $k \neq 6$. If $k = 6$, then the equations represent the same line, in which case, there are infinitely many solutions. Since this covers all of the possibilities, there is never a unique solution.

11. **(a)** If the system of equations fails to have a solution, then there are three possibilities: The three lines intersect in (i) 3 distinct points, (ii) 2 distinct points, or (iii) not at all.

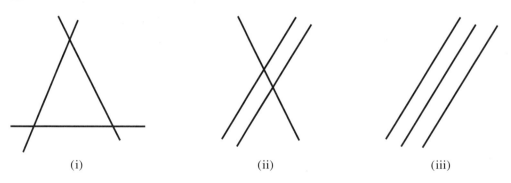

(i)   (ii)   (iii)

(b) If the system of equations has exactly one solution, then all of the lines must pass through a common point. Moreover, at least two of the lines must be distinct.

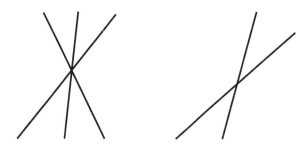

(c) If the system has infinitely many solutions, then the three lines coincide; that is, the three equations represent the same line.

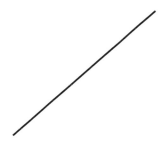

13. If $k = \ell = m = 0$, then all 3 lines must intersect at the origin.

EXERCISE SET 1.2

1. **(e)** Not in reduced row-echelon form because Property 2 is not satisfied.

   (f) Not in reduced row-echelon form because Property 3 is not satisfied.

   (g) Not in reduced row-echelon form because Property 4 is not satisfied.

2. **(c)** Not in row-echelon form because Property 1 is not satisfied.

   (d) In row-echelon form.

   (f) Not in row-echelon form because Property 1 is not satisfied.

4. **(b)** Let $x_4 = t$. Then we can read off the solution

$$x_1 = 8 + 7t$$
$$x_2 = 2 - 3t$$
$$x_3 = -5 - t$$
$$x_4 = t$$

   (c) Let $x_5 = t$. Then $x_4 = 8 - 5t$ and $x_3 = 7 - 4t$. Let $x_2 = s$. Then $x_1 = -2 - 3t + 6s$.

5. **(a)** The solution is

$$x_3 = 5$$
$$x_2 = 2 - 2x_3 = -8$$
$$x_1 = 7 - 4x_3 + 3x_2 = -37$$

5. **(b)** Let $x_4 = t$. Then $x_3 = 2 - t$. Therefore

$$x_2 = 3 + 9t - 4x_3 = 3 + 9t - 4(2 - t) = -5 + 13t$$
$$x_1 = 6 + 5t - 8x_3 = 6 + 5t - 8(2 - t) = -10 + 13t$$

6. **(a)** The augmented matrix is

$$\begin{bmatrix} 1 & 1 & 2 & 8 \\ -1 & -2 & 3 & 1 \\ 3 & -7 & 4 & 10 \end{bmatrix}$$

Replace Row 2 by Row 1 plus Row 2 and replace Row 3 by Row 3 plus -3 times Row 1.

$$\begin{bmatrix} 1 & 1 & 2 & 8 \\ 0 & -1 & 5 & 9 \\ 0 & -10 & -2 & -14 \end{bmatrix}$$

Multiply Row 2 by -1.

$$\begin{bmatrix} 1 & 1 & 2 & 8 \\ 0 & 1 & -5 & -9 \\ 0 & -10 & -2 & -14 \end{bmatrix}$$

Note that we resist the temptation to multiply Row 3 by $-1/2$. While this is sensible, it is not part of the formal elimination procedure. Replace Row 3 by Row 3 plus 10 times Row 2.

$$\begin{bmatrix} 1 & 1 & 2 & 8 \\ 0 & 1 & -5 & -9 \\ 0 & 0 & -52 & -104 \end{bmatrix}$$

Multiply Row 3 by $-1/52$.

$$
\begin{bmatrix}
1 & 1 & 2 & 8 \\
0 & 1 & -5 & -9 \\
0 & 0 & 1 & 2
\end{bmatrix}
$$

Now multiply Row 3 by 5 and add the result to Row 2.

$$
\begin{bmatrix}
1 & 1 & 2 & 8 \\
0 & 1 & 0 & 1 \\
0 & 0 & 1 & 2
\end{bmatrix}
$$

Next multiply Row 3 by -2 and add the result to Row 1.

$$
\begin{bmatrix}
1 & 1 & 0 & 4 \\
0 & 1 & 0 & 1 \\
0 & 0 & 1 & 2
\end{bmatrix}
$$

Finally, multiply Row 2 by -1 and add to Row 1.

$$
\begin{bmatrix}
1 & 0 & 0 & 3 \\
0 & 1 & 0 & 1 \\
0 & 0 & 1 & 2
\end{bmatrix}
$$

Thus the solution is $x_1 = 3$, $x_2 = 1$, $x_3 = 2$.

(c) The augmented matrix is

$$
\begin{bmatrix}
1 & -1 & 2 & -1 & -1 \\
2 & 1 & -2 & -2 & -2 \\
-1 & 2 & -4 & 1 & 1 \\
3 & 0 & 0 & -3 & -3
\end{bmatrix}
$$

Replace Row 2 by Row 2 plus -2 times Row 1, replace Row 3 by Row 3 plus Row 1, and replace Row 4 by Row 4 plus -3 times Row 1.

$$
\begin{bmatrix}
1 & -1 & 2 & -1 & -1 \\
0 & 3 & -6 & 0 & 0 \\
0 & 1 & -2 & 0 & 0 \\
0 & 3 & -6 & 0 & 0
\end{bmatrix}
$$

Multiply Row 2 by $1/3$ and then add -1 times the new Row 2 to Row 3 and add -3 times the new Row 2 to Row 4.

$$
\begin{bmatrix}
1 & -1 & 2 & -1 & -1 \\
0 & 1 & -2 & 0 & 0 \\
0 & 0 & 0 & 0 & 0 \\
0 & 0 & 0 & 0 & 0
\end{bmatrix}
$$

Finally, replace Row 1 by Row 1 plus Row 2.

$$
\begin{bmatrix}
1 & 0 & 0 & -1 & -1 \\
0 & 1 & -2 & 0 & 0 \\
0 & 0 & 0 & 0 & 0 \\
0 & 0 & 0 & 0 & 0
\end{bmatrix}
$$

Thus if we let $z = s$ and $w = t$, then we have $x = -1 + w = -1 + t$ and $y = 2z = 2s$. The solution is therefore

$$x = -1 + t$$
$$y = 2s$$
$$z = s$$
$$w = t$$

7. **(a)** In Problem 6(a), we reduced the augmented matrix to the following row-echelon matrix:

$$
\begin{bmatrix}
1 & 1 & 2 & 8 \\
0 & 1 & -5 & -9 \\
0 & 0 & 1 & 2
\end{bmatrix}
$$

By Row 3, $x_3 = 2$. Thus by Row 2, $x_2 = 5x_3 - 9 = 1$. Finally, Row 1 implies that $x_1 = -x_2 - 2x_3 + 8 = 3$. Hence the solution is

$$x_1 = 3$$
$$x_2 = 1$$
$$x_3 = 2$$

(c) According to the solution to Problem 6(c), one row-echelon form of the augmented matrix is

$$\begin{bmatrix} 1 & -1 & 2 & -1 & -1 \\ 0 & 1 & -2 & 0 & 0 \\ 0 & 0 & 0 & 0 & 0 \\ 0 & 0 & 0 & 0 & 0 \end{bmatrix}$$

Row 2 implies that $y = 2z$. Thus if we let $z = s$, we have $y = 2s$. Row 1 implies that $x = -1 + y - 2z + w$. Thus if we let $w = t$, then $x = -1 + 2s - 2s + t$ or $x = -1 + t$. Hence the solution is

$$x = -1 + t$$
$$y = 2s$$
$$z = s$$
$$w = t$$

8. **(a)** The augmented matrix is

$$\begin{bmatrix} 2 & -3 & -2 \\ 2 & 1 & 1 \\ 3 & 2 & 1 \end{bmatrix}$$

Divide Row 1 by 2.

$$\begin{bmatrix} 1 & -3/2 & -1 \\ 2 & 1 & 1 \\ 3 & 2 & 1 \end{bmatrix}$$

Replace Row 2 by Row 2 plus -2 times Row 1 and replace Row 3 by Row 3 plus -3 times Row 1.

$$\begin{bmatrix} 1 & -3/2 & -1 \\ 0 & 4 & 3 \\ 0 & 13/2 & 4 \end{bmatrix}$$

Divide Row 2 by 4. Then multiply the new Row 2 by $-13/2$ and add the result to Row 3.

$$\begin{bmatrix} 1 & -3/2 & -1 \\ 0 & 1 & 3/4 \\ 0 & 0 & -7/8 \end{bmatrix}$$

Thus, if we multiply Row 3 by $-8/7$, we obtain the following row-echelon matrix:

$$\begin{bmatrix} 1 & -3/2 & -1 \\ 0 & 1 & 3/4 \\ 0 & 0 & 1 \end{bmatrix}$$

Finally, if we complete the reduction to the reduced row-echelon form, we obtain

$$\begin{bmatrix} 1 & 0 & 0 \\ 0 & 1 & 0 \\ 0 & 0 & 1 \end{bmatrix}$$

The system of equations represented by this matrix is inconsistent because Row 3 implies that $0x_1 + 0x_2 = 0 = 1$.

8. (c) The augmented matrix is

$$\begin{bmatrix} 4 & -8 & 12 \\ 3 & -6 & 9 \\ -2 & 4 & -6 \end{bmatrix}$$

Divide Row 1 by 4.

$$\begin{bmatrix} 1 & -2 & 3 \\ 3 & -6 & 9 \\ -2 & 4 & -6 \end{bmatrix}$$

Replace Row 2 by Row 2 plus -3 times Row 1 and replace Row 3 by Row 3 plus 2 times Row 1.

$$\begin{bmatrix} 1 & -2 & 3 \\ 0 & 0 & 0 \\ 0 & 0 & 0 \end{bmatrix}$$

If we let $x_2 = t$, then $x_1 = 3 + 2x_2 = 3 + 2t$. Hence, the solution is

$$x_1 = 3 + 2t$$
$$x_2 = t$$

9. **(a)** In Problem 8(a), we reduced the augmented matrix of this system to row-echelon form, obtaining the matrix

$$\begin{bmatrix} 1 & -3/2 & -1 \\ 0 & 1 & 3/4 \\ 0 & 0 & 1 \end{bmatrix}$$

Row 3 again yields the equation $0 = 1$ and hence the system is inconsistent.

(c) In Problem 8(c), we found that one row-echelon form of the augmented matrix is

$$\begin{bmatrix} 1 & -2 & 3 \\ 0 & 0 & 0 \\ 0 & 0 & 0 \end{bmatrix}$$

Again if we let $x_2 = t$, then $x_1 = 3 + 2x_2 = 3 + 2t$.

10. (a) The augmented matrix of the system is

$$\begin{bmatrix} 5 & -2 & 6 & 0 \\ -2 & 1 & 3 & 1 \end{bmatrix}$$

Divide Row 1 by 5.

$$\begin{bmatrix} 1 & -2/5 & 6/5 & 0 \\ -2 & 1 & 3 & 1 \end{bmatrix}$$

Replace Row 2 by Row 2 plus 2 times Row 1.

$$\begin{bmatrix} 1 & -2/5 & 6/5 & 0 \\ 0 & 1/5 & 27/5 & 1 \end{bmatrix}$$

Multiply Row 2 by 5.

$$\begin{bmatrix} 1 & -2/5 & 6/5 & 0 \\ 0 & 1 & 27 & 5 \end{bmatrix}$$

Replace Row 1 by Row 1 plus 2/5 times Row 2.

$$\begin{bmatrix} 1 & 0 & 12 & 2 \\ 0 & 1 & 27 & 5 \end{bmatrix}$$

Thus, if we let $x_3 = t$, then $x_2 = 5 - 27x_3 = 5 - 27t$ and $x_1 = 2 - 12x_3 = 2 - 12t$. Therefore, the solution is

$$x_1 = 2 - 12t$$
$$x_2 = 5 - 27t$$
$$x_3 = t$$

(c) The augmented matrix of the system is

$$
\begin{bmatrix}
0 & 0 & 1 & 2 & -1 & 4 \\
0 & 0 & 0 & 1 & -1 & 3 \\
0 & 0 & 1 & 3 & -2 & 7 \\
2 & 4 & 1 & 7 & 0 & 7
\end{bmatrix}
$$

Interchange Rows 1 and 4 and multiply the new Row 1 by 1/2.

$$
\begin{bmatrix}
1 & 2 & 1/2 & 7/2 & 0 & 7/2 \\
0 & 0 & 0 & 1 & -1 & 3 \\
0 & 0 & 1 & 3 & -2 & 7 \\
0 & 0 & 1 & 2 & -1 & 4
\end{bmatrix}
$$

Interchange Rows 2 and 4. Then replace Row 3 by Row 3 plus -1 times the new Row 2.

$$
\begin{bmatrix}
1 & 2 & 1/2 & 7/2 & 0 & 7/2 \\
0 & 0 & 1 & 2 & -1 & 4 \\
0 & 0 & 0 & 1 & -1 & 3 \\
0 & 0 & 0 & 1 & -1 & 3
\end{bmatrix}
$$

Replace Row 4 by Row 4 plus -1 times Row 3.

$$
\begin{bmatrix}
1 & 2 & 1/2 & 7/2 & 0 & 7/2 \\
0 & 0 & 1 & 2 & -1 & 4 \\
0 & 0 & 0 & 1 & -1 & 3 \\
0 & 0 & 0 & 0 & 0 & 0
\end{bmatrix}
$$

Replace Row 1 by Row 1 plus $-1/2$ times Row 2.

$$
\begin{bmatrix}
1 & 2 & 0 & 5/2 & 1/2 & 3/2 \\
0 & 0 & 1 & 2 & -1 & 4 \\
0 & 0 & 0 & 1 & -1 & 3 \\
0 & 0 & 0 & 0 & 0 & 0
\end{bmatrix}
$$

Replace Row 2 by Row 2 plus -2 times Row 3 and replace Row 1 by Row 1 plus $-5/2$ times Row 3.

$$\begin{bmatrix} 1 & 2 & 0 & 0 & 3 & -6 \\ 0 & 0 & 1 & 0 & 1 & -2 \\ 0 & 0 & 0 & 1 & -1 & 3 \\ 0 & 0 & 0 & 0 & 0 & 0 \end{bmatrix}$$

Thus, if we let $y = t$ and $v = s$, we have

$$u = -6 - 2s - 3t$$
$$v = s$$
$$w = -2 - t$$
$$x = 3 + t$$
$$y = t$$

11. (a) From Problem 10(a), a row-echelon form of the augmented matrix is

$$\begin{bmatrix} 1 & -2/5 & 6/5 & 0 \\ 0 & 1 & 27 & 5 \end{bmatrix}$$

If we let $x_3 = t$, then Row 2 implies that $x_2 = 5 - 27t$. Row 1 then implies that $x_1 = (-6/5)x_3 + (2/5)x_2 = 2 - 12t$. Hence the solution is

$$x_1 = 2 - 12t$$
$$x_2 = 5 - 27t$$
$$x_3 = t$$

(c) From Problem 10(c), a row-echelon form of the augmented matrix is

$$\begin{bmatrix} 1 & 2 & 1/2 & 7/2 & 0 & 7/2 \\ 0 & 0 & 1 & 2 & -1 & 4 \\ 0 & 0 & 0 & 1 & -1 & 3 \\ 0 & 0 & 0 & 0 & 0 & 0 \end{bmatrix}$$

If we let $y = t$, then Row 3 implies that $x = 3 + t$. Row 2 then implies that

$$w = 4 - 2x + t = -2 - t.$$

Now let $v = s$. By Row 1, $u = 7/2 - 2s - (1/2)w - (7/2)x = -6 - 2s - 3t$. Thus we have the same solution which we obtained in Problem 10(c).

12. **(a)** There are more unknowns than equations. By Theorem 1.2.1, the system has infinitely many solutions.

  (d) If we let $x_2 = t$, then $x_1 = \frac{2}{3}t$. Thus the system has infinitely many solutions.

13. **(b)** The augmented matrix of the homogenous system is

$$\begin{bmatrix} 3 & 1 & 1 & 1 & 0 \\ 5 & -1 & 1 & -1 & 0 \end{bmatrix}$$

This matrix may be reduced to

$$\begin{bmatrix} 3 & 1 & 1 & 1 & 0 \\ 0 & 4 & 1 & 4 & 0 \end{bmatrix}$$

If we let $x_3 = 4s$ and $x_4 = t$, then Row 2 implies that

$$4x_2 = -4t - 4s \quad \text{or} \quad x_2 = -t - s$$

Now Row 1 implies that

$$3x_1 = -x_2 - 4s - t = t + s - 4s - t = -3s \quad \text{or} \quad x_1 = -s$$

Therefore the solution is

$$x_1 = -s$$
$$x_2 = -(t + s)$$
$$x_3 = 4s$$
$$x_4 = t$$

14. (b) The augmented matrix of the homogenous system is

$$\begin{bmatrix} 0 & 1 & 3 & -2 & 0 \\ 2 & 1 & -4 & 3 & 0 \\ 2 & 3 & 2 & -1 & 0 \\ -4 & -3 & 5 & -4 & 0 \end{bmatrix}$$

This matrix can be reduced to

$$\begin{bmatrix} 1 & 0 & -\dfrac{7}{2} & \dfrac{5}{2} & 0 \\ 0 & 1 & 3 & -2 & 0 \\ 0 & 0 & 0 & 0 & 0 \\ 0 & 0 & 0 & 0 & 0 \end{bmatrix}$$

If we let $w = 2s$ and $x = 2t$, then the above matrix yields the solution

$$u = 7s - 5t$$
$$v = -6s + 4t$$
$$w = 2s$$
$$x = 2t$$

15. (a) The augmented matrix of this system is

$$\begin{bmatrix} 2 & -1 & 3 & 4 & 9 \\ 1 & 0 & -2 & 7 & 11 \\ 3 & -3 & 1 & 5 & 8 \\ 2 & 1 & 4 & 4 & 10 \end{bmatrix}$$

Its reduced row-echelon form is

$$\begin{bmatrix} 1 & 0 & 0 & 0 & -1 \\ 0 & 1 & 0 & 0 & 0 \\ 0 & 0 & 1 & 0 & 1 \\ 0 & 0 & 0 & 1 & 2 \end{bmatrix}$$

Hence the solution is

$$I_1 = -1$$
$$I_2 = 0$$
$$I_3 = 1$$
$$I_4 = 2$$

(b) The reduced row-echelon form of the augmented matrix is

$$\begin{bmatrix} 1 & 1 & 0 & 0 & 1 & 0 \\ 0 & 0 & 1 & 0 & 1 & 0 \\ 0 & 0 & 0 & 1 & 0 & 0 \\ 0 & 0 & 0 & 0 & 0 & 0 \end{bmatrix}$$

If we let $Z_2 = s$ and $Z_5 = t$, then we obtain the solution

$$Z_1 = -s - t$$
$$Z_2 = s$$
$$Z_3 = -t$$
$$Z_4 = 0$$
$$Z_5 = t$$

16. **(a)** The augmented matrix of the system is

$$\begin{bmatrix} 2 & 1 & a \\ 3 & 6 & b \end{bmatrix}$$

Divide Row 1 by 2 and Row 2 by 3.

$$\begin{bmatrix} 1 & \dfrac{1}{2} & \dfrac{a}{2} \\[3mm] 1 & 2 & \dfrac{b}{3} \end{bmatrix}$$

Subtract Row 1 from Row 2.

$$\begin{bmatrix} 1 & \dfrac{1}{2} & \dfrac{a}{2} \\[3mm] 0 & \dfrac{3}{2} & -\dfrac{a}{2} + \dfrac{b}{3} \end{bmatrix}$$

Multiply Row 2 by 2/3.

$$\begin{bmatrix} 1 & \dfrac{1}{2} & \dfrac{a}{2} \\[3mm] 0 & 1 & -\dfrac{a}{3} + \dfrac{2b}{9} \end{bmatrix}$$

Thus

$$y = -\frac{a}{3} + \frac{2b}{9}$$

Row 1 then implies that

$$x = \frac{a}{2} - \frac{y}{2}$$

or

$$x = \frac{a}{2} - \frac{1}{2}\left(-\frac{a}{3} + \frac{2b}{9}\right) = \frac{2a}{3} - \frac{b}{9}$$

17.   The Gauss-Jordan process will reduce this system to the equations

$$x + 2y - 3z = 4$$
$$y - 2z = 10/7$$
$$(a^2 - 16)z = a - 4$$

If $a = 4$, then the last equation becomes $0 = 0$, and hence there will be infinitely many solutions — for instance, $z = t$, $y = 2t + \frac{10}{7}$, $x = -2\left(2t + \frac{10}{7}\right) + 3t + 4$. If $a = -4$, then the last equation becomes $0 = -8$, and so the system will have no solutions. Any other value of a will yield a unique solution for z and hence also for y and x.

19.   One possibility is

$$\begin{bmatrix} 1 & 3 \\ 2 & 7 \end{bmatrix} \rightarrow \begin{bmatrix} 1 & 3 \\ 0 & 1 \end{bmatrix}$$

Another possibility is

$$\begin{bmatrix} 1 & 3 \\ 2 & 7 \end{bmatrix} \rightarrow \begin{bmatrix} 2 & 7 \\ 1 & 3 \end{bmatrix} \rightarrow \begin{bmatrix} 1 & 7/2 \\ 1 & 3 \end{bmatrix} \rightarrow \begin{bmatrix} 1 & 7/2 \\ 0 & 1 \end{bmatrix}$$

20.   Let $x_1 = \sin \alpha$, $x_2 = \cos \beta$, and $x_3 = \tan \gamma$. If we solve the given system of equations for x_1, x_2, and x_3, we find that $x_1 = 1$, $x_2 = -1$, and $x_3 = 0$. Thus $\alpha = \frac{\pi}{2}$, $\beta = \pi$, and $\gamma = 0$.

21.   If we treat the given system as linear in the variables $\sin \alpha$, $\cos \beta$, and $\tan \gamma$, then the augmented matrix is

$$\begin{bmatrix} 1 & 2 & 3 & 0 \\ 2 & 5 & 3 & 0 \\ -1 & -5 & 5 & 0 \end{bmatrix}$$

This reduces to

$$\begin{bmatrix} 1 & 0 & 0 & 0 \\ 0 & 1 & 0 & 0 \\ 0 & 0 & 1 & 0 \end{bmatrix}$$

so that the solution (for α, β, γ between 0 and 2π) is

$$\sin \alpha = 0 \quad \Rightarrow \quad \alpha = 0, \ \pi, \ 2\pi$$
$$\cos \beta = 0 \quad \Rightarrow \quad \beta = \pi/2, \ 3\pi/2$$
$$\tan \gamma = 0 \quad \Rightarrow \quad \gamma = 0, \ \pi, \ 2\pi$$

That is, there are $3 \cdot 2 \cdot 3 = 18$ possible triples α, β, γ which satisfy the system of equations.

22. If (x, y) is a solution, then the first equation implies that $y = -(\lambda - 3)x$. If we then substitute this value for y into the second equation, we obtain $[1 - (\lambda - 3)^2]x = 0$. But $[1 - (\lambda - 3)^2]x = 0$ implies that either $x = 0$ or $1 - (\lambda - 3)^2 = 0$. Now if $x = 0$, then either of the given equations implies that $y = 0$; that is, if $x = 0$, then the system of equations has only the trivial solution. On the other hand, if $1 - (\lambda - 3)^2 = 0$, then either $\lambda = 2$ or $\lambda = 4$. For either of these values of λ, the two given equations are identical — that is, they represent exactly the same line — and, hence, there are infinitely many solutions.

23. If $\lambda = 2$, the system becomes

$$- x_2 = 0$$
$$2x_1 - 3x_2 + x_3 = 0$$
$$-2x_1 + 2x_2 - x_3 = 0$$

Thus $x_2 = 0$ and the third equation becomes -1 times the second. If we let $x_1 = t$, then $x_3 = -2t$.

24. We regard this as a system of linear equations in the variables $1/x$, $1/y$, and $1/z$. The augmented matrix for this system of equations is therefore

$$\begin{bmatrix} 1 & 2 & -4 & 1 \\ 2 & 3 & 8 & 0 \\ -1 & 9 & 10 & 5 \end{bmatrix}$$

which reduces to

$$
\begin{bmatrix}
1 & 0 & 0 & -7/13 \\
0 & 1 & 0 & 54/91 \\
0 & 0 & 1 & -8/91
\end{bmatrix}
$$

Hence, the solution is $x = -13/7$, $y = 91/54$, and $z = -91/8$.

25. Using the given points, we obtain the equations

$$
\begin{aligned}
d &= 10 \\
a + \quad b + \quad c + d &= 7 \\
27a + \quad 9b + 3c + d &= -11 \\
64a + 16b + 4c + d &= -14
\end{aligned}
$$

If we solve this system, we find that $a = 1$, $b = -6$, $c = 2$, and $d = 10$.

26. Using the given points, we obtain the equations

$$
\begin{aligned}
41a - 4b + 5c + d &= 0 \\
53a - 2b + 7c + d &= 0 \\
25a + 4b - 3c + d &= 0
\end{aligned}
$$

This homogeneous system with more unknowns than equations will have infinitely many solutions. Since we know that $a \neq 0$ (otherwise we'd have a line and not a circle), we can expect to solve for the other variables in terms of a. Since the augmented matrix reduces to

$$
\begin{bmatrix}
29 & 0 & 0 & 1 & 0 \\
4 & 0 & 1 & 0 & 0 \\
2 & 1 & 0 & 0 & 0
\end{bmatrix}
$$

we have $d = -29a$, $c = -4a$, and $b = -2a$. If we let $a = 1$, the equation of the circle becomes $x^2 + y^2 - 2x - 4y - 29 = 0$. Any other choice of a will yield a multiple of this equation.

27. **(a)** If $a \neq 0$, then the reduction can be accomplished as follows:

$$\begin{bmatrix} a & b \\ c & d \end{bmatrix} \rightarrow \begin{bmatrix} 1 & \dfrac{b}{a} \\ c & d \end{bmatrix} \rightarrow \begin{bmatrix} 1 & \dfrac{b}{a} \\ 0 & \dfrac{ad-bc}{a} \end{bmatrix} \rightarrow \begin{bmatrix} 1 & \dfrac{b}{a} \\ 0 & 1 \end{bmatrix} \rightarrow \begin{bmatrix} 1 & 0 \\ 0 & 1 \end{bmatrix}$$

If $a = 0$, then $b \neq 0$ and $c \neq 0$, so the reduction can be carried out as follows:

$$\begin{bmatrix} 0 & b \\ c & d \end{bmatrix} \rightarrow \begin{bmatrix} c & d \\ 0 & b \end{bmatrix} \rightarrow \begin{bmatrix} 1 & \dfrac{d}{c} \\ 0 & b \end{bmatrix} \rightarrow \begin{bmatrix} 1 & \dfrac{d}{c} \\ 0 & 1 \end{bmatrix} \rightarrow \begin{bmatrix} 1 & 0 \\ 0 & 1 \end{bmatrix}$$

Where did you use the fact that $ad - bc \neq 0$? (This proof uses it twice.)

29. There are eight possibilities. They are

$$\begin{bmatrix} 1 & 0 & 0 \\ 0 & 1 & 0 \\ 0 & 0 & 1 \end{bmatrix} \begin{bmatrix} 1 & 0 & r \\ 0 & 1 & s \\ 0 & 0 & 0 \end{bmatrix} \begin{bmatrix} 1 & r & 0 \\ 0 & 0 & 1 \\ 0 & 0 & 0 \end{bmatrix} \begin{bmatrix} 1 & r & s \\ 0 & 0 & 0 \\ 0 & 0 & 0 \end{bmatrix}$$

$$\begin{bmatrix} 0 & 1 & 0 \\ 0 & 0 & 1 \\ 0 & 0 & 0 \end{bmatrix} \begin{bmatrix} 0 & 1 & r \\ 0 & 0 & 0 \\ 0 & 0 & 0 \end{bmatrix} \begin{bmatrix} 0 & 0 & 1 \\ 0 & 0 & 0 \\ 0 & 0 & 0 \end{bmatrix} \begin{bmatrix} 0 & 0 & 0 \\ 0 & 0 & 0 \\ 0 & 0 & 0 \end{bmatrix}$$

where r and s can be any real numbers.

30. **(a)** If the system of equations has only the trivial solution, then the three lines all pass through $(0,0)$ and at least two of them are distinct.

(b) Nontrivial solutions can exist only if the three lines coincide; that is, the three equations represent the same line and that line passes through $(0,0)$.

31. **(a)** False. The reduced row-echelon form of a matrix is unique, as stated in the remark in this section.

(b) True. The row-echelon form of a matrix is not unique, as shown in the following example:

$$\begin{bmatrix} 1 & 2 \\ 1 & 3 \end{bmatrix} \rightarrow \begin{bmatrix} 1 & 2 \\ 0 & 1 \end{bmatrix}$$

but

$$\begin{bmatrix} 1 & 2 \\ 1 & 3 \end{bmatrix} \rightarrow \begin{bmatrix} 1 & 3 \\ 1 & 2 \end{bmatrix} \rightarrow \begin{bmatrix} 1 & 3 \\ 0 & -1 \end{bmatrix} \rightarrow \begin{bmatrix} 1 & 3 \\ 0 & 1 \end{bmatrix}$$

(c) False. If the reduced row-echelon form of the augmented matrix for a system of 3 equations in 2 unknowns is

$$\begin{bmatrix} 1 & 0 & a \\ 0 & 1 & b \\ 0 & 0 & 0 \end{bmatrix}$$

then the system has a unique solution. If the augmented matrix of a system of 3 equations in 3 unknowns reduces to

$$\begin{bmatrix} 1 & 1 & 1 & 0 \\ 0 & 0 & 0 & 1 \\ 0 & 0 & 0 & 0 \end{bmatrix}$$

then the system has no solutions.

(d) False. The system can have a solution only if the 3 lines meet in at least one point which is common to all 3.

32. **(a)** False. Let 2 of the equations be $x_1+x_2+x_3+x_4+x_5 = 0$ and $x_1+x_2+x_3+x_4+x_5 = 1$ and the system cannot be consistent.

(b) False. Let all 5 equations be the same.

(c) False. The nth row might be n zeros followed by a 1. In that case, the system would have no solution.

(d) False. If one equation is a multiple of another, those 2 equations represent the same solution set. Thus, 1 of the 2 equations can be disregarded and we will have a system of $n-1$ equations in n unknowns. Such a system need not be inconsistent.

EXERCISE SET 1.3

1. **(c)** The matrix AE is 4×4. Since B is 4×5, $AE + B$ is not defined.

   (e) The matrix $A + B$ is 4×5. Since E is 5×4, $E(A + B)$ is 5×5.

   (h) Since A^T is 5×4 and E is 5×4, their sum is also 5×4. Thus $(A^T + E)D$ is 5×2.

2. Since two matrices are equal if and only if their corresponding entries are equal, we have the system of equations

$$
\begin{aligned}
a - b & & = 8 \\
b + c & & = 1 \\
c + 3d & = 7 \\
2a & - 4d & = 6
\end{aligned}
$$

   The augmented matrix of this system may be reduced to

$$
\begin{bmatrix}
1 & 0 & 0 & 0 & 5 \\
0 & 1 & 0 & 0 & -3 \\
0 & 0 & 1 & 0 & 4 \\
0 & 0 & 0 & 1 & 1
\end{bmatrix}
$$

   Hence, $a = 5$, $b = -3$, $c = 4$, and $d = 1$.

3. **(e)** Since $2B$ is a 2×2 matrix and C is a 2×3 matrix, $2B - C$ is not defined.

3. **(g)** We have

$$-3(D + 2E) = -3 \left(\begin{bmatrix} 1 & 5 & 2 \\ -1 & 0 & 1 \\ 3 & 2 & 4 \end{bmatrix} + \begin{bmatrix} 12 & 2 & 6 \\ -2 & 2 & 4 \\ 8 & 2 & 6 \end{bmatrix} \right)$$

$$= -3 \begin{bmatrix} 13 & 7 & 8 \\ -3 & 2 & 5 \\ 11 & 4 & 10 \end{bmatrix} = \begin{bmatrix} -39 & -21 & -24 \\ 9 & -6 & -15 \\ -33 & -12 & -30 \end{bmatrix}$$

(j) We have $tr(D - 3E) = (1 - 3(6)) + (0 - 3(1)) + (4 - 3(3)) = -25$.

4. **(b)** We have

$$D^T - E^T = \begin{bmatrix} 1 & -1 & 3 \\ 5 & 0 & 2 \\ 2 & 1 & 4 \end{bmatrix} - \begin{bmatrix} 6 & -1 & 4 \\ 1 & 1 & 1 \\ 3 & 2 & 3 \end{bmatrix} = \begin{bmatrix} -5 & 0 & -1 \\ 4 & -1 & 1 \\ -1 & -1 & 1 \end{bmatrix}$$

(d) Since B^T is a 2×2 matrix and $5C^T$ is a 3×2 matrix, the indicated addition is impossible.

(h) We have

$$(2E^T - 3D^T)^T = \left(\begin{bmatrix} 12 & -2 & 8 \\ 2 & 2 & 2 \\ 6 & 4 & 6 \end{bmatrix} - \begin{bmatrix} 3 & -3 & 9 \\ 15 & 0 & 6 \\ 6 & 3 & 12 \end{bmatrix} \right)^T$$

$$= \begin{bmatrix} 9 & 1 & -1 \\ -13 & 2 & -4 \\ 0 & 1 & -6 \end{bmatrix} = \begin{bmatrix} 9 & -13 & 0 \\ 1 & 2 & 1 \\ -1 & -4 & -6 \end{bmatrix}$$

5. **(b)** Since B is a 2×2 matrix and A is a 3×2 matrix, BA is not defined (although AB is).

(d) We have

$$AB = \begin{bmatrix} 12 & -3 \\ -4 & 5 \\ 4 & 1 \end{bmatrix}$$

Hence

$$(AB)C = \begin{bmatrix} 3 & 45 & 9 \\ 11 & -11 & 17 \\ 7 & 17 & 13 \end{bmatrix}$$

(e) We have

$$A(BC) = \begin{bmatrix} 3 & 0 \\ -1 & 2 \\ 1 & 1 \end{bmatrix} \begin{bmatrix} 1 & 15 & 3 \\ 6 & 2 & 10 \end{bmatrix} = \begin{bmatrix} 3 & 45 & 9 \\ 11 & -11 & 17 \\ 7 & 17 & 13 \end{bmatrix}$$

(f) We have

$$CC^T = \begin{bmatrix} 1 & 4 & 2 \\ 3 & 1 & 5 \end{bmatrix} \begin{bmatrix} 1 & 3 \\ 4 & 1 \\ 2 & 5 \end{bmatrix} = \begin{bmatrix} 21 & 17 \\ 17 & 35 \end{bmatrix}$$

(j) We have $tr(4E^T - D) = tr(4E - D) = (4(6) - 1) + (4(1) - 0) + (4(3) - 4) = 35$.

6. **(a)** We have

$$(2D^T - E)A = \left(\begin{bmatrix} 2 & -2 & 6 \\ 10 & 0 & 4 \\ 4 & 2 & 8 \end{bmatrix} - \begin{bmatrix} 6 & 1 & 3 \\ -1 & 1 & 2 \\ 4 & 1 & 3 \end{bmatrix} \right) \begin{bmatrix} 3 & 0 \\ -1 & 2 \\ 1 & 1 \end{bmatrix}$$

$$= \begin{bmatrix} -4 & -3 & 3 \\ 11 & -1 & 2 \\ 0 & 1 & 5 \end{bmatrix} \begin{bmatrix} 3 & 0 \\ -1 & 2 \\ 1 & 1 \end{bmatrix}$$

$$= \begin{bmatrix} -6 & -3 \\ 36 & 0 \\ 4 & 7 \end{bmatrix}$$

6. (b) Since $(4B)C$ is a 2×3 matrix and $2B$ is a 2×2 matrix, then $(4B)C + 2B$ is undefined.

(d)

$$BA^T - 2C = \begin{bmatrix} 4 & -1 \\ 0 & 2 \end{bmatrix} \begin{bmatrix} 3 & -1 & 1 \\ 0 & 2 & 1 \end{bmatrix} - 2 \begin{bmatrix} 1 & 4 & 2 \\ 3 & 1 & 5 \end{bmatrix}$$

$$= \begin{bmatrix} 12 & -6 & 3 \\ 0 & 4 & 2 \end{bmatrix} - \begin{bmatrix} 2 & 8 & 4 \\ 6 & 2 & 10 \end{bmatrix}$$

$$= \begin{bmatrix} 10 & -14 & -1 \\ -6 & 2 & -8 \end{bmatrix}$$

Thus,

$$(BA^T - 2C)^T = \begin{bmatrix} 10 & -6 \\ -14 & 2 \\ -1 & -8 \end{bmatrix}$$

(f) Each of the matrices $D^T E^T$ and $(ED)^T$ is equal to

$$\begin{bmatrix} 14 & 4 & 12 \\ 36 & -1 & 26 \\ 25 & 7 & 21 \end{bmatrix}$$

Thus their difference is the 3×3 matrix with all zero entries.

7. (a) The first row of A is

$$A_1 = \begin{bmatrix} 3 & -2 & 7 \end{bmatrix}$$

Thus, the first row of AB is

$$A_1 B = \begin{bmatrix} 3 & -2 & 7 \end{bmatrix} \begin{bmatrix} 6 & -2 & 4 \\ 0 & 1 & 3 \\ 7 & 7 & 5 \end{bmatrix}$$

$$= \begin{bmatrix} 67 & 41 & 41 \end{bmatrix}$$

(c) The second column of B is

$$B_2 = \begin{bmatrix} -2 \\ 1 \\ 7 \end{bmatrix}$$

Thus, the second column of AB is

$$AB_2 = \begin{bmatrix} 3 & -2 & 7 \\ 6 & 5 & 4 \\ 0 & 4 & 9 \end{bmatrix} \begin{bmatrix} -2 \\ 1 \\ 7 \end{bmatrix} = \begin{bmatrix} 41 \\ 21 \\ 67 \end{bmatrix}$$

(e) The third row of A is

$$A_3 = \begin{bmatrix} 0 & 4 & 9 \end{bmatrix}$$

Thus, the third row of AA is

$$A_3 A = \begin{bmatrix} 0 & 4 & 9 \end{bmatrix} \begin{bmatrix} 3 & -2 & 7 \\ 6 & 5 & 4 \\ 0 & 4 & 9 \end{bmatrix}$$

$$= \begin{bmatrix} 24 & 56 & 97 \end{bmatrix}$$

8. (b) For instance, the 2nd column of BA is the matrix

$$-2 \begin{bmatrix} 6 \\ 0 \\ 7 \end{bmatrix} + 5 \begin{bmatrix} -2 \\ 1 \\ 7 \end{bmatrix} + 4 \begin{bmatrix} 4 \\ 3 \\ 5 \end{bmatrix}$$

which is obtained by using the entries from the 2nd column of A as coefficients in the linear combination of the column matrices of B.

9. The product $\mathbf{y}A$ is the matrix

$$[y_1 a_{11} + y_2 a_{21} + \cdots + y_m a_{m1} \qquad y_1 a_{12} + y_2 a_{22} + \cdots + y_m a_{m2} \qquad \cdots$$
$$y_1 a_{1n} + y_2 a_{2n} + \cdots + y_m a_{mn}]$$

We can rewrite this matrix in the form

$$y_1 \begin{bmatrix} a_{11} & a_{12} & \cdots & a_{1n} \end{bmatrix} + y_2 \begin{bmatrix} a_{21} & a_{22} & \cdots & a_{2n} \end{bmatrix} + \cdots + y_m \begin{bmatrix} a_{m1} & a_{m2} & \cdots & a_{mn} \end{bmatrix}$$

which is, indeed, a linear combination of the row matrices of A with the scalar coefficients of $\mathbf{y}$.

10. (a) By the result of Exercise 9, the rows of AB are the matrices

$$3 \begin{bmatrix} 6 & -2 & 4 \end{bmatrix} - 2 \begin{bmatrix} 0 & 1 & 3 \end{bmatrix} + 7 \begin{bmatrix} 7 & 7 & 5 \end{bmatrix}$$
$$6 \begin{bmatrix} 6 & -2 & 4 \end{bmatrix} + 5 \begin{bmatrix} 0 & 1 & 3 \end{bmatrix} + 4 \begin{bmatrix} 7 & 7 & 5 \end{bmatrix}$$
$$0 \begin{bmatrix} 6 & -2 & 4 \end{bmatrix} + 4 \begin{bmatrix} 0 & 1 & 3 \end{bmatrix} + 9 \begin{bmatrix} 7 & 7 & 5 \end{bmatrix}$$

11. Let f_{ij} denote the entry in the ith row and j^{th} column of $C(DE)$. We are asked to find f_{23}. In order to compute f_{23}, we must calculate the elements in the second row of C and the third column of DE. According to Equation (3), we can find the elements in the third column of DE by computing DE_3 where E_3 is the third column of E. That is,

$$f_{23} = \begin{bmatrix} 3 & 1 & 5 \end{bmatrix} \left(\begin{bmatrix} 1 & 5 & 2 \\ -1 & 0 & 1 \\ 3 & 2 & 4 \end{bmatrix} \begin{bmatrix} 3 \\ 2 \\ 3 \end{bmatrix} \right)$$

$$= \begin{bmatrix} 3 & 1 & 5 \end{bmatrix} \begin{bmatrix} 19 \\ 0 \\ 25 \end{bmatrix} = 182$$

12. (a) Suppose that A is $m \times n$ and B is $r \times s$. If AB is defined, then $n = r$. That is, B is $n \times s$. On the other hand, if BA is defined, then $s = m$. That is, B is $n \times m$. Hence AB is $m \times m$ and BA is $n \times n$, so that both are square matrices.

(b) Suppose that A is $m \times n$ and B is $r \times s$. If BA is defined, then $s = m$ and BA is $r \times n$. If $A(BA)$ is defined, then $n = r$. Thus B is an $n \times m$ matrix.

15. **(a)** By block multiplication,

$$AB = \left[\begin{array}{c|c} \begin{bmatrix} -1 & 2 \\ 0 & -3 \end{bmatrix}\begin{bmatrix} 2 & 1 \\ -3 & 5 \end{bmatrix} + \begin{bmatrix} 1 & 5 \\ 4 & 2 \end{bmatrix}\begin{bmatrix} 7 & -1 \\ 0 & 3 \end{bmatrix} & \begin{bmatrix} -1 & 2 \\ 0 & -3 \end{bmatrix}\begin{bmatrix} 4 \\ 2 \end{bmatrix} + \begin{bmatrix} 1 & 5 \\ 4 & 2 \end{bmatrix}\begin{bmatrix} 5 \\ -3 \end{bmatrix} \\ \hline \begin{bmatrix} 1 & 5 \end{bmatrix}\begin{bmatrix} 2 & 1 \\ -3 & 5 \end{bmatrix} + \begin{bmatrix} 6 & 1 \end{bmatrix}\begin{bmatrix} 7 & -1 \\ 0 & 3 \end{bmatrix} & \begin{bmatrix} 1 & 5 \end{bmatrix}\begin{bmatrix} 4 \\ 2 \end{bmatrix} + \begin{bmatrix} 6 & 1 \end{bmatrix}\begin{bmatrix} 5 \\ -3 \end{bmatrix} \end{array} \right]$$

$$= \left[\begin{array}{c|c} \begin{bmatrix} -8 & 9 \\ 9 & -15 \end{bmatrix} + \begin{bmatrix} 7 & 14 \\ 28 & 2 \end{bmatrix} & \begin{bmatrix} 0 \\ -6 \end{bmatrix} + \begin{bmatrix} -10 \\ 14 \end{bmatrix} \\ \hline \begin{bmatrix} -13 & 26 \end{bmatrix} + \begin{bmatrix} 42 & -3 \end{bmatrix} & \begin{bmatrix} 14 \end{bmatrix} + \begin{bmatrix} 27 \end{bmatrix} \end{array} \right]$$

$$= \left[\begin{array}{cc|c} -1 & 23 & -10 \\ 37 & -13 & 8 \\ \hline 29 & 23 & 41 \end{array} \right]$$

16. **(b)** Call the matrices A and B. Then

$$AB = \left[\begin{array}{c|c} \begin{bmatrix} 2 & -5 \\ 1 & 3 \\ 0 & 5 \end{bmatrix}\begin{bmatrix} 2 & -1 & 3 \\ 0 & 1 & 5 \end{bmatrix} & \begin{bmatrix} 2 & -5 \\ 1 & 3 \\ 0 & 5 \end{bmatrix}\begin{bmatrix} -4 \\ 7 \end{bmatrix} \\ \hline \begin{bmatrix} 1 & 4 \end{bmatrix}\begin{bmatrix} 2 & -1 & 3 \\ 0 & 1 & 5 \end{bmatrix} & \begin{bmatrix} 1 & 4 \end{bmatrix}\begin{bmatrix} -4 \\ 7 \end{bmatrix} \end{array} \right]$$

$$= \left[\begin{array}{ccc|c} 4 & -7 & -19 & -43 \\ 2 & 2 & 18 & 17 \\ 0 & 5 & 25 & 35 \\ \hline 2 & 3 & 23 & 24 \end{array} \right]$$

17. **(a)** The partitioning of A and B makes them each effectively 2×2 matrices, so block multiplication might be possible. However, if

$$A = \begin{bmatrix} A_{11} & A_{12} \\ A_{21} & A_{22} \end{bmatrix} \quad \text{and} \quad B = \begin{bmatrix} B_{11} & B_{12} \\ B_{21} & B_{22} \end{bmatrix}$$

then the products $A_{11}B_{11}$, $A_{12}B_{21}$, $A_{11}B_{12}$, $A_{12}B_{22}$, $A_{21}B_{11}$, $A_{22}B_{21}$, $A_{21}B_{12}$, and $A_{22}B_{22}$ are *all* undefined. If even *one* of these is undefined, block multiplication is impossible.

18. **(a)** The entry in the ith row and j^{th} column of AB is

$$a_{i1}b_{1j} + a_{i2}b_{2j} + \cdots + a_{in}b_{nj}$$

Thus, if the ith row of A consists entirely of zeros, i.e., if $a_{ik} = 0$ for i fixed and for $k = 1, 2, \ldots, n$, then the entry in the ith row and the jth column of AB is zero for fixed i and for $j = 1, 2, \ldots, n$. That is, if the ith row of A consists entirely of zeros, then so does the ith row of AB.

20. The element in the ith row and j^{th} column of I is

$$\delta_{ij} = \begin{cases} 0 & \text{if} \quad i \neq j \\ 1 & \text{if} \quad i = j \end{cases}$$

The element in the ith row and j^{th} column of AI is

$$a_{i1}\delta_{1j} + a_{i2}\delta_{2j} + \cdots + a_{in}\delta_{nj}$$

But the above sum equals a_{ij} because $\delta_{kj} = 0$ if $k \neq j$ and $\delta_{jj} = 1$. This proves that $AI = A$.

The proof that $IA = A$ is similar.

21. **(b)** If $i > j$, then the entry a_{ij} has row number larger than column number; that is, it lies below the matrix diagonal. Thus $[a_{ij}]$ has all zero elements below the diagonal.

(d) If $|i - j| > 1$, then either $i - j > 1$ or $i - j < -1$; that is, either $i > j + 1$ or $j > i + 1$. The first of these inequalities says that the entry a_{ij} lies below the diagonal and also below the "subdiagonal" consisting of all entries immediately below the diagonal ones. The second inequality says that the entry a_{ij} lies above the diagonal and also above the entries immediately above the diagonal ones. Thus we have

$$[a_{ij}] = \begin{bmatrix} a_{11} & a_{12} & 0 & 0 & 0 & 0 \\ a_{21} & a_{22} & a_{23} & 0 & 0 & 0 \\ 0 & a_{32} & a_{33} & a_{34} & 0 & 0 \\ 0 & 0 & a_{43} & a_{44} & a_{45} & 0 \\ 0 & 0 & 0 & a_{54} & a_{55} & a_{56} \\ 0 & 0 & 0 & 0 & a_{65} & a_{66} \end{bmatrix}$$

22. **(b)** For example, $a_{11} = 1^{1-1} = 1^0 = 1$, $a_{13} = 1^{3-1} = 1^2 = 1$, $a_{23} = 2^{3-1} = 2^2 = 4$, and $a_{32} = 3^{2-1} = 3^1 = 3$. Thus the matrix is

$$\begin{bmatrix} 1 & 1 & 1 & 1 \\ 1 & 2 & 4 & 8 \\ 1 & 3 & 9 & 27 \\ 1 & 4 & 16 & 64 \end{bmatrix}$$

23. Let $A = [a_{ij}]$ and $B = [b_{ij}]$. Then $\mathrm{tr}(A) = a_{11} + a_{22} + \cdots + a_{nn}$ and $\mathrm{tr}(B) = b_{11} + b_{22} + \cdots + b_{nn}$. Since $A + B = [a_{ij} + b_{ij}]$, then $\mathrm{tr}(A + B) = (a_{11} + b_{11}) + (a_{22} + b_{22}) + \cdots + (a_{nn} + b_{nn})$. Thus $\mathrm{tr}(A + B) = \mathrm{tr}(A) + \mathrm{tr}(B)$.

25.   If $A = [a_{ij}]$, then we have

$$a_{11}x + a_{12}y + a_{13}z = x + y$$
$$a_{21}x + a_{22}y + a_{23}z = x - y$$
$$a_{31}x + a_{32}y + a_{33}z = 0$$

The only solution to this system of equations is, by inspection,

$$A = \begin{bmatrix} 1 & 1 & 0 \\ 1 & -1 & 0 \\ 0 & 0 & 0 \end{bmatrix}$$

27.   (a) Let $B = \begin{bmatrix} a & b \\ c & d \end{bmatrix}$. Then $B^2 = A$ implies that

$$(*) \qquad \begin{matrix} a^2 + bc = 2 & \qquad ab + bd = 2 \\ ac + cd = 2 & \qquad bc + d^2 = 2 \end{matrix}$$

One might note that $a = b = c = d = 1$ and $a = b = c = d = -1$ satisfy $(*)$. Solving the first and last of the above equations simultaneously yields $a^2 = d^2$. Thus $a = \pm d$. Solving the remaining 2 equations yields $c(a + d) = b(a + d) = 2$. Therefore $a \neq -d$ and a and d cannot both be zero. Hence we have $a = d \neq 0$, so that $ac = ab = 1$, or $b = c = 1/a$. The first equation in $(*)$ then becomes $a^2 + 1/a^2 = 2$ or $a^4 - 2a^2 + 1 = 0$. Thus $a = \pm 1$. That is,

$$\begin{bmatrix} 1 & 1 \\ 1 & 1 \end{bmatrix} \quad \text{and} \quad \begin{bmatrix} -1 & -1 \\ -1 & -1 \end{bmatrix}$$

are the only square roots of A.

(b) Using the reasoning and the notation of Part (a), show that either $a = -d$ or $b = c = 0$. If $a = -d$, then $a^2 + bc = 5$ and $bc + a^2 = 9$. This is impossible, so we have $b = c = 0$. This implies that $a^2 = 5$ and $d^2 = 9$. Thus

$$\begin{bmatrix} \sqrt{5} & 0 \\ 0 & 3 \end{bmatrix} \begin{bmatrix} -\sqrt{5} & 0 \\ 0 & 3 \end{bmatrix} \begin{bmatrix} \sqrt{5} & 0 \\ 0 & -3 \end{bmatrix} \begin{bmatrix} -\sqrt{5} & 0 \\ 0 & -3 \end{bmatrix}$$

are the 4 square roots of A.

Note that if A were $\begin{bmatrix} 5 & 0 \\ 0 & 5 \end{bmatrix}$, say, then $B = \begin{bmatrix} 1 & r \\ 4/r & -1 \end{bmatrix}$ would be a square root of A for every nonzero real number r and there would be infinitely many other square roots as well.

(c) By an argument similar to the above, show that if, for instance,

$$A = \begin{bmatrix} -1 & 0 \\ 0 & 1 \end{bmatrix} \quad \text{and} \quad B = \begin{bmatrix} a & b \\ c & d \end{bmatrix}$$

where $BB = A$, then either $a = -d$ or $b = c = 0$. Each of these alternatives leads to a contradiction. Why?

29. **(a)** True. If A is an $m \times n$ matrix, then A^T is $n \times m$. Thus AA^T is $m \times m$ and $A^T A$ is $n \times n$. Since the trace is defined for every square matrix, the result follows.

(b) True. Partition A into its row matrices, so that

$$A = \begin{bmatrix} r_1 \\ r_2 \\ \vdots \\ r_m \end{bmatrix} \quad \text{and} \quad A^T = \begin{bmatrix} r_1^T & r_2^T & \cdots & r_m^T \end{bmatrix}$$

Then

$$AA^T = \begin{bmatrix} r_1 r_1^T & r_1 r_2^T & \cdots & r_1 r_m^T \\ r_2 r_1^T & r_2 r_2^T & \cdots & r_2 r_m^T \\ \vdots & \vdots & & \vdots \\ r_m r_1^T & r_m r_2^T & \cdots & r_m r_m^T \end{bmatrix}$$

Since each of the rows r_i is a $1 \times n$ matrix, each r_i^T is an $n \times 1$ matrix, and therefore each matrix $r_i\, r_j^T$ is a 1×1 matrix. Hence

$$\text{tr}(AA^T) = r_1\, r_1^T + r_2\, r_2^T + \cdots + r_m r_m^T$$

Note that since $r_i\, r_i^T$ is just the sum of the squares of the entries in the ith row of A, $r_1\, r_1^T + r_2\, r_2^T + \cdots + r_m r_m^T$ is the sum of the squares of *all* of the entries of A.

A similar argument works for $A^T A$, and since the sum of the squares of the entries of A^T is the same as the sum of the squares of the entries of A, the result follows.

29. **(c)** False. For instance, let $A = \begin{bmatrix} 0 & 1 \\ 0 & 1 \end{bmatrix}$ and $B = \begin{bmatrix} 1 & 1 \\ 1 & 1 \end{bmatrix}$.

(d) True. Every entry in the first row of AB is the matrix product of the first row of A with a column of B. If the first row of A has all zeros, then this product is zero.

30. **(a)** True. If the rows of A are $r_1, \cdots, r_n$ and the columns are $c_1, \cdots, c_n$, then the ith row of AA is given by $r_i c_1 \cdots r_i c_n$. Thus if $r_i = r_j$ for some $i \neq j$, then the ith row of AA will equal its jth row.

(b) False. For instance if

$$A = \begin{bmatrix} 1 & -1 & 1 \\ 0 & 0 & 1 \\ 0 & 0 & 0 \end{bmatrix} \quad \text{then} \quad A^2 = \begin{bmatrix} 1 & -1 & 0 \\ 0 & 0 & 0 \\ 0 & 0 & 0 \end{bmatrix}$$

(c) True. If every entry of B is a positive even integer, we can write $B = [b_{ij}] = [2c_{ij}]$ where each entry c_{ij} is a positive integer. If $A = [a_{ij}]$ where each a_{ij} is also a positive integer, then each entry of AB and of BA is the sum of numbers of the form $2c_{ij}a_{kl}$. Thus each will be an even positive integer.

(d) True. If A is $m \times n$ and B is $r \times s$, then for AB to be defined we must have $n = r$ and for BA to be defined, we must have $s = m$. Thus AB is $m \times m$ and BA is $n \times n$. But for $AB + BA$ to be defined, we must have $m = n$. Thus A and B must both be square $m \times m$ matrices.

EXERCISE SET 1.4

1. (a) We have

$$A + B = \begin{bmatrix} 10 & -4 & -2 \\ 0 & 5 & 7 \\ 2 & -6 & 10 \end{bmatrix}$$

Hence,

$$(A + B) + C = \begin{bmatrix} 10 & -4 & -2 \\ 0 & 5 & 7 \\ 2 & -6 & 10 \end{bmatrix} + \begin{bmatrix} 0 & -2 & 3 \\ 1 & 7 & 4 \\ 3 & 5 & 9 \end{bmatrix}$$

$$= \begin{bmatrix} 10 & -6 & 1 \\ 1 & 12 & 11 \\ 5 & -1 & 19 \end{bmatrix}$$

On the other hand,

$$B + C = \begin{bmatrix} 8 & -5 & -2 \\ 1 & 8 & 6 \\ 7 & -2 & 15 \end{bmatrix}$$

Hence,

$$A + (B + C) = \begin{bmatrix} 2 & -1 & 3 \\ 0 & 4 & 5 \\ -2 & 1 & 4 \end{bmatrix} + \begin{bmatrix} 8 & -5 & -2 \\ 1 & 8 & 6 \\ 7 & -2 & 15 \end{bmatrix} = \begin{bmatrix} 10 & -6 & 1 \\ 1 & 12 & 11 \\ 5 & -1 & 19 \end{bmatrix}$$

1. **(c)** Since $a + b = -3$, we have

$$(a + b)C = (-3) \begin{bmatrix} 0 & -2 & 3 \\ 1 & 7 & 4 \\ 3 & 5 & 9 \end{bmatrix} = \begin{bmatrix} 0 & 6 & -9 \\ -3 & -21 & -12 \\ -9 & -15 & -27 \end{bmatrix}$$

Also

$$aC + bC = \begin{bmatrix} 0 & -8 & 12 \\ 4 & 28 & 16 \\ 12 & 20 & 36 \end{bmatrix} + \begin{bmatrix} 0 & 14 & -21 \\ -7 & -49 & -28 \\ -21 & -35 & -63 \end{bmatrix} = \begin{bmatrix} 0 & 6 & -9 \\ -3 & -21 & -12 \\ -9 & -15 & -27 \end{bmatrix}$$

2. **(a)** We have

$$a(BC) = (4) \left(\begin{bmatrix} 8 & -3 & -5 \\ 0 & 1 & 2 \\ 4 & -7 & 6 \end{bmatrix} \begin{bmatrix} 0 & -2 & 3 \\ 1 & 7 & 4 \\ 3 & 5 & 9 \end{bmatrix} \right)$$

$$= (4) \begin{bmatrix} -18 & -62 & -33 \\ 7 & 17 & 22 \\ 11 & -27 & 38 \end{bmatrix} = \begin{bmatrix} -72 & -248 & -132 \\ 28 & 68 & 88 \\ 44 & -108 & 152 \end{bmatrix}$$

On the other hand

$$(aB)C = \begin{bmatrix} 32 & -12 & -20 \\ 0 & 4 & 8 \\ 16 & -28 & 24 \end{bmatrix} \begin{bmatrix} 0 & -2 & 3 \\ 1 & 7 & 4 \\ 3 & 5 & 9 \end{bmatrix} = \begin{bmatrix} -72 & -248 & -132 \\ 28 & 68 & 88 \\ 44 & -108 & 152 \end{bmatrix}$$

Finally,

$$
B(aC) = \begin{bmatrix} 8 & -3 & 5 \\ 0 & 1 & 2 \\ 4 & -7 & 6 \end{bmatrix} \begin{bmatrix} 0 & -8 & 12 \\ 4 & 28 & 16 \\ 12 & 20 & 36 \end{bmatrix} = \begin{bmatrix} -72 & -248 & -132 \\ 28 & 68 & 88 \\ 44 & -108 & 152 \end{bmatrix}
$$

(c) Since

$$
(B+C)A = \begin{bmatrix} 8 & -5 & -2 \\ 1 & 8 & 6 \\ 7 & -2 & 15 \end{bmatrix} \begin{bmatrix} 2 & -1 & 3 \\ 0 & 4 & 5 \\ -2 & 1 & 4 \end{bmatrix} = \begin{bmatrix} 20 & -30 & -9 \\ -10 & 37 & 67 \\ -16 & 0 & 71 \end{bmatrix}
$$

and

$$
BA + CA = \begin{bmatrix} 26 & -25 & -11 \\ -4 & 6 & 13 \\ -4 & -26 & 1 \end{bmatrix} + \begin{bmatrix} -6 & -5 & 2 \\ -6 & 31 & 54 \\ -12 & 26 & 70 \end{bmatrix} = \begin{bmatrix} 20 & -30 & -9 \\ -10 & 37 & 67 \\ -16 & 0 & 71 \end{bmatrix}
$$

the two matrices are equal.

3. (b) Since

$$
(A+B)^T = \begin{bmatrix} 10 & -4 & -2 \\ 0 & 5 & 7 \\ 2 & -6 & 10 \end{bmatrix}^T = \begin{bmatrix} 10 & 0 & 2 \\ -4 & 5 & -6 \\ -2 & 7 & 10 \end{bmatrix}
$$

and

$$
A^T + B^T = \begin{bmatrix} 2 & 0 & -2 \\ -1 & 4 & 1 \\ 3 & 5 & 4 \end{bmatrix} + \begin{bmatrix} 8 & 0 & 4 \\ -3 & 1 & -7 \\ -5 & 2 & 6 \end{bmatrix} = \begin{bmatrix} 10 & 0 & 2 \\ -4 & 5 & -6 \\ -2 & 7 & 10 \end{bmatrix}
$$

the two matrices are equal.

3.   **(d)** Since

$$(AB)^T = \begin{bmatrix} 28 & -28 & 6 \\ 20 & -31 & 38 \\ 0 & -21 & 36 \end{bmatrix}^T = \begin{bmatrix} 28 & 20 & 0 \\ -28 & -31 & -21 \\ 6 & 38 & 36 \end{bmatrix}$$

and

$$B^T A^T = \begin{bmatrix} 8 & 0 & 4 \\ -3 & 1 & -7 \\ -5 & 2 & 6 \end{bmatrix} \begin{bmatrix} 2 & 0 & -2 \\ -1 & 4 & 1 \\ 3 & 5 & 4 \end{bmatrix} = \begin{bmatrix} 28 & 20 & 0 \\ -28 & -31 & -21 \\ 6 & 38 & 36 \end{bmatrix}$$

the two matrices are equal.

4.   For A, we have $ad - bc = 1$. Hence,

$$A^{-1} = \begin{bmatrix} 2 & -1 \\ -5 & 3 \end{bmatrix}$$

For B, we have $ad - bc = 20$. Hence,

$$B^{-1} = \frac{1}{20} \begin{bmatrix} 4 & 3 \\ -4 & 2 \end{bmatrix}$$

For C, we have $ad - bc = 2$. Hence,

$$C^{-1} = \frac{1}{2} \begin{bmatrix} -1 & -4 \\ 2 & 6 \end{bmatrix}$$

5.   **(b)**

$$(B^T)^{-1} = \begin{bmatrix} 2 & 4 \\ -3 & 4 \end{bmatrix}^{-1} = \frac{1}{20} \begin{bmatrix} 4 & -4 \\ 3 & 2 \end{bmatrix}$$

$$(B^{-1})^T = \begin{bmatrix} \frac{1}{20} \begin{bmatrix} 4 & 3 \\ -4 & 2 \end{bmatrix} \end{bmatrix}^T = \frac{1}{20} \begin{bmatrix} 4 & 3 \\ -4 & 2 \end{bmatrix}^T = \frac{1}{20} \begin{bmatrix} 4 & -4 \\ 3 & 2 \end{bmatrix}$$

6.   (a)

$$(AB)^{-1} = \begin{bmatrix} 10 & -5 \\ 18 & -7 \end{bmatrix}^{-1} = \frac{1}{20} \begin{bmatrix} -7 & 5 \\ -18 & 10 \end{bmatrix}$$

$$B^{-1}A^{-1} = \frac{1}{20} \begin{bmatrix} 4 & 3 \\ -4 & 2 \end{bmatrix} \begin{bmatrix} 2 & -1 \\ -5 & 3 \end{bmatrix} = \frac{1}{20} \begin{bmatrix} -7 & 5 \\ -18 & 10 \end{bmatrix}$$

7.   (b)   We are given that $(7A)^{-1} = \begin{bmatrix} -3 & 7 \\ 1 & -2 \end{bmatrix}$. Therefore

$$7A = ((7A)^{-1})^{-1} = \begin{bmatrix} -3 & 7 \\ 1 & -2 \end{bmatrix}^{-1} = \begin{bmatrix} 2 & 7 \\ 1 & 3 \end{bmatrix}$$

Thus,

$$A = \begin{bmatrix} 2/7 & 1 \\ 1/7 & 3/7 \end{bmatrix}$$

(d)   If $(I + 2A)^{-1} = \begin{bmatrix} -1 & 2 \\ 4 & 5 \end{bmatrix}$, then $I + 2A = \begin{bmatrix} -1 & 2 \\ 4 & 5 \end{bmatrix}^{-1} = \begin{bmatrix} -\dfrac{5}{13} & \dfrac{2}{13} \\ \dfrac{4}{13} & \dfrac{1}{13} \end{bmatrix}$. Hence

$$2A = \begin{bmatrix} -\dfrac{5}{13} & \dfrac{2}{13} \\ \dfrac{4}{13} & \dfrac{1}{13} \end{bmatrix} - \begin{bmatrix} 1 & 0 \\ 0 & 1 \end{bmatrix} = \begin{bmatrix} -\dfrac{18}{13} & \dfrac{2}{13} \\ \dfrac{4}{13} & -\dfrac{12}{13} \end{bmatrix}, \text{ so that } A = \begin{bmatrix} -\dfrac{9}{13} & \dfrac{1}{13} \\ \dfrac{2}{13} & -\dfrac{6}{13} \end{bmatrix}.$$

8.   To compute A^{-3}, we can either compute A^3 and then take its inverse or else we can find A^{-1} and then find $(A^{-1})^3$. We choose the first alternative. Thus

$$A^3 = \left(\begin{bmatrix} 2 & 0 \\ 4 & 1 \end{bmatrix} \begin{bmatrix} 2 & 0 \\ 4 & 1 \end{bmatrix} \right) \begin{bmatrix} 2 & 0 \\ 4 & 1 \end{bmatrix} = \begin{bmatrix} 4 & 0 \\ 12 & 1 \end{bmatrix} \begin{bmatrix} 2 & 0 \\ 4 & 1 \end{bmatrix} = \begin{bmatrix} 8 & 0 \\ 28 & 1 \end{bmatrix}$$

so that

$$A^{-3} = \begin{bmatrix} \dfrac{1}{8} & 0 \\ -\dfrac{7}{2} & 1 \end{bmatrix}$$

9.　**(b)**　We have

$$p(A) = 2 \begin{bmatrix} 3 & 1 \\ 2 & 1 \end{bmatrix}^2 - \begin{bmatrix} 3 & 1 \\ 2 & 1 \end{bmatrix} + 1 \begin{bmatrix} 1 & 0 \\ 0 & 1 \end{bmatrix}$$

$$= 2 \begin{bmatrix} 11 & 4 \\ 8 & 3 \end{bmatrix} - \begin{bmatrix} 3 & 1 \\ 2 & 1 \end{bmatrix} + \begin{bmatrix} 1 & 0 \\ 0 & 1 \end{bmatrix}$$

$$= \begin{bmatrix} 22 & 8 \\ 16 & 6 \end{bmatrix} - \begin{bmatrix} 3 & 1 \\ 2 & 1 \end{bmatrix} + \begin{bmatrix} 1 & 0 \\ 0 & 1 \end{bmatrix}$$

$$= \begin{bmatrix} 20 & 7 \\ 14 & 6 \end{bmatrix}$$

10.　**(b)**　Let A be any square matrix. Then

$$p_2(A)p_3(A) = (A+3)(A-3) = A^2 + 3A - 3A - 9I = A^2 - 9 = p_1(A)$$

11.　Call the matrix A. By Theorem 1.4.5,

$$A^{-1} = \frac{1}{\cos^2\theta + \sin^2\theta} \begin{bmatrix} \cos\theta & -\sin\theta \\ \sin\theta & \cos\theta \end{bmatrix}$$

$$= \begin{bmatrix} \cos\theta & -\sin\theta \\ \sin\theta & \cos\theta \end{bmatrix}$$

since $\cos^2\theta + \sin^2\theta = 1$.

12. Since the determinant is

$$\frac{1}{2}(e^x + e^{-x})\frac{1}{2}(e^x + e^{-x}) - \frac{1}{2}(e^x - e^{-x})\frac{1}{2}(e^x - e^{-x})$$

$$= \frac{1}{4}(e^{2x} + e^{-2x} + 2) - \frac{1}{4}(e^{2x} + e^{-2x} - 2)$$

$$= 1 \neq 0$$

Theorem 1.4.5 applies.

13. If $a_{11}a_{22}\cdots a_{nn} \neq 0$, then $a_{ii} \neq 0$, and hence $1/a_{ii}$ is defined for $i = 1, 2, \ldots, n$. It is now easy to verify that

$$A^{-1} = \begin{bmatrix} 1/a_{11} & 0 & \cdots & 0 \\ 0 & 1/a_{22} & \cdots & 0 \\ \vdots & \vdots & & \vdots \\ 0 & 0 & \cdots & 1/a_{nn} \end{bmatrix}$$

14. We are given that $3A - A^2 = I$. This implies that $A(3I - A) = (3I - A)A = I$; hence $A^{-1} = 3I - A$.

15. Let A denote a matrix which has an entire row or an entire column of zeros. Then if B is any matrix, either AB has an entire row of zeros or BA has an entire column of zeros, respectively. (See Exercise 18, Section 1.3.) Hence, neither AB nor BA can be the identity matrix; therefore, A cannot have an inverse.

16. Not necessarily; for example, both I and $-I$ are invertible, but their sum is not.

17. Suppose that $AB = 0$ and A is invertible. Then $A^{-1}(AB) = A^{-1}0$ or $IB = 0$. Hence, $B = 0$.

18. By Exercise 15 of Section 1.3, we are looking for a matrix C such that

$$\left[\begin{array}{c:c} A & O \\ \hdashline B & A \end{array}\right] \left[\begin{array}{c:c} A^{-1} & O \\ \hdashline C & A^{-1} \end{array}\right] = \left[\begin{array}{c:c} AA^{-1}+O & O+O \\ \hdashline BA^{-1}+AC & O+AA^{-1} \end{array}\right]$$

$$= \left[\begin{array}{c:c} I_2 & O \\ \hdashline BA^{-1}+AC & I_2 \end{array}\right]$$

$$= I_4$$

That is, $BA^{-1} + AC = O$. Thus

$$AC = -BA^{-1}$$

and hence

$$C = -A^{-1}BA^{-1}$$

19. **(a)** Using the notation of Exercise 18, let

$$A = \left[\begin{array}{cc} 1 & 1 \\ -1 & 1 \end{array}\right] \quad \text{and} \quad B = \left[\begin{array}{cc} 1 & 1 \\ 1 & 1 \end{array}\right]$$

Then

$$A^{-1} = \frac{1}{2}\left[\begin{array}{cc} 1 & -1 \\ 1 & 1 \end{array}\right]$$

so that

$$C = -\frac{1}{4}\left[\begin{array}{cc} 1 & -1 \\ 1 & 1 \end{array}\right]\left[\begin{array}{cc} 1 & 1 \\ 1 & 1 \end{array}\right]\left[\begin{array}{cc} 1 & -1 \\ 1 & 1 \end{array}\right]$$

$$= -\frac{1}{4}\left[\begin{array}{cc} 0 & 0 \\ 4 & 0 \end{array}\right]$$

$$= \left[\begin{array}{cc} 0 & 0 \\ -1 & 0 \end{array}\right]$$

Thus the inverse of the given matrix is

$$
\begin{bmatrix}
\dfrac{1}{2} & -\dfrac{1}{2} & 0 & 0 \\[2ex]
\dfrac{1}{2} & \dfrac{1}{2} & 0 & 0 \\[2ex]
0 & 0 & \dfrac{1}{2} & -\dfrac{1}{2} \\[2ex]
-1 & 0 & \dfrac{1}{2} & \dfrac{1}{2}
\end{bmatrix}
$$

20. **(a)** One such example is

$$
A = \begin{bmatrix}
1 & 0 & -1 \\
0 & 2 & -2 \\
-1 & -2 & 3
\end{bmatrix}
$$

In general, any matrix of the form

$$
\begin{bmatrix}
a & d & e \\
d & b & f \\
e & f & c
\end{bmatrix}
$$

will work.

(b) One such example is

$$
A = \begin{bmatrix}
0 & 0 & -1 \\
0 & 0 & -2 \\
1 & 2 & 0
\end{bmatrix}
$$

In general, any matrix of the form

$$
\begin{bmatrix}
0 & b & c \\
-b & 0 & f \\
-c & -f & 0
\end{bmatrix}
$$

will work.

21. We use Theorem 1.4.9.

    (a) If $A = BB^T$, then

    $$A^T = (BB^T)^T = (B^T)^T B^T = BB^T = A$$

    Thus A is symmetric. On the other hand, if $A = B + B^T$, then

    $$A^T = (B + B^T)^T = B^T + (B^T)^T = B^T + B = B + B^T = A$$

    Thus A is symmetric.

    (b) If $A = B - B^T$, then

    $$A^T = (B - B^T)^T = [B + (-1)B^T]^T = B^T + [(-1)B^T]^T$$

    $$= B^T + (-1)(B^T)^T = B^T + (-1)B = B^T - B = -A$$

    Thus A is skew-symmetric.

22. By Theorem 1.4.9, $(A^n)^T = (A(A^{n-1}))^T = (A^{n-1})^T A^T$. If we repeat this process $n - 1$ times, we get that $(A^n)^T = (A^T)^n$ which justifies the equality.

23. Let

    $$A^{-1} = \begin{bmatrix} x_{11} & x_{12} & x_{13} \\ x_{21} & x_{22} & x_{23} \\ x_{31} & x_{32} & x_{33} \end{bmatrix}$$

    Then

    $$AA^{-1} = \begin{bmatrix} 1 & 0 & 1 \\ 1 & 1 & 0 \\ 0 & 1 & 1 \end{bmatrix} \begin{bmatrix} x_{11} & x_{12} & x_{13} \\ x_{21} & x_{22} & x_{23} \\ x_{31} & x_{32} & x_{33} \end{bmatrix}$$

    $$= \begin{bmatrix} x_{11} + x_{31} & x_{12} + x_{32} & x_{13} + x_{33} \\ x_{11} + x_{21} & x_{12} + x_{22} & x_{13} + x_{23} \\ x_{21} + x_{31} & x_{22} + x_{32} & x_{23} + x_{33} \end{bmatrix}$$

Since $AA^{-1} = I$, we equate corresponding entries to obtain the system of equations

$$
\begin{array}{rcrcrcl}
x_{11} & & & + & x_{31} & & = 1 \\
& x_{12} & & & + & x_{32} & = 0 \\
& & x_{13} & & & + & x_{33} = 0 \\
x_{11} & + & x_{21} & & & & = 0 \\
& x_{12} & + & x_{22} & & & = 1 \\
& & x_{13} & + & x_{23} & & = 0 \\
& & x_{21} & & + & x_{31} & = 0 \\
& & & x_{22} & & + & x_{32} = 0 \\
& & & & x_{23} & + & x_{33} = 1
\end{array}
$$

The solution to this system of equations gives

$$
A^{-1} = \begin{bmatrix} 1/2 & 1/2 & -1/2 \\ -1/2 & 1/2 & 1/2 \\ 1/2 & -1/2 & 1/2 \end{bmatrix}
$$

24. **(a)** Let ℓ_{ij} denote the ij^{th} entry of the matrix on the left and r_{ij} be the corresponding entry for the matrix on the right. Then

$$
\ell_{ij} = a_{ij} + (b_{ij} + c_{ij})
$$

Since the associative rule holds for real numbers, this implies that

$$
\ell_{ij} = (a_{ij} + b_{ij}) + c_{ij} = r_{ij}
$$

(c) Suppose that B is $m \times n$ and C is $n \times p$. Then

$$
[a(BC)]_{ij} = a(b_{i1}c_{1j} + b_{i2}c_{2j} + \cdots + b_{in}c_{nj})
$$

$$
= (ab_{i1})c_{1j} + (ab_{i2})c_{2j} + \cdots + (ab_{in})c_{nj} = [(aB)C]_{ij}
$$

$$
= b_{i1}(ac_{ij}) + b_{i2}(ac_{2j}) + \cdots + b_{in}(ac_{nj}) = [B(aC)]_{ij}
$$

25.   We wish to show that $A(B - C) = AB - AC$. By Part (d) of Theorem 1.4.1, we have $A(B - C) = A(B + (-C)) = AB + A(-C)$. Finally by Part (m), we have $A(-C) = -AC$ and the desired result can be obtained by substituting this result in the above equation.

26.   Let a_{ij} denote the ij^{th} entry of the $m \times n$ matrix A. All the entries of 0 are 0. In the proofs below, we need only show that the ij^{th} entries of all the matrices involved are equal.

(a) $a_{ij} + 0 = 0 + a_{ij} = a_{ij}$

(b) $a_{ij} - a_{ij} = 0$

(c) $0 - a_{ij} = -a_{ij}$

(d) $a_{i1}0 + a_{i2}0 + \cdots + a_{in}0 = 0$ and $0a_{1j} + 0a_{2j} + \cdots + a_{mj} = 0$

27.   **(a)**   We have

$$A^r A^s = \underbrace{(AA \cdots A)}_{\substack{r \\ \text{factors}}} \underbrace{(AA \cdots A)}_{\substack{s \\ \text{factors}}}$$

$$= \underbrace{AA \cdots A}_{\substack{r + s \\ \text{factors}}} = A^{r+s}$$

On the other hand,

$$(A^r)^s = \underbrace{\underbrace{(AA \cdots A)}_{\substack{r \\ \text{factors}}} \underbrace{(AA \cdots A)}_{\substack{r \\ \text{factors}}} \cdots \underbrace{(AA \cdots A)}_{\substack{r \\ \text{factors}}}}_{\substack{s \\ \text{factors}}}$$

$$= \underbrace{AA \cdots A}_{\substack{rs \\ \text{factors}}}$$

(b) Suppose that $r < 0$ and $s < 0$; let $\rho = -r$ and $\sigma = -s$, so that

$$A^r A^s = A^{-\rho} A^{-\sigma}$$

$$= (A^{-1})^\rho (A^{-1})^\sigma \qquad \text{(by the definition)}$$

$$= (A^{-1})^{\rho+\sigma} \qquad \text{(by Part (a))}$$

$$= A^{-(\rho+\sigma)} \qquad \text{(by the definition)}$$

$$= A^{-\rho-\sigma}$$

$$= A^{r+s}$$

Also

$$(A^r)^s = (A^{-\rho})^{-\sigma}$$

$$= \left[(A^{-1})^\rho\right]^{-\sigma} \qquad \text{(by the definition)}$$

$$= \left([(A^{-1})^\rho]^{-1}\right)^\sigma \qquad \text{(by the definition)}$$

$$= \left([(A^{-1})^{-1}]^\rho\right)^\sigma \qquad \text{(by Theorem 1.4.8b)}$$

$$= ([A]^\rho)^\sigma \qquad \text{(by Theorem 1.4.8a)}$$

$$= A^{\rho\sigma} \qquad \text{(by Part (a))}$$

$$= A^{(-\rho)(-\sigma)}$$

$$= A^{rs}$$

29. **(a)** If $AB = AC$, then

$$A^{-1}(AB) = A^{-1}(AC)$$

or

$$(A^{-1}A)B = (A^{-1}A)C$$

or

$$B = C$$

(b) The matrix A in Example 3 is not invertible.

30. Following the hint, let $\ell_{ij} = [A(BC)]_{ij}$ and $r_{ij} = [(AB)C]_{ij}$ and suppose that A is $m \times n$, B is $n \times p$, and C is $p \times q$. Then

$$\ell_{ij} = a_{i1}[BC]_{1j} + a_{i2}[BC]_{2j} + \cdots + a_{in}[BC]_{nj}$$

$$= a_{i1}(b_{11}c_{1j} + \cdots + b_{1p}c_{pi}) + a_{i2}(b_{21}c_{1j} + \cdots + b_{2p}c_{pj}) + \cdots + a_{in}(b_{n1}c_{1j} + \cdots + b_{np}c_{pj})$$

$$= (a_{i1}b_{11} + a_{i2}b_{21} + \cdots + a_{in}b_{n1})c_{1j} + \cdots + (a_{i1}b_{1p} + a_{i2}b_{2p} + \cdots + a_{in}b_{np})c_{pj}$$

$$= [AB]_{i1}c_{1j} + \cdots + [AB]_{ip}c_{pj}$$

$$= r_{ij}$$

For anyone familiar with sigma notation, this proof can be simplified as follows:

$$\ell_{ij} = \sum_{s=1}^{n} a_{is}[BC]_{sj} = \sum_{s=1}^{n} a_{is} \sum_{t=1}^{p} b_{st}c_{tj}$$

$$= \sum_{t=1}^{p} \left(\sum_{s=1}^{n} a_{is}b_{st} \right) c_{tj} = \sum_{t=1}^{p} [AB]_{it}c_{tj}$$

$$= r_{ij}$$

31. **(a)** Any pair of matrices that do not commute will work. For example, if we let

$$A = \begin{bmatrix} 1 & 0 \\ 0 & 0 \end{bmatrix} \qquad B = \begin{bmatrix} 0 & 1 \\ 0 & 1 \end{bmatrix}$$

then

$$(A+B)^2 = \begin{bmatrix} 1 & 1 \\ 0 & 1 \end{bmatrix}^2 = \begin{bmatrix} 1 & 2 \\ 0 & 1 \end{bmatrix}$$

whereas

$$A^2 + 2AB + B^2 = \begin{bmatrix} 1 & 3 \\ 0 & 1 \end{bmatrix}$$

(b) In general,

$$(A + B)^2 = (A + B)(A + B) = A^2 + AB + BA + B^2$$

32. **(a)** As in Exercise 31, let A and B be any appropriate matrices which do not commute.

33. If

$$A = \begin{bmatrix} a_{11} & 0 & 0 \\ 0 & a_{22} & 0 \\ 0 & 0 & a_{33} \end{bmatrix} \quad \text{then} \quad A^2 = \begin{bmatrix} a_{11}^2 & 0 & 0 \\ 0 & a_{22}^2 & 0 \\ 0 & 0 & a_{33}^2 \end{bmatrix}$$

Thus, $A^2 = I$ if and only if $a_{11}^2 = a_{22}^2 = a_{33}^2 = 1$, or $a_{11} = \pm 1$, $a_{22} = \pm 1$, and $a_{33} = \pm 1$. There are exactly eight possibilities:

$$\begin{bmatrix} 1 & 0 & 0 \\ 0 & 1 & 0 \\ 0 & 0 & 1 \end{bmatrix} \quad \begin{bmatrix} 1 & 0 & 0 \\ 0 & 1 & 0 \\ 0 & 0 & -1 \end{bmatrix} \quad \begin{bmatrix} 1 & 0 & 0 \\ 0 & -1 & 0 \\ 0 & 0 & 1 \end{bmatrix} \quad \begin{bmatrix} 1 & 0 & 0 \\ 0 & -1 & 0 \\ 0 & 0 & -1 \end{bmatrix}$$

$$\begin{bmatrix} -1 & 0 & 0 \\ 0 & 1 & 0 \\ 0 & 0 & 1 \end{bmatrix} \quad \begin{bmatrix} -1 & 0 & 0 \\ 0 & 1 & 0 \\ 0 & 0 & -1 \end{bmatrix} \quad \begin{bmatrix} -1 & 0 & 0 \\ 0 & -1 & 0 \\ 0 & 0 & 1 \end{bmatrix} \quad \begin{bmatrix} -1 & 0 & 0 \\ 0 & -1 & 0 \\ 0 & 0 & -1 \end{bmatrix}$$

34. **(a)** The logical contrapositive is "If A is invertible, then A^T is invertible."

(b) The statement is true by Theorem 1.4.10.

35. **(b)** The statement is true, since $(A - B)^2 = (-(B - A))^2 = (B - A)^2$.

(c) The statement is true only if A^{-1} and B^{-1} exist, in which case

$$(AB^{-1})(BA^{-1}) = A(B^{-1}B)A^{-1} = AI_n A^{-1} = AA^{-1} = I_n$$

EXERCISE SET 1.5

1. (a) The matrix may be obtained from I_2 by adding -5 times Row 1 to Row 2. Thus, it is elementary.

   (c) The matrix may be obtained from I_2 by multiplying Row 2 of I_2 by $\sqrt{3}$. Thus it is elementary.

   (e) This is not an elementary matrix because it is not invertible.

   (g) The matrix may be obtained from I_4 only by performing two elementary row operations such as replacing Row 1 of I_4 by Row 1 plus Row 4, and then multiplying Row 1 by 2. Thus it is not an elementary matrix.

3. (a) If we interchange Rows 1 and 3 of A, then we obtain B. Therefore, E_1 must be the matrix obtained from I_3 by interchanging Rows 1 and 3 of I_3, i.e.,

$$E_1 = \begin{bmatrix} 0 & 0 & 1 \\ 0 & 1 & 0 \\ 1 & 0 & 0 \end{bmatrix}$$

   (c) If we multiply Row 1 of A by -2 and add it to Row 3, then we obtain C. Therefore, E_3 must be the matrix obtained from I_3 by replacing its third row by -2 times Row 1 plus Row 3, i.e.,

$$E_3 = \begin{bmatrix} 1 & 0 & 0 \\ 0 & 1 & 0 \\ -2 & 0 & 1 \end{bmatrix}$$

5. (a)

$$\left[\begin{array}{cc|cc} 1 & 4 & 1 & 0 \\ 2 & 7 & 0 & 1 \end{array}\right]$$

$$\left[\begin{array}{cc|cc} 1 & 4 & 1 & 0 \\ 0 & -1 & -3 & 1 \end{array}\right]$$

> Add -2 times Row 1 to Row 2.

$$\left[\begin{array}{cc|cc} 1 & 0 & -7 & 4 \\ 0 & 1 & 2 & -1 \end{array}\right]$$

> Multiply Row 2 by -1; then multiply Row 2 by -4 and add to Row 1.

Therefore

$$\left[\begin{array}{cc} 1 & 4 \\ 2 & 7 \end{array}\right]^{-1} = \left[\begin{array}{cc} -7 & 4 \\ 2 & -1 \end{array}\right]$$

(c)

$$\left[\begin{array}{cc|cc} 6 & -4 & 1 & 0 \\ -3 & 2 & 0 & 1 \end{array}\right]$$

$$\left[\begin{array}{cc|cc} 1 & -2/3 & 1/6 & 0 \\ -3 & 2 & 0 & 1 \end{array}\right]$$

> Divide Row 1 by 6.

$$\left[\begin{array}{cc|cc} 1 & -2/3 & 1/6 & 0 \\ 0 & 0 & 1/2 & 1 \end{array}\right]$$

> Multiply Row 1 by 3 and add to Row 2.

Thus, the given matrix is not invertible because we have obtained a row of zeros on the left side.

6. (a)

$$\left[\begin{array}{ccc|ccc} 3 & 4 & -1 & 1 & 0 & 0 \\ 1 & 0 & 3 & 0 & 1 & 0 \\ 2 & 5 & -4 & 0 & 0 & 1 \end{array}\right]$$

$$\left[\begin{array}{ccc|ccc} 1 & 0 & 3 & 0 & 1 & 0 \\ 3 & 4 & -1 & 1 & 0 & 0 \\ 2 & 5 & -4 & 0 & 0 & 1 \end{array}\right]$$

Interchange Rows 1 and 2.

$$\left[\begin{array}{ccc|ccc} 1 & 0 & 3 & 0 & 1 & 0 \\ 0 & 4 & -10 & 1 & -3 & 0 \\ 0 & 5 & -10 & 0 & -2 & 1 \end{array}\right]$$

Add −3 times Row 1 to Row 2 and −2 times Row 1 to Row 3.

$$\left[\begin{array}{ccc|ccc} 1 & 0 & 3 & 0 & 1 & 0 \\ 0 & 4 & -10 & 1 & -3 & 0 \\ 0 & 1 & 0 & -1 & 1 & 1 \end{array}\right]$$

Add −1 times Row 2 to Row 3.

$$\left[\begin{array}{ccc|ccc} 1 & 0 & 3 & 0 & 1 & 0 \\ 0 & 1 & 0 & -1 & 1 & 1 \\ 0 & 0 & -10 & 5 & -7 & -4 \end{array}\right]$$

Add −4 times Row 3 to Row 2 and interchange Rows 2 and 3.

$$\left[\begin{array}{ccc|ccc} 1 & 0 & 0 & \dfrac{3}{2} & -\dfrac{11}{10} & -\dfrac{6}{5} \\ 0 & 1 & 0 & -1 & 1 & 1 \\ 0 & 0 & 1 & -\dfrac{1}{2} & \dfrac{7}{10} & \dfrac{2}{5} \end{array}\right]$$

Multiply Row 3 by −1/10. Then add −3 times Row 3 to Row 1.

Thus, the desired inverse is

$$\left[\begin{array}{ccc} \dfrac{3}{2} & -\dfrac{11}{10} & -\dfrac{6}{5} \\ -1 & 1 & 1 \\ -\dfrac{1}{2} & \dfrac{7}{10} & \dfrac{2}{5} \end{array}\right]$$

6. **(c)**

$$\left[\begin{array}{ccc|ccc} 1 & 0 & 1 & 1 & 0 & 0 \\ 0 & 1 & 1 & 0 & 1 & 0 \\ 1 & 1 & 0 & 0 & 0 & 1 \end{array}\right]$$

$$\left[\begin{array}{ccc|ccc} 1 & 0 & 1 & 1 & 0 & 0 \\ 0 & 1 & 1 & 0 & 1 & 0 \\ 0 & 1 & -1 & -1 & 0 & 1 \end{array}\right]$$

Subtract Row 1 from Row 3.

$$\left[\begin{array}{ccc|ccc} 1 & 0 & 1 & 1 & 0 & 0 \\ 0 & 1 & 1 & 0 & 1 & 0 \\ 0 & 0 & 1 & \dfrac{1}{2} & \dfrac{1}{2} & -\dfrac{1}{2} \end{array}\right]$$

Subtract Row 2 from Row 3 and multiply Row 3 by $-\dfrac{1}{2}$.

$$\left[\begin{array}{ccc|ccc} 1 & 0 & 0 & \dfrac{1}{2} & -\dfrac{1}{2} & \dfrac{1}{2} \\ 0 & 1 & 0 & -\dfrac{1}{2} & \dfrac{1}{2} & \dfrac{1}{2} \\ 0 & 0 & 1 & \dfrac{1}{2} & \dfrac{1}{2} & -\dfrac{1}{2} \end{array}\right]$$

Subtract Row 3 from Rows 1 and 2.

Thus

$$\left[\begin{array}{ccc} 1 & 0 & 1 \\ 0 & 1 & 1 \\ 1 & 1 & 0 \end{array}\right]^{-1} = \left[\begin{array}{ccc} \dfrac{1}{2} & -\dfrac{1}{2} & \dfrac{1}{2} \\ -\dfrac{1}{2} & \dfrac{1}{2} & \dfrac{1}{2} \\ \dfrac{1}{2} & \dfrac{1}{2} & -\dfrac{1}{2} \end{array}\right]$$

(e)

$$\left[\begin{array}{ccc|ccc} 1 & 0 & 1 & 1 & 0 & 0 \\ -1 & 1 & 1 & 0 & 1 & 0 \\ 0 & 1 & 0 & 0 & 0 & 1 \end{array}\right]$$

$$\left[\begin{array}{ccc|ccc} 1 & 0 & 1 & 1 & 0 & 0 \\ 0 & 1 & 2 & 1 & 1 & 0 \\ 0 & 0 & -2 & -1 & -1 & 1 \end{array}\right]$$

Add Row 1 to Row 2 and subtract the new Row 2 from Row 3.

$$\left[\begin{array}{ccc|ccc} 1 & 0 & 1 & 1 & 0 & 0 \\ 0 & 1 & 0 & 0 & 0 & 1 \\ 0 & 0 & 1 & \dfrac{1}{2} & \dfrac{1}{2} & -\dfrac{1}{2} \end{array}\right]$$

Add Row 3 to Row 2 and then multiply Row 3 by $-\dfrac{1}{2}$.

$$\left[\begin{array}{ccc|ccc} 1 & 0 & 0 & \dfrac{1}{2} & -\dfrac{1}{2} & \dfrac{1}{2} \\ 0 & 1 & 0 & 0 & 0 & 1 \\ 0 & 0 & 1 & \dfrac{1}{2} & \dfrac{1}{2} & -\dfrac{1}{2} \end{array}\right]$$

Subtract Row 3 from Row 1.

Thus

$$\left[\begin{array}{ccc} 1 & 0 & 1 \\ -1 & 1 & 1 \\ 0 & 1 & 0 \end{array}\right]^{-1} = \left[\begin{array}{ccc} \dfrac{1}{2} & -\dfrac{1}{2} & \dfrac{1}{2} \\ 0 & 0 & 1 \\ \dfrac{1}{2} & \dfrac{1}{2} & -\dfrac{1}{2} \end{array}\right]$$

7. (a)

$$\left[\begin{array}{ccc|ccc} \frac{1}{5} & \frac{1}{5} & -\frac{2}{5} & 1 & 0 & 0 \\ \frac{1}{5} & \frac{1}{5} & \frac{1}{10} & 0 & 1 & 0 \\ \frac{1}{5} & -\frac{4}{5} & \frac{1}{10} & 0 & 0 & 1 \end{array}\right]$$

$$\left[\begin{array}{ccc|ccc} 1 & 1 & -2 & 5 & 0 & 0 \\ 1 & 1 & \frac{1}{2} & 0 & 5 & 0 \\ 1 & -4 & \frac{1}{2} & 0 & 0 & 5 \end{array}\right] \quad \boxed{\begin{array}{l}\text{Multiply all three}\\ \text{Rows by 5.}\end{array}}$$

$$\left[\begin{array}{ccc|ccc} 1 & 1 & -2 & 5 & 0 & 0 \\ 0 & 0 & \frac{5}{2} & -5 & 5 & 0 \\ 0 & -5 & \frac{5}{2} & -5 & 0 & 5 \end{array}\right] \quad \boxed{\begin{array}{l}\text{Subtract Row 1}\\ \text{from both Row 2}\\ \text{and Row 3.}\end{array}}$$

$$\left[\begin{array}{ccc|ccc} 1 & 1 & -2 & 5 & 0 & 0 \\ 0 & 0 & 1 & -2 & 2 & 0 \\ 0 & 1 & -\frac{1}{2} & 1 & 0 & -1 \end{array}\right] \quad \boxed{\begin{array}{l}\text{Multiply Row 2 by}\\ \text{2/5 and Row 3 by}\\ -1/5.\end{array}}$$

$$\left[\begin{array}{ccc|ccc} 1 & 1 & 0 & 1 & 4 & 0 \\ 0 & 0 & 1 & -2 & 2 & 0 \\ 0 & 1 & 0 & 0 & 1 & -1 \end{array}\right] \quad \boxed{\begin{array}{l}\text{Add 2 times Row 2}\\ \text{to Row 1 and 1/2}\\ \text{times Row 2 to Row 3.}\end{array}}$$

$$\left[\begin{array}{ccc|ccc} 1 & 0 & 0 & 1 & 3 & 1 \\ 0 & 1 & 0 & 0 & 1 & -1 \\ 0 & 0 & 1 & -2 & 2 & 0 \end{array}\right] \quad \boxed{\begin{array}{l}\text{Subtract Row 3 from}\\ \text{Row 1 and interchange}\\ \text{Rows 2 and 3.}\end{array}}$$

Thus the desired inverse is

$$
\begin{bmatrix}
1 & 3 & 1 \\
0 & 1 & -1 \\
-2 & 2 & 0
\end{bmatrix}
$$

(b) Hint: First multiply Rows 1 and 2 by $1/\sqrt{2}$.

(c)

$$
\left[\begin{array}{cccc|cccc}
1 & 0 & 0 & 0 & 1 & 0 & 0 & 0 \\
1 & 3 & 0 & 0 & 0 & 1 & 0 & 0 \\
1 & 3 & 5 & 0 & 0 & 0 & 1 & 0 \\
1 & 3 & 5 & 7 & 0 & 0 & 0 & 1
\end{array}\right]
$$

$$
\left[\begin{array}{cccc|cccc}
1 & 0 & 0 & 0 & 1 & 0 & 0 & 0 \\
0 & 3 & 0 & 0 & -1 & 1 & 0 & 0 \\
0 & 3 & 5 & 0 & -1 & 0 & 1 & 0 \\
0 & 3 & 5 & 7 & -1 & 0 & 0 & 1
\end{array}\right]
$$

Subtract Row 1 from Rows 2-4.

$$
\left[\begin{array}{cccc|cccc}
1 & 0 & 0 & 0 & 1 & 0 & 0 & 0 \\
0 & 1 & 0 & 0 & -\dfrac{1}{3} & \dfrac{1}{3} & 0 & 0 \\
0 & 0 & 5 & 0 & 0 & -1 & 1 & 0 \\
0 & 0 & 5 & 7 & 0 & -1 & 0 & 1
\end{array}\right]
$$

Subtract Row 2 from Rows 3 and 4 and multiply Row 2 by 1/3.

$$
\left[\begin{array}{cccc|cccc}
1 & 0 & 0 & 0 & 1 & 0 & 0 & 0 \\
0 & 1 & 0 & 0 & -\dfrac{1}{3} & \dfrac{1}{3} & 0 & 0 \\
0 & 0 & 1 & 0 & 0 & -\dfrac{1}{5} & \dfrac{1}{5} & 0 \\
0 & 0 & 0 & 1 & 0 & 0 & -\dfrac{1}{7} & \dfrac{1}{7}
\end{array}\right]
$$

Subtract Row 3 from Row 4 and multiply Row 3 by 1/5 and Row 4 by 1/7.

Thus the desired inverse is

$$\begin{bmatrix} 1 & 0 & 0 & 0 \\ -\dfrac{1}{3} & \dfrac{1}{3} & 0 & 0 \\ 0 & -\dfrac{1}{5} & \dfrac{1}{5} & 0 \\ 0 & 0 & -\dfrac{1}{7} & \dfrac{1}{7} \end{bmatrix}$$

7. **(d)** Since the third row consists entirely of zeros, this matrix is not invertible.

8. **(b)** Multiplying Row i of

$$\left[\begin{array}{cccc|cccc} 0 & 0 & 0 & k_1 & 1 & 0 & 0 & 0 \\ 0 & 0 & k_2 & 0 & 0 & 1 & 0 & 0 \\ 0 & k_3 & 0 & 0 & 0 & 0 & 1 & 0 \\ k_4 & 0 & 0 & 0 & 0 & 0 & 0 & 1 \end{array}\right]$$

by $1/k_i$ for $i = 1, 2, 3, 4$ and then reversing the order of the rows yields I_4 on the left and the desired inverse

$$\begin{bmatrix} 0 & 0 & 0 & 1/k_4 \\ 0 & 0 & 1/k_3 & 0 \\ 0 & 1/k_2 & 0 & 0 \\ 1/k_1 & 0 & 0 & 0 \end{bmatrix}$$

on the right.

(c) To reduce

$$\left[\begin{array}{cccc|cccc} k & 0 & 0 & 0 & 1 & 0 & 0 & 0 \\ 1 & k & 0 & 0 & 0 & 1 & 0 & 0 \\ 0 & 1 & k & 0 & 0 & 0 & 1 & 0 \\ 0 & 0 & 1 & k & 0 & 0 & 0 & 1 \end{array}\right]$$

we multiply Row i by $1/k$ and then subtract Row i from Row $(i + 1)$ for $i = 1, 2, 3$. Then multiply Row 4 by $1/k$. This produces I_4 on the left and the inverse,

$$
\begin{bmatrix}
1/k & 0 & 0 & 0 \\
-1/k^2 & 1/k & 0 & 0 \\
1/k^3 & -1/k^2 & 1/k & 0 \\
1/k^4 & 1/k^3 & -1/k^2 & 1/k
\end{bmatrix}
$$

on the right.

9. (a) It takes 2 elementary row operations to convert A to I: first multiply Row 1 by 5 and add to Row 2, then multiply Row 2 by $1/2$. If we apply each of these operations to I, we obtain the 2 elementary matrices

$$
E_1 = \begin{bmatrix} 1 & 0 \\ 5 & 1 \end{bmatrix} \quad \text{and} \quad E_2 = \begin{bmatrix} 1 & 0 \\ 0 & 1/2 \end{bmatrix}
$$

From Theorem 1.5.1, we conclude that

$$
E_2 E_1 A = I
$$

(b) Since the inverse of a matrix is unique, the above equation guarantees that (with E_1 and E_2 as defined above)

$$
A^{-1} = E_2 E_1
$$

(c) From the equation $E_2 E_1 A = I$, we conclude that

$$
E_2^{-1} E_2 E_1 A = E_2^{-1} I
$$

or

$$
E_1 A = E_2^{-1}
$$

or

$$
E_1^{-1} E_1 A = E_1^{-1} E_2^{-1}
$$

or

$$
A = E_1^{-1} E_2^{-1}
$$

so that

$$
A = \begin{bmatrix} 1 & 0 \\ -5 & 1 \end{bmatrix} \begin{bmatrix} 1 & 0 \\ 0 & 2 \end{bmatrix}
$$

11. We first reduce A to row-echelon form, keeping track as we do so of the elementary row operations which are required. Here

$$A = \begin{bmatrix} 0 & 1 & 7 & 8 \\ 1 & 3 & 3 & 8 \\ -2 & -5 & 1 & -8 \end{bmatrix}$$

1. If we interchange Rows 1 and 2, we obtain the matrix

$$\begin{bmatrix} 1 & 3 & 3 & 8 \\ 0 & 1 & 7 & 8 \\ -2 & -5 & 1 & -8 \end{bmatrix}$$

2. If we add twice Row 1 to Row 3, we obtain the matrix

$$\begin{bmatrix} 1 & 3 & 3 & 8 \\ 0 & 1 & 7 & 8 \\ 0 & 1 & 7 & 8 \end{bmatrix}$$

3. If we subtract Row 2 from Row 3, we obtain the matrix

$$R = \begin{bmatrix} 1 & 3 & 3 & 8 \\ 0 & 1 & 7 & 8 \\ 0 & 0 & 0 & 0 \end{bmatrix}$$

Note that the above matrix R is in row-echelon form. Thus, Theorem 1.5.1 implies that $E_3 E_2 E_1 A = R$, where

$$E_1 = \begin{bmatrix} 0 & 1 & 0 \\ 1 & 0 & 0 \\ 0 & 0 & 1 \end{bmatrix}, E_2 = \begin{bmatrix} 1 & 0 & 0 \\ 0 & 1 & 0 \\ 2 & 0 & 1 \end{bmatrix}, \text{ and } E_3 = \begin{bmatrix} 1 & 0 & 0 \\ 0 & 1 & 0 \\ 0 & -1 & 1 \end{bmatrix}$$

Thus, $A = E_1^{-1} E_2^{-1} E_3^{-1} R$. We need only let

$$E = E_1^{-1} = \begin{bmatrix} 0 & 1 & 0 \\ 1 & 0 & 0 \\ 0 & 0 & 1 \end{bmatrix}$$

$$F = E_2^{-1} = \begin{bmatrix} 1 & 0 & 0 \\ 0 & 1 & 0 \\ -2 & 0 & 1 \end{bmatrix}$$

and

$$G = E_3^{-1} = \begin{bmatrix} 1 & 0 & 0 \\ 0 & 1 & 0 \\ 0 & 1 & 1 \end{bmatrix}$$

and we're done.

12. If A is an elementary matrix, then it can be obtained from the identity matrix I by a single elementary row operation. If we start with I and multiply Row 3 by a nonzero constant, then $a = b = 0$. If we interchange Row 1 or Row 2 with Row 3, then $c = 0$. If we add a nonzero multiple of Row 1 or Row 2 to Row 3, then either $b = 0$ or $a = 0$. Finally, if we operate only on the first two rows, then $a = b = 0$. Thus at least one entry in Row 3 must equal zero.

14. Every $m \times n$ matrix A can be transformed into reduced row-echelon form B by a sequence of row operations. From Theorem 1.5.1,

$$B = E_k E_{k-1} \cdots E_1 A$$

where $E_1, E_2, \ldots, E_k$ are the elementary matrices corresponding to the row operations. If we take $C = E_k E_{k-1} \cdots E_1$, then C is invertible by Theorem 1.5.2 and the rule following Theorem 1.4.6.

15. The hypothesis that B is row equivalent to A can be expressed as follows:

$$B = E_k E_{k-1} \cdots E_1 A$$

where $E_1, E_2, \ldots, E_k$ are elementary matrices. Since A is invertible, then (by the rule following Theorem 1.5.3) B is invertible and

$$B^{-1} = A^{-1} E_1^{-1} E_2^{-1} \cdots E_k^{-1}$$

16. **(a)** First suppose that A and B are row equivalent. Then there are elementary matrices $E_1, \ldots, E_p$ such that $A = E_1 \cdots E_p B$. There are also elementary matrices $E_{p+1}, \ldots, E_{p+q}$ such that $E_{p+1} \cdots E_{p+q} A$ is in reduced row-echelon form. Therefore, the matrix $E_{p+1} \cdots E_{p+q} E_1 \cdots E_p B$ is also in (the same) reduced row-echelon form. Hence we have found, via elementary matrices, a sequence of elementary row operations which will put B in the same reduced row-echelon form as A.

Now suppose that A and B have the same reduced row-echelon form. Then there are elementary matrices $E_1, \ldots, E_p$ and $E_{p+1}, \ldots, E_{p+q}$ such that $E_1 \cdots E_p A = E_{p+1} \cdots E_{p+q} B$. Since elementary matrices are invertible, this equation implies that $A = E_p^{-1} \cdots E_1^{-1} E_{p+1} \cdots E_{p+q} B$. Since the inverse of an elementary matrix is also an elementary matrix, we have that A and B are row equivalent.

17. There are 3 kinds of elementary row operations, and therefore 3 basic kinds of elementary matrices. First, suppose that we multiply row k by the constant $c \neq 0$. This results in the elementary matrix E with $e_{kk} = c$ and all other entries identical to those of I_m. Now consider the product EA. Since

$$[EA]_{ij} = e_{i1} a_{1j} + e_{i2} a_{2j} + \cdots + e_{im} a_{mj}$$

where $e_{ij} = 0$ unless $i = j$, we have that $[EA]_{ij} = a_{ij}$ unless $i = k$, in which case $[EA]_{kj} = c a_{kj}$. That is, EA is just A with the kth row multiplied by c.

Next, suppose that we interchange rows k and ℓ. This results in the elementary matrix E in which $e_{k\ell} = e_{\ell k} = 1$, all other entries in rows k and ℓ are zero, and otherwise E coincides with I_m. Thus $[EA]_{ij} = a_{ij}$ unless $i = k$ or $i = \ell$, in which case $[EA]_{kj} = a_{\ell j}$ and $[EA]_{\ell j} = a_{kj}$. That is, EA is just A with rows k and ℓ interchanged.

Finally, suppose that we multiply row k by c and add the result to row ℓ where $k \neq \ell$. This results in the elementary matrix E in which $e_{\ell k} = c$ and all other entries coincide with those of I_m. Thus $[EA]_{ij} = a_{ij}$ unless $i = \ell$, in which case $[EA]_{\ell j} = c a_{kj} + a_{\ell j}$, so that EA is just A with c times row k added to row ℓ.

18. The matrix A, by hypothesis, can be reduced to the identity matrix via a sequence of elementary row operations. We can therefore find elementary matrices $E_1, E_2, \ldots E_k$ such that

$$E_k \cdots E_2 \cdot E_1 \cdot A = I_n$$

Since every elementary matrix is invertible, it follows that

$$A = E_1^{-1} E_2^{-1} \cdots E_k^{-1} I_n$$

19. **(a)** False. A square matrix need not be invertible, but the product of elementary matrices is. Thus the zero matrix provides a counterexample.

(b) False. The product of two elementary matrices is a matrix which represents two successive elementary row operations. These operations can usually not be accomplished by a single elementary row operation. For instance, let

$$E_1 = \begin{bmatrix} 2 & 0 \\ 0 & 1 \end{bmatrix} \quad \text{and} \quad E_2 = \begin{bmatrix} 1 & 0 \\ 0 & 2 \end{bmatrix}$$

Then

$$E_1 E_2 = \begin{bmatrix} 2 & 0 \\ 0 & 2 \end{bmatrix}$$

which is clearly not an elementary matrix since it cannot be obtained from I_2 via a single elementary row operation.

(c) True. If A is invertible and we apply an elementary row operation to A, then we have, in effect, multiplied A by an elementary matrix E. The resulting matrix EA is invertible as the product of invertible matrices.

20. **(a)** True. Suppose we reduce A to its reduced row-echelon form via a sequence of elementary row operations. The resulting matrix must have at least one row of zeros, since otherwise we would obtain the identity matrix and A would be invertible. Thus at least one of the variables in $\mathbf{x}$ must be arbitrary and the system of equations will have infinitely many solutions.

(b) See Part (a).

(d) False. If $B = EA$ for any elementary matrix E, then $A = E^{-1}B$. Thus, if B were invertible, then A would also be invertible, contrary to hypothesis.

21. There is not. For instance, let $b = 1$ and $a = c = d = 0$. Then there is no matrix A which will satisfy the equation.

EXERCISE SET 1.6

1. This system of equations is of the form $A\mathbf{x} = \mathbf{b}$, where

$$A = \begin{bmatrix} 1 & 1 \\ 5 & 6 \end{bmatrix} \qquad \mathbf{x} = \begin{bmatrix} x_1 \\ x_2 \end{bmatrix} \quad \text{and} \quad \mathbf{b} = \begin{bmatrix} 2 \\ 9 \end{bmatrix}$$

By Theorem 1.4.5,

$$A^{-1} = \begin{bmatrix} 6 & -1 \\ -5 & 1 \end{bmatrix}$$

Thus

$$\mathbf{x} = A^{-1}\mathbf{b} = \begin{bmatrix} 6 & -1 \\ -5 & 1 \end{bmatrix} \begin{bmatrix} 2 \\ 9 \end{bmatrix} = \begin{bmatrix} 3 \\ -1 \end{bmatrix}$$

That is,

$$x_1 = 3 \qquad \text{and} \qquad x_2 = -1$$

3. This system is of the form $A\mathbf{x} = \mathbf{b}$, where

$$A = \begin{bmatrix} 1 & 3 & 1 \\ 2 & 2 & 1 \\ 2 & 3 & 1 \end{bmatrix} \qquad \mathbf{x} = \begin{bmatrix} x_1 \\ x_2 \\ x_3 \end{bmatrix} \quad \text{and} \quad \mathbf{b} = \begin{bmatrix} 4 \\ -1 \\ 3 \end{bmatrix}$$

By direct computation we obtain

$$A^{-1} = \begin{bmatrix} -1 & 0 & 1 \\ 0 & -1 & 1 \\ 2 & 3 & -4 \end{bmatrix}$$

so that

$$\mathbf{x} = A^{-1}\mathbf{b} = \begin{bmatrix} -1 \\ 4 \\ -7 \end{bmatrix}$$

That is,

$$x_1 = -1, \quad x_2 = 4, \quad \text{and} \quad x_3 = -7$$

5.   The system is of the form $A\mathbf{x} = \mathbf{b}$, where

$$A = \begin{bmatrix} 1 & 1 & 1 \\ 1 & 1 & -4 \\ -4 & 1 & 1 \end{bmatrix} \qquad \mathbf{x} = \begin{bmatrix} x_1 \\ x_2 \\ x_3 \end{bmatrix} \quad \text{and} \quad \mathbf{b} = \begin{bmatrix} 5 \\ 10 \\ 0 \end{bmatrix}$$

By direct computation, we obtain

$$A^{-1} = \left(\frac{1}{5}\right) \begin{bmatrix} 1 & 0 & -1 \\ 3 & 1 & 1 \\ 1 & -1 & 0 \end{bmatrix}$$

Thus,

$$\mathbf{x} = A^{-1}\mathbf{b} = \begin{bmatrix} 1 \\ 5 \\ -1 \end{bmatrix}$$

That is, $x_1 = 1$, $x_2 = 5$, and $x_3 = -1$.

7. The system is of the form $A\mathbf{x} = \mathbf{b}$ where

$$A - \begin{bmatrix} 3 & 5 \\ 1 & 2 \end{bmatrix} \qquad \mathbf{x} = \begin{bmatrix} x_1 \\ x_2 \end{bmatrix} \qquad \text{and} \qquad \mathbf{b} = \begin{bmatrix} b_1 \\ b_2 \end{bmatrix}$$

By Theorem 1.4.5, we have

$$A^{-1} = \begin{bmatrix} 2 & -5 \\ -1 & 3 \end{bmatrix}$$

Thus

$$\mathbf{x} = A^{-1}\mathbf{b} = \begin{bmatrix} 2b_1 - 5b_2 \\ -b_1 + 3b_2 \end{bmatrix}$$

That is,

$$x_1 = 2b_1 - 5b_2 \qquad \text{and} \qquad x_2 = -b_1 + 3b_2$$

9. The system is of the form $A\mathbf{x} = \mathbf{b}$, where

$$A = \begin{bmatrix} 1 & 2 & 1 \\ 1 & -1 & 1 \\ 1 & 1 & 0 \end{bmatrix} \qquad \mathbf{x} = \begin{bmatrix} x_1 \\ x_2 \\ x_3 \end{bmatrix} \qquad \text{and} \qquad \mathbf{b} = \begin{bmatrix} b_1 \\ b_2 \\ b_3 \end{bmatrix}$$

We compute

$$A^{-1} = \begin{bmatrix} -1/3 & 1/3 & 1 \\ 1/3 & -1/3 & 0 \\ 2/3 & 1/3 & -1 \end{bmatrix}$$

so that

$$\mathbf{x} = A^{-1}\mathbf{b} = \begin{bmatrix} -(1/3)b_1 + (1/3)b_2 + b_3 \\ (1/3)b_1 - (1/3)b_2 \\ (2/3)b_1 + (1/3)b_2 - b_3 \end{bmatrix}$$

9. **(a)** In this case, we let

$$\mathbf{b} = \begin{bmatrix} -1 \\ 3 \\ 4 \end{bmatrix}$$

Then

$$\mathbf{x} = A^{-1}\mathbf{b} = \begin{bmatrix} 16/3 \\ -4/3 \\ -11/3 \end{bmatrix}$$

That is, $x_1 = 16/3$, $x_2 = -4/3$, and $x_3 = -11/3$.

(c) In this case, we let

$$\mathbf{b} = \begin{bmatrix} -1 \\ -1 \\ 3 \end{bmatrix}$$

Then

$$\mathbf{x} = A^{-1}\mathbf{b} = \begin{bmatrix} 3 \\ 0 \\ -4 \end{bmatrix}$$

That is, $x_1 = 3$, $x_2 = 0$, and $x_3 = -4$.

10. We apply the method of Example 2 to the two systems (a) and (c) solved above. The coefficient matrix augmented by the two **b** matrices yields

$$\left[\begin{array}{ccc|c|c} 1 & 2 & 1 & -1 & -1 \\ 1 & -1 & 1 & 3 & -1 \\ 1 & 1 & 0 & 4 & 3 \end{array}\right]$$

The reduced row-echelon form may be obtained as follows:

$$\left[\begin{array}{ccc|c|c} 1 & 2 & 1 & -1 & -1 \\ 0 & -3 & 0 & 4 & 0 \\ 0 & 2 & -1 & 1 & 4 \end{array}\right]$$

Add -1 times Row 2 to Row 3 and -1 times Row 1 to Row 2.

$$\left[\begin{array}{ccc|c|c} 1 & 0 & 2 & -2 & -5 \\ 0 & 1 & 0 & -4/3 & 0 \\ 0 & 0 & -1/2 & 11/6 & 2 \end{array}\right]$$

Add -1 times Row 3 to Row 2, divide Row 2 by -3 and Row 3 by 2, add -1 times Row 2 to Row 3.

$$\left[\begin{array}{ccc|c|c} 1 & 0 & 0 & 16/3 & 3 \\ 0 & 1 & 0 & -4/3 & 0 \\ 0 & 0 & 1 & -11/3 & -4 \end{array}\right]$$

Multiply Row 3 by -2 and then add -2 times Row 3 to Row 1.

This, fortunately, yields the same results as in Exercise 9.

11. The coefficient matrix, augmented by the two **b** matrices, yields

$$\left[\begin{array}{cc|c|c} 1 & -5 & 1 & -2 \\ 3 & 2 & 4 & 5 \end{array}\right]$$

This reduces to

$$\left[\begin{array}{cc|c|c} 1 & -5 & 1 & -2 \\ 0 & 17 & 1 & 11 \end{array}\right]$$

Add -3 times Row 1 to Row 2.

or

$$\left[\begin{array}{cc|c|c} 1 & 0 & 22/17 & 21/17 \\ 0 & 1 & 1/17 & 11/17 \end{array}\right]$$

Divide Row 2 by 17 and add 5 times Row 2 to Row 1.

Thus the solution to Part (a) is $x_1 = 22/17$, $x_2 = 1/17$, and to Part (b) is $x_1 = 21/17$, $x_2 = 11/17$.

15. As above, we set up the matrix

$$\left[\begin{array}{ccc|c|c} 1 & -2 & 1 & -2 & 1 \\ 2 & -5 & 1 & 1 & -1 \\ 3 & -7 & 2 & -1 & 0 \end{array}\right]$$

This reduces to

$$\left[\begin{array}{ccc|c|c} 1 & -2 & 1 & -2 & 1 \\ 0 & -1 & -1 & 5 & -3 \\ 0 & -1 & -1 & 5 & -3 \end{array}\right]$$

Add appropriate multiples of Row 1 to Rows 2 and 3.

or

$$\left[\begin{array}{ccc|c|c} 1 & -2 & 1 & -2 & 1 \\ 0 & 1 & 1 & -5 & 3 \\ 0 & 0 & 0 & 0 & 0 \end{array}\right]$$

Add -1 times Row 2 to Row 3 and multiply Row 2 by -1.

or

$$\left[\begin{array}{ccc|c|c} 1 & 0 & 3 & -12 & 7 \\ 0 & 1 & 1 & -5 & 3 \\ 0 & 0 & 0 & 0 & 0 \end{array}\right]$$

Add twice Row 2 to Row 3.

Thus if we let $x_3 = t$, we have for Part (a) $x_1 = -12 - 3t$ and $x_2 = -5 - t$, while for Part (b) $x_1 = 7 - 3t$ and $x_2 = 3 - t$.

16.   The augmented matrix for this sytem is

$$\begin{bmatrix} 6 & -4 & b_1 \\ 3 & -2 & b_2 \end{bmatrix}$$

which, when reduced to row-echelon form, becomes

$$\begin{bmatrix} 1 & -2/3 & b_2/3 \\ 0 & 0 & b_1 - 2b_2 \end{bmatrix}$$

The system is consistent if and only if $b_1 - 2b_2 = 0$ or $b_1 = 2b_2$. Thus $A\mathbf{x} = \mathbf{b}$ is consistent if and only if $\mathbf{b}$ has the form

$$\mathbf{b} = \begin{bmatrix} 2b_2 \\ b_2 \end{bmatrix}$$

17.   The augmented matrix for this system of equations is

$$\begin{bmatrix} 1 & -2 & 5 & b_1 \\ 4 & -5 & 8 & b_2 \\ -3 & 3 & -3 & b_3 \end{bmatrix}$$

If we reduce this matrix to row-echelon form, we obtain

$$\begin{bmatrix} 1 & -2 & 5 & b_1 \\ 0 & 1 & -4 & \frac{1}{3}(b_2 - 4b_1) \\ 0 & 0 & 0 & -b_1 + b_2 + b_3 \end{bmatrix}$$

The third row implies that $b_3 = b_1 - b_2$. Thus, $A\mathbf{x} = \mathbf{b}$ is consistent if and only if $\mathbf{b}$ has the form

$$\mathbf{b} = \begin{bmatrix} b_1 \\ b_2 \\ b_1 - b_2 \end{bmatrix}$$

20. **(a)** In general, $A\mathbf{x} = \mathbf{x}$ if and only if $A\mathbf{x} - \mathbf{x} = \mathbf{0}$ if and only if $(A - I)\mathbf{x} = \mathbf{0}$. In this case,

$$A = \begin{bmatrix} 2 & 1 & 2 \\ 2 & 2 & -2 \\ 3 & 1 & 1 \end{bmatrix}$$

and hence,

$$A - I = \begin{bmatrix} 1 & 1 & 2 \\ 2 & 1 & -2 \\ 3 & 1 & 0 \end{bmatrix}$$

The reduced row-echelon form of A–I is

$$\begin{bmatrix} 1 & 0 & 0 \\ 0 & 1 & 0 \\ 0 & 0 & 1 \end{bmatrix}$$

Therefore, the system of equations $(A - I)\mathbf{x} = \mathbf{0}$ has only the trivial solution $x_1 = x_2 = x_3 = 0$; that is, $\mathbf{x} = \mathbf{0}$ is the only solution to $A\mathbf{x} = \mathbf{x}$.

22. **(a)** It is clear that $\mathbf{x} = \mathbf{0}$ is the only solution to the equation $A\mathbf{x} = \mathbf{0}$. We may then invoke Theorem 1.6.4 to conclude that A is invertible.

23.   Since $A\mathbf{x} = \mathbf{0}$ has only $\mathbf{x} = \mathbf{0}$ as a solution, Theorem 1.6.4 guarantees that A is invertible. By Theorem 1.4.8 (b), A^k is also invertible. In fact,

$$(A^k)^{-1} = (A^{-1})^k$$

Since the proof of Theorem 1.4.8 (b) was omitted, we note that

$$\underbrace{A^{-1}A^{-1}\cdots A^{-1}}_{\substack{k \\ \text{factors}}} \underbrace{AA\cdots A}_{\substack{k \\ \text{factors}}} = I$$

Because A^k is invertible, Theorem 1.6.4 allows us to conclude that $A^k\mathbf{x} = \mathbf{0}$ has only the trivial solution.

24.   First, assume that $A\mathbf{x} = \mathbf{0}$ has only the trivial solution. It then follows from Theorem 1.6.4 that A is invertible. Now Q is invertible by assumption. Thus, QA is invertible because the product of two invertible matrices is invertible. If we apply Theorem 1.6.4 again, we see that $(QA)\mathbf{x} = \mathbf{0}$ has only the trivial solution.

   Conversely, assume that $(QA)\mathbf{x} = \mathbf{0}$ has only the trivial solution; thus, QA is invertible by Theorem 1.6.4. But if QA is invertible, then so is A because $A = Q^{-1}(QA)$ is the product of invertible matrices. But if A is invertible, then $A\mathbf{x} = \mathbf{0}$ has only the trivial solution.

25.   Suppose that $\mathbf{x}_1$ is a fixed matrix which satisfies the equation $A\mathbf{x}_1 = \mathbf{b}$. Further, let $\mathbf{x}$ be any matrix whatsoever which satisfies the equation $A\mathbf{x} = \mathbf{b}$. We must then show that there is a matrix $\mathbf{x}_0$ which satisfies both of the equations $\mathbf{x} = \mathbf{x}_1 + \mathbf{x}_0$ and $A\mathbf{x}_0 = \mathbf{0}$.

   Clearly, the first equation implies that

$$\mathbf{x}_0 = \mathbf{x} - \mathbf{x}_1$$

This candidate for $\mathbf{x}_0$ will satisfy the second equation because

$$A\mathbf{x}_0 = A(\mathbf{x} - \mathbf{x}_1) = A\mathbf{x} - A\mathbf{x}_1 = \mathbf{b} - \mathbf{b} = \mathbf{0}$$

We must also show that if both $A\mathbf{x}_1 = \mathbf{b}$ and $A\mathbf{x}_0 = \mathbf{0}$, then $A(\mathbf{x}_1 + \mathbf{x}_0) = \mathbf{b}$. But

$$A(\mathbf{x}_1 + \mathbf{x}_0) = A\mathbf{x}_1 + A\mathbf{x}_0 = \mathbf{b} + \mathbf{0} = \mathbf{b}$$

26. We are given that $AB = I$. By Part (a) of Theorem 1.6.3, A must be the inverse of B. But then

$$A = B^{-1} \quad \Rightarrow \quad A^{-1} = (B^{-1})^{-1} = B$$

That is, $A^{-1} = B$.

27. **(a)** For the equation $\mathbf{x} = A\mathbf{x} + \mathbf{b}$ to have a unique solution for $\mathbf{x}$, the equation $(I - A)\mathbf{x} = \mathbf{b}$ must also have a unique solution. We are assured of this if the matrix $I - A$ is invertible.

28. No. The system of equations $A\mathbf{x} = \mathbf{x}$ is equivalent to the system $(A - I)\mathbf{x} = \mathbf{0}$. For this system to have a unique solution, $A - I$ must be invertible. If, for instance, $A = I$, then any vector $\mathbf{x}$ will be a solution to the system of equations $A\mathbf{x} = \mathbf{x}$.

Note that if $\mathbf{x} \neq \mathbf{0}$ is a solution to the equation $A\mathbf{x} = \mathbf{x}$, then so is $k\mathbf{x}$ for any real number k. A unique solution can only exist if $A - I$ is invertible, in which case, $\mathbf{x} = \mathbf{0}$.

29. Yes. The matrices A and B need not be square. For example, if

$$A = \begin{bmatrix} 1 & 0 & 0 \\ 0 & 0 & 1 \end{bmatrix} \quad \text{and} \quad B = \begin{bmatrix} 1 & 0 \\ 0 & 0 \\ 0 & 1 \end{bmatrix}$$

then $AB = I_2$ but

$$BA = \begin{bmatrix} 1 & 0 & 0 \\ 0 & 0 & 0 \\ 0 & 0 & 1 \end{bmatrix}$$

In fact, only square matrices can be invertible.

30. Let A and B be square matrices of the same size. If either A or B is singular, then AB is singular.

EXERCISE SET 1.7

6. For A to be symmetric, the following equations must hold:

$$
\begin{aligned}
a - 2b + 2c &= 3 \\
2a + b + c &= 0 \\
a + c &= -2
\end{aligned}
$$

The solution to this system of equations is $a = 11$, $b = -9$, $c = -13$. This is the only set of values for which A is symmetric.

7. The matrix A fails to be invertible if and only if $a + b - 1 = 0$ and the matrix B fails to be invertible if and only if $2a - 3b - 7 = 0$. For both of these conditions to hold, we must have $a = 2$ and $b = -1$.

8. **(a)** Since the product is not symmetric, the matrices do not commute.

(b) Since the product is symmetric, the matrices commute.

9. We know that A and B will commute if and only if

$$
AB = \begin{bmatrix} 2 & 1 \\ 1 & -5 \end{bmatrix} \begin{bmatrix} a & b \\ b & d \end{bmatrix} = \begin{bmatrix} 2a + b & 2b + d \\ a - 5b & b - 5d \end{bmatrix}
$$

is symmetric. So $2b + d = a - 5b$, from which it follows that $a - d = 7b$.

10.   (b)   If

$$A^{-2} = \begin{bmatrix} 9 & 0 & 0 \\ 0 & 4 & 0 \\ 0 & 0 & 1 \end{bmatrix} \quad \text{then} \quad A^{-1} = \begin{bmatrix} \pm 3 & 0 & 0 \\ 0 & \pm 2 & 0 \\ 0 & 0 & \pm 1 \end{bmatrix}$$

so that

$$A = \begin{bmatrix} \pm\dfrac{1}{3} & 0 & 0 \\ 0 & \pm\dfrac{1}{2} & 0 \\ 0 & 0 & \pm 1 \end{bmatrix}$$

Thus there are 8 possible matrices A which satisfy the given equation.

11.   (b)   Clearly

$$A = \begin{bmatrix} ka_{11} & ka_{12} & ka_{13} \\ ka_{21} & ka_{22} & ka_{23} \\ ka_{31} & ka_{32} & ka_{33} \end{bmatrix} \begin{bmatrix} 3/k & 0 & 0 \\ 0 & 5/k & 0 \\ 0 & 0 & 7/k \end{bmatrix}$$

for any real number $k \neq 0$.

13.   We verify the result for the matrix A by finding its inverse.

$$\left[\begin{array}{ccc|ccc} -1 & 2 & 5 & 1 & 0 & 0 \\ 0 & 1 & 3 & 0 & 1 & 0 \\ 0 & 0 & -4 & 0 & 0 & 1 \end{array}\right]$$

$$\left[\begin{array}{ccc|ccc} 1 & -2 & -5 & -1 & 0 & 0 \\ 0 & 1 & 3 & 0 & 1 & 0 \\ 0 & 0 & 1 & 0 & 0 & -1/4 \end{array}\right]$$

> Multiply Row 1 by -1
> and Row 3 by $-1/4$.

$$\left[\begin{array}{ccc|ccc} 1 & 0 & 1 & -1 & 2 & 0 \\ 0 & 1 & 0 & 0 & 1 & 3/4 \\ 0 & 0 & 1 & 0 & 0 & 1/4 \end{array} \right] \quad \boxed{\begin{array}{l} \text{Add 2 times Row 2 to} \\ \text{Row 1 and } -3 \text{ times} \\ \text{Row 3 to Row 2.} \end{array}}$$

$$\left[\begin{array}{ccc|ccc} 1 & 0 & 0 & -1 & 2 & 1/4 \\ 0 & 1 & 0 & 0 & 1 & 3/4 \\ 0 & 0 & 1 & 0 & 0 & -1/4 \end{array} \right] \quad \boxed{\begin{array}{l} \text{Add } -1 \text{ times} \\ \text{Row 3 to Row 1.} \end{array}}$$

Thus A^{-1} is indeed upper triangular.

14. **(a)** Since $A^{-1} = \left[\begin{array}{cc} 3/5 & 1/5 \\ 1/5 & 2/5 \end{array} \right]$, the inverse is indeed symmetric.

15. **(b)** By Theorem 1.4.9,

$$(2A^2 - 3A + I)^T = (2A^2)^T + (-3A)^T + I^T$$
$$= 2A^T A^T - 3A^T + I^T$$
$$= 2A^2 - 3A + I$$

because A and I are both symmetric. Therefore $2A^2 - 3A + I$ is also symmetric.

16. **(a)** We prove this result by induction on k. If A is symmetric, then so is A^k when $k = 1$. (Since $A^0 = I$, the result is even true for $k = 0$.) Suppose that A^k is symmetric for $k = 1, \ldots, N$ for some $N \geq 1$. Then $A^{N+1} = A(A^N)$ is the product of two symmetric matrices. Hence, by Theorem 1.4.9,

$$(A^{N+1})^T = (A^N)^T A^T = A^N A = A^{N+1}$$

and therefore A^{N+1} is also symmetric. Thus A^k is symmetric for all integers $k \geq 0$.

(b) From Theorem 1.7.2 and Part (a), above, we know that A^k is symmetric if A is symmetric. So also is $a_k A^k$ for any constant a_k and so is any such sum or difference. Thus, $p(A)$ is symmetric.

18. If $A = A^T A$, then $A^T = (A^T A)^T = A^T (A^T)^T = A^T A = A$, so A is symmetric. This implies that $A = A^T$, so $A = A^2$.

19. Let

$$A = \begin{bmatrix} x & 0 & 0 \\ 0 & y & 0 \\ 0 & 0 & z \end{bmatrix}$$

Then if $A^2 - 3A - 4I = O$, we have

$$\begin{bmatrix} x^2 & 0 & 0 \\ 0 & y^2 & 0 \\ 0 & 0 & z^2 \end{bmatrix} - 3 \begin{bmatrix} x & 0 & 0 \\ 0 & y & 0 \\ 0 & 0 & z \end{bmatrix} - 4 \begin{bmatrix} 1 & 0 & 0 \\ 0 & 1 & 0 \\ 0 & 0 & 1 \end{bmatrix} = O$$

This leads to the system of equations

$$x^2 - 3x - 4 = 0$$
$$y^2 - 3y - 4 = 0$$
$$z^2 - 3z - 4 = 0$$

which has the solutions $x = 4, -1$, $y = 4, -1$, $z = 4, -1$. Hence, there are 8 possible choices for x, y, and z, respectively, namely $(4, 4, 4)$, $(4, 4, -1)$, $(4, -1, 4)$, $(4, -1, -1)$, $(-1, 4, 4)$, $(-1, 4, -1)$, $(-1, -1, 4)$, and $(-1, -1, -1)$.

20. **(a)** Since $a_{ij} = i^2 + j^2 = j^2 + i^2 = a_{ji}$, A is symmetric.

(d) Since $2i^2 + 2j^3$ need not equal $2j^2 + 2i^3$, A is not symmetric. For instance $a_{21} \neq a_{12}$. In fact, A will be symmetric only in the trivial case where $n = 1$.

22. **(a)** If A is invertible and skew-symmetric, then by Theorem 1.4.10,

$$(A^{-1})^T = (A^T)^{-1} = (-A)^{-1} = -(A^{-1})$$

so A^{-1} is skew-symmetric.

(c) From the hint,

$$A = \frac{1}{2}(A + A^T) + \frac{1}{2}(A - A^T)$$

so we need only prove that $\frac{1}{2}(A + A^T)$ is symmetric and that $\frac{1}{2}(A - A^T)$ is skew-symmetric. To this end, note that

$$\frac{1}{2}(A + A^T)^T = \frac{1}{2}(A^T + (A^T)^T) = \frac{1}{2}(A + A^T)$$

and that

$$\frac{1}{2}(A - A^T)^T = \frac{1}{2}(A^T - (A^T)^T) = \frac{1}{2}(A^T - A) = -\frac{1}{2}(A - A^T)$$

23. The matrix

$$A = \begin{bmatrix} 0 & 1 \\ -1 & 0 \end{bmatrix}$$

is skew-symmetric but

$$AA = A^2 = \begin{bmatrix} -1 & 0 \\ 0 & -1 \end{bmatrix}$$

is not skew-symmetric. Therefore, the result does not hold.

In general, suppose that A and B are commuting skew-symmetric matrices. Then $(AB)^T = (BA)^T = A^T B^T = (-A)(-B) = AB$, so that AB is symmetric rather than skew-symmetric. [We note that if A and B are skew-symmetric and their product is symmetric, then $AB = (AB)^T = B^T A^T = (-B)(-A) = BA$, so the matrices commute and thus skew-symmetric matrices, too, commute if and only if their product is symmetric.]

24. **(a)** Here

$$Ly = \begin{bmatrix} 1 & 0 & 0 \\ -2 & 3 & 0 \\ 2 & 4 & 1 \end{bmatrix} \begin{bmatrix} y_1 \\ y_2 \\ y_3 \end{bmatrix} = \begin{bmatrix} 1 \\ -2 \\ 0 \end{bmatrix} = b$$

has the solution $y_1 = 1$, $y_2 = 0$, $y_3 = -2$. Thus

$$Ux = \begin{bmatrix} 2 & -1 & 3 \\ 0 & 1 & 2 \\ 0 & 0 & 4 \end{bmatrix} \begin{bmatrix} x_1 \\ x_2 \\ x_3 \end{bmatrix} = \begin{bmatrix} 1 \\ 0 \\ -2 \end{bmatrix} = y$$

so that $x_1 = 7/4$, $x_2 = 1$, $x_3 = -1/2$.

25. Let

$$A = \begin{bmatrix} x & y \\ 0 & z \end{bmatrix}$$

Then

$$A^3 = \begin{bmatrix} x^3 & y(x^2 + xz + z^2) \\ 0 & z^3 \end{bmatrix} = \begin{bmatrix} 1 & 30 \\ 0 & -8 \end{bmatrix}$$

Hence, $x^3 = 1$ which implies that $x = 1$, and $z^3 = -8$ which implies that $z = -2$. Therefore, $3y = 30$ and thus $y = 10$.

26. An $n \times n$ matrix has n^2 entries, n of them along the diagonal and $(n^2 - n)/2$ of them above the diagonal. Thus we could assign distinct values to $n + (n^2 - n)/2 = (n^2 + n)/2$ entries. The remaining $(n^2 - n)/2$ entries below the diagonal would coincide with the corresponding entries above the diagonal.

27. To multiply two diagonal matrices, multiply their corresponding diagonal elements to obtain a new diagonal matrix. Thus, if D_1 and D_2 are diagonal matrices with diagonal elements $d_1, \ldots, d_n$ and $e_1, \ldots, e_n$ respectively, then $D_1 D_2$ is a diagonal matrix with diagonal elements $d_1 e_1, \ldots, d_n e_n$. The proof follows directly from the definition of matrix multiplication.

28. Suppose that D has diagonal elements $d_1, \ldots, d_n$. Since A is a square matrix and $AD = I$, Theorem 1.6.3 guarantees that $A = D^{-1}$. Thus, A is a diagonal matrix with diagonal elements $1/d_1, \ldots, 1/dn$. Note that since D is invertible, it follows that all of its diagonal elements must be nonzero.

29. In general, let $A = [a_{ij}]_{n \times n}$ denote a lower triangular matrix with no zeros on or below the diagonal and let $A\mathbf{x} = \mathbf{b}$ denote the system of equations where $\mathbf{b} = [b_1, b_2, \ldots, b_n]^T$. Since A is lower triangular, the first row of A yields the equation $a_{11}x_1 = b_1$. Since $a_{11} \neq 0$, we can solve for x_1. Next, the second row of A yields the equation $a_{21}x_1 + a_{22}x_2 = b_2$. Since we know x_1 and since $a_{22} \neq 0$, we can solve for x_2. Continuing in this way, we can solve for successive values of x_i by back substituting all of the previously found values $x_1, x_2, \ldots, x_{i-1}$.

30. **(a)** False. If A were invertible, then A^T would also be invertible, and hence so would AA^T.

(c) True. Use Theorems 1.4.10 and 1.5.3.

(d) False. Let $A = \begin{bmatrix} a & b \\ c & d \end{bmatrix}$. Then if A^2 is symmetric, we must have $ab + bd = ac + cd$ so that either $a = -d$ or $b = c$. If $b = c$, then A is symmetric. However, if $b \neq c$ but $a = -d$, then A^2 is symmetric, but A is not. For instance, let $a = \begin{bmatrix} 1 & 0 \\ 1 & -1 \end{bmatrix}$.

SUPPLEMENTARY EXERCISES 1

1.

$$\begin{bmatrix} \dfrac{3}{5} & -\dfrac{4}{5} & x \\[2ex] \dfrac{4}{5} & \dfrac{3}{5} & y \end{bmatrix}$$

$$\begin{bmatrix} 1 & -\dfrac{4}{3} & \dfrac{5}{3}x \\[2ex] \dfrac{4}{5} & \dfrac{3}{5} & y \end{bmatrix}$$

Multiply Row 1 by 5/3.

$$\begin{bmatrix} 1 & -\dfrac{4}{3} & \dfrac{5}{3}x \\[2ex] 0 & \dfrac{5}{3} & -\dfrac{4}{3}x + y \end{bmatrix}$$

Add −4/5 times Row 1 to Row 2.

$$\begin{bmatrix} 1 & -\dfrac{4}{3} & \dfrac{5}{3}x \\[2ex] 0 & 1 & -\dfrac{4}{5}x + \dfrac{3}{5}y \end{bmatrix}$$

Multiply Row 2 by 3/5.

$$\begin{bmatrix} 1 & 0 & \dfrac{3}{5}x + \dfrac{4}{5}y \\[2ex] 0 & 1 & -\dfrac{4}{5}x + \dfrac{3}{5}y \end{bmatrix}$$

Add 4/3 times Row 2 to Row 1.

Thus,

$$x' = \frac{3}{5}x + \frac{4}{5}y$$

$$y' = -\frac{4}{5}x + \frac{3}{5}y$$

3. We denote the system of equations by

$$a_{11}x_1 + a_{12}x_2 + a_{13}x_3 + a_{14}x_4 = 0$$

$$a_{21}x_1 + a_{22}x_2 + a_{23}x_3 + a_{24}x_4 = 0$$

If we substitute both sets of values for x_1, x_2, x_3, and x_4 into the first equation, we obtain

$$a_{11} - a_{12} + a_{13} + 2a_{14} = 0$$

$$2a_{11} + 3a_{13} - 2a_{14} = 0$$

where a_{11}, a_{12}, a_{13}, and a_{14} are variables. If we substitute both sets of values for x_1, x_2, x_3, and x_4 into the second equation, we obtain

$$a_{21} - a_{22} + a_{23} + 2a_{24} = 0$$

$$2a_{21} + 3a_{23} - a_{24} = 0$$

where a_{21}, a_{22}, a_{23}, and a_{24} are again variables. The two systems above both yield the matrix

$$\begin{bmatrix} 1 & -1 & 1 & 2 & 0 \\ 2 & 0 & 3 & -1 & 0 \end{bmatrix}$$

which reduces to

$$\begin{bmatrix} 1 & 0 & 3/2 & -1/2 & 0 \\ 0 & 1 & 1/2 & -5/2 & 0 \end{bmatrix}$$

This implies that

$$a_{11} = -(3/2)a_{13} + (1/2)a_{14}$$

$$a_{12} = -(1/2)a_{13} + (5/2)a_{14}$$

and similarly,

$$a_{21} = (-3/2)a_{23} + (1/2)a_{24}$$

$$a_{22} = (-1/2)a_{23} + (5/2)a_{24}$$

As long as our choice of values for the numbers a_{ij} is consistent with the above, then the system will have a solution. For simplicity, and to insure that neither equation is a multiple of the other, we let $a_{13} = a_{14} = -1$ and $a_{23} = 0$, $a_{24} = 2$. This means that $a_{11} = 1$, $a_{12} = -2$, $a_{21} = 1$, and $a_{22} = 5$, so that the system becomes

$$x_1 - 2x_2 - x_3 - x_4 = 0$$

$$x_1 + 5x_2 \qquad + 2x_4 = 0$$

Of course, this is just one of infinitely many possibilities.

4. Suppose that the box contains x pennies, y nickels, and z dimes. We know that

$$x + 5y + 10z = 83$$

$$x + y + z = 13$$

The augmented matrix of this system is

$$\begin{bmatrix} 1 & 5 & 10 & 83 \\ 1 & 1 & 1 & 13 \end{bmatrix}$$

which can be reduced to the matrix

$$\begin{bmatrix} 1 & 0 & -\dfrac{5}{4} & -\dfrac{9}{2} \\ 0 & 1 & \dfrac{9}{4} & \dfrac{35}{2} \end{bmatrix}$$

by Gauss-Jordan elimination. Thus,

$$x = \frac{5}{4}z - \frac{9}{2} \quad \text{and} \quad y = -\frac{9}{4}z + \frac{35}{2}$$

Since x, y, and z must all be integers between 0 and 13, then z must be either 2, 6, or 10 because those are the only values of z for which x and y are both integers. However, $z = 2$ implies that $x < 0$ while $z = 10$ implies that $y < 0$. Therefore, the only solution is $x = 3$, $y = 4$, and $z = 6$.

5.   As in Exercise 4, we reduce the system to the equations

$$x = \frac{1 + 5z}{4}$$

$$y = \frac{35 - 9z}{4}$$

Since x, y, and z must all be positive integers, we have $z > 0$ and $35 - 9z > 0$ or $4 > z$. Thus we need only check the three values $z = 1, 2, 3$ to see whether or not they produce integer solutions for x and y. This yields the unique solution $x = 4$, $y = 2$, $z = 3$.

6.   The augmented matrix of this system is

$$\begin{bmatrix} 1 & 1 & 1 & 4 \\ 0 & 0 & 1 & 2 \\ 0 & 0 & a^2 - 4 & a - 2 \end{bmatrix}$$

which can be reduced to the matrix

$$\begin{bmatrix} 1 & 1 & 0 & 2 \\ 0 & 0 & 1 & 2 \\ 0 & 0 & 0 & 2a^2 - a - 6 \end{bmatrix}$$

Row 3 of the above matrix implies that the system of equations will have a solution if and only if $2a^2 - a - 6 = 0$; that is, if and only if $a = 2$ or $a = -3/2$. For either of those values of a, the solution is $x_1 = 2 - t$, $x_2 = t$, and $x_3 = 2$. Hence, the system has infinitely many solutions provided $a = 2$ or $a = -3/2$. For all other values of a, there will be no solution.

8.   Let $s = xy$, $t = \sqrt{y}$, and $u = zy$. Then we have a system of linear equations in s, t, and u with augmented matrix

$$\begin{bmatrix} 1 & -2 & 3 & 8 \\ 2 & -3 & 2 & 7 \\ -1 & 1 & 2 & 4 \end{bmatrix}$$

This reduces to

$$\begin{bmatrix} 1 & 0 & 0 & 5 \\ 0 & 1 & 0 & 3 \\ 0 & 0 & 1 & 3 \end{bmatrix}$$

so that

$$xy = 5$$

$$\sqrt{y} = 3$$

$$zy = 3$$

Thus from the second equation, $y = 9$, and from the first and third equations, $x = 5/9$ and $z = 3/9 = 1/3$.

9. Note that K must be a 2×2 matrix. Let

$$K = \begin{bmatrix} a & b \\ c & d \end{bmatrix}$$

Then

$$\begin{bmatrix} 1 & 4 \\ -2 & 3 \\ 1 & -2 \end{bmatrix} \begin{bmatrix} a & b \\ c & d \end{bmatrix} \begin{bmatrix} 2 & 0 & 0 \\ 0 & 1 & -1 \end{bmatrix} = \begin{bmatrix} 8 & 6 & -6 \\ 6 & -1 & 1 \\ -4 & 0 & 0 \end{bmatrix}$$

or

$$\begin{bmatrix} 1 & 4 \\ -2 & 3 \\ 1 & -2 \end{bmatrix} \begin{bmatrix} 2a & b & -b \\ 2c & d & -d \end{bmatrix} = \begin{bmatrix} 8 & 6 & -6 \\ 6 & -1 & 1 \\ -4 & 0 & 0 \end{bmatrix}$$

or

$$\begin{bmatrix} 2a + 8c & b + 4d & -b - 4d \\ -4a + 6c & -2b + 3d & 2b - 3d \\ 2a - 4c & b - 2d & -b + 2d \end{bmatrix} = \begin{bmatrix} 8 & 6 & -6 \\ 6 & -1 & 1 \\ -4 & 0 & 0 \end{bmatrix}$$

Thus

$$
\begin{aligned}
2a & & +8c & & &= 8 \\
& b & & +4d &&= 6 \\
-4a & & +6c & & &= 6 \\
& -2b & & +3d &&= -1 \\
2a & & -4c & & &= -4 \\
& b & & -2d &&= 0
\end{aligned}
$$

Note that we have omitted the 3 equations obtained by equating elements of the last columns of these matrices because the information so obtained would be just a repeat of that gained by equating elements of the second columns. The augmented matrix of the above system is

$$
\begin{bmatrix}
2 & 0 & 8 & 0 & 8 \\
0 & 1 & 0 & 4 & 6 \\
-4 & 0 & 6 & 0 & 6 \\
0 & -2 & 0 & 3 & -1 \\
2 & 0 & -4 & 0 & -4 \\
0 & 1 & 0 & -2 & 0
\end{bmatrix}
$$

The reduced row-echelon form of this matrix is

$$
\begin{bmatrix}
1 & 0 & 0 & 0 & 0 \\
0 & 1 & 0 & 0 & 2 \\
0 & 0 & 1 & 0 & 1 \\
0 & 0 & 0 & 1 & 1 \\
0 & 0 & 0 & 0 & 0 \\
0 & 0 & 0 & 0 & 0
\end{bmatrix}
$$

Thus $a = 0$, $b = 2$, $c = 1$, and $d = 1$.

10. If we substitute the given values for x, y, and z into the system of equations, we obtain

$$a - b - 6 = -3$$
$$-2 + b + 2c = -1$$
$$a - 3 - 2c = -3$$

or

$$a - b \quad\;\; = 3$$
$$b + 2c = 1$$
$$a \quad\; - 2c = 0$$

This system of equations in a, b, and c yields the matrix

$$\begin{bmatrix} 1 & -1 & 0 & 3 \\ 0 & 1 & 2 & 1 \\ 1 & 0 & -2 & 0 \end{bmatrix}$$

which reduces to

$$\begin{bmatrix} 1 & 0 & 0 & 2 \\ 0 & 1 & 0 & -1 \\ 0 & 0 & 1 & 1 \end{bmatrix}$$

Hence, $a = 2$, $b = -1$, and $c = 1$.

11. The matrix X in Part (a) must be 2×3 for the operations to make sense. The matrices in Parts (b) and (c) must be 2×2.

(b) Let $X = \begin{bmatrix} x & y \\ z & w \end{bmatrix}$. Then

$$X \begin{bmatrix} 1 & -1 & 2 \\ 3 & 0 & 1 \end{bmatrix} = \begin{bmatrix} x + 3y & -x & 2x + y \\ z + 3w & -z & 2z + w \end{bmatrix}$$

If we equate matrix entries, this gives us the equations

$$x + 3y = -5 \qquad x + 3w = 6$$

$$-x = -1 \qquad -z = -3$$

$$2x + y = 0 \qquad 2z + w = 7$$

Thus $x = 1$ and $z = 3$, so that the top two equations give $y = -2$ and $w = 1$. Since these values are consistent with the bottom two equations, we have that

$$X = \begin{bmatrix} 1 & -2 \\ 3 & 1 \end{bmatrix}$$

11. **(c)** As above, let $X = \begin{bmatrix} x & y \\ z & w \end{bmatrix}$, so that the matrix equation becomes

$$\begin{bmatrix} 3x + z & 3y + w \\ -x + 2z & -y + 2w \end{bmatrix} - \begin{bmatrix} x + 2y & 4x \\ z + 2w & 4z \end{bmatrix} = \begin{bmatrix} 2 & -2 \\ 5 & 4 \end{bmatrix}$$

This yields the system of equations

$$2x - 2y + z \qquad = 2$$

$$-4x + 3y \qquad + w = -2$$

$$-x \qquad + z - 2w = 5$$

$$-y - 4z + 2w = 4$$

with matrix

$$\begin{bmatrix} 2 & -2 & 1 & 0 & 2 \\ -4 & 3 & 0 & 1 & -2 \\ -1 & 0 & 1 & -2 & 5 \\ 0 & -1 & -4 & 2 & 4 \end{bmatrix}$$

which reduces to

$$\begin{bmatrix} 1 & 0 & 0 & 0 & -113/37 \\ 0 & 1 & 0 & 0 & -160/37 \\ 0 & 0 & 1 & 0 & -20/37 \\ 0 & 0 & 0 & 1 & -46/37 \end{bmatrix}$$

Hence, $x = -113/37$, $y = -160/37$, $z = -20/37$, and $w = -46/37$.

12. **(a)** By inspection

$$\begin{bmatrix} y_1 \\ y_2 \\ y_3 \end{bmatrix} = \begin{bmatrix} 1 & -1 & 1 \\ 3 & 1 & -4 \\ -2 & -2 & 3 \end{bmatrix} \begin{bmatrix} x_1 \\ x_2 \\ x_3 \end{bmatrix}$$

and

$$\begin{bmatrix} z_1 \\ z_2 \end{bmatrix} = \begin{bmatrix} 4 & -1 & 1 \\ -3 & 5 & -1 \end{bmatrix} \begin{bmatrix} y_1 \\ y_2 \\ y_3 \end{bmatrix}$$

Thus,

$$\begin{bmatrix} z_1 \\ z_2 \end{bmatrix} = \begin{bmatrix} 4 & -1 & 1 \\ -3 & 5 & -1 \end{bmatrix} \begin{bmatrix} 1 & -1 & 1 \\ 3 & 1 & -4 \\ -2 & -2 & 3 \end{bmatrix} \begin{bmatrix} x_1 \\ x_2 \\ x_3 \end{bmatrix}$$

$$= \begin{bmatrix} -1 & -7 & 11 \\ 14 & 10 & -26 \end{bmatrix} \begin{bmatrix} x_1 \\ x_2 \\ x_3 \end{bmatrix}$$

14. **(a)** If $A^4 = 0$, then

$$(I - A)(I + A + A^2 + A^3) = I + A + A^2 + A^3 - A - A^2 - A^3 - 0 = I$$

Therefore, $(I - A)^{-1} = I + A + A^2 + A^3$ by Theorem 1.6.3.

(b) If $A^{n+1} = 0$, then

$$(I - A)(I + A + A^2 + \cdots + A^n) = I - A^{n+1} = I + 0 = I$$

Thus Theorem 1.6.3 again supplies the desired result.

15. Since the coordinates of the given points must satisfy the polynomial, we have

$$\begin{aligned} p(1) = 2 &\Rightarrow & a + b + c &= 2 \\ p(-1) = 6 &\Rightarrow & a - b + c &= 6 \\ p(2) = 3 &\Rightarrow & 4a + 2b + c &= 3 \end{aligned}$$

The reduced row-echelon form of the augmented matrix of this system of equations is

$$\begin{bmatrix} 1 & 0 & 0 & 1 \\ 0 & 1 & 0 & -2 \\ 0 & 0 & 1 & 3 \end{bmatrix}$$

Thus, $a = 1$, $b = -2$, and $c = 3$.

16. If $p(x) = ax^2 + bx + c$, then $p'(x) = 2ax + b$. Hence,

$$
\begin{aligned}
p(-1) &= 0 &\Rightarrow\quad & a - b + c = 0 \\
p(2) &= -9 &\Rightarrow\quad & 4a + 2b + c = -9 \\
p'(2) &= 0 &\Rightarrow\quad & 4a + b = 0
\end{aligned}
$$

The augmented matrix of this system reduces to

$$\begin{bmatrix} 1 & 0 & 0 & 1 \\ 0 & 1 & 0 & -4 \\ 0 & 0 & 1 & -5 \end{bmatrix}$$

so that $a = 1$, $b = -4$, and $c = -5$.

17. We must show that $(I - J_n)\left(I - \frac{1}{n-1}J_n\right) = I$ or that $\left(I - \frac{1}{n-1}J_n\right)(I - J_n) = I$. (By virtue of Theorem 1.6.3, we need only demonstrate one of these equalities.) We have

$$(I - J_n)\left(I - \frac{1}{n-1}J_n\right) = I^2 - \frac{1}{n-1}IJ_n - J_nI + \frac{1}{n-1}J_n^2$$

$$= I - \frac{n}{n-1}J_n + \frac{1}{n-1}J_n^2$$

But $J_n^2 = nJ_n$ (think about actually squaring J_n), so that the right-hand side of the above equation is just I, as desired.

18. If $A^3 + 4A^2 - 2A + 7I = 0$, then $(A^3 + 4A^2 - 2A + 7I)^T = 0$ as well. Now repeated applications of Theorem 1.4.9 will finish the problem.

19. First suppose that $AB^{-1} = B^{-1}A$. Note that all matrices must be square and of the same size. Therefore

$$(AB^{-1})B = (B^{-1}A)B$$

or

$$A = B^{-1}AB$$

so that

$$BA = B(B^{-1}AB) = (BB^{-1})(AB) = AB$$

It remains to show that if $AB = BA$ then $AB^{-1} = B^{-1}A$. An argument similar to the one given above will serve, and we leave the details to you.

20. Suppose that A and $A + B$ are both invertible. Then there is a matrix C such that $(A + B)C = I$. We must show that $I + BA^{-1}$ is also invertible. Consider the matrix

$$(I + BA^{-1})(AC) = AC + (BA^{-1})(AC)$$
$$= AC + BIC$$
$$= AC + BC$$
$$= (A + B)C$$
$$= I$$

Thus AC is the inverse of $I + BA^{-1}$.

Now suppose that A is invertible but $A + B$ is not. We must show that $(I + BA^{-1})$ is also not invertible. Suppose that $(I + BA^{-1})D = I$ for some matrix D. Then

$$(A + B)A^{-1}D = AA^{-1}D + BA^{-1}D$$
$$= D + BA^{-1}D$$
$$= (I + BA^{-1})D$$
$$= I$$

This implies that $A + B$ is invertible. Since this is contrary to our hypothesis, $I + BA^{-1}$ is not invertible.

21.   (b) Let the ij^{th} entry of A be a_{ij}. Then $\text{tr}(A) = a_{11} + a_{22} + \cdots + a_{nn}$, so that

$$\text{tr}(kA) = ka_{11} + ka_{22} + \cdots + ka_{nn}$$
$$= k(a_{11} + a_{22} + \cdots + a_{nn})$$
$$= k\text{tr}(A)$$

(d) Let the ij^{th} entries of A and B be a_{ij} and b_{ij}, respectively. Then

$$\text{tr}(AB) = a_{11}b_{11} + a_{12}b_{21} + \cdots + a_{1n}b_{n1}$$
$$+ \; a_{21}b_{12} + a_{22}b_{22} + \cdots + a_{2n}b_{n2}$$
$$+ \; \cdots$$
$$+ \; a_{n1}b_{1n} + a_{n2}b_{2n} + \cdots + a_{nn}b_{nn}$$

and

$$\text{tr}(BA) = b_{11}a_{11} + b_{12}a_{21} + \cdots + b_{1n}a_{n1}$$
$$+ \; b_{21}a_{12} + b_{22}a_{22} + \cdots + b_{2n}a_{n2}$$
$$+ \; \cdots$$
$$+ \; b_{n1}a_{1n} + b_{n2}a_{2n} + \cdots + b_{nn}a_{nn}$$

If we rewrite each of the terms $b_{ij}a_{ji}$ in the above expression as $a_{ji}b_{ij}$ and list the terms in the order indicated by the arrows below,

$$\text{tr}(BA) = a_{11}b_{11} + a_{21}b_{12} + \cdots + a_{n1}b_{1n}$$
$$+ \; a_{12}b_{21} + a_{22}b_{22} + \cdots + a_{n2}b_{2n}$$
$$+ \; \cdots$$
$$+ \; a_{1n}b_{n1} + a_{2n}b_{n2} + \cdots + a_{nn}b_{nn}$$

then we have $\text{tr}(AB) = \text{tr}(BA)$.

22.   Suppose that A and B are square matrices such that $AB - BA = I$. Then

$$\text{tr}(AB - BA) = \text{tr}(AB) - \text{tr}(BA) = \text{tr}(I).$$

But $\text{tr}(I) = n$ for any $n \times n$ identity matrix I and $\text{tr}(AB) - \text{tr}(BA) = 0$ by Part (d) of Problem 21.

24. **(c)** Let $A = [a_{ij}(x)]$ be an $m \times n$ matrix and let $B = [b_{ij}(x)]$ be an $n \times s$ matrix. Then the ij^{th} entry of AB is

$$a_{i1}(x)b_{1j}(x) + \cdots + a_{in}(x)b_{nj}(x)$$

Thus the ij^{th} entry of $\dfrac{d}{dx}(AB)$ is

$$\frac{d}{dx}\Big(a_{i1}(x)b_{1j}(x) + \cdots + a_{in}(x)b_{nj}(x)\Big)$$

$$= a'_{i1}(x)b_{1j}(x) + a_{i1}(x)b'_{1j}(x) + \cdots + a'_{in}(x)b_{nj}(x) + a_{in}(x)b'_{nj}(x)$$

$$= [a'_{i1}(x)b_{1j}(x) + \cdots + a'_{in}(x)b_{nj}(x)] + [a_{i1}(x)b'_{1j}(x) + \cdots + a_{in}(x)b'_{nj}(x)]$$

This is just the ij^{th} entry of $\dfrac{dA}{dx}B + A\dfrac{dB}{dx}$.

25. Suppose that A is a square matrix whose entries are differentiable functions of x. Suppose also that A has an inverse, A^{-1}. Then we shall show that A^{-1} also has entries which are differentiable functions of x and that

$$\frac{dA^{-1}}{dx} = -A^{-1}\frac{dA}{dx}A^{-1}$$

Since we can find A^{-1} by the method used in Chapter 1, its entries are functions of x which are obtained from the entries of A by using only addition together with multiplication and division by constants or entries of A. Since sums, products, and quotients of differentiable functions are differentiable wherever they are defined, the resulting entries in the inverse will be differentiable functions except, perhaps, for values of x where their denominators are zero. (Note that we never have to divide by a function which is identically zero.) That is, the entries of A^{-1} are differentiable wherever they are defined. But since we are assuming that A^{-1} is defined, its entries must be differentiable. Moreover,

$$\frac{d}{dx}(AA^{-1}) = \frac{d}{dx}(I) = 0$$

or

$$\frac{dA}{dx}A^{-1} + A\frac{dA^{-1}}{dx} = 0$$

Therefore

$$A\frac{dA^{-1}}{dx} = -\frac{dA}{dx}A^{-1}$$

so that

$$\frac{dA^{-1}}{dx} = -A^{-1}\frac{dA}{dx}A^{-1}$$

26. Following the hint, we obtain the equation

$$x^2 + x - 2 = A(x^2 + 1) + (Bx + C)(3x - 1)$$

or

$$x^2 + x - 2 = (A + 3B)x^2 + (3C - B)x + (A - C)$$

If we equate coefficients of like powers of x, we obtain the system of equations

$$A + 3B \qquad = \quad 1$$
$$- \quad B + 3C = \quad 1$$
$$A \qquad - \quad C = -2$$

the solution to this system is

$$A = -7/5, \qquad B = 4/5, \qquad C = 3/5$$

27. **(b)** Let H be a Householder matrix, so that $H = I - 2PP^T$ where P is an $n \times 1$ matrix. Then using Theorem 1.4.9,

$$H^T = (I - 2PP^T)^T$$
$$= I^T - (2PP^T)^T$$
$$= I - 2(P^T)^T P^T$$
$$= I - 2PP^T$$
$$= H$$

and (using Theorem 1.4.1)

$$H^T H = H^2 \qquad \text{(by the above result)}$$

$$= (I - 2PP^T)^2$$

$$= I^2 - 2PP^T - 2PP^T + (-2PP^T)^2$$

$$= I - 4PP^T + 4PP^T PP^T$$

$$= I - 4PP^T + 4PP^T \qquad \text{(because } P^T P = I\text{)}$$

$$= I$$

28. **(a)** We are asked to show that $(C^{-1} + D^{-1})^{-1} = C(C + D)^{-1}D$. Since all the matrices in sight are invertible and the inverse of a matrix is unique, it will suffice to show that the inverses of each side are equal, or that $(C(C+D)^{-1}D)^{-1} = C^{-1} + D^{-1}$. We have

$$(C(C + D)^{-1}D)^{-1} = D^{-1}(C + D)C^{-1}$$

$$= D^{-1}CC^{-1} + D^{-1}DC^{-1}$$

$$= D^{-1} + C^{-1}$$

$$= C^{-1} + D^{-1}$$

(b) Since we don't know whether or not C is invertible, we can't use the technique of Part (a). Instead, we observe that

$$(*) \qquad\qquad (I + CD)C = C(I + DC)$$

since each side of $(*)$ is just $C + CDC$. If we multiply both sides of $(*)$ on the left by $(I + CD)^{-1}$ and on the right by $(I + DC)^{-1}$, we obtain the desired result.

(c) Start with $D + DD^T C^{-1}D$, write it in two different ways, and proceed more or less as in Part (b).

29. **(b)** A bit of experimenting and an application of Part (a) indicates that

$$A^n = \begin{bmatrix} a^n & 0 & 0 \\ 0 & b^n & 0 \\ d & 0 & c^n \end{bmatrix}$$

where

$$d = a^{n-1} + a^{n-2}c + \cdots + ac^{n-2} + c^{n-1} = \frac{a^n - c^n}{a - c} \quad \text{if } a \neq c$$

If $a = c$, then $d = na^{n-1}$. We prove this by induction. Observe that the result holds when $n = 1$. Suppose that it holds when $n = N$. Then

$$A^{N+1} = AA^N = A \begin{bmatrix} a^N & 0 & 0 \\ 0 & b^N & 0 \\ d & 0 & c^N \end{bmatrix} = \begin{bmatrix} a^{N+1} & 0 & 0 \\ 0 & b^{N+1} & 0 \\ a^N + cd & 0 & c^{N+1} \end{bmatrix}$$

Here

$$a^N + cd = \begin{cases} a^N + c\dfrac{a^N - c^N}{a - c} = \dfrac{a^{N+1} - a^N c + a^N c - c^{N+1}}{a - c} = \dfrac{a^{N+1} - c^{N+1}}{a - c} & \text{if } a \neq c \\[3mm] a^N + a(Na^{N-1}) = (N+1)a^N & \text{if } a = c. \end{cases}$$

Thus the result holds when $n = N + 1$ and so must hold for all values of n.

EXERCISE SET 2.1

1. **(a)** The number of inversions in $(4,1,3,5,2)$ is $3 + 0 + 1 + 1 = 5$.

   (d) The number of inversions in $(5,4,3,2,1)$ is $4 + 3 + 2 + 1 = 10$.

2. **(a)** The permutation is odd because 5 is odd.

   (d) The permutation is even because 10 is even.

3. $\begin{vmatrix} 3 & 5 \\ -2 & 4 \end{vmatrix} = 12 - (-10) = 22$

5. $\begin{vmatrix} -5 & 6 \\ -7 & -2 \end{vmatrix} = (-5)(-2) - (-7)(6) = 52$

7. $\begin{vmatrix} a - 3 & 5 \\ -3 & a - 2 \end{vmatrix} = (a - 3)(a - 2) - (-3)(5) = a^2 - 5a + 21$

9. $\begin{vmatrix} -2 & 1 & 4 \\ 3 & 5 & -7 \\ 1 & 6 & 2 \end{vmatrix} = (-20 - 7 + 72) - (20 + 84 + 6) = -65$

11. $\begin{vmatrix} 3 & 0 & 0 \\ 2 & -1 & 5 \\ 1 & 9 & -4 \end{vmatrix} = (12 + 0 + 0) - (0 + 135 + 0) = -123$

13. **(a)**

$$\det(A) = \begin{vmatrix} \lambda - 2 & 1 \\ -5 & \lambda + 4 \end{vmatrix} = (\lambda - 2)(\lambda + 4) + 5$$

$$= \lambda^2 + 2\lambda - 3 = (\lambda - 1)(\lambda + 3)$$

Hence, $\det(A) = 0$ if and only if $\lambda = 1$ or $\lambda = -3$.

14. The following permutations are even:

$$(1,2,3,4) \quad (1,3,4,2) \quad (1,4,2,3)$$
$$(2,1,4,3) \quad (2,3,1,4) \quad (2,4,3,1)$$
$$(3,1,2,4) \quad (3,2,4,1) \quad (3,4,1,2)$$
$$(4,1,3,2) \quad (4,2,1,3) \quad (4,3,2,1)$$

The following permutations are odd:

$$(1,2,4,3) \quad (1,3,2,4) \quad (1,4,3,2)$$
$$(2,1,3,4) \quad (2,3,4,1) \quad (2,4,1,3)$$
$$(3,1,4,2) \quad (3,2,1,4) \quad (3,4,2,1)$$
$$(4,1,2,3) \quad (4,2,3,1) \quad (4,3,1,2)$$

15. If A is a 4×4 matrix, then

$$\det(A) = \sum (-1)^p a_{1i_1} a_{2i_2} a_{3i_3} a_{4i_4}$$

where $p = 1$ if (i_1, i_2, i_3, i_4) is an odd permutation of $\{1, 2, 3, 4\}$ and $p = 2$ otherwise. There are 24 terms in this sum.

17. **(a)** The only nonzero product in the expansion of the determinant is

$$a_{15}a_{24}a_{33}a_{42}a_{51} = (-3)(-4)(-1)(2)(5) = -120$$

Since (5,4,3,2,1) is even, $\det(A) = -120$.

(b) The only nonzero product in the expansion of the determinant is

$$a_{11}a_{25}a_{33}a_{44}a_{52} = (5)(-4)(3)(1)(-2) = 120$$

Since (1,5,3,4,2) is odd, $\det(A) = -120$.

18. We have

$$\begin{vmatrix} x & -1 \\ 3 & 1-x \end{vmatrix} = x(1-x) - 3(-1) = 3 + x - x^2$$

and

$$\begin{vmatrix} 1 & 0 & -3 \\ 2 & x & -6 \\ 1 & 3 & x-5 \end{vmatrix} = x(x-5) + 0 + (-3)(2)(3) - (-3x + (-6)(3) + 0) = x^2 - 2x$$

Thus $x^2 - 2x = 3 + x - x^2$ or $2x^2 - 3x - 3 = 0$. By the quadratic formula, $x = (3 \pm \sqrt{33})/4$.

19. The value of the determinant is

$$\sin^2\theta - (-\cos^2\theta) = \sin^2\theta + \cos^2\theta = 1$$

The identity $\sin^2\theta + \cos^2\theta = 1$ holds for all values of θ.

20. We have

$$AB = \begin{bmatrix} a & b \\ 0 & c \end{bmatrix} \begin{bmatrix} d & e \\ 0 & f \end{bmatrix} = \begin{bmatrix} ad & ae+bf \\ 0 & cf \end{bmatrix}$$

and

$$BA = \begin{bmatrix} d & e \\ 0 & f \end{bmatrix} \begin{bmatrix} a & b \\ 0 & c \end{bmatrix} = \begin{bmatrix} ad & bd+ce \\ 0 & cf \end{bmatrix}$$

Thus $AB = BA$ if and only if $ae + bf = bd + ce$, which is just the condition that

$$\begin{vmatrix} b & a-e \\ e & d-f \end{vmatrix} = bd - bf - ae + ce = 0.$$

21. Since the product of integers is always an integer, each elementary product is an integer. The result then follows from the fact that the sum of integers is always an integer.

22. The signed elementary products of an $n \times n$ matrix $(n > 1)$ all of whose entries are 1 will all be either 1 or -1. There are $n!$ such elementary products, half of which are 1 and the other half of which are -1. Thus the determinant of such a matrix will always be zero.

23. **(a)** Since each elementary product in the expansion of the determinant contains a factor from each row, each elementary product must contain a factor from the row of zeros. Thus, each signed elementary product is zero and $\det(A) = 0$.

24. Suppose that $A = [a_{ij}]$ denotes an n by n diagonal matrix. That is $a_{ij} = 0$ unless $i = j$. If any elementary product $a_{1j_1} a_{2j_2} \cdots a_{nj_n}$ contains a factor of the form a_{ij_i} with $i \neq j_i$, then that elementary product is zero. But, the only elementary product where $i = j_i$ for all $i = 1, 2, \ldots, n$ is $a_{11} a_{22} \cdots a_{nn}$, that is, the product of elements from the diagonal of A. Since the column indices for this elementary product are in natural order, the sign of this elementary product is positive. Hence, the determinant of any diagonal matrix is the product of its diagonal elements.

25. Let $U = [a_{ij}]$ be an n by n upper triangular matrix. That is, suppose that $a_{ij} = 0$ whenever $i > j$. Now consider any elementary product $a_{1j_1} a_{2j_2} \cdots a_{nj_n}$. If $k > j_k$ for any factor a_{kj_k} in this product, then the product will be zero. But if $k \leq j_k$ for all $k = 1, 2, \ldots, n$, then $k = j_k$ for all k because $j_1, j_2, \ldots, j_n$ is just a permutation of the integers $1, 2, \ldots, n$. Hence, $a_{11} a_{22} \cdots a_{nn}$ is the only elementary product which is not guaranteed to be zero. Since the column indices in this product are in natural order, the product appears with a plus sign. Thus, the determinant of U is the product of its diagonal elements. A similar argument works for lower triangular matrices.

EXERCISE SET 2.2

1. **(b)** We have

$$\det(A) = \begin{vmatrix} 2 & -1 & 3 \\ 1 & 2 & 4 \\ 5 & -3 & 6 \end{vmatrix} = \begin{vmatrix} 0 & -5 & -5 \\ 1 & 2 & 4 \\ 0 & -13 & -14 \end{vmatrix}$$

> Add -2 times Row 2 to Row 1 and -5 times Row 2 to Row 3.

$$= (-1)(-5) \begin{vmatrix} 1 & 2 & 4 \\ 0 & 1 & 1 \\ 0 & -13 & -14 \end{vmatrix}$$

> Factor -5 from Row 1 and interchange Row 1 and Row 2.

$$= (-1)(-5) \begin{vmatrix} 1 & 2 & 4 \\ 0 & 1 & 1 \\ 0 & 0 & -1 \end{vmatrix}$$

> Add 13 times Row 2 to Row 3.

$$= (-1)(-5)(-1) = -5$$

> By Theorem 2.2.2.

$$\det(A^T) = \begin{vmatrix} 2 & 1 & 5 \\ -1 & 2 & -3 \\ 3 & 4 & 6 \end{vmatrix} = \begin{vmatrix} 0 & 5 & -1 \\ -1 & 2 & -3 \\ 0 & 10 & -3 \end{vmatrix}$$

> Add 2 times Row 2 to Row 1 and 3 times Row 2 to Row 3.

$$= (-1) \begin{vmatrix} -1 & 2 & -3 \\ 0 & 5 & -1 \\ 0 & 0 & -1 \end{vmatrix}$$

> Add -2 times Row 1 to Row 3, and interchange Row 1 and Row 2.

$$= (-1)(-1)(5)(-1) = -5$$

> By Theorem 2.2.2.

105

2. **(a)** Since the matrix is upper triangular, its determinant equals the product of the diagonal elements. Hence, if we call the matrix A, then $\det(A) = -30$.

(c) The first and third rows of the given matrix are proportional — in fact, they are equal — and hence, by Theorem 2.2.5, $\det(A) = 0$.

3. **(b)** Since this matrix is just I_4 with Row 2 and Row 3 interchanged, its determinant is -1.

5.

$$\det(A) = \begin{vmatrix} 0 & 3 & 1 \\ 1 & 1 & 2 \\ 3 & 2 & 4 \end{vmatrix} = (-1) \begin{vmatrix} 1 & 1 & 2 \\ 0 & 3 & 1 \\ 3 & 2 & 4 \end{vmatrix}$$

Interchange Row 1 and Row 2.

$$= (-1) \begin{vmatrix} 1 & 1 & 2 \\ 0 & 3 & 1 \\ 0 & -1 & -2 \end{vmatrix}$$

Add -3 times Row 1 to Row 3.

$$= (-1)(3) \begin{vmatrix} 1 & 1 & 2 \\ 0 & 1 & 1/3 \\ 0 & -1 & -2 \end{vmatrix}$$

Factor 3 from Row 2.

$$= -3 \begin{vmatrix} 1 & 1 & 2 \\ 0 & 1 & 1/3 \\ 0 & 0 & -5/3 \end{vmatrix}$$

Add Row 2 to Row 3.

If we factor $-5/3$ from Row 3 and apply Theorem 2.2.2 we find that

$$\det(A) = -3(-5/3)(1) = 5$$

7.

$$\det(A) = \begin{vmatrix} 3 & -6 & 9 \\ -2 & 7 & -2 \\ 0 & 1 & 5 \end{vmatrix} = 3 \begin{vmatrix} 1 & -2 & 3 \\ 0 & 3 & 4 \\ 0 & 1 & 5 \end{vmatrix}$$

Factor 3 from Row 1 and Add twice Row 1 to Row 2.

$$= (3)(3) \begin{vmatrix} 1 & -2 & 3 \\ 0 & 1 & 4/3 \\ 0 & 0 & 11/3 \end{vmatrix}$$

Factor 3 from Row 2 and subtract Row 2 from Row 3.

$$= 9\left(\frac{11}{3}\right) \begin{vmatrix} 1 & -2 & 3 \\ 0 & 1 & 4/3 \\ 0 & 0 & 1 \end{vmatrix}$$

Factor 11/3 from Row 3.

$$= 9(11/3)(1) = 33$$

9.

$$\det(A) = \begin{vmatrix} 2 & 1 & 3 & 1 \\ 1 & 0 & 1 & 1 \\ 0 & 2 & 1 & 0 \\ 0 & 1 & 2 & 3 \end{vmatrix} = (-1) \begin{vmatrix} 1 & 0 & 1 & 1 \\ 2 & 1 & 3 & 1 \\ 0 & 2 & 1 & 0 \\ 0 & 1 & 2 & 3 \end{vmatrix}$$

Interchange Row 1 and Row 2.

$$= (-1) \begin{vmatrix} 1 & 0 & 1 & 1 \\ 0 & 1 & 1 & -1 \\ 0 & 2 & 1 & 0 \\ 0 & 1 & 2 & 3 \end{vmatrix}$$

Add -2 times Row 1 to Row 2.

$$= (-1) \begin{vmatrix} 1 & 0 & 1 & 1 \\ 0 & 1 & 1 & -1 \\ 0 & 0 & -1 & 2 \\ 0 & 0 & 1 & 4 \end{vmatrix}$$

Add -2 times Row 2 to Row 3; subtract Row 2 from Row 4.

$$= (-1) \begin{vmatrix} 1 & 0 & 1 & 1 \\ 0 & 1 & 1 & -1 \\ 0 & 0 & -1 & 2 \\ 0 & 0 & 0 & 6 \end{vmatrix}$$

Add Row 3 to Row 4.

$$= (-1)(-1)(6)(1) = 6$$

11.

$$\det(A) = \begin{vmatrix} 1 & 3 & 1 & 5 & 3 \\ -2 & -7 & 0 & -4 & 2 \\ 0 & 0 & 1 & 0 & 1 \\ 0 & 0 & 2 & 1 & 1 \\ 0 & 0 & 0 & 1 & 1 \end{vmatrix}$$

$$= \begin{vmatrix} 1 & 3 & 1 & 5 & 3 \\ 0 & -1 & 2 & 6 & 8 \\ 0 & 0 & 1 & 0 & 1 \\ 0 & 0 & 0 & 1 & -1 \\ 0 & 0 & 0 & 1 & 1 \end{vmatrix}$$

> Add 2 times Row 1
> to Row 2; add -2
> times Row 3 to Row 4.

$$= \begin{vmatrix} 1 & 3 & 1 & 5 & 3 \\ 0 & -1 & 2 & 6 & 8 \\ 0 & 0 & 1 & 0 & 1 \\ 0 & 0 & 0 & 1 & -1 \\ 0 & 0 & 0 & 0 & 2 \end{vmatrix}$$

> Add -1 times
> Row 4 to Row 5.

Hence, $\det(A) = (-1)(2)(1) = -2$.

12. Let

$$A = \begin{bmatrix} a & b & c \\ d & e & f \\ g & h & i \end{bmatrix}$$

We are given that $\det(A) = -6$.

(a)

$$\det(B) = \begin{vmatrix} d & e & f \\ g & h & i \\ a & b & c \end{vmatrix} = (-1) \begin{vmatrix} d & e & f \\ a & b & c \\ g & h & i \end{vmatrix} \qquad \boxed{\begin{array}{l} \text{Interchange} \\ \text{Row 2 and} \\ \text{Row 3.} \end{array}}$$

$$= (-1)^2 \begin{vmatrix} a & b & c \\ d & e & f \\ g & h & i \end{vmatrix} \qquad \boxed{\begin{array}{l} \text{Interchange} \\ \text{Row 1 and Row 2.} \end{array}}$$

$$= \det(A) = -6$$

(b)

$$\det(C) = \begin{vmatrix} 3a & 3b & 3c \\ -d & -e & -f \\ 4g & 4h & 4i \end{vmatrix}$$

$$= (3)(-1)(4) \begin{vmatrix} a & b & c \\ d & e & f \\ g & h & i \end{vmatrix} \qquad \boxed{\begin{array}{l} \text{Factor 3 from Row 1,} \\ -1 \text{ from Row 2, and} \\ 4 \text{ from Row 3.} \end{array}}$$

$$= -12 \det(A) = 72$$

(c) Let

$$D = \begin{bmatrix} a+g & b+h & c+i \\ d & e & f \\ g & h & i \end{bmatrix}$$

Then $\det(D) = \det(A) = -6$ because the matrix D may be obtained from A by replacing the first row by the sum of the first and third rows.

12. **(d)** Let

$$E = \begin{bmatrix} -3a & -3b & -3c \\ d & e & f \\ g-4d & h-4e & i-4f \end{bmatrix}$$

If we factor -3 from Row 1, we obtain

$$\det(E) = (-3) \begin{vmatrix} a & b & c \\ d & e & f \\ g-4d & h-4e & i-4f \end{vmatrix}$$

But the above matrix can be obtained from A by adding -4 times Row 2 to Row 3. Hence $\det(E) = (-3)\det(A) = 18$.

13.

$$\det(A) = \begin{vmatrix} 1 & 1 & 1 \\ a & b & c \\ a^2 & b^2 & c^2 \end{vmatrix}$$

$$= \begin{vmatrix} 1 & 1 & 1 \\ 0 & b-a & c-a \\ 0 & b^2-a^2 & c^2-a^2 \end{vmatrix} \qquad \boxed{\begin{array}{l} \text{Add } -a \text{ times Row 1} \\ \text{to Row 2; add } -a^2 \\ \text{times Row 1 to Row 3.} \end{array}}$$

Since $b^2 - a^2 = (b-a)(b+a)$, we add $-(b+a)$ times Row 2 to Row 3 to obtain

$$\det(A) = \begin{vmatrix} 1 & 1 & 1 \\ 0 & b-a & c-a \\ 0 & 0 & (c^2-a^2)-(c-a)(b+a) \end{vmatrix}$$

$$= (b-a)[(c^2-a^2)-(c-a)(b+a)]$$

$$= (b-a)(c-a)[(c+a)-(b+a)]$$

$$= (b-a)(c-a)(c-b)$$

14. **(a)** Consider the elementary product $p = a_{1j_1}a_{2j_2}a_{3j_3}$. If $j_1 = 1$ or $j_1 = 2$, then $p = 0$ because $a_{11} = a_{12} = 0$. If $j_1 = 3$, then the only choices for j_2 are $j_2 = 1$ or $j_2 = 2$. But $a_{21} = 0$, so that $p = 0$ unless $j_2 = 2$. Finally, if $j_1 = 3$ and $j_2 = 2$, then the only choice for j_3 is $j_3 = 1$. Therefore, $a_{13}a_{22}a_{31}$ is the only elementary product which can be nonzero. Since the permutation (3,2,1) is odd, then

$$\det(A) = -a_{13}a_{22}a_{31}$$

(b) By an argument which is completely analogous to that given in Part (a), we can argue that $a_{14}a_{23}a_{32}a_{41}$ is the only nonzero elementary product. Since (4,3,2,1) is an even permutation, then

$$\det(A) = a_{14}a_{23}a_{32}a_{41}$$

15. In each case, d will denote the determinant on the left and, as usual, $\det(A) = \sum \pm a_{1j_1}a_{2j_2}a_{3j_3}$, where $\sum$ denotes the sum of all such elementary products.

(a) $d = \sum \pm (ka_{1j_1})a_{2j_2}a_{3j_3} = k\sum \pm a_{1j_1}a_{2j_2}a_{3j_3} = k\ \det(A)$

(b) $d = \sum \pm a_{2j_1}a_{1j_2}a_{3j_3} = \sum \pm a_{1j_2}a_{2j_1}a_{3j_3}$

The sign of a given elementary product on the left depends upon whether (j_1, j_2, j_3) is even or odd; the sign of the same elementary product appearing on the right depends upon whether (j_2, j_1, j_3) is even or odd. By counting the number of inversions in each of the six permutations of $\{1, 2, 3\}$, it can be verified that (j_1, j_2, j_3) is even if and only if (j_2, j_1, j_3) is odd. Therefore every elementary product in A appears in d with opposite sign, and $d = -\det(A)$.

16. **(a)** Since A can be obtained from I_3 by interchanging the first and third rows, it follows that $\det(A) = -1$.

17. Since the given matrix is upper triangular, its determinant is the product of the diagonal elements. That is, the determinant is $x(x+1)(2x-1)$. This product is zero if and only if $x = 0$, $x = -1$, or $x = 1/2$.

18. **(a)** If $x = 1$ or if $x = -3$, then two rows of the given matrix will be the same and the determinant will be zero.

(b) Since the determinant is $12 - 8x - 4x^2$ and since a quadratic polynomial has exactly two roots, it follows that the values of x which we found in Exercise 18(a) are the only possibilities.

EXERCISE SET 2.3

1. **(a)** We have

$$\det(A) = \begin{vmatrix} -1 & 2 \\ 3 & 4 \end{vmatrix} = -4 - 6 = -10$$

and

$$\det(2A) = \begin{vmatrix} -2 & 4 \\ 6 & 8 \end{vmatrix} = (-2)(8) - (4)(6) = -40 = 2^2(-10)$$

2.

$$\det(A) = \begin{vmatrix} 2 & 1 & 0 \\ 3 & 4 & 0 \\ 0 & 0 & 2 \end{vmatrix} = (2)\begin{vmatrix} 1 & 1/2 & 0 \\ 0 & 5/2 & 0 \\ 0 & 0 & 2 \end{vmatrix} = 10$$

$$\det(B) = \begin{vmatrix} 1 & -1 & 3 \\ 7 & 1 & 2 \\ 5 & 0 & 1 \end{vmatrix} = \begin{vmatrix} -14 & -1 & 0 \\ -3 & 1 & 0 \\ 5 & 0 & 1 \end{vmatrix}$$

$$= \begin{vmatrix} -17 & 0 & 0 \\ -3 & 1 & 0 \\ 5 & 0 & 1 \end{vmatrix} = -17$$

Thus $\det(A) = 10$ and $\det(B) = -17$. Now by direct computation,

$$AB = \begin{bmatrix} 9 & -1 & 8 \\ 31 & 1 & 17 \\ 10 & 0 & 2 \end{bmatrix}$$

If we add 9 times Column 2 to Column 1, we find that

$$\det(AB) = \begin{vmatrix} 0 & -1 & 8 \\ 40 & 1 & 17 \\ 10 & 0 & 2 \end{vmatrix} = (10) \begin{vmatrix} 0 & -1 & 8 \\ 4 & 1 & 17 \\ 1 & 0 & 2 \end{vmatrix}$$

$$= (10) \begin{vmatrix} 0 & -1 & 8 \\ 0 & 1 & 9 \\ 1 & 0 & 2 \end{vmatrix} = -(10) \begin{vmatrix} 1 & 0 & 2 \\ 0 & 1 & 9 \\ 0 & 0 & 17 \end{vmatrix}$$

$$= -170 = \det(A)\det(B)$$

4. (a)

$$\begin{vmatrix} 1 & 0 & -1 \\ 9 & -1 & 4 \\ 8 & 9 & -1 \end{vmatrix} = \begin{vmatrix} 1 & 0 & 0 \\ 9 & -1 & 13 \\ 8 & 9 & 7 \end{vmatrix} = \begin{vmatrix} 1 & 0 & 0 \\ 9 & -1 & 0 \\ 8 & 9 & 124 \end{vmatrix} = -124$$

Since the determinant of this matrix is not zero, the matrix is invertible.

(c) Since the first two rows of this matrix are proportional, its determinant is zero. Hence, it is not invertible.

5. (a) By Equation (1),

$$\det(3A) = 3^3 \det(A) = (27)(-7) = -189$$

(c) Again, by Equation (1), $\det(2A^{-1}) = 2^3 \det(A^{-1})$. By Theorem 2.3.5, we have

$$\det(2A^{-1}) = \frac{8}{\det(A)} = -\frac{8}{7}$$

(d) Again, by Equation (1), $\det(2A) = 2^3 \det(A) = -56$. By Theorem 2.3.5, we have

$$\det[(2A)^{-1}] = \frac{1}{\det(2A)} = -\frac{1}{56}$$

(e)

$$\begin{vmatrix} a & g & d \\ b & h & e \\ c & i & f \end{vmatrix} = - \begin{vmatrix} a & d & g \\ b & e & h \\ c & f & i \end{vmatrix} \qquad \boxed{\begin{array}{l} \text{Interchange} \\ \text{Columns 2 and 3.} \end{array}}$$

$$= - \begin{vmatrix} a & b & c \\ d & e & f \\ g & h & i \end{vmatrix} \qquad \boxed{\begin{array}{l} \text{Take the transpose} \\ \text{of the matrix.} \end{array}}$$

$$= 7$$

7. If we replace Row 1 by Row 1 plus Row 2, we obtain

$$\begin{vmatrix} b+c & c+a & b+a \\ a & b & c \\ 1 & 1 & 1 \end{vmatrix} = \begin{vmatrix} a+b+c & b+c+a & c+b+a \\ a & b & c \\ 1 & 1 & 1 \end{vmatrix} = 0$$

because the first and third rows are proportional.

8. If we add -1 times Column 1 to Column 3 and -1 times Column 2 to Column 3 in the determinant on the left, we obtain the determinant on the right.

10. If we add $-t$ times Row 1 to Row 2 in the determinant on the left, we get

$$\begin{vmatrix} a_1 + b_1 t & a_2 + b_2 t & a_3 + b_3 t \\ (1-t^2)b_1 & (1-t^2)b_2 & (1-t^2)b_3 \\ c_1 & c_2 & c_3 \end{vmatrix}$$

or

$$(1 - t^2) \begin{vmatrix} a_1 + b_1 t & a_2 + b_2 t & a_3 + b_3 t \\ b_1 & b_2 & b_3 \\ c_1 & c_2 & c_3 \end{vmatrix}$$

Adding $-t$ times Row 2 to Row 1 in the above determinant yields the desired result.

12. **(a)** We have

$$\det(A) = (k - 3)(k - 2) - 4$$

$$= k^2 - 5k + 2$$

From Theorem 2.3.3, A will be invertible if and only if $\det(A) \neq 0$. But $\det(A) = 0$ if and only if $k = (5 \pm \sqrt{17})/2$. Thus A is invertible unless $k = (5 \pm \sqrt{17})/2$.

13. By adding Row 1 to Row 2 and using the identity $\sin^2 x + \cos^2 x = 1$, we see that the determinant of the given matrix can be written as

$$\begin{vmatrix} \sin^2 \alpha & \sin^2 \beta & \sin^2 \gamma \\ 1 & 1 & 1 \\ 1 & 1 & 1 \end{vmatrix}$$

But this is zero because two of its rows are identical. Therefore the matrix is not invertible.

14. **(b)** The system can be written as

$$\begin{bmatrix} 2 & 3 \\ 4 & 3 \end{bmatrix} \begin{bmatrix} x_1 \\ x_2 \end{bmatrix} = \lambda \begin{bmatrix} x_1 \\ x_2 \end{bmatrix}$$

or

$$\left(\lambda \begin{bmatrix} 1 & 0 \\ 0 & 1 \end{bmatrix} - \begin{bmatrix} 2 & 3 \\ 4 & 3 \end{bmatrix} \right) \begin{bmatrix} x_1 \\ x_2 \end{bmatrix} = \begin{bmatrix} 0 \\ 0 \end{bmatrix}$$

This expression can be simplified —

$$\begin{bmatrix} \lambda - 2 & -3 \\ -4 & \lambda - 3 \end{bmatrix} \begin{bmatrix} x_1 \\ x_2 \end{bmatrix} = \begin{bmatrix} 0 \\ 0 \end{bmatrix}$$

15.   We work with the system from Part (b).

(i)   Here

$$\det(\lambda I - A) = \begin{vmatrix} \lambda - 2 & -3 \\ -4 & \lambda - 3 \end{vmatrix} = (\lambda - 2)(\lambda - 3) - 12 = \lambda^2 - 5\lambda - 6$$

so the characteristic equation is $\lambda^2 - 5\lambda - 6 = 0$.

(ii)   The eigenvalues are just the solutions to this equation, or $\lambda = 6$ and $\lambda = -1$.

(iii)   If $\lambda = 6$, then the corresponding eigenvectors are the nonzero solutions $\mathbf{x} = \begin{bmatrix} x_1 \\ x_2 \end{bmatrix}$ to the equation

$$\begin{bmatrix} 6 - 2 & -3 \\ -4 & 6 - 3 \end{bmatrix} \begin{bmatrix} x_1 \\ x_2 \end{bmatrix} = \begin{bmatrix} 4 & -3 \\ -4 & 3 \end{bmatrix} \begin{bmatrix} x_1 \\ x_2 \end{bmatrix} = \begin{bmatrix} 0 \\ 0 \end{bmatrix}$$

The solution to this system is $x_1 = (3/4)t$, $x_2 = t$, so $\mathbf{x} = \begin{bmatrix} (3/4)t \\ t \end{bmatrix}$ is an eigenvector whenever $t \neq 0$.

If $\lambda = -1$, then the corresponding eigenvectors are the nonzero solutions $\mathbf{x} = \begin{bmatrix} x_1 \\ x_2 \end{bmatrix}$ to the equation

$$\begin{bmatrix} -3 & -3 \\ -4 & -4 \end{bmatrix} \begin{bmatrix} x_1 \\ x_2 \end{bmatrix} = \begin{bmatrix} 0 \\ 0 \end{bmatrix}$$

If we let $x_1 = t$, then $x_2 = -t$, so $\mathbf{x} = \begin{bmatrix} t \\ -t \end{bmatrix}$ is an eigenvector whenever $t \neq 0$.

It is easy to check that these eigenvalues and their corresponding eigenvectors satisfy the original system of equations by substituting for x_1, x_2, and λ. The solution is valid for all values of t.

16.   Since A is invertible, $\det(A) \neq 0$. Thus

$$\det(A^{-1}BA) = \det(A^{-1})\det(B)\det(A)$$

$$= \frac{1}{\det(A)}\det(B)\det(A)$$

$$= \det(B)$$

17.   **(a)** We have, for instance,

$$\begin{vmatrix} a_1 + b_1 & c_1 + d_1 \\ a_2 + b_2 & c_2 + d_2 \end{vmatrix} = \begin{vmatrix} a_1 + b_1 & c_1 + d_1 \\ a_2 & c_2 \end{vmatrix} + \begin{vmatrix} a_1 + b_1 & c_1 + d_1 \\ b_2 & d_2 \end{vmatrix}$$

$$= \begin{vmatrix} a_1 & c_1 \\ a_2 & c_2 \end{vmatrix} + \begin{vmatrix} b_1 & d_1 \\ a_2 & c_2 \end{vmatrix} + \begin{vmatrix} a_1 & c_1 \\ b_2 & d_2 \end{vmatrix} + \begin{vmatrix} b_1 & d_1 \\ b_2 & d_2 \end{vmatrix}$$

The answer is clearly not unique.

18.   We know that

A is invertible if and only if $\det(A) \neq 0$

if and only if $\det(A^T) \neq 0$ (since $\det(A) = \det(A^T)$)

if and only if $\det(A^T)\det(A) \neq 0$

if and only if $\det(A^T A) \neq 0$

if and only if $A^T A$ is invertible

19.   Let B be an $n \times n$ matrix and E be an $n \times n$ elementary matrix.

Case 2: Let E be obtained by interchanging two rows of I_n. Then $\det(E) = -1$ and EA is just A with (the same) two rows interchanged. By Theorem 2.2.3, $\det(EA) = -\det(A) = \det(E)\det(A)$.

Case 3: Let E be obtained by adding a multiple of one row of I_n to another. Then $\det(E) = 1$ and $\det(EA) = \det(A)$. Hence $\det(EA) = \det(A) = \det(E)\det(A)$.

20. No, since $\det(AB) = \det(A)\det(B) = \det(B)\det(A) = \det(BA)$. Hence, the 2 determinants are always equal.

21. If either A or B is singular, then either $\det(A)$ or $\det(B)$ is zero. Hence, $\det(AB) = \det(A)\det(B) = 0$. Thus AB is also singular.

22. **(a)** False. If A is an n by n matrix, then $\det(2A) = 2^n \det(A)$. Hence, the statement holds only in the trivial cases when $n = 1$ or when $\det(A) = 0$.

(c) False. For instance, let $A = \begin{bmatrix} 1 & 0 \\ 0 & 0 \end{bmatrix}$.

(d) True. By Theorem 1.6.1, any system of linear equations has no solution, one solution, or infinitely many solutions. Since $A\mathbf{0} = \mathbf{0}$, there is at least one solution. Since $\det(A) = 0$, it follows from Theorem 2.3.6 that there cannot be just one solution. Hence, there must be infinitely many. Also see the argument used in Exercise 20(a) of Section 1.5.

23. **(a)** False. If $\det(A) = 0$, then A cannot be expressed as the product of elementary matrices. If it could, then it would be invertible as the product of invertible matrices.

(b) True. The reduced row echelon form of A is the product of A and elementary matrices, all of which are invertible. Thus for the reduced row echelon form to have a row of zeros and hence zero determinant, we must also have $\det(A) = 0$.

(c) False. Consider the 2×2 identity matrix. In general, reversing the order of the columns may change the sign of the determinant.

(d) True. Since $\det(AA^T) = \det(A)\det(A^T) = [\det(A)]^2$, $\det(AA^T)$ cannot be negative.

EXERCISE SET 2.4

1. For example,

$$M_{11} = \begin{vmatrix} 7 & -1 \\ 1 & 4 \end{vmatrix} = 29 \quad \Rightarrow \quad C_{11} = 29$$

$$M_{12} = \begin{vmatrix} 6 & -1 \\ -3 & 4 \end{vmatrix} = 21 \quad \Rightarrow \quad C_{12} = -21$$

$$M_{22} = \begin{vmatrix} 1 & 3 \\ -3 & 4 \end{vmatrix} = 13 \quad \Rightarrow \quad C_{22} = 13$$

2. (a)

$$M_{13} = \begin{vmatrix} 0 & 0 & 3 \\ 4 & 1 & 14 \\ 4 & 1 & 2 \end{vmatrix} = 0 \quad \Rightarrow \quad C_{13} = 0$$

(b)

$$M_{23} = \begin{vmatrix} 4 & -1 & 6 \\ 4 & 1 & 14 \\ 4 & 1 & 2 \end{vmatrix} = \begin{vmatrix} 8 & -1 & 6 \\ 0 & 1 & 14 \\ 0 & 1 & 2 \end{vmatrix} = 8(-12) = -96 \quad \Rightarrow \quad C_{23} = 96$$

3. (a)

$$\det(A) = (1) \begin{vmatrix} 7 & -1 \\ 1 & 4 \end{vmatrix} - (-2) \begin{vmatrix} 6 & -1 \\ -3 & 4 \end{vmatrix} + (3) \begin{vmatrix} 6 & 7 \\ -3 & 1 \end{vmatrix}$$

$$= (1)(29) + (2)(21) + (3)(27)$$

$$= 152$$

(b)

$$\det(A) = (1) \begin{vmatrix} 7 & -1 \\ 1 & 4 \end{vmatrix} - (6) \begin{vmatrix} -2 & 3 \\ 1 & 4 \end{vmatrix} + (-3) \begin{vmatrix} -2 & 3 \\ 7 & -1 \end{vmatrix}$$

$$= (1)(29) - (6)(-11) + (-3)(-19)$$

$$= 152$$

(c)

$$\det(A) = -(6) \begin{vmatrix} -2 & 3 \\ 1 & 4 \end{vmatrix} + (7) \begin{vmatrix} 1 & 3 \\ -3 & 4 \end{vmatrix} - (-1) \begin{vmatrix} 1 & -2 \\ -3 & 1 \end{vmatrix}$$

$$= -(6)(-11) + (7)(13) - (-1)(-5)$$

$$= 152$$

5. Since Column 2 contains 2 zeros, we expand using cofactors of that column. Hence,

$$\det(A) = (5) \begin{vmatrix} -3 & 7 \\ -1 & 5 \end{vmatrix} = 5(-15 + 7) = -40$$

7. We factor k from the second column to obtain

$$\begin{vmatrix} 1 & k & k^2 \\ 1 & k & k^2 \\ 1 & k & k^2 \end{vmatrix} = k \begin{vmatrix} 1 & 1 & k^2 \\ 1 & 1 & k^2 \\ 1 & 1 & k^2 \end{vmatrix} = 0$$

because Column 1 and Column 2 are identical.

9.   Let A denote the given matrix. Note that the third column of A has two zero entries. If we add Row 3 to Row 4, we won't change the value of $\det(A)$, but Column 3 will then have three zero entries. Thus

$$\det(A) = \begin{vmatrix} 3 & 3 & 0 & 5 \\ 2 & 2 & 0 & -2 \\ 4 & 1 & -3 & 0 \\ 6 & 11 & 0 & 2 \end{vmatrix}$$

$$= (-3) \begin{vmatrix} 3 & 3 & 5 \\ 2 & 2 & -2 \\ 6 & 11 & 2 \end{vmatrix} \qquad \boxed{\begin{array}{l} \text{Take cofactors} \\ \text{of Column 3.} \end{array}}$$

$$= (-3) \begin{vmatrix} 0 & 3 & 5 \\ 0 & 2 & -2 \\ -5 & 11 & 2 \end{vmatrix} \qquad \boxed{\begin{array}{l} \text{Subtract Column 2} \\ \text{from Column 1.} \end{array}}$$

$$= (-3)(-5) \begin{vmatrix} 3 & 5 \\ 2 & -2 \end{vmatrix} \qquad \boxed{\begin{array}{l} \text{Take cofactors} \\ \text{of Column 1.} \end{array}}$$

$$= (-3)(-5)(-16)$$

$$= -240$$

11.   The determinant of A is -1. The matrix of cofactors is

$$\begin{bmatrix} -3 & 3 & -2 \\ 5 & -4 & 2 \\ 5 & -5 & 3 \end{bmatrix}$$

Thus

$$-\operatorname{adj}(A) = A^{-1} = \begin{bmatrix} 3 & -5 & -5 \\ -3 & 4 & 5 \\ 2 & -2 & -3 \end{bmatrix}$$

13. The determinant of A is 4. The matrix of cofactors is

$$\begin{bmatrix} 2 & 0 & 0 \\ 6 & 4 & 0 \\ 4 & 6 & 2 \end{bmatrix}$$

Thus

$$\text{adj}(A) = \begin{bmatrix} 2 & 6 & 4 \\ 0 & 4 & 6 \\ 0 & 0 & 2 \end{bmatrix}$$

and

$$A^{-1} = \begin{bmatrix} 1/2 & 3/2 & 1 \\ 0 & 1 & 3/2 \\ 0 & 0 & 1/2 \end{bmatrix}$$

15. (a) We must first compute $\text{adj}(A)$. By direct calculation, we find that the matrix of cofactors is

$$\begin{bmatrix} -4 & 2 & -7 & 6 \\ 3 & -1 & 0 & 0 \\ 0 & 0 & -1 & 1 \\ -1 & 0 & 8 & -7 \end{bmatrix}$$

Thus

$$\text{adj}(A) = \begin{bmatrix} -4 & 3 & 0 & -1 \\ 2 & -1 & 0 & 0 \\ -7 & 0 & -1 & 8 \\ 6 & 0 & 1 & -7 \end{bmatrix}$$

We must now compute $\det(A)$. We replace Row 2 by Row 2 plus (-2) times Row 1 to obtain

$$\det(A) = \begin{vmatrix} 1 & 3 & 1 & 1 \\ 0 & -1 & 0 & 0 \\ 1 & 3 & 8 & 9 \\ 1 & 3 & 2 & 2 \end{vmatrix}$$

$$= (-1)\begin{vmatrix} 1 & 1 & 1 \\ 1 & 8 & 9 \\ 1 & 2 & 2 \end{vmatrix} = (-1)(-1) = 1$$

Thus, $A^{-1} = \text{adj}(A)$.

(b)

$$\left[\begin{array}{cccc|cccc} 1 & 3 & 1 & 1 & 1 & 0 & 0 & 0 \\ 2 & 5 & 2 & 2 & 0 & 1 & 0 & 0 \\ 1 & 3 & 8 & 9 & 0 & 0 & 1 & 0 \\ 1 & 3 & 2 & 2 & 0 & 0 & 0 & 1 \end{array}\right]$$

$$\downarrow$$

$$\left[\begin{array}{cccc|cccc} 1 & 3 & 1 & 1 & 1 & 0 & 0 & 0 \\ 0 & -1 & 0 & 0 & -2 & 1 & 0 & 0 \\ 0 & 0 & 7 & 8 & -1 & 0 & 1 & 0 \\ 0 & 0 & 1 & 1 & -1 & 0 & 0 & 1 \end{array}\right]$$

$$\downarrow$$

$$\left[\begin{array}{cccc|cccc} 1 & 0 & 1 & 1 & -5 & 3 & 0 & 0 \\ 0 & 1 & 0 & 0 & 2 & -1 & 0 & 0 \\ 0 & 0 & 7 & 8 & -1 & 0 & 1 & 0 \\ 0 & 0 & 1 & 1 & -1 & 0 & 0 & 1 \end{array}\right]$$

$$\downarrow$$

$$\begin{bmatrix} 1 & 0 & 0 & 0 & \bigm| & -4 & 3 & 0 & -1 \\ 0 & 1 & 0 & 0 & \bigm| & 2 & -1 & 0 & 0 \\ 0 & 0 & 7 & 8 & \bigm| & -1 & 0 & 1 & 0 \\ 0 & 0 & 1 & 1 & \bigm| & -1 & 0 & 0 & 1 \end{bmatrix}$$

$$\downarrow$$

$$\begin{bmatrix} 1 & 0 & 0 & 0 & \bigm| & -4 & 3 & 0 & -1 \\ 0 & 1 & 0 & 0 & \bigm| & 2 & -1 & 0 & 0 \\ 0 & 0 & 7 & 8 & \bigm| & -1 & 0 & 1 & 0 \\ 0 & 0 & 0 & 1 & \bigm| & 6 & 0 & 1 & -7 \end{bmatrix}$$

$$\downarrow$$

$$\begin{bmatrix} 1 & 0 & 0 & 0 & \bigm| & -4 & 3 & 0 & -1 \\ 0 & 1 & 0 & 0 & \bigm| & 2 & -1 & 0 & 0 \\ 0 & 0 & 7 & 0 & \bigm| & -49 & 0 & -7 & 56 \\ 0 & 0 & 0 & 1 & \bigm| & 6 & 0 & 1 & -7 \end{bmatrix}$$

$$\downarrow$$

$$\begin{bmatrix} 1 & 0 & 0 & 0 & \bigm| & -4 & 3 & 0 & -1 \\ 0 & 1 & 0 & 0 & \bigm| & 2 & -1 & 0 & 0 \\ 0 & 0 & 1 & 0 & \bigm| & -7 & 0 & -1 & 8 \\ 0 & 0 & 0 & 1 & \bigm| & 6 & 0 & 1 & -7 \end{bmatrix}$$

Thus

$$A^{-1} = \begin{bmatrix} -4 & 3 & 0 & -1 \\ 2 & -1 & 0 & 0 \\ -7 & 0 & -1 & 8 \\ 6 & 0 & 1 & -7 \end{bmatrix}$$

and the answers to (a) and (b) agree.

17. By direct computation, we obtain

$$\det(A) = \begin{vmatrix} 4 & 5 & 0 \\ 11 & 1 & 2 \\ 1 & 5 & 2 \end{vmatrix} - -132$$

$$\det(A_1) = \begin{vmatrix} 2 & 5 & 0 \\ 3 & 1 & 2 \\ 1 & 5 & 2 \end{vmatrix} = -36$$

$$\det(A_2) = \begin{vmatrix} 4 & 2 & 0 \\ 11 & 3 & 2 \\ 1 & 1 & 2 \end{vmatrix} = -24$$

$$\det(A_3) = \begin{vmatrix} 4 & 5 & 2 \\ 11 & 1 & 3 \\ 1 & 5 & 1 \end{vmatrix} = 12$$

Thus, $x = 3/11$, $y = 2/11$, and $z = -1/11$.

19. By direct calculation, we have

$$\det(A) = \begin{vmatrix} 1 & -3 & 1 \\ 2 & -1 & 0 \\ 4 & 0 & -3 \end{vmatrix} = -11$$

$$\det(A_1) = \begin{vmatrix} 4 & -3 & 1 \\ -2 & -1 & 0 \\ 0 & 0 & -3 \end{vmatrix} = 30$$

$$\det(A_2) = \begin{vmatrix} 1 & 4 & 1 \\ 2 & -2 & 0 \\ 4 & 0 & -3 \end{vmatrix} = 38$$

$$\det(A_3) = \begin{vmatrix} 1 & -3 & 4 \\ 2 & -1 & -2 \\ 4 & 0 & 0 \end{vmatrix} = 40$$

Hence, $x_1 = -30/11$, $x_2 = -38/11$, and $x_3 = -40/11$.

21. By direct calculation, we have

$$\det(A) = \begin{vmatrix} 3 & -1 & 1 \\ -1 & 7 & -2 \\ 2 & 6 & -1 \end{vmatrix} = 0$$

Since $\det(A) = 0$, Cramer's Rule cannot be used to find the solution to this system of equations.

22. Since

$$\det(A) = \begin{vmatrix} \cos\theta & \sin\theta \\ -\sin\theta & \cos\theta \end{vmatrix} = \cos^2\theta + \sin^2\theta = 1$$

the matrix is invertible for all values of θ. The matrix of cofactors is

$$\begin{bmatrix} \cos\theta & \sin\theta & 0 \\ -\sin\theta & \cos\theta & 0 \\ 0 & 0 & 1 \end{bmatrix}$$

so that

$$A^{-1} = \mathrm{adj}(A) = \begin{bmatrix} \cos\theta & -\sin\theta & 0 \\ \sin\theta & \cos\theta & 0 \\ 0 & 0 & 1 \end{bmatrix}$$

24. (a) By direct computation, we have

$$\det(A) = -424$$

$$\det(A_1) = -424$$

$$\det(A_2) = 0$$

$$\det(A_3) = -848$$

$$\det(A_4) = 0$$

Hence, $x = 1$, $z = 2$, and $y = w = 0$.

(b) The augmented matrix is

$$\begin{bmatrix} 4 & 1 & 1 & 1 & 6 \\ 3 & 7 & -1 & 1 & 1 \\ 7 & 3 & -5 & 8 & -3 \\ 1 & 1 & 1 & 2 & 3 \end{bmatrix} \rightarrow \begin{bmatrix} 1 & 1 & 1 & 2 & 3 \\ 3 & 7 & -1 & 1 & 1 \\ 7 & 3 & -5 & 8 & -3 \\ 4 & 1 & 1 & 1 & 6 \end{bmatrix} \rightarrow$$

$$\begin{bmatrix} 1 & 1 & 1 & 2 & 3 \\ 0 & 4 & -4 & -5 & -8 \\ 0 & -4 & -12 & -6 & -24 \\ 0 & -3 & -3 & -7 & -6 \end{bmatrix} \rightarrow \begin{bmatrix} 1 & 1 & 1 & 2 & 3 \\ 0 & 1 & 3 & 3/2 & 6 \\ 0 & 4 & -4 & -5 & -8 \\ 0 & 3 & 3 & 7 & 6 \end{bmatrix} \rightarrow$$

$$\begin{bmatrix} 1 & 0 & -2 & 1/2 & -3 \\ 0 & 1 & 3 & 3/2 & 6 \\ 0 & 0 & -16 & -11 & -32 \\ 0 & 0 & -6 & 5/2 & -12 \end{bmatrix} \rightarrow \begin{bmatrix} 1 & 0 & -2 & 1/2 & -3 \\ 0 & 1 & 3 & 3/2 & 6 \\ 0 & 0 & 1 & -5/12 & 2 \\ 0 & 0 & 16 & 11 & 32 \end{bmatrix} \rightarrow$$

$$\begin{bmatrix} 1 & 0 & 0 & -1/3 & 1 \\ 0 & 1 & 0 & 11/4 & 0 \\ 0 & 0 & 1 & -5/12 & 2 \\ 0 & 0 & 0 & 53/3 & 0 \end{bmatrix} \rightarrow \begin{bmatrix} 1 & 0 & 0 & -1/3 & 1 \\ 0 & 1 & 0 & 11/4 & 0 \\ 0 & 0 & 1 & -5/12 & 2 \\ 0 & 0 & 0 & 1 & 0 \end{bmatrix} \rightarrow$$

$$\begin{bmatrix} 1 & 0 & 0 & 0 & 1 \\ 0 & 1 & 0 & 0 & 0 \\ 0 & 0 & 1 & 0 & 2 \\ 0 & 0 & 0 & 1 & 0 \end{bmatrix}$$

Hence, $x = 1$, $z = 2$, and $y = w = 0$.

25. In case $\det(A) = 1$, then A is invertible and, by Theorem 2.4.2, $A^{-1} = \text{adj}(A)$. If A has only integer entries, then each of its cofactors will be a sum of products of integers. Hence, each cofactor will be an integer. Since $\text{adj}(A)$ is the transpose of a matrix with integer entries, then its entries will also be integers. Thus, A^{-1} can have only integer entries.

26. Since $\det(A) = 1$, then $x_i = \det(A_i)$. But since the coefficients and the constants are all integers, it follows that $\det(A_i)$ is an integer.

27. If A is a lower triangular matrix, then its jth column starts with at least $j - 1$ zeros and its $(j + 1)^{st}$ column starts with at least j zeros. If we delete the jth column of A and any row beyond the jth row, we are left with an upper triangular matrix M_{ij} whose j^{th} column is the old $(j + 1)^{st}$ column of A with the ith entry deleted. Thus M_{ij} has j zeros at the beginning of the jth column; that is, M_{ij} has a zero on the main diagonal. Since the determinant of a lower triangular matrix is just the product of the diagonal elements, then $\det(M_{ij}) = 0$. Because this is true whenever $j < i$, then the matrix of cofactors of A is upper triangular. Therefore, $\text{adj}(A)$ and hence A^{-1} will be lower triangular.

28. In order to prove the last identity, we rearrange the terms in Equation (1) as follows:

$$\det(A) = a_{13}(a_{21}a_{32} - a_{22}a_{31}) - a_{23}(a_{11}a_{32} - a_{12}a_{31}) + a_{33}(a_{11}a_{22} - a_{12}a_{21})$$

$$= a_{13}C_{13} + a_{23}C_{23} + a_{33}C_{33}$$

29.   The point (x, y) lies on the line determined by (a_1, b_1) and (a_2, b_2) if and only if there are constants α, β, and γ, not all zero, such that

$$\alpha x + \beta y + \gamma = 0$$

$$\alpha a_1 + \beta b_1 + \gamma = 0$$

$$\alpha a_2 + \beta b_2 + \gamma = 0$$

or equivalently, if and only if

$$\begin{bmatrix} x & y & 1 \\ a_1 & b_1 & 1 \\ a_2 & b_2 & 1 \end{bmatrix} \begin{bmatrix} \alpha \\ \beta \\ \gamma \end{bmatrix} = \begin{bmatrix} 0 \\ 0 \\ 0 \end{bmatrix}$$

has a nontrivial solution for α, β, and γ. This is so if and only if the determinant of the coefficient matrix vanishes, that is, if and only if

$$\begin{vmatrix} x & y & 1 \\ a_1 & b_1 & 1 \\ a_2 & b_2 & 1 \end{vmatrix} = 0$$

Alternatively, we can notice that the above equation is

$$(*) \qquad x(b_1 - b_2) + y(a_2 - a_1) + a_1 b_2 - a_2 b_1 = 0$$

If (a_1, b_1) and (a_2, b_2) are distinct points, then $(*)$ is the equation of a line. Moreover, it is easy to verify that the coordinates of these two points satisfy $(*)$.

30.   The points (x_1, y_1), (x_2, y_2) and (x_3, y_3) are collinear if and only if for some α, β, and γ, not all zero, the equations

$$\alpha x_1 + \beta y_1 + \gamma = 0$$

$$\alpha x_2 + \beta y_2 + \gamma = 0$$

$$\alpha x_3 + \beta y_3 + \gamma = 0$$

are satisfied.   If we use the same argument that we employed in the solution to Exercise 29, we see that the existence of a solution to the above system of equations for α, β, and γ not all zero is equivalent to the condition that

$$
\begin{vmatrix}
x_1 & y_1 & 1 \\
x_2 & y_2 & 1 \\
x_3 & y_3 & 1
\end{vmatrix} = 0
$$

32.   Let A be upper triangular. Then $a_{kj} = 0$ for $j < k$. If we remove Row i, then the resulting matrix will still have $k - 1$ leading zeros in Row k for $k < i$, but it will have k leading zeros for $i \leq k$. If we then remove Column j where $i < j$ to obtain B_{ij}, the resulting matrix will have $k - 1$ leading zeros in Row k for $1 \leq k < i$, k leading zeros for $i \leq k < j$, and $k - 1$ leading zeros for $j \leq k$. Thus the kth row of B_{ij} has at least $k - 1$ leading zeros, and hence B_{ij} is upper triangular.

33.   A 4×4 matrix has a nonzero determinant only if at least one elementary product is nonzero. Thus it must have at least 4 nonzero entries, or at most 12 zeros.

34.   We expand by cofactors along the first row to obtain the formula

$$
\det(A) = \star C_{11} + \star C_{12}
$$

where

$$
C_{ij} = \pm \det \left(\begin{bmatrix}
\star & 0 & 0 & 0 \\
\star & 0 & 0 & 0 \\
\star & \star & \star & \star \\
\star & \star & \star & \star
\end{bmatrix} \right)
$$

Expanding the above determinants by cofactors along their first rows yields

$$
C_{ij} = \pm \star \det \left(\begin{bmatrix}
0 & 0 & 0 \\
\star & \star & \star \\
\star & \star & \star
\end{bmatrix} \right) = 0
$$

Thus $\det(A) = 0$ no matter what values are assigned to the starred quantities.

35. **(a)** True. In the proof of Theorem 2.4.2, it is shown that $A \operatorname{adj}(A) = \det(A)I$.

(b) False. It is true provided the system has a solution and is square.

(c) True. The inverse of $\operatorname{adj}(A)$ is $A/\det(A)$.

(d) False. If the ith row of A consists entirely of zeros, then every row in the matrix of cofactors *except* the ith row must consist entirely of zeros. Since the adjoint is the transpose of the matrix of cofactors, then all of the *columns* of the adjoint except possibly the jth column will consist entirely of zeros. For instance, let $A = \begin{bmatrix} 0 & 0 \\ 1 & 1 \end{bmatrix}$. Then $\operatorname{adj}(A) = \begin{bmatrix} 1 & 0 \\ -1 & 0 \end{bmatrix}$.

SUPPLEMENTARY EXERCISES 2

1.

$$x' = \frac{\begin{vmatrix} x & -\dfrac{4}{5} \\[2ex] y & \dfrac{3}{5} \end{vmatrix}}{\begin{vmatrix} \dfrac{3}{5} & -\dfrac{4}{5} \\[2ex] \dfrac{4}{5} & \dfrac{3}{5} \end{vmatrix}} = \frac{\dfrac{3}{5}x + \dfrac{4}{5}y}{\dfrac{9}{25} + \dfrac{16}{25}} = \frac{3}{5}x + \frac{4}{5}y$$

$$y' = \frac{\begin{vmatrix} \dfrac{3}{5} & x \\[2ex] \dfrac{4}{5} & y \end{vmatrix}}{\begin{vmatrix} \dfrac{3}{5} & -\dfrac{4}{5} \\[2ex] \dfrac{4}{5} & \dfrac{3}{5} \end{vmatrix}} = \frac{\dfrac{3}{5}y - \dfrac{4}{5}x}{1} = -\frac{4}{5}x + \frac{3}{5}y$$

3. The determinant of the coefficient matrix is

$$\begin{vmatrix} 1 & 1 & \alpha \\ 1 & 1 & \beta \\ \alpha & \beta & 1 \end{vmatrix} = \begin{vmatrix} 1 & 1 & \alpha \\ 0 & 0 & \beta - \alpha \\ \alpha & \beta & 1 \end{vmatrix} = -(\beta - \alpha)\begin{vmatrix} 1 & 1 \\ \alpha & \beta \end{vmatrix} = -(\beta - \alpha)(\beta - \alpha)$$

The system of equations has a nontrivial solution if and only if this determinant is zero; that is, if and only if $\alpha = \beta$. (See Theorem 2.3.6.)

4. Recall that $\det(A)$ is the sum of 6 signed elementary products, 3 of which are positive and 3 of which are negative. Since each of the factors in these products is either 0 or 1, then the only values which the products can assume are 0, +1, or −1. If $\det(A)$ has 3 signed elementary products which are equal to +1, then the remaining 3 products must equal −1. (Why?) In this case, $\det(A) = 0$. However, $\det(A)$ can have 2 signed elementary products which are equal to +1 and the remaining 4 products equal to 0. For instance, this is true of

$$\det \begin{bmatrix} 1 & 0 & 1 \\ 1 & 1 & 0 \\ 0 & 1 & 1 \end{bmatrix}$$

Thus, the largest possible value for $\det(A)$ is 2.

5. **(a)** If the perpendicular from the vertex of angle α to side a meets side a between angles β and γ, then we have the following picture:

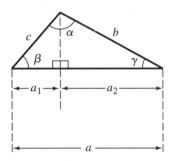

Thus $\cos\beta = \dfrac{a_1}{c}$ and $\cos\gamma = \dfrac{a_2}{b}$ and hence

$$a = a_1 + a_2 = c\cos\beta + b\cos\gamma$$

This is the first equation which you are asked to derive. If the perpendicular intersects side a outside of the triangle, the argument must be modified slightly, but the same result holds. Since there is nothing sacred about starting at angle α, the same argument starting at angles β and γ will yield the second and third equations.

Cramer's Rule applied to this system of equations yields the following results:

$$\cos\alpha = \frac{\begin{vmatrix} a & c & b \\ b & 0 & a \\ c & a & 0 \\ 0 & c & b \\ c & 0 & a \\ b & a & 0 \end{vmatrix}}{} = \frac{a(-a^2 + b^2 + c^2)}{2abc} = \frac{b^2 + c^2 - a^2}{2bc}$$

$$\cos\beta = \frac{\begin{vmatrix} 0 & a & b \\ c & b & a \\ b & c & 0 \end{vmatrix}}{2abc} = \frac{b(a^2 - b^2 + c^2)}{2abc} = \frac{a^2 + c^2 - b^2}{2ac}$$

$$\cos\gamma = \frac{\begin{vmatrix} 0 & c & a \\ c & 0 & b \\ b & a & c \end{vmatrix}}{2abc} = \frac{c(a^2 + b^2 - c^2)}{2abc} = \frac{a^2 + b^2 - c^2}{2ab}$$

6. The system of equations

$$(1 - \lambda)x - 2y = 0$$

$$x - (1 + \lambda)y = 0$$

always has the trivial solution $x = y = 0$. There can only be a nontrivial solution in case the determinant of the coefficient matrix is zero. But this determinant is just

$$-(1 - \lambda)(1 + \lambda) + 2 = \lambda^2 + 1$$

and $\lambda^2 + 1$ can never equal zero for any real value of λ.

7. If A is invertible, then $A^{-1} = \dfrac{1}{\det(A)} \operatorname{adj}(A)$, or $\operatorname{adj}(A) = [\det(A)]A^{-1}$. Thus

$$\operatorname{adj}(A)\frac{A}{\det(A)} = I$$

That is, adj(A) is invertible and

$$[\text{adj}(A)]^{-1} = \frac{1}{\det(A)} A$$

It remains only to prove that $A = \det(A)\text{adj}(A^{-1})$. This follows from Theorem 2.4.2 and Theorem 2.3.5 as shown:

$$A = [A^{-1}]^{-1} = \frac{1}{\det(A^{-1})}\text{adj}(A^{-1}) = \det(A)\text{adj}(A^{-1})$$

8.   From the proof of Theorem 2.4.2, we have

$(*)$ $A[\text{adj}(A)] = [\det(A)]I$

First, assume that A is invertible. Then $\det(A) \neq 0$. Notice that $[\det(A)]I$ is a diagonal matrix, each of whose diagonal elements is $\det(A)$. Thus, if we take the determinant of the matrices on the right- and left-hand sides of $(*)$, we obtain

$$\det(A[\text{adj}(A)]) = \det([\det(A)]I)$$

or

$$\det(A) \cdot \det(\text{adj}(A)) = [\det(A)]^n$$

and the result follows.

Now suppose that A is not invertible. Then $\det(A) = 0$ and hence we must show that $\det(\text{adj}(A)) = 0$. But if $\det(A) = 0$, then, by $(*)$, $A[\text{adj}(A)] = \mathit{0}$. Certainly, if $A = \mathit{0}$, then $\text{adj}(A) = \mathit{0}$; thus, $\det(\text{adj}(A)) = 0$ and the result holds. On the other hand, if $A \neq \mathit{0}$, then A must have a nonzero row $\mathbf{x}$. Hence $A[\text{adj}(A)] = \mathit{0}$ implies that $\mathbf{x}\,\text{adj}(A) = \mathit{0}$ or

$$[\text{adj}(A)]^T \mathbf{x}^T = \mathit{0}^T$$

Since $\mathbf{x}^T$ is a nontrivial solution to the above system of homogeneous equations, it follows that the determinant of $[\text{adj}(A)]^T$ is zero. Hence, $\det(\text{adj}(A)) = 0$ and the result is proved.

9. We simply expand W. That is,

$$\frac{dW}{dx} = \frac{d}{dx} \begin{vmatrix} f_1(x) & f_2(x) \\ g_1(x) & g_2(x) \end{vmatrix}$$

$$= \frac{d}{dx}(f_1(x)g_2(x) - f_2(x)g_1(x))$$

$$= f_1'(x)g_2(x) + f_1(x)g_2'(x) - f_2'(x)g_1(x) - f_2(x)g_1'(x)$$

$$= [f_1'(x)g_2(x) - f_2'(x)g_1(x)] + [f_1(x)g_2'(x) - f_2(x)g_1'(x)]$$

$$= \begin{vmatrix} f_1'(x) & f_2'(x) \\ g_1(x) & g_2(x) \end{vmatrix} + \begin{vmatrix} f_1(x) & f_2(x) \\ g_1'(x) & g_2'(x) \end{vmatrix}$$

10. (a) As suggested in the figure and the hints,

$$\text{area } ABC = \text{area } ADEC + \text{ area } CEFB - \text{ area } ADFB$$

$$= \frac{1}{2}(x_3 - x_1)(y_1 + y_3) + \frac{1}{2}(x_2 - x_3)(y_3 + y_2)$$

$$-\frac{1}{2}(x_2 - x_1)(y_1 + y_2)$$

$$= \frac{1}{2}[x_3y_1 - x_1y_1 + x_3y_3 - x_1y_3 + x_2y_3 - x_3y_3 + x_2y_2$$

$$- x_3y_2 - x_2y_1 + x_1y_1 - x_2y_2 + x_1y_2]$$

$$= \frac{1}{2}[x_1y_2 - x_1y_3 - x_2y_1 + x_2y_3 + x_3y_1 - x_3y_2]$$

$$= \frac{1}{2} \begin{vmatrix} x_1 & y_1 & 1 \\ x_2 & y_2 & 1 \\ x_3 & y_3 & 1 \end{vmatrix}$$

10.   **(b)**   The area of the triangle is

$$\frac{1}{2} \begin{vmatrix} 3 & 3 & 1 \\ -2 & -1 & 1 \\ 4 & 0 & 1 \end{vmatrix} = \frac{1}{2}[3(-1) + 2(3) + 4(3+1)] = \frac{19}{2}$$

11.   Let A be an $n \times n$ matrix for which the entries in each row add up to zero and let $\mathbf{x}$ be the $n \times 1$ matrix each of whose entries is one. Then all of the entries in the $n \times 1$ matrix $A\mathbf{x}$ are zero since each of its entries is the sum of the entries of one of the rows of A. That is, the homogeneous system of linear equations

$$A\mathbf{x} = \begin{bmatrix} 0 \\ \vdots \\ 0 \end{bmatrix}$$

has a nontrivial solution. Hence $\det(A) = 0$. (See Theorem 2.3.6.)

12.   To obtain B from A, we first interchange Row 1 and Row n. This changes the sign of $\det(A)$. If $n = 2$ or 3, we are done. If $n > 3$, then we interchange Rows 2 and $n - 1$, which again reverses the sign of the determinant. In general, we interchange Rows $k + 1$ and $n - k$ until all of the rows have been interchanged.

   If n is even, then we will have made $n/2$ interchanges and the sign of $\det(A)$ will have changed by a factor of $(-1)^{n/2}$. On the other hand, if n is odd, then we will have made $(n-1)/2$ interchanges and the sign of $\det(A)$ will have changed by a factor of $(-1)^{(n-1)/2}$. All of this can be summarized by the formula

$$\det(B) = (-1)^{\frac{n(n-1)}{2}} \det(A)$$

Can you see why?

13.   **(a)**   If we interchange the ith and j^{th} rows of A, then we claim that we must interchange the ith and j^{th} columns of A^{-1}. To see this, let

$$A = \begin{bmatrix} \text{Row } 1 \\ \text{Row } 2 \\ \vdots \\ \text{Row } n \end{bmatrix} \quad \text{and} \quad A^{-1} = [\text{Col. 1, Col. 2}, \ldots, \text{ Col. } n]$$

where $AA^{-1} = I$. Thus, the sum of the products of corresponding entries from Row s in A and from Column r in A^{-1} must be 0 unless $s = r$, in which case it is 1. That is, if Rows i and j are interchanged in A, then Columns i and j must be interchanged in A^{-1} in order to insure that only 1's will appear on the diagonal of the product AA^{-1}.

(b) If we multiply the ith row of A by a nonzero scalar c, then we must divide the ith column of A^{-1} by c. This will insure that the sum of the products of corresponding entries from the ith row of A and the ith column of A^{-1} will remain equal to 1.

(c) Suppose we add c times the ith row of A to the jth row of A. Call that matrix B. Now suppose that we add $-c$ times the jth column of A^{-1} to the ith column of A^{-1}. Call that matrix C. We claim that $C = B^{-1}$. To see that this is so, consider what happens when

$$\text{Row } j \rightarrow \text{ Row } j + c \text{ Row } i \qquad [\text{in } A]$$

$$\text{Column } i \rightarrow \text{ Column } i - c \text{ Column } j \quad [\text{in } A^{-1}]$$

The sum of the products of corresponding entries from the jth row of B and any k^{th} column of C will clearly be 0 unless $k = i$ or $k = j$. If $k = i$, then the result will be $c - c = 0$. If $k = j$, then the result will be 1. The sum of the products of corresponding entries from any other row of B — say the rth row — and any column of C — say the kth column — will be 1 if $r = k$ and 0 otherwise. This follows because there have been no changes unless $k = i$. In case $k = i$, the result is easily checked.

14. Let C be the matrix which is the same as A except for its ith row which is the sum of the ith rows of B_1 and B_2, or twice the ith row of A. Then by Theorem 2.3.1,

$$\det(B_1) + \det(B_2) = \det(C)$$

But $\det(C) = 2 \det(A)$, so that

$$\det(A) = \frac{1}{2} \left[\det(B_1) + \det(B_2) \right]$$

15. (a) We have

$$\det(\lambda I - A) = \begin{vmatrix} \lambda - a_{11} & -a_{12} & -a_{13} \\ -a_{21} & \lambda - a_{22} & -a_{23} \\ -a_{31} & -a_{32} & \lambda - a_{33} \end{vmatrix}$$

If we calculate this determinant by any method, we find that

$$\det(\lambda I - A) = (\lambda - a_{11})(\lambda - a_{22})(\lambda - a_{33}) - a_{23}a_{32}(\lambda - a_{11})$$

$$-a_{13}a_{31}(\lambda - a_{22}) - a_{12}a_{21}(\lambda - a_{33})$$

$$-a_{13}a_{21}a_{32} - a_{12}a_{23}a_{31}$$

$$= \lambda^3 + (-a_{11} - a_{22} - a_{33})\lambda^2$$

$$+(a_{11}a_{22} + a_{11}a_{33} + a_{22}a_{33} - a_{12}a_{21} - a_{13}a_{31} - a_{23}a_{32})\lambda$$

$$+(a_{11}a_{23}a_{32} + a_{12}a_{21}a_{33} + a_{13}a_{22}a_{31}$$

$$-a_{11}a_{22}a_{33} - a_{12}a_{23}a_{31} - a_{13}a_{21}a_{32})$$

15. **(b)** From Part (a) we see that $b = -\operatorname{tr}(A)$ and $d = -\det(A)$. (It is less obvious that c is the trace of the matrix of minors of the entries of A; that is, the sum of the minors of the diagonal entries of A.)

16. Recall that $\sin(a + b) = \sin a \cos b + \cos a \sin b$. If we multiply the 1st column of the determinant by $\cos \delta$ and the 2nd column by $\sin \delta$ and then add the results, we get the 3rd column. Therefore the determinant is 0.

17. If we multiply Column 1 by 10^4, Column 2 by 10^3, Column 3 by 10^2, Column 4 by 10, and add the results to Column 5, we obtain a new Column 5 whose entries are just the 5 numbers listed in the problem. Since each is divisible by 19, so is the resulting determinant.

18. **(b)** Here we have

$$A = \begin{bmatrix} 0 & 1 & 1 \\ 1 & 0 & -1 \\ 1 & 5 & 3 \end{bmatrix} \quad \text{so that} \quad \lambda I - A = \begin{bmatrix} \lambda & -1 & -1 \\ -1 & \lambda & 1 \\ -1 & -5 & \lambda - 3 \end{bmatrix}$$

If we add λ times Row 2 to Row 1, we have

$$\det(\lambda I - A) = \begin{vmatrix} 0 & \lambda^2 - 1 & \lambda - 1 \\ -1 & \lambda & 1 \\ -1 & -5 & \lambda - 3 \end{vmatrix} = (\lambda - 1) \begin{vmatrix} 0 & \lambda + 1 & 1 \\ -1 & \lambda & 1 \\ -1 & -5 & \lambda - 3 \end{vmatrix}$$

(Factor $\lambda - 1$ from Row 1.) Now add -1 times Row 2 to Row 3 and interchange Rows 1 and 2 to obtain the equation

$$\det(\lambda I - A) = -(\lambda - 1) \begin{vmatrix} -1 & \lambda & 1 \\ 0 & \lambda + 1 & 1 \\ 0 & -\lambda - 5 & \lambda - 4 \end{vmatrix}$$

To simplify, add Row 2 to Row 3 and then add $(\lambda + 1)/4$ times Row 3 to Row 2, getting

$$\det(\lambda I - A) = -(\lambda - 1) \begin{vmatrix} -1 & \lambda & 1 \\ 0 & \lambda + 1 & 1 \\ 0 & -4 & \lambda - 3 \end{vmatrix} = -(\lambda - 1) \begin{vmatrix} -1 & \lambda & 1 \\ 0 & 0 & \dfrac{\lambda^2 - 2\lambda + 1}{4} \\ 0 & -4 & \lambda - 3 \end{vmatrix}$$

If we interchange Rows 2 and 3, the last determinant becomes lower triangular, so that

$$\det(\lambda I - A) = -(\lambda - 1)(-1)(-1)(-4)\left(\frac{\lambda^2 - 2\lambda + 1}{4}\right)$$

$$= (\lambda - 1)(\lambda^2 - 2\lambda + 1) = (\lambda - 1)^3$$

Thus, the characteristic equation is $(\lambda - 1)^3 = 0$ and so $\lambda = 1$ is the only eigenvalue. If we let $\lambda = 1$, then the eigenvectors are the nonzero solutions to the equation

$$\begin{bmatrix} 1 & -1 & -1 \\ -1 & 1 & 1 \\ -1 & -5 & -2 \end{bmatrix} \begin{bmatrix} x_1 \\ x_2 \\ x_3 \end{bmatrix} = \begin{bmatrix} 0 \\ 0 \\ 0 \end{bmatrix}$$

The augmented matrix reduces to

$$\begin{bmatrix} 1 & -1 & -1 & 0 \\ 0 & 1 & \dfrac{1}{2} & 0 \\ 0 & 0 & 0 & 0 \end{bmatrix}$$

If we put $x_3 = 2t$, then $x_2 = -t$ and $x_1 = t$, so the eigenvectors are

$$\mathbf{x} = \begin{bmatrix} t & -t & 2t \end{bmatrix}^T \quad \text{where } t \neq 0$$

EXERCISE SET 3.1

1.   (a)

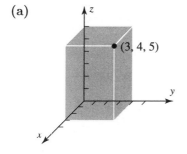

(c)

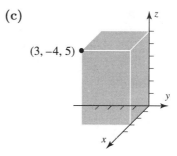

(e) (−3, −4, 5)

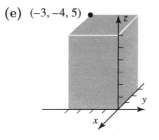

(j)

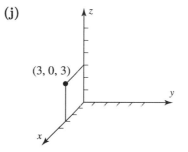

2.   (a)

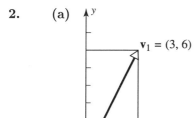

(c)

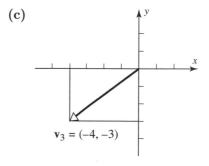

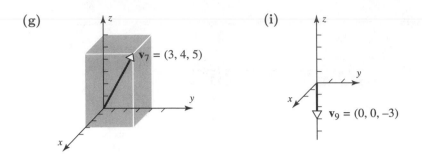

(g) $\mathbf{v}_7 = (3, 4, 5)$

(i) $\mathbf{v}_9 = (0, 0, -3)$

3. (a) $\overrightarrow{P_1 P_2} = (3 - 4,\ 7 - 8) = (-1, -1)$

    (e) $\overrightarrow{P_1 P_2} = (-2 - 3,\ 5 + 7,\ -4 - 2) = (-5, 12, -6)$

5. (a) Let $P = (x, y, z)$ be the initial point of the desired vector and assume that this vector has the same length as $\mathbf{v}$. Since $\overrightarrow{PQ}$ has the same direction as $\mathbf{v} = (4, -2, -1)$, we have the equation

$$\overrightarrow{PQ} = (3 - x,\ 0 - y,\ -5 - z) = (4, -2, -1)$$

If we equate components in the above equation, we obtain

$$x = -1, \quad y = 2, \quad \text{and} \quad z = -4$$

Thus, we have found a vector $\overrightarrow{PQ}$ which satisfies the given conditions. Any positive multiple $k\,\overrightarrow{PQ}$ will also work provided the terminal point remains fixed at Q. Thus, P could be any point $(3 - 4k, 2k, k - 5)$ where $k > 0$.

(b) Let $P = (x, y, z)$ be the initial point of the desired vector and assume that this vector has the same length as $\mathbf{v}$. Since $\overrightarrow{PQ}$ is oppositely directed to $\mathbf{v} = (4, -2, -1)$, we have the equation

$$\overrightarrow{PQ} = (3 - x,\ 0 - y,\ -5 - z) = (-4, 2, 1)$$

If we equate components in the above equation, we obtain

$$x = 7,\ y = -2,\ \text{and}\ z = -6$$

Thus, we have found a vector $\overrightarrow{PQ}$ which satisfies the given conditions. Any positive multiple $k\,\overrightarrow{PQ}$ will also work, provided the terminal point remains fixed at Q. Thus, P could be any point $(3 + 4k, -2k, -k - 5)$ where $k > 0$.

6. **(a)** $\mathbf{v} - \mathbf{w} = (4 - 6,\ 0 - (-1),\ -8 - (-4)) = (-2, 1, -4)$

    (c) $-\mathbf{v} + \mathbf{u} = (-4 - 3,\ 0 + 1,\ 8 + 2) = (-7, 1, 10)$

    (e) $-3(\mathbf{v} - 8\mathbf{w}) = -3(4 - 48,\ 0 + 8,\ -8 + 32) = (132, -24, -72)$

7. Let $\mathbf{x} = (x_1, x_2, x_3)$. Then

    $$2\mathbf{u} - \mathbf{v} + \mathbf{x} = (-6, 2, 4) - (4, 0, -8) + (x_1, x_2, x_3)$$
    $$= (-10 + x_1,\ 2 + x_2,\ 12 + x_3)$$

    On the other hand,

    $$7\mathbf{x} + \mathbf{w} = 7(x_1, x_2, x_3) + (6, -1, -4)$$
    $$= (7x_1 + 6,\ 7x_2 - 1,\ 7x_3 - 4)$$

    If we equate the components of these two vectors, we obtain

    $$7x_1 + 6 = x_1 - 10$$
    $$7x_2 - 1 = x_2 + 2$$
    $$7x_3 - 4 = x_3 + 12$$

    Hence, $\mathbf{x} = (-8/3, 1/2, 8/3)$.

9. Suppose there are scalars c_1, c_2, and c_3 which satisfy the given equation. If we equate components on both sides, we obtain the following system of equations:

    $$-2c_1 - 3c_2 + c_3 = 0$$
    $$9c_1 + 2c_2 + 7c_3 = 5$$
    $$6c_1 + c_2 + 5c_3 = 4$$

    The augmented matrix of this system of equations can be reduced to

    $$\begin{bmatrix} 2 & 3 & -1 & 0 \\ 0 & 2 & -2 & -1 \\ 0 & 0 & 0 & -1 \end{bmatrix}$$

    The third row of the above matrix implies that $0c_1 + 0c_2 + 0c_3 = -1$. Clearly, there do not exist scalars c_1, c_2, and c_3 which satisfy the above equation, and hence the system is inconsistent.

10. If we equate components on both sides of the given equation, we obtain

$$c_1 + 2c_2 \qquad = 0$$
$$2c_1 + \ c_2 + 3c_3 = 0$$
$$c_2 + \ c_3 = 0$$

The augmented matrix of this system of equations can be reduced to

$$\begin{bmatrix} 1 & 0 & 0 & 0 \\ 0 & 1 & 0 & 0 \\ 0 & 0 & 1 & 0 \end{bmatrix}$$

Thus the only solution is $c_1 = c_2 = c_3 = 0$.

11. We work in the plane determined by the three points $O = (0,0,0)$, $P = (2,3,-2)$, and $Q = (7,-4,1)$. Let X be a point on the line through P and Q and let $t\,\overrightarrow{PQ}$ (where t is a positive, real number) be the vector with initial point P and terminal point X. Note that the length of $t\,\overrightarrow{PQ}$ is t times the length of $\overrightarrow{PQ}$. Referring to the figure below, we see that

$$\overrightarrow{OP} + t\,\overrightarrow{PQ} = \overrightarrow{OX}$$

and

$$\overrightarrow{OP} + \overrightarrow{PQ} = \overrightarrow{OQ}$$

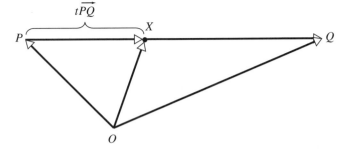

Therefore,

$$\overrightarrow{OX} = \overrightarrow{OP} + t(\overrightarrow{OQ} - \overrightarrow{OP})$$

$$= (1-t)\,\overrightarrow{OP} + t\,\overrightarrow{OQ}$$

(a) To obtain the midpoint of the line segment connecting P and Q, we set $t = 1/2$. This gives

$$\overrightarrow{OX} = \frac{1}{2}\,\overrightarrow{OP} + \frac{1}{2}\,\overrightarrow{OQ}$$

$$= \frac{1}{2}(2,3,-2) + \frac{1}{2}(7,-4,1)$$

$$= \left(\frac{9}{2}, -\frac{1}{2}, -\frac{1}{2}\right)$$

(b) Now set $t = 3/4$. This gives

$$\overrightarrow{OX} = \frac{1}{4}(2,3,-2) + \frac{3}{4}(7,-4,1) = \left(\frac{23}{4}, -\frac{9}{4}, \frac{1}{4}\right)$$

12. The relationship between the xy-coordinate system and the $x'y'$-coordinate system is given by

$$x' = x - 2, \quad y' = y - (-3) = y + 3$$

(a) $x' = 7 - 2 = 5$ and $y' = 5 + 3 = 8$

(b) $x = x' + 2 = -1$ and $y = y' - 3 = 3$

(c)

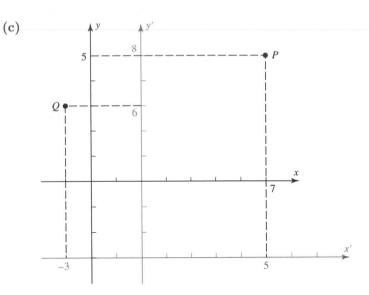

13. Let (x_0, y_0, z_0) denote the origin in the $x'y'z'$-coordinate system with respect to the xyz-coordinate system. Suppose that P_1 and P_2 are the initial and terminal points, respectively, for a vector $\mathbf{v}$. Let (x_1, y_1, z_1) and (x_2, y_2, z_2) be the coordinates of P_1 and P_2 in the xyz-coordinate system, and let (x_1', y_1', z_1') and (z_2', y_2', z_2') be the coordinates of P_1 and P_2 in the $x'y'z'$-coordinate system. Further, let (v_1, v_2, v_3) and (v_1', v_2', v_3') denote the coordinates of $\mathbf{v}$ with respect to the xyz- and $x'y'z'$-coordinate systems, respectively. Then

$$v_1 = x_2 - x_1, \quad v_2 = y_2 - y_1, \quad v_3 = z_2 - z_1$$

and

$$v_1' = x_2' - x_1', \quad v_2' = y_2' - y_1', \quad v_3' = z_2' - z_1'$$

However,

$$x_1' = x_1 - x_0, \quad y_1' = y_1 - y_0, \quad z_1' = z_1 - z_0$$

and

$$x_2' = x_2 - x_0, \quad y_2' = y_2 - y_0, \quad z_2' = z_2 - z_0$$

If we substitute the above expressions for x_1' and x_2' into the equation for v_1', we obtain

$$v_1' = (x_2 - x_0) - (x_1 - x_0) = x_2 - x_1 = v_1$$

In a similar way, we can show that $v_2' = v_2$ and $v_3' = v_3$.

15. Let $\mathbf{v} = (v_1, v_2)$ and $\mathbf{u} = k\mathbf{v} = (u_1, u_2)$. Let P and Q be the points (v_1, v_2) and (u_1, u_2), respectively. Since the triangles OPR and OQS are similar (see the diagram),

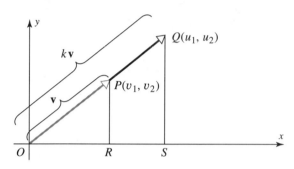

we have

$$\frac{u_1}{v_1} = \frac{\text{length of } \overrightarrow{OS}}{\text{length of } \overrightarrow{OR}} = \frac{\text{length of } \overrightarrow{OQ}}{\text{length of } \overrightarrow{OP}} = k$$

Thus, $u_1 = kv_1$. Similarly, $u_2 = kv_2$.

16. The vector $\mathbf{u}$ has terminal point Q which is the midpoint of the line segment connecting P_1 and P_2.

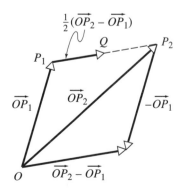

17.   We have $\mathbf{x} + \mathbf{y} + \mathbf{z} + \mathbf{w} = \mathbf{0}$ where

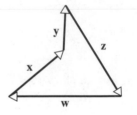

18.   Geometrically, given 4 nonzero vectors attach the "tail" of one to the "head" of another and continue until all 4 have been strung together. The vector from the "tail" of the first vector to the "head" of the last one will be their sum.

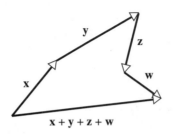

EXERCISE SET 3.2

1. **(a)** $\|\mathbf{v}\| = \left(4^2 + (-3)^2\right)^{1/2} = 5$

   (c) $\|\mathbf{v}\| = \left[(-5)^2 + 0^2\right]^{1/2} = 5$

   (e) $\|\mathbf{v}\| = \left[(-7)^2 + 2^2 + (-1)^2\right]^{1/2} = \sqrt{54}$

2. **(a)** $d = \left[(3-5)^2 + (4-7)^2\right]^{1/2} = \sqrt{13}$

   (c) $d = \left[7 - (-7)^2 + (-5) - (-2)^2 + 1 - (-1)^2\right]^{1/2} = \sqrt{209}$

3. **(a)** Since $\mathbf{u} + \mathbf{v} = (3, -5, 7)$, then

   $$\|\mathbf{u} + \mathbf{v}\| = \left[3^2 + (-5)^2 + 7^2\right]^{1/2} = \sqrt{83}$$

   (c) Since

   $$\| - 2\mathbf{u}\| = \left[(-4)^2 + 4^2 + (-6)^2\right]^{1/2} = 2\sqrt{17}$$

   and

   $$2\|\mathbf{u}\| = 2\left[2^2 + (-2)^2 + 3^2\right]^{1/2} = 2\sqrt{17}$$

   then

   $$\| - 2\mathbf{u}\| + 2\|\mathbf{u}\| = 4\sqrt{17}$$

   (e) Since $\|\mathbf{w}\| = [3^2 + 6^2 + (-4)^2]^{1/2} = \sqrt{61}$, then

   $$\frac{1}{\|\mathbf{w}\|}\mathbf{w} = \left(\frac{3}{\sqrt{61}}, \ \frac{6}{\sqrt{61}}, \ \frac{-4}{\sqrt{61}}\right)$$

4.   Since $k\mathbf{v} = (-k, 2k, 5k)$, then

$$\|k\mathbf{v}\| = \left[k^2 + 4k^2 + 25k^2\right]^{1/2} = |k|\sqrt{30}$$

If $\|k\mathbf{v}\| = 4$, it follows that $|k|\sqrt{30} = 4$ or $k = \pm 4/\sqrt{30}$.

6.   **(b)**   From Part (a), we know that the norm of $\mathbf{v}/\|\mathbf{v}\|$ is 1. But if $\mathbf{v} = (3, 4)$, then $\|\mathbf{v}\| = 5$. Hence $\mathbf{u} = \mathbf{v}/\|\mathbf{v}\| = (3/5, 4/5)$ has norm 1 and has the same direction as $\mathbf{v}$.

7.   **(b)**   By the result of Part (a), we have $4\mathbf{u} = 4(\|\mathbf{u}\| \cos 30°, \|\mathbf{u}\| \sin 30°) = 4(3(\sqrt{3}/2), 3(1/2)) = (6\sqrt{3},\ 6)$ and $5\mathbf{v} = 5(\|\mathbf{v}\| \cos(135°), \|\mathbf{v}\| \sin(135°)) = 5(2(-1/\sqrt{2}),\ 2(1/\sqrt{2})) = (-10/\sqrt{2},\ 10/\sqrt{2}) = (-5\sqrt{2},\ 5\sqrt{2})$. Thus $4\mathbf{u} - 5\mathbf{v} = (6\sqrt{3} + 5\sqrt{2},\ 6 - 5\sqrt{2})$.

8.   Note that $\|\mathbf{p} - \mathbf{p}_0\| = 1$ if and only if $\|\mathbf{p} - \mathbf{p}_0\|^2 = 1$. Thus

$$(x - x_0)^2 + (y - y_0)^2 + (z - z_0)^2 = 1$$

The points (x, y, z) which satisfy these equations are just the points on the sphere of radius 1 with center (x_0, y_0, z_0); that is, they are all the points whose distance from (x_0, y_0, z_0) is 1.

9.   First, suppose that $\mathbf{u}$ and $\mathbf{v}$ are neither similarly nor oppositely directed and that neither is the zero vector.

   If we place the initial point of $\mathbf{v}$ at the terminal point of $\mathbf{u}$, then the vectors $\mathbf{u}$, $\mathbf{v}$, and $\mathbf{u} + \mathbf{v}$ form a triangle, as shown in (i) below.

(i)   (ii)

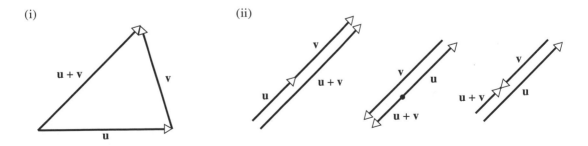

Since the length of one side of a triangle — say $\|\mathbf{u} + \mathbf{v}\|$ — cannot exceed the sum of the lengths of the other two sides, then $\|\mathbf{u} + \mathbf{v}\| < \|\mathbf{u}\| + \|\mathbf{v}\|$.

Now suppose that $\mathbf{u}$ and $\mathbf{v}$ have the same direction. From diagram (ii), we see that $\|\mathbf{u} + \mathbf{v}\| = \|\mathbf{u}\| + \|\mathbf{v}\|$. If $\mathbf{u}$ and $\mathbf{v}$ have opposite directions, then (again, see diagram (ii)) $\|\mathbf{u} + \mathbf{v}\| < \|\mathbf{u}\| + \|\mathbf{v}\|$.

Finally, if either vector is zero, then $\|\mathbf{u} + \mathbf{v}\| = \|\mathbf{u}\| + \|\mathbf{v}\|$.

10. These proofs are for vectors in 3-space. To obtain proofs in 2-space, just delete the 3rd component. Let $\mathbf{u} = (u_1, u_2, u_3)$ and $\mathbf{v} = (v_1, v_2, v_3)$. Then

(a) $\mathbf{u} + \mathbf{v} = (u_1 + v_1,\ u_2 + v_2,\ u_3 + v_3)$

$\qquad = (v_1 + u_1,\ v_2 + u_2,\ v_3 + u_3) = \mathbf{v} + \mathbf{u}$

(c) $\mathbf{u} + \mathbf{0} = (u_1 + 0,\ u_2 + 0,\ u_3 + 0)$

$\qquad = (0 + u_1,\ 0 + u_2,\ 0 + u_3)$

$\qquad = (u_1, u_2, u_3) = \mathbf{0} + \mathbf{u} = \mathbf{u}$

(e) $k(l\mathbf{u}) = k(lu_1, lu_2, lu_3) = (klu_1, klu_2, klu_3) = (kl)\mathbf{u}$

11. Again, we work in 3-space. Let $\mathbf{u} = (u_1, u_2, u_3)$.

(d) $\mathbf{u} + (-\mathbf{u}) = (u_1 + (-u_1),\ u_2 + (-u_2),\ u_3 + (-u_3)) = (0, 0, 0) = \mathbf{0}$

(g) $(k + l)\mathbf{u} = ((k + l)u_1,\ (k + l)u_2,\ (k + l)u_3)$

$\qquad = (ku_1 + lu_1,\ ku_2 + lu_2,\ ku_3 + lu_3)$

$\qquad = (ku_1, ku_2, ku_3) + (lu_1, lu_2, lu_3)$

$\qquad = k\mathbf{u} + l\mathbf{u}$

(h) $1\mathbf{u} = (1u_1, 1u_2, 1u_3) = (u_1, u_2, u_3) = \mathbf{u}$

12. See Exercise 9. Equality occurs only when $\mathbf{u}$ and $\mathbf{v}$ have the same direction or when one is the zero vector.

13. (a) For $\mathbf{p} = (a, b, c)$ to be equidistant from the origin and the xz-plane, we must have $\sqrt{a^2 + b^2 + c^2} = |b|$. Thus, $a^2 + b^2 + c^2 = b^2$ or $a^2 + c^2 = 0$, so that $a = c = 0$. That is, $\mathbf{p}$ must lie on the y-axis.

(b) For $\mathbf{p} = (a, b, c)$ to be farther from the origin than the xz-plane, we must have $\sqrt{a^2 + b^2 + c^2} > |b|$. Thus $a^2 + b^2 + c^2 > b^2$ or $a^2 + c^2 > 0$, so that a and c are not both zero. That is, $\mathbf{p}$ cannot lie on the y-axis.

14.   (a) If $\|\mathbf{x}\| < 1$, then the point $\mathbf{x}$ lies inside the circle or sphere of radius one with center at the origin.

       (b) Such points $\mathbf{x}$ must satisfy the inequality $\|\mathbf{x} - \mathbf{x}_0\| > 1$.

15.   The two triangles pictured are similar since $\mathbf{u}$ and $k\mathbf{u}$ and $\mathbf{v}$ and $k\mathbf{v}$ are parallel and their lengths are proportional; i.e., $\|k\mathbf{u}\| = k\|\mathbf{u}\|$ and $\|k\mathbf{v}\| = k\|\mathbf{v}\|$. Therefore $\|k\mathbf{u} + k\mathbf{v}\| = k\|\mathbf{u} + \mathbf{v}\|$. That is, the corresponding sides of the two triangles are all proportional, where k is the constant of proportionality. Thus, the vectors $k(\mathbf{u} + \mathbf{v})$ and $k\mathbf{u} + k\mathbf{v}$ have the same length. Moreover, $\mathbf{u} + \mathbf{v}$ and $k\mathbf{u} + k\mathbf{v}$ have the same direction because corresponding sides of the two triangles are parallel. Hence, $k(\mathbf{u} + \mathbf{v})$ and $k\mathbf{u} + k\mathbf{v}$ have the same direction. Therefore, $k(\mathbf{u} + \mathbf{v}) = k\mathbf{u} + k\mathbf{v}$.

EXERCISE SET 3.3

1. **(a)** $\mathbf{u} \cdot \mathbf{v} = (2)(5) + (3)(-7) = -11$

   (c) $\mathbf{u} \cdot \mathbf{v} = (1)(3) + (-5)(3) + (4)(3) = 0$

2. **(a)** We have $\|\mathbf{u}\| = [2^2 + 3^2]^{1/2} = \sqrt{13}$ and $\|\mathbf{v}\| = [5^2 + (-7)^2]^{1/2} = \sqrt{74}$. From Problem 1(a), we know that $\mathbf{u} \cdot \mathbf{v} = -11$. Hence,

$$\cos\theta = \frac{-11}{\sqrt{13}\sqrt{74}} \approx 110.8°$$

   (c) From Problem 1(c), we know that $\mathbf{u} \cdot \mathbf{v} = 0$. Since neither $\mathbf{u}$ nor $\mathbf{v}$ is the zero vector, this implies that $\cos\theta = 0$.

3. **(a)** $\mathbf{u} \cdot \mathbf{v} = (6)(2) + (1)(0) + (4)(-3) = 0$. Thus the vectors are orthogonal.

   (b) $\mathbf{u} \cdot \mathbf{v} = -1 < 0$. Thus θ is obtuse.

4. **(a)** Since $\mathbf{u} \cdot \mathbf{a} = 0$ and $\|\mathbf{a}\| \neq 0$, then $\mathbf{w}_1 = (0,0)$.

   (c) Since $\mathbf{u} \cdot \mathbf{a} = (3)(1) + (1)(0) + (-7)(5) = -32$ and $\|\mathbf{a}\|^2 = 1^2 + 0^2 + 5^2 = 26$, we have

$$\mathbf{w}_1 = -\frac{32}{26}(1,\ 0,\ 5) = \left(-\frac{16}{13},\ 0,\ -\frac{80}{13}\right)$$

5. **(a)** From Problem 4(a), we have

$$\mathbf{w}_2 = \mathbf{u} - \mathbf{w}_1 = \mathbf{u} = (6, 2)$$

5. **(c)** From Problem 4(c), we have

$$\mathbf{w}_2 = (3, 1, -7) - (-16/13, \ 0, \ -80/13) = (55/13, \ 1, \ -11/13)$$

6. **(a)** Since $|\mathbf{u} \cdot \mathbf{a}| = |-4 + 6| = |2| = 2$ and $\|\mathbf{a}\| = 5$, it follows from Formula (10) that $\|\text{proj}_\mathbf{a}\mathbf{u}\| = 2/5$.

 (d) Since $|\mathbf{u} \cdot \mathbf{a}| = |3 - 4 - 42| = |-43| = 43$ and $\|\mathbf{a}\| = \sqrt{54}$, Formula (10) implies that $\|\text{proj}_\mathbf{a}\mathbf{u}\| = 43/\sqrt{54}$.

8. **(c)** From Part (a) it follows that the vectors $(-4, -3)$ and $(4, 3)$ are orthogonal to $(-3, 4)$. Divide each component by the norm, 5, to obtain the desired unit vectors.

10. We look for vectors (x, y, z) such that $(x, y, z) \cdot (5, -2, 3) = 0$. That is, such that $5x - 2y + 3z = 0$. One such vector is $(0, 3, 2)$. Another is $(1, 1, -1)$. To find others, assign arbitrary values, not both zero, to any two of the variables x, y, and z and solve for the remaining variable.

12. Since the three points are not collinear (Why?), they do form a triangle. Since

$$\overrightarrow{AB} \ \cdot \ \overrightarrow{BC} = (1, 3, -2) \cdot (4, -2, 1) = 0$$

the right angle is at B.

13. Let $\mathbf{w} = (x, y, z)$ be orthogonal to both $\mathbf{u}$ and $\mathbf{v}$. Then $\mathbf{u} \cdot \mathbf{w} = 0$ implies that $x + z = 0$ and $\mathbf{v} \cdot \mathbf{w} = 0$ implies that $y + z = 0$. That is $\mathbf{w} = (x, x, -x)$. To transform $\mathbf{w}$ into a unit vector, we divide each component by $\|\mathbf{w}\| = \sqrt{3x^2}$. Thus either $(1/\sqrt{3}, \ 1/\sqrt{3}, \ -1/\sqrt{3})$ or $(-1/\sqrt{3}, \ -1/\sqrt{3}, \ 1/\sqrt{3})$ will work.

14. **(a)** We look for a value of k which will make $\mathbf{p}$ a nonzero multiple of $\mathbf{q}$. If $\mathbf{p} = c\mathbf{q}$, then $(2, k) = c(3, 5)$, so that $k/5 = 2/3$ or $k = 10/3$.

 (c) We want k such that

$$\mathbf{p} \cdot \mathbf{q} = \|\mathbf{p}\| \, \|\mathbf{q}\| \cos \frac{\pi}{3}$$

or

$$6 + 5k = \sqrt{4 + k^2} \ \sqrt{3^2 + 5^2} \left(\frac{1}{2}\right)$$

If we square both sides and use the quadratic formula, we find that

$$k = \frac{-60 \pm 34\sqrt{3}}{33}$$

The minus sign in the above equation is extraneous because it yields an angle of $2\pi/3$.

15. **(b)** Here

$$D = \frac{|4(2) + 1(-5) - 2|}{\sqrt{(4)^2 + (1)^2}} = \frac{1}{\sqrt{17}}$$

16. From the definition of the norm, we have

$$\|\mathbf{u} + \mathbf{v}\|^2 = (\mathbf{u} + \mathbf{v}) \cdot (\mathbf{u} + \mathbf{v})$$

Using Theorem 3.3.2, we obtain

$$\|\mathbf{u} + \mathbf{v}\|^2 = (\mathbf{u} \cdot \mathbf{u}) + (\mathbf{u} \cdot \mathbf{v}) + (\mathbf{v} \cdot \mathbf{u}) + (\mathbf{v} \cdot \mathbf{v})$$

or

(∗) $$\|\mathbf{u} + \mathbf{v}\|^2 = \|\mathbf{u}\|^2 + 2(\mathbf{u} \cdot \mathbf{v}) + \|\mathbf{v}\|^2$$

Similarly,

(∗∗) $$\|\mathbf{u} - \mathbf{v}\|^2 = \|\mathbf{u}\|^2 - 2(\mathbf{u} \cdot \mathbf{v}) + \|\mathbf{v}\|^2$$

If we add Equations (∗) and (∗∗), we obtain the desired result.

17. If we subtract Equation (∗∗) from Equation (∗) in the solution to Problem 16, we obtain

$$\|\mathbf{u} + \mathbf{v}\|^2 - \|\mathbf{u} - \mathbf{v}\|^2 = 4(\mathbf{u} \cdot \mathbf{v})$$

If we then divide both sides by 4, we obtain the desired result.

18. Let $\mathbf{u}_1$, $\mathbf{u}_2$, and $\mathbf{u}_3$ be three sides of the cube as shown. The diagonal of the face of the cube determined by $\mathbf{u}_1$ and $\mathbf{u}_2$ is $\mathbf{b} = \mathbf{u}_1 + \mathbf{u}_2$; the diagonal of the cube itself is $\mathbf{a} = \mathbf{u}_1 + \mathbf{u}_2 + \mathbf{u}_3$. The angle θ between $\mathbf{a}$ and $\mathbf{b}$ is given by

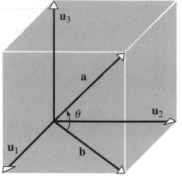

$$\cos\theta = \frac{\mathbf{a} \cdot \mathbf{b}}{\|\mathbf{a}\| \|\mathbf{b}\|}$$

$$= \frac{(\mathbf{u}_1 + \mathbf{u}_2 + \mathbf{u}_3) \cdot (\mathbf{u}_1 + \mathbf{u}_2)}{\sqrt{(\mathbf{u}_1 + \mathbf{u}_2 + \mathbf{u}_3) \cdot (\mathbf{u}_1 + \mathbf{u}_2 + \mathbf{u}_3)} \sqrt{(\mathbf{u}_1 + \mathbf{u}_2) \cdot (\mathbf{u}_1 + \mathbf{u}_2)}}$$

Because $\mathbf{u}_1$, $\mathbf{u}_2$, and $\mathbf{u}_3$ are mutually orthogonal, we have $\mathbf{u}_i \cdot \mathbf{u}_j = 0$ whenever $i \neq j$. Also $\mathbf{u}_i \cdot \mathbf{u}_i = \|\mathbf{u}_i\|^2$ and $\|\mathbf{u}_1\| = \|\mathbf{u}_2\| = \|\mathbf{u}_3\|$. Thus

$$\cos\theta = \frac{\|\mathbf{u}_1\|^2 + \|\mathbf{u}_2\|^2}{\sqrt{\|\mathbf{u}_1\|^2 + \|\mathbf{u}_2\|^2 + \|\mathbf{u}_3\|^2} \sqrt{\|\mathbf{u}_1\|^2 + \|\mathbf{u}_2\|^2}}$$

$$= \frac{2\|\mathbf{u}_1\|^2}{\sqrt{3\|\mathbf{u}_1\|^2} \sqrt{2\|\mathbf{u}_1\|^2}} = \frac{2}{\sqrt{6}}$$

That is, $\theta = \arccos(2/\sqrt{6})$.

19. (a) Let $\mathbf{i} = (1,0,0)$, $\mathbf{j} = (0,1,0)$, and $\mathbf{k} = (0,0,1)$ denote the unit vectors along the x, y, and z axes, respectively. If $\mathbf{v}$ is the arbitrary vector (a,b,c), then we can write $\mathbf{v} = a\mathbf{i} + b\mathbf{j} + c\mathbf{k}$. Hence, the angle α between $\mathbf{v}$ and $\mathbf{i}$ is given by

$$\cos\alpha = \frac{\mathbf{v} \cdot \mathbf{i}}{\|\mathbf{v}\| \|\mathbf{i}\|} = \frac{a}{\sqrt{a^2 + b^2 + c^2}} = \frac{a}{\|\mathbf{v}\|}$$

since $\|\mathbf{i}\| = 1$ and $\mathbf{i} \cdot \mathbf{j} = \mathbf{i} \cdot \mathbf{k} = 0$.

21. By the results of Exercise 19, we have that if $\mathbf{v}_i = (a_i, b_i, c_i)$ for $i = 1$ and 2, then $\cos\alpha_i = a_i/\|\mathbf{v}_i\|$, $\cos\beta_i = b_i/\|\mathbf{v}_i\|$, and $\cos\gamma_i = c_i/\|\mathbf{v}_i\|$. Now

$$\mathbf{v}_1 \text{ and } \mathbf{v}_2 \text{ are orthogonal} \quad \Longleftrightarrow \quad \mathbf{v}_1 \cdot \mathbf{v}_2 = 0$$

$$\Longleftrightarrow \quad a_1 a_2 + b_1 b_2 + c_1 c_2 = 0$$

$$\Longleftrightarrow \quad \frac{a_1 a_2}{\|\mathbf{v}_1\| \, \|\mathbf{v}_2\|} + \frac{b_1 b_2}{\|\mathbf{v}_1\| \, \|\mathbf{v}_2\|} + \frac{c_1 c_2}{\|\mathbf{v}_1\| \, \|\mathbf{v}_2\|} = 0$$

$$\Longleftrightarrow \quad \cos \alpha_1 \cos \alpha_2 + \cos \beta_1 \cos \beta_2 + \cos \gamma_1 \cos \gamma_2 = 0$$

22. Note that

$$\mathbf{v} \cdot (k_1 \mathbf{w}_1 + k_2 \mathbf{w}_2) = k_1 (\mathbf{v} \cdot \mathbf{w}_1) + k_2 (\mathbf{v} \cdot \mathbf{w}_2) = 0$$

because, by hypothesis, $\mathbf{v} \cdot \mathbf{w_1} = \mathbf{v} \cdot \mathbf{w_2} = 0$. Therefore $\mathbf{v}$ is orthogonal to $k_1 \mathbf{w}_1 + k_2 \mathbf{w}_2$ for any scalars k_1 and k_2.

23. Note that $\mathbf{w}$ is a multiple of $\mathbf{u}$ plus a multiple of $\mathbf{v}$ and thus lies in the plane determined by $\mathbf{u}$ and $\mathbf{v}$. Let α be the angle between $\mathbf{u}$ and $\mathbf{w}$ and let β be the angle between $\mathbf{v}$ and $\mathbf{w}$. Then

$$\cos \alpha = \frac{\mathbf{u} \cdot \mathbf{w}}{\|\mathbf{u}\| \, \|\mathbf{v}\|} = \frac{[k\mathbf{u} \cdot \mathbf{v} + \ell \mathbf{u} \cdot \mathbf{u}]}{k \|\mathbf{w}\|}$$

$$= \frac{[k\mathbf{u} \cdot \mathbf{v} + \ell k^2]}{k \|\mathbf{w}\|}$$

$$= \frac{\mathbf{u} \cdot \mathbf{v} + \ell k}{\|\mathbf{w}\|}$$

Similarly, we can show that

$$\cos \beta = \frac{\mathbf{v} \cdot \mathbf{w}}{\|\mathbf{v}\| \, \|\mathbf{w}\|} = \frac{\mathbf{u} \cdot \mathbf{v} + kl}{\|\mathbf{w}\|}$$

Since both α and β lie between 0 and π, then $\cos \alpha = \cos \beta$ implies that $\alpha = \beta$.

24. (a) The inner product $\mathbf{x} \cdot \mathbf{y}$ is defined only if both $\mathbf{x}$ and $\mathbf{y}$ are vectors, but here $\mathbf{v} \cdot \mathbf{w}$ is a scalar.

(b) We can add two vectors or two scalars, but not one of each.

24. **(c)** The norm of $\mathbf{x}$ is defined only for $\mathbf{x}$ a vector, but $\mathbf{u} \cdot \mathbf{v}$ is a scalar.

(d) Again, the dot product of a scalar and a vector is undefined.

25. Assume that neither $\mathbf{a}$ nor $\mathbf{u}$ is the zero vector. If $\text{proj}_{\mathbf{a}}\mathbf{u} = \text{proj}_{\mathbf{u}}\mathbf{a}$, then

$$\frac{\mathbf{u} \cdot \mathbf{a}}{\|\mathbf{a}\|^2}\mathbf{a} = \frac{\mathbf{a} \cdot \mathbf{u}}{\|\mathbf{u}\|^2}\mathbf{u} = \frac{\mathbf{u} \cdot \mathbf{a}}{\|\mathbf{u}\|^2}\mathbf{u}$$

Thus, either $\mathbf{u} \cdot \mathbf{a} = 0$ and the two vectors are orthogonal, or

$$\frac{\mathbf{a}}{\|\mathbf{a}\|^2} = \frac{\mathbf{u}}{\|\mathbf{u}\|^2}$$

In this case, $\mathbf{a}$ is clearly a scalar multiple of $\mathbf{u}$. If we let $\mathbf{a} = k\mathbf{u}$, then

$$\frac{\mathbf{a}}{\|\mathbf{a}\|^2} = \frac{k\mathbf{u}}{k^2\|\mathbf{u}\|^2} = \frac{\mathbf{u}}{k\|\mathbf{u}\|^2}.$$

Hence, $k = 1$, so that $\mathbf{a} = \mathbf{u}$.

26. If, for instance, $\mathbf{u} = (1, 0, 0)$, $\mathbf{v} = (0, 1, 0)$ and $\mathbf{w} = (0, 0, 1)$, we have $\mathbf{u} \cdot \mathbf{v} = \mathbf{u} \cdot \mathbf{w} = 0$, but $\mathbf{v} \neq \mathbf{w}$.

27. Suppose that $\mathbf{r} = c_1\mathbf{u} + c_2\mathbf{v} + c_3\mathbf{w}$. Then

$$\mathbf{u} \cdot \mathbf{r} = c_1\mathbf{u} \cdot \mathbf{u} + c_2\mathbf{u} \cdot \mathbf{v} + c_3\mathbf{u} \cdot \mathbf{w} = c_1\|\mathbf{u}\|^2$$

since $\mathbf{u} \cdot \mathbf{v} = \mathbf{u} \cdot \mathbf{w} = 0$. Similarly, $\mathbf{v} \cdot \mathbf{r} = c_2\|\mathbf{v}\|^2$ and $\mathbf{w} \cdot \mathbf{r} = c_3\|\mathbf{w}\|^2$. Since we know $\mathbf{u} \cdot \mathbf{r}$, $\mathbf{v} \cdot \mathbf{r}$, and $\mathbf{w} \cdot \mathbf{r}$, this allows us to solve for c_1, c_2, and c_3. We find that

$$\mathbf{r} = \frac{\mathbf{u} \cdot \mathbf{r}}{\|\mathbf{u}\|^2}\mathbf{u} + \frac{\mathbf{v} \cdot \mathbf{r}}{\|\mathbf{v}\|^2}\mathbf{v} + \frac{\mathbf{w} \cdot \mathbf{r}}{\|\mathbf{w}\|^2}\mathbf{w}$$

28. This is just the Pythagorean Theorem.

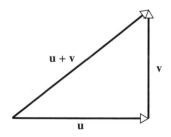

EXERCISE SET 3.4

1. **(a)** $\mathbf{v} \times \mathbf{w} = \left(\begin{vmatrix} 2 & -3 \\ 6 & 7 \end{vmatrix}, \ - \begin{vmatrix} 0 & -3 \\ 2 & 7 \end{vmatrix}, \ \begin{vmatrix} 0 & 2 \\ 2 & 6 \end{vmatrix} \right) = (32, -6, -4)$

(c) Since

$$\mathbf{u} \times \mathbf{v} = \left(\begin{vmatrix} 2 & -1 \\ 2 & -3 \end{vmatrix}, \ - \begin{vmatrix} 3 & -1 \\ 0 & -3 \end{vmatrix}, \ \begin{vmatrix} 3 & 2 \\ 0 & 2 \end{vmatrix} \right) = (-4, 9, 6)$$

we have

$$(\mathbf{u} \times \mathbf{v}) \times \mathbf{w} = \left(\begin{vmatrix} 9 & 6 \\ 6 & 7 \end{vmatrix}, \ - \begin{vmatrix} -4 & 6 \\ 2 & 7 \end{vmatrix}, \ \begin{vmatrix} -4 & 9 \\ 2 & 6 \end{vmatrix} \right) = (27, 40, -42)$$

(e) Since

$$\mathbf{v} - 2\mathbf{w} = (0, 2, -3) - (4, 12, 14) = (-4, -10, -17)$$

we have

$$\mathbf{u} \times (\mathbf{v} - 2\mathbf{w}) = \left(\begin{vmatrix} 2 & -1 \\ -10 & -17 \end{vmatrix}, \ - \begin{vmatrix} 3 & -1 \\ -4 & -17 \end{vmatrix}, \ \begin{vmatrix} 3 & 2 \\ -4 & -10 \end{vmatrix} \right) = (-44, 55, -22)$$

2. **(a)** By Theorem 3.4.1, $\mathbf{u} \times \mathbf{v}$ will be orthogonal to both $\mathbf{u}$ and $\mathbf{v}$ where

$$\mathbf{u} \times \mathbf{v} = \left(\begin{vmatrix} 4 & 2 \\ 1 & 5 \end{vmatrix}, \ - \begin{vmatrix} -6 & 2 \\ 3 & 5 \end{vmatrix}, \ \begin{vmatrix} -6 & 4 \\ 3 & 1 \end{vmatrix} \right) = (18, 36, -18)$$

so $(1, 2, -1)$, for instance, is orthogonal to both $\mathbf{u}$ and $\mathbf{v}$.

3. **(a)** Since $\mathbf{u} \times \mathbf{v} = (-7, -1, 3)$, the area of the parallelogram is $\|\mathbf{u} \times \mathbf{v}\| = \sqrt{59}$.

3. **(c)** Since **u** and **v** are proportional, they lie on the same line and hence the area of the parallelogram they determine is zero, which is, of course, $\|\mathbf{u} \times \mathbf{v}\|$.

4. **(a)** We have $\mathbf{u} = \overrightarrow{PQ} = (-1, -5, 2)$ and $\mathbf{v} = \overrightarrow{PR} = (2, 0, 3)$. Thus

$$\mathbf{u} \times \mathbf{v} = \left(\begin{vmatrix} -5 & 2 \\ 0 & 3 \end{vmatrix}, -\begin{vmatrix} -1 & 2 \\ 2 & 3 \end{vmatrix}, \begin{vmatrix} -1 & -5 \\ 2 & 0 \end{vmatrix} \right) = (-15, 7, 10)$$

so that

$$\|\mathbf{u} \times \mathbf{v}\| = \left[(-15)^2 + 7^2 + 10^2\right]^{1/2} = \sqrt{374}$$

Thus the area of the triangle is $\sqrt{374}/2$.

6. For Part (a), we have

$$\mathbf{u} \times \mathbf{v} = \left(\begin{vmatrix} -1 & 2 \\ 0 & -2 \end{vmatrix}, -\begin{vmatrix} 5 & 2 \\ 6 & -2 \end{vmatrix}, \begin{vmatrix} 5 & -1 \\ 6 & 0 \end{vmatrix} \right) = (2, 22, 6)$$

and

$$\mathbf{v} \times \mathbf{u} = \left(\begin{vmatrix} 0 & -2 \\ -1 & 2 \end{vmatrix}, -\begin{vmatrix} 6 & -2 \\ 5 & 2 \end{vmatrix}, \begin{vmatrix} 6 & 0 \\ 5 & -1 \end{vmatrix} \right) = (-2, -22, -6)$$

Hence

$$\mathbf{u} \times \mathbf{v} = -(\mathbf{v} \times \mathbf{u})$$

For Part (b), we have $\mathbf{v} + \mathbf{w} = (7, 2, -3)$. Hence

$$\mathbf{u} \times (\mathbf{v} + \mathbf{w}) = \left(\begin{vmatrix} -1 & 2 \\ 2 & -3 \end{vmatrix}, -\begin{vmatrix} 5 & 2 \\ 7 & -3 \end{vmatrix}, \begin{vmatrix} 5 & -1 \\ 7 & 2 \end{vmatrix} \right) = (-1, 29, 17)$$

On the other hand, from Part (a) we have

$$\mathbf{u} \times \mathbf{v} = (2, 22, 6)$$

Also

$$\mathbf{u} \times \mathbf{w} = \left(\begin{vmatrix} -1 & 2 \\ 2 & -1 \end{vmatrix}, -\begin{vmatrix} 5 & 2 \\ 1 & -1 \end{vmatrix}, \begin{vmatrix} 5 & -1 \\ 1 & 2 \end{vmatrix} \right) = (-3, 7, 11)$$

Thus

$$(\mathbf{u} \times \mathbf{v}) + (\mathbf{u} \times \mathbf{w}) = (2, 22, 6) + (-3, 7, 11)$$
$$= (-1, 29, 17)$$

and we have that

$$\mathbf{u} \times (\mathbf{v} + \mathbf{w}) = (\mathbf{u} \times \mathbf{v}) + (\mathbf{u} \times \mathbf{w})$$

The proof of Part (c) is similar to the proof of Part (b).

For Part (d), we already have that $\mathbf{u} \times \mathbf{v} = (2, -22, 6)$. Hence $k(\mathbf{u} \times \mathbf{v}) = (-10, 110, -30)$. But $k\mathbf{u} = (-25, 5, -10)$ and $k\mathbf{v} = (-30, 0, 10)$. Thus

$$(k\mathbf{u}) \times \mathbf{v} = \left(\begin{vmatrix} 5 & -10 \\ 0 & -2 \end{vmatrix}, -\begin{vmatrix} -25 & -10 \\ 6 & -2 \end{vmatrix}, \begin{vmatrix} -25 & 5 \\ 6 & 0 \end{vmatrix} \right)$$

$$= (-10, -110, -30)$$

and

$$\mathbf{u} \times (k\mathbf{v}) = \left(\begin{vmatrix} -1 & 2 \\ 0 & 10 \end{vmatrix}, -\begin{vmatrix} 5 & 2 \\ -30 & 10 \end{vmatrix}, \begin{vmatrix} 5 & -1 \\ -30 & 0 \end{vmatrix} \right)$$

$$= (-10, -110, -30)$$

Thus

$$k(\mathbf{u} \times \mathbf{v}) = (k\mathbf{u}) \times \mathbf{v} = \mathbf{u} \times (k\mathbf{v})$$

For Part (e), we have

$$\mathbf{u} \times \mathbf{0} = \left(\begin{vmatrix} -1 & 2 \\ 0 & 0 \end{vmatrix}, -\begin{vmatrix} 5 & 2 \\ 0 & 0 \end{vmatrix}, \begin{vmatrix} 5 & -1 \\ 0 & 0 \end{vmatrix} \right) = (0, 0, 0) = \mathbf{0}$$

Similarly $\mathbf{0} \times \mathbf{u} = \mathbf{0}$.

Finally, for Part (f), we have

$$\mathbf{u} \times \mathbf{u} = \left(\begin{vmatrix} -1 & 2 \\ -1 & 2 \end{vmatrix}, -\begin{vmatrix} 5 & 2 \\ 5 & 2 \end{vmatrix}, \begin{vmatrix} 5 & -1 \\ 5 & -1 \end{vmatrix} \right) = (0, 0, 0) = \mathbf{0}$$

7. Choose any nonzero vector **w** which is not parallel to **u**. For instance, let **w** = $(1, 0, 0)$ or $(0, 1, 0)$. Then **v** = **u** × **w** will be orthogonal to **u**. Note that if **u** and **w** were parallel, then **v** = **u** × **w** would be the zero vector.

   Alternatively, let **w** = (x, y, z). Then **w** orthogonal to **u** implies $2x - 3y + 5z = 0$. Now assign nonzero values to any two of the variables x, y, and z and solve for the remaining variable.

8. **(a)** We have

$$\mathbf{u} \cdot (\mathbf{v} \times \mathbf{w}) = \begin{vmatrix} -1 & 2 & 4 \\ 3 & 4 & -2 \\ -1 & 2 & 5 \end{vmatrix} = -(20 + 4) - 2(15 - 2) + 4(6 + 4) = -10$$

9. **(e)** Since $(\mathbf{u} \times \mathbf{w}) \cdot \mathbf{v} = \mathbf{v} \cdot (\mathbf{u} \times \mathbf{w})$ is a determinant whose rows are the components of **v**, **u**, and **w**, respectively, we interchange Rows 1 and 2 to obtain the determinant which represents $\mathbf{u} \cdot (\mathbf{v} \times \mathbf{w})$. Since the value of this determinant is 3, we have $(\mathbf{u} \times \mathbf{w}) \cdot \mathbf{v} = -3$.

10. **(a)** Call this volume V. Then, since

$$\begin{vmatrix} 2 & -6 & 2 \\ 0 & 4 & -2 \\ 2 & 2 & -4 \end{vmatrix} = 2 \begin{vmatrix} 4 & -2 \\ 2 & -4 \end{vmatrix} + 2 \begin{vmatrix} -6 & 2 \\ 4 & -2 \end{vmatrix} = -16$$

   we have $V = 16$.

11. **(a)** Since the determinant

$$\begin{vmatrix} -1 & -2 & 1 \\ 3 & 0 & -2 \\ 5 & -4 & 0 \end{vmatrix} = 16 \neq 0$$

   the vectors do not lie in the same plane.

12. For a vector to be parallel to the yz-plane, it must be perpendicular to the vector $(1, 0, 0)$, so we are looking for a vector which is perpendicular to both $(1, 0, 0)$ and $(3, -1, 2)$. Such a vector is $(1, 0, 0) \times (3, -1, 2) = (0, -2, -1)$. Since we want a unit vector, we divide through by the norm to obtain the vector $(0, -2/\sqrt{5}, -1/\sqrt{5})$. Obviously any vector parallel to this one, such as $(0, 2/\sqrt{5}, 1/\sqrt{5})$, will also work.

15.   By Theorem 3.4.2, we have

$$(\mathbf{u}+\mathbf{v}) \times (\mathbf{u}-\mathbf{v}) = \mathbf{u} \times (\mathbf{u}-\mathbf{v}) + \mathbf{v} \times (\mathbf{u}-\mathbf{v})$$
$$= (\mathbf{u} \times \mathbf{u}) + (\mathbf{u} \times (-\mathbf{v})) + (\mathbf{v} \times \mathbf{u}) + (\mathbf{v} \times (-\mathbf{v}))$$
$$= \mathbf{0} - (\mathbf{u} \times \mathbf{v}) - (\mathbf{u} \times \mathbf{v}) - (\mathbf{v} \times \mathbf{v})$$
$$= -2(\mathbf{u} \times \mathbf{v})$$

17.   **(a)**   The area of the triangle with sides $\overrightarrow{AB}$ and $\overrightarrow{AC}$ is the same as the area of the triangle with sides $(-1, 2, 2)$ and $(1, 1, -1)$ where we have "moved" A to the origin and translated B and C accordingly. This area is $\frac{1}{2}\|(-1,2,2) \times (1,1,-1)\| = \frac{1}{2}\|(-4,1,-3)\| = \sqrt{26}/2$.

18.   If the vector $\mathbf{u}$ and the given line make an angle θ, then the distance D between the point and the line is given by

$$D = \|\mathbf{u}\| \sin \theta \qquad \qquad \text{(Why?)}$$
$$= \|\mathbf{u} \times \mathbf{v}\|/\|\mathbf{v}\| \qquad \text{(Why?)}$$

19.   **(a)**   Let $\mathbf{u} = \overrightarrow{AP} = (-4, 0, 2)$ and $\mathbf{v} = \overrightarrow{AB} = (-3, 2, -4)$. Then the distance we want is

$$\|(-4,0,2) \times (-3,2,-4)\|/\|(-3,2,-4)\| = \|(-4,-22,-8)\|/\sqrt{29} = 2\sqrt{141}/\sqrt{29}$$

20.   We have that $\mathbf{u} \cdot \mathbf{v} = \|\mathbf{u}\| \, \|\mathbf{v}\| \cos\theta$ and $\|\mathbf{u} \times \mathbf{v}\| = \|\mathbf{u}\| \, \|\mathbf{v}\| \sin\theta$. Thus if $\mathbf{u} \cdot \mathbf{v} \neq 0$ then $\|\mathbf{u}\|$, $\|\mathbf{v}\|$, and $\cos\theta$ are all nonzero, and

$$\tan\theta = \frac{\sin\theta}{\cos\theta} = \frac{\|\mathbf{u} \times \mathbf{v}\|}{\|\mathbf{u}\| \, \|\mathbf{v}\|} \Big/ \frac{\mathbf{u} \cdot \mathbf{v}}{\|\mathbf{u}\| \, \|\mathbf{v}\|} = \frac{\|\mathbf{u} \times \mathbf{v}\|}{\mathbf{u} \cdot \mathbf{v}}$$

21.   **(b)**   One vector $\mathbf{n}$ which is perpendicular to the plane containing $\mathbf{v}$ and $\mathbf{w}$ is given by

$$\mathbf{n} = \mathbf{w} \times \mathbf{v} = (1, 3, 3) \times (1, 1, 2) = (3, 1, -2)$$

Therefore the angle ϕ between $\mathbf{u}$ and $\mathbf{n}$ is given by

$$\phi = \cos^{-1}\left(\frac{\mathbf{u} \cdot \mathbf{n}}{\|\mathbf{u}\| \, \|\mathbf{n}\|}\right) = \cos^{-1}\left(\frac{9}{14}\right)$$

$$\approx 0.8726 \text{ radians (or } 49.99°)$$

Hence the angle θ between $\mathbf{u}$ and the plane is given by

$$\theta = \frac{\pi}{2} - \phi \approx .6982 \text{ radians (or } 40°19'')$$

If we had interchanged the rôles of $\mathbf{v}$ and $\mathbf{w}$ in the formula for $\mathbf{n}$ so that $\mathbf{n} = \mathbf{v} \times \mathbf{w} = (-3, -1, 2)$, then we would have obtained $\phi = \cos^{-1}(-\frac{9}{14}) \approx 2.269$ radians or $130.0052°$. In this case, $\theta = \phi - \frac{\pi}{2}$.

In either case, note that θ may be computed using the formula

$$\theta = \left|\sin^{-1}\left(\frac{\mathbf{u} \cdot \mathbf{n}}{\|\mathbf{u}\| \, \|\mathbf{n}\|}\right)\right|.$$

24. **(a)** By Theorems 3.4.2 and 3.2.1 and the definition of $k\mathbf{0}$,

$$(\mathbf{u} + k\mathbf{v}) \times \mathbf{v} = (\mathbf{u} \times \mathbf{v}) + (k\mathbf{v} \times \mathbf{v})$$

$$= (\mathbf{u} \times \mathbf{v}) + k(\mathbf{v} \times \mathbf{v})$$

$$= (\mathbf{u} \times \mathbf{v}) + k\mathbf{0}$$

$$= \mathbf{u} \times \mathbf{v}$$

(b) Let $\mathbf{u} = (u_1, u_2, u_3)$, $\mathbf{v} = (v_1, v_2, v_3)$, and $\mathbf{z} = (z_1, z_2, z_3)$. Then we know from the text that

$$\mathbf{u} \cdot (\mathbf{v} \times \mathbf{z}) = \begin{vmatrix} u_1 & u_2 & u_3 \\ v_1 & v_2 & v_3 \\ z_1 & \mathbf{z}_2 & \mathbf{z}_3 \end{vmatrix}$$

Since $\mathbf{x} \cdot \mathbf{y} = \mathbf{y} \cdot \mathbf{x}$ for any vectors $\mathbf{x}$ and $\mathbf{y}$, we have

$$(\mathbf{u} \times \mathbf{z}) \cdot \mathbf{v} = \mathbf{v} \cdot (\mathbf{u} \times \mathbf{z}) = \begin{vmatrix} v_1 & v_2 & v_3 \\ u_1 & u_2 & u_3 \\ z_1 & z_2 & z_3 \end{vmatrix}$$

$$= - \begin{vmatrix} u_1 & u_2 & u_3 \\ v_1 & v_2 & v_3 \\ z_1 & z_2 & z_3 \end{vmatrix}$$

$$= -\mathbf{u} \cdot (\mathbf{v} \times \mathbf{z})$$

25. **(a)** By Theorem 3.4.1, we know that the vector $\mathbf{v} \times \mathbf{w}$ is perpendicular to both $\mathbf{v}$ and $\mathbf{w}$. Hence $\mathbf{v} \times \mathbf{w}$ is perpendicular to every vector in the plane determined by $\mathbf{v}$ and $\mathbf{w}$; moreover the only vectors perpendicular to $\mathbf{v} \times \mathbf{w}$ which share its initial point must be in this plane. But also by Theorem 3.4.1, $\mathbf{u} \times (\mathbf{v} \times \mathbf{w})$ is perpendicular to $\mathbf{v} \times \mathbf{w}$ for any vector $\mathbf{u} \neq \mathbf{0}$ and hence must lie in the plane determined by $\mathbf{v}$ and $\mathbf{w}$.

(b) The argument is completely similar to Part (a), above.

26. Let $\mathbf{u} = (u_1, u_2, u_3)$, $\mathbf{v} = (v_1, v_2, v_3)$, and $\mathbf{w} = (w_1, w_2, w_3)$. Following the hint, we find

$$\mathbf{v} \times \mathbf{i} = (0, v_3, -v_2)$$

and

$$\mathbf{u} \times (\mathbf{v} \times \mathbf{i}) = (-u_2 v_2 - u_3 v_3, u_1 v_2, u_1 v_3)$$

But

$$(\mathbf{u} \cdot \mathbf{i})\mathbf{v} - (\mathbf{u} \cdot \mathbf{v})\mathbf{i} = u_1(v_1, v_2, v_3) - (u_1 v_1 + u_2 v_2 + u_3 v_3)(1, 0, 0)$$

$$= (u_1 v_1, \ u_1 v_2, \ u_1 v_3) - (u_1 v_1 + u_2 v_2 + u_3 v_3, \ 0, \ 0)$$

$$= (-u_2 v_2 - u_3 v_3, \ u_1 v_2, \ u_1 v_3)$$

$$= \mathbf{u} \times (\mathbf{v} \times \mathbf{i})$$

Similarly,

$$\mathbf{u} \times (\mathbf{v} \times \mathbf{j}) = (\mathbf{u} \cdot \mathbf{j})\mathbf{v} - (\mathbf{u} \cdot \mathbf{v})\mathbf{j}$$

and

$$\mathbf{u} \times (\mathbf{v} \times \mathbf{k}) = (\mathbf{u} \cdot \mathbf{k})\mathbf{v} - (\mathbf{u} \cdot \mathbf{v})\mathbf{k}$$

Now write $\mathbf{w} = (w_1, w_2, w_3) = w_1\mathbf{i} + w_2\mathbf{j} + w_3\mathbf{k}$. Then

$$\mathbf{u} \times (\mathbf{v} \times \mathbf{w}) = \mathbf{u} \times (\mathbf{v} \times (w_1\mathbf{i} + w_2\mathbf{j} + w_3\mathbf{k}))$$

$$= w_1[\mathbf{u} \times (\mathbf{v} \times \mathbf{i})] + w_2[\mathbf{u} \times (\mathbf{v} \times \mathbf{j})] + w_3[\mathbf{u} \times (\mathbf{v} \times \mathbf{k})]$$

$$= w_1[(\mathbf{u} \cdot \mathbf{i})\mathbf{v} - (\mathbf{u} \cdot \mathbf{v})\mathbf{i}] + w_2[(\mathbf{u} \cdot \mathbf{j})\mathbf{v} - (\mathbf{u} \cdot \mathbf{v})\mathbf{j}] + w_3[(\mathbf{u} \cdot \mathbf{k})\mathbf{v} - (\mathbf{u} \cdot \mathbf{v})\mathbf{k}]$$

$$= (\mathbf{u} \cdot (w_1\mathbf{i} + w_2\mathbf{j} + w_3\mathbf{k}))\mathbf{v} - (\mathbf{u} \cdot \mathbf{v})(w_1\mathbf{i} + w_2\mathbf{j} + w_3\mathbf{k})$$

$$= (\mathbf{u} \cdot \mathbf{w})\mathbf{v} - (\mathbf{u} \cdot \mathbf{v})\mathbf{w}$$

29. If $\mathbf{a}$, $\mathbf{b}$, $\mathbf{c}$, and $\mathbf{d}$ lie in the same plane, then $(\mathbf{a} \times \mathbf{b})$ and $(\mathbf{c} \times \mathbf{d})$ are both perpendicular to this plane, and are therefore parallel. Hence, their cross-product is zero.

30. Recall that $|\mathbf{a} \cdot (\mathbf{b} \times \mathbf{c})|$ represents the volume of the parallelepiped with sides $\mathbf{a}$, $\mathbf{b}$, and $\mathbf{c}$, which is the area of its base times its height. The volume of the tetrahedron is just $\frac{1}{3}$ (area of its base) times (its height). The two heights are the same, but the area of the base of the tetrahedron is half of the area of the base of the parallelepiped. Hence, the volume of the tetrahedron is $\frac{1}{3}\left(\frac{1}{2}|\mathbf{a} \cdot (\mathbf{b} \times \mathbf{c})|\right)$.

31. **(a)** The required volume is

$$\frac{1}{6}\left(|(-1 - 3,\ 2 + 2,\ 0 - 3)\ \cdot\ ((2 - 3,\ 1 + 2,\ -3 - 3) \times (1 - 3,\ 0 + 2,\ 1 - 3))|\right)$$

$$= \frac{1}{6}\left(|(-4, 4, -3) \cdot (6, 10, 4)|\right)$$

$$= 2/3$$

33. Let $\mathbf{u} = (u_1, u_2, u_3)$, $\mathbf{v} = (v_1, v_2, v_3)$, and $\mathbf{w} = (w_1, w_2, w_3)$.

For Part (c), we have

$$\mathbf{u} \times \mathbf{w} = (u_2 w_3 - u_3 w_2,\ u_3 w_1 - u_1 w_3,\ u_1 w_2 - u_2 w_1)$$

and

$$\mathbf{v} \times \mathbf{w} = (v_2 w_3 - v_3 w_2,\ v_3 w_1 - v_1 w_3,\ v_1 w_2 - v_2 w_1)$$

Thus

$$(\mathbf{u} \times \mathbf{w}) + (\mathbf{v} \times \mathbf{w})$$
$$= ([u_2 + v_2]w_3 - [u_3 + v_3]w_2, [u_3 + v_3]w_1 - [u_1 + v_1]w_3, [u_1 + v_1]w_2 - [u_2 + v_2]w_1)$$

But, by definition, this is just $(\mathbf{u} + \mathbf{v}) \times \mathbf{w}$.

For Part (d), we have

$$k(\mathbf{u} \times \mathbf{v}) = (k[u_2v_3 - u_3v_2], \ k[u_3v_1 - u_1v_3], \ k[u_1v_2 - u_2v_1])$$

and

$$(k\mathbf{u}) \times \mathbf{v} = (ku_2v_3 - ku_3v_2, \ ku_3v_1 - ku_1v_3, \ ku_1v_2 - ku_2v_1)$$

Thus, $k(\mathbf{u} \times \mathbf{v}) = (k\mathbf{u}) \times \mathbf{v}$. The identity $k(\mathbf{u} \times \mathbf{v}) = \mathbf{u} \times (k\mathbf{v})$ may be proved in an analogous way.

35. **(a)** Observe that $\mathbf{u} \times \mathbf{v}$ is perpendicular to both $\mathbf{u}$ and $\mathbf{v}$, and hence to all vectors in the plane which they determine. Similarly, $\mathbf{w} = \mathbf{v} \times (\mathbf{u} \times \mathbf{v})$ is perpendicular to both $\mathbf{v}$ and to $\mathbf{u} \times \mathbf{v}$. Hence, it must lie on the line through the origin perpendicular to $\mathbf{v}$ and in the plane determined by $\mathbf{u}$ and $\mathbf{v}$.

(b) From the above, $\mathbf{v} \cdot \mathbf{w} = 0$. Applying Part (d) of Theorem 3.7.1, we have

$$\mathbf{w} = \mathbf{v} \times (\mathbf{u} \times \mathbf{v}) = (\mathbf{v} \cdot \mathbf{v})\mathbf{u} - (\mathbf{v} \cdot \mathbf{u})\mathbf{v}$$

so that

$$\mathbf{u} \cdot \mathbf{w} = (\mathbf{v} \cdot \mathbf{v})(\mathbf{u} \cdot \mathbf{u}) - (\mathbf{v} \cdot \mathbf{u})(\mathbf{u} \cdot \mathbf{v})$$
$$= \|\mathbf{v}\|^2 \|\mathbf{u}\|^2 - (\mathbf{u} \cdot \mathbf{v})^2$$

36. It is not valid. For instance, let $\mathbf{u} = (1, 0, 0)$, $\mathbf{v} = (0, 1, 0)$, and $\mathbf{w} = (1, 1, 0)$. Then $\mathbf{u} \times \mathbf{v} = \mathbf{u} \times \mathbf{w} = (0, 0, 1)$, which is a vector perpendicular to $\mathbf{u}$, $\mathbf{v}$, and $\mathbf{w}$. However, $\mathbf{v} \neq \mathbf{w}$.

37. The expression $\mathbf{u} \cdot (\mathbf{v} \times \mathbf{w})$ is clearly well-defined.

Since the cross product is not associative, the expression $\mathbf{u} \times \mathbf{v} \times \mathbf{w}$ is not well-defined because the result is dependent upon the order in which we compute the cross products, i.e., upon the way in which we insert the parentheses. For example, $(\mathbf{i} \times \mathbf{j}) \times \mathbf{j} = \mathbf{k} \times \mathbf{j} = -\mathbf{i}$ but $\mathbf{i} \times (\mathbf{j} \times \mathbf{j}) = \mathbf{i} \times \mathbf{0} = \mathbf{0}$.

The expression $\mathbf{u} \cdot \mathbf{v} \times \mathbf{w}$ may be deemed to be acceptable because there is only one meaningful way to insert parenthesis, namely, $\mathbf{u} \cdot (\mathbf{v} \times \mathbf{w})$. The alternative, $(\mathbf{u} \cdot \mathbf{v}) \times \mathbf{w}$, does not make sense because it is the cross product of a scalar with a vector.

38. If either $\mathbf{u}$ or $\mathbf{v}$ is the zero vector, then $\mathbf{u} \times \mathbf{v} = \mathbf{0}$. If neither $\mathbf{u}$ nor $\mathbf{v}$ is the zero vector, then $\|\mathbf{u} \times \mathbf{v}\| = \|\mathbf{u}\| \, \|\mathbf{v}\| \sin \theta$ where θ is the angle between $\mathbf{u}$ and $\mathbf{v}$. Thus if $\|\mathbf{u} \times \mathbf{v}\| = 0$, then $\sin \theta = 0$, so that $\mathbf{u}$ and $\mathbf{v}$ are parallel.

EXERCISE SET 3.5

4. **(a)** We have

$$\overrightarrow{PQ} = (2, 1, 2) \quad \text{and} \quad \overrightarrow{PR} = (3, -1, -2)$$

Thus $\overrightarrow{PQ} \times \overrightarrow{PR} = (0, 10, -5)$ is perpendicular to the plane determined by $\overrightarrow{PQ}$ and $\overrightarrow{PR}$ and P, say, is a point in that plane. Hence, an equation for the plane is

$$0(x + 4) + 10(y + 1) - 5(z + 1) = 0$$

or

$$2y - z + 1 = 0$$

5. **(a)** Normal vectors for the planes are $(4, -1, 2)$ and $(7, -3, 4)$. Since these vectors are not multiples of one another, the planes are not parallel.

    (b) The normal vectors are $(1, -4, -3)$ and $(3, -12, -9)$. Since one vector is three times the other, the planes are parallel.

6. **(a)** A normal vector for the plane is $(1, 2, 3)$ and a direction vector for the line is $(-4, -1, 2)$. The inner product of these two vectors is $-4 - 2 + 6 = 0$, and therefore they are perpendicular. This guarantees that the line and the plane are parallel.

7. **(a)** Normal vectors for the planes are $(3, -1, 1)$ and $(1, 0, 2)$. Since the inner product of these two vectors is not zero, the planes are not perpendicular.

8. **(a)** A normal vector for the plane is $(2, 1, -1)$ and a direction vector for the line is $(-4, -2, 2)$. Since one of these vectors is a multiple of the other, the line and the plane are perpendicular.

10. **(a)** Call the points P and Q and call the line ℓ. Then the vector $\overrightarrow{PQ} = (2, 4, -8)$ is parallel to ℓ and the point $P = (5, -2, 4)$ lies on ℓ. Hence, one set of parametric equations for ℓ is: $x = 5 + t$, $y = -2 + 2t$, $z = 4 - 4t$ where t is any real number.

11. **(a)** As in Example 6, we solve the two equations simultaneously. If we eliminate y, we have $x + 7z + 12 = 0$. Let, say, $z = t$, so that $x = -12 - 7t$, and substitute these values into the equation for either plane to get $y = -41 - 23t$.

Alternatively, recall that a direction vector for the line is just the cross-product of the normal vectors for the two planes, i.e.,

$$(7, -2, 3) \times (-3, 1, 2) = (-7, -23, 1)$$

Thus if we can find a point which lies on the line (that is, any point whose coordinates satisfy the equations for both planes), we are done. If we set $z = 0$ and solve the two equations simultaneously, we get $x = -12$ and $y = -41$, so that $x = -12 - 7t$, $y = -41 - 23t$, $z = 0 + t$ is one set of equations for the line (see above).

13. **(a)** Since the normal vectors $(-1, 2, 4)$ and $(2, -4, -8)$ are parallel, so are the planes.

(b) Since the normal vectors $(3, 0, -1)$ and $(-1, 0, 3)$ are not parallel, neither are the planes.

14. **(a)** Since $(-2, 1, 4) \cdot (1, -2, 1) = 0$, the planes are perpendicular.

(b) Since $(3, 0, -2) \cdot (1, 1, 1) = 1 \neq 0$, the planes are not perpendicular.

16. **(a)** Since $6(0) + 4t - 4t = 0$ for all t, it follows that every point on the line also lies in the plane.

(b) The normal to the plane is $\mathbf{n} = (5, -3, 3)$; the line is parallel to $\mathbf{v} = (0, 1, 1)$. But $\mathbf{n} \cdot \mathbf{v} = 0$, so $\mathbf{n}$ and $\mathbf{v}$ are perpendicular. Thus $\mathbf{v}$, and therefore the line, are parallel to the plane. To conclude that the line lies below the plane, simply note that $(0, 0, 1/3)$ is in the plane and $(0, 0, 0)$ is on the line.

(c) Here $\mathbf{n} = (6, 2, -2)$ and $\mathbf{v}$ is the same as before. Again, $\mathbf{n} \cdot \mathbf{v} = 0$, so the line and the plane are parallel. Since $(0, 0, 0)$ lies on the line and $(0, 0, -3/2)$ lies in the plane, then the line is above the plane.

17. Since the plane is perpendicular to a line with direction $(2, 3, -5)$, we can use that vector as a normal to the plane. The point-normal form then yields the equation $2(x + 2) + 3(y - 1) - 5(z - 7) = 0$, or $2x + 3y - 5z + 36 = 0$.

19. **(a)** Since the vector $(0,0,1)$ is perpendicular to the xy-plane, we can use this as the normal for the plane. The point-normal form then yields the equation $z - z_0 = 0$. This equation could just as well have been derived by inspection, since it represents the set of all points with fixed z and x and y arbitrary.

20. The plane will have normal $(7, 4, -2)$, so the point-normal form yields $7x + 4y - 2z = 0$.

21. A normal to the plane is $\mathbf{n} = (5, -2, 1)$ and the point $(3, -6, 7)$ is in the desired plane. Hence, an equation for the plane is $5(x-3) - 2(y+6) + (z-7) = 0$ or $5x - 2y + z - 34 = 0$.

22. If we substitute $x = 9 - 5t$, $y = -1 - t$, and $z = 3 + t$ into the equation for the plane, we find that $t = 40/3$. Substituting this value for t into the parametric equations for the line yields $x = -173/3$, $y = -43/3$, and $z = 49/3$.

24. The two planes intersect at points given by $(-5 - 2t, \ -7 - 6t, \ t)$. Two such points are $(-5, -7, 0)$ and $(-3, -1, -1)$. The plane $ax + by + cz + d = 0$ through these points and also through $(2, 4, -1)$ will satisfy the equations

$$-5a - 7b \qquad + d = 0$$

$$-3a - \ b - c + d = 0$$

$$2a + 4b - c + d = 0$$

This system of equations has the solution $a = (-1/2)t$, $b = (1/2)t$, $c = 2t$, $d = t$ where t is arbitrary. Thus, if we let $t = -2$, we obtain the equation

$$x - y - 4z - 2 = 0$$

for the desired plane.

Alternatively, note that the line of intersection of the two planes has direction given by the vector $\mathbf{v} = (-2, -6, 1)$. It also contains the point $(-5, -7, 0)$. The direction of the line connecting this point and $(2, 4, -1)$ is given by $\mathbf{w} = (7, 11, -1)$. Thus the vector $\mathbf{v} \times \mathbf{w} = (5, -5, -20)$ is normal to the desired plane, so that a point-normal form for the plane is

$$(x - 2) - (y - 4) - 4(z + 1) = 0$$

or

$$x - y - 4z - 2 = 0$$

25. Call the points A, B, C, and D, respectively. Since the vectors $\overrightarrow{AB} = (-1, 2, 4)$ and $\overrightarrow{BC} = (-2, -1, -2)$ are not parallel, then the points A, B, and C do determine a plane (and not just a line). The normal to this plane is $\overrightarrow{AB} \times \overrightarrow{BC} = (0, -10, 5)$. Therefore an equation for the plane is

$$2y - z + 1 = 0$$

Since the coordinates of the point D satisfy this equation, all four points must lie in the same plane.

Alternatively, it would suffice to show that (for instance) $\overrightarrow{AB} \times \overrightarrow{BC}$ and $\overrightarrow{AD} \times \overrightarrow{DC}$ are parallel, so that the planes determined by A, B, and C and A, D, and C are parallel. Since they have points in common, they must coincide.

27. The normals to the two planes are $(4, -2, 2)$ and $(3, 3, -6)$ or, simplifying, $\mathbf{n}_1 = (2, -1, 1)$ and $\mathbf{n}_2 = (1, 1, -2)$. The normal $\mathbf{n}$ to a plane which is perpendicular to both of the given planes must be perpendicular to both $\mathbf{n}_1$ and $\mathbf{n}_2$. That is, $\mathbf{n} = \mathbf{n}_1 \times \mathbf{n}_2 = (1, 5, 3)$. The plane with this normal which passes through the point $(-2, 1, 5)$ has the equation

$$(x + 2) + 5(y - 1) + 3(z - 5) = 0$$

or

$$x + 5y + 3z - 18 = 0$$

28. The line of intersection of the given planes has the equations $x = -\dfrac{10}{9} - \dfrac{t}{3}$, $y = \dfrac{31}{18} - \dfrac{t}{3}$, $z = t$ and hence has direction $(1, 1, -3)$. The plane with this normal vector which passes through the point $(2, -1, 4)$ is the one we are looking for. Its equation is

$$(x - 2) + (y + 1) - 3(z - 4) = 0$$

Note that the normal vector for the desired plane can also be obtained by computing the cross product of the normal vectors of the two given planes.

30. The directions of the two lines are given by the vectors $(-2, 1, -1)$ and $(2, -1, 1)$. Since each is a multiple of the other, they represent the same direction. The first line passes (for example) through the point $(3, 4, 1)$ and the second line passes through the points $(5, 1, 7)$ and $(7, 0, 8)$, among others. Either of the methods of Example 2 will yield an equation for the plane determined by these points.

31. If, for instance, we set $t = 0$ and $t = -1$ in the line equation, we obtain the points $(0, 1, -3)$ and $(-1, 0, -5)$. These, together with the given point and the methods of Example 2, will yield an equation for the desired plane.

33. The plane we are looking for is just the set of all points $P = (x, y, z)$ such that the distances from P to the two fixed points are equal. If we equate the squares of these distances, we have

$$(x+1)^2 + (y+4)^2 + (z+2)^2 = (x-0)^2 + (y+2)^2 + (z-2)^2$$

or

$$2x + 1 + 8y + 16 + 4z + 4 = 4y + 4 - 4z + 4$$

or

$$2x + 4y + 8z + 13 = 0$$

34. The vector $\mathbf{v} = (-1, 2, -5)$ is parallel to the line and the vector $\mathbf{n} = (-3, 1, 1)$ is perpendicular to the plane. But $\mathbf{n} \cdot \mathbf{v} = 0$ and the result follows.

35. We change the parameter in the equations for the second line from t to s. The two lines will then intersect if we can find values of s and t such that the x, y, and z coordinates for the two lines are equal; that is, if there are values for s and t such that

$$4t + 3 = 12s - 1$$
$$t + 4 = 6s + 7$$
$$1 = 3s + 5$$

This system of equations has the solution $t = -5$ and $s = -4/3$. If we then substitute $t = -5$ into the equations for the first line or $s = -4/3$ into the equations for the second line, we find that $x = -17$, $y = -1$, and $z = 1$ is the point of intersection.

36. The vector $\mathbf{v}_1 = (4, 1, 0)$ is parallel to the first line and $\mathbf{v}_2 = (4, 2, 1)$ is parallel to the second line. Hence $\mathbf{n} = \mathbf{v}_1 \times \mathbf{v}_2 = (1, -4, 4)$ is perpendicular to both lines and is therefore a normal vector for the plane determined by the lines. If we substitute $t = 0$ into the parametric equations for the first line, we see that $(3, 4, 1)$ must lie in the plane. Hence, an equation for the plane is

$$(x - 3) - 4(y - 4) + 4(z - 1) = 0$$

or

$$x - 4y + 4z + 9 = 0$$

37.   **(a)** If we set $z = t$ and solve for x and y in terms of z, then we find that

$$x = \frac{11}{23} + \frac{7}{23}t, \quad y = -\frac{41}{23} - \frac{1}{23}t, \quad z = t$$

38.   Call the plane $Ax + By + Cz + D = 0$. Since the points $(a, 0, 0)$, $(0, b, 0)$, and $(0, 0, c)$ lie in this plane, we have

$$aA + D = 0$$
$$bB + D = 0$$
$$cC + D = 0$$

Thus $A = -D/a$, $B = -D/b$, and $C = -D/c$ and an equation for the plane is

$$\frac{x}{a} + \frac{y}{b} + \frac{z}{c} = 1$$

Alternatively, let P, Q, and R denote the points $(a, 0, 0)$, $(0, b, 0)$, and $(0, 0, c)$, respectively. Then

$$\overrightarrow{PQ} = (-a, b, 0) \quad \text{and} \quad \overrightarrow{QR} = (0, -b, c)$$

So

$$\mathbf{n} = \overrightarrow{PQ} \times \overrightarrow{QR} = (bc, ac, ab)$$

is a normal vector for the plane. Since P, say, lies on the plane, we have

$$bc(x - a) + acy + abz = 0$$

Dividing the above equation by abc yields the desired result.

39.   **(b)** By Theorem 3.5.2, the distance is

$$D = \frac{|2(-1) + 3(2) - 4(1) - 1|}{\sqrt{2^2 + 3^2 + (-4)^2}} = \frac{1}{\sqrt{29}}$$

40. **(a)** The point $(0, 0, 1)$ lies in the first plane. The distance between this point and the second plane is

$$D = \frac{|2 - 3|}{\sqrt{6^2 + (-8)^2 + (2)^2}} = \frac{1}{2\sqrt{26}}$$

(b) The two planes coincide, so the distance between them is zero.

41. First observe that if we substitute the values for x, y, and z from the line equations into the symmetric equations, we obtain the equations $t = t = t$, which hold for all values of t. Thus every point on the line satisfies the symmetric equations.

It remains to show that every point (x, y, z) which satisfies the symmetric equations also lies on the line. That is, we must find a value of t for which the line equations hold. Clearly

$$t = \frac{x - x_0}{a} = \frac{y - y_0}{b} = \frac{z - z_0}{c}$$

is the desired value of t.

43. **(a)** The symmetric equations for the line are

$$\frac{x - 7}{4} = \frac{y + 5}{2} = \frac{z - 5}{-1}$$

Thus, two of the planes are given by

$$\frac{x - 7}{4} = \frac{y + 5}{2} \quad \text{and} \quad \frac{x - 7}{4} = \frac{z - 5}{-1}$$

or, equivalently,

$$x - 2y - 17 = 0 \quad \text{and} \quad x + 4z - 27 = 0$$

44. **(a)** The normals to the two planes are $(1, 0, 0)$ and $(2, -1, 1)$. The angle between them is given by

$$\cos \theta = \frac{(1, 0, 0) \cdot (2, -1, 1)}{\sqrt{1} \, \sqrt{4 + 1 + 1}} = \frac{2}{\sqrt{6}}$$

Thus $\theta = \cos^{-1}(2/\sqrt{6}) \approx 35°15'52''$.

46. If we substitute any value of the parameter—say t_0—into $\mathbf{r} = \mathbf{r}_0 + t\mathbf{v}$ and $-t_0$ into $\mathbf{r} = \mathbf{r}_0 - t\mathbf{v}$, we clearly obtain the same point. Hence, the two lines coincide. They both pass through the point $\mathbf{r}_0$ and both are parallel to $\mathbf{v}$.

47. Since the vector (a, b, c) is both the normal vector to the plane and the direction vector for the line, the line must be perpendicular to the plane.

48. The equation $\mathbf{r} = (1 - t)\mathbf{r}_1 + t\mathbf{r}_2$ can be rewritten as $\mathbf{r} = \mathbf{r}_1 + t(\mathbf{r}_2 - \mathbf{r}_1)$. This represents a line through the point P_1 with direction $\mathbf{r}_2 - \mathbf{r}_1$. If $t = 0$, we have the point P_1. If $t = 1$, we have the point P_2. If $0 < t < 1$, we have a point on the line segment connecting P_1 and P_2. Hence the given equation represents this line segment.

EXERCISE SET 4.1

3. We must find numbers c_1, c_2, c_3, and c_4 such that

$$c_1(-1, 3, 2, 0) + c_2(2, 0, 4, -1) + c_3(7, 1, 1, 4) + c_4(6, 3, 1, 2) = (0, 5, 6, -3)$$

If we equate vector components, we obtain the following system of equations:

$$-c_1 + 2c_2 + 7c_3 + 6c_4 = 0$$
$$3c_1 \qquad + c_3 + 3c_4 = 5$$
$$2c_1 + 4c_2 + c_3 + c_4 = 6$$
$$- c_2 + 4c_3 + 2c_4 = 3$$

The augmented matrix of this system is

$$\begin{bmatrix} -1 & 2 & 7 & 6 & 0 \\ 3 & 0 & 1 & 3 & 5 \\ 2 & 4 & 1 & 1 & 6 \\ 0 & -1 & 4 & 2 & -3 \end{bmatrix}$$

The reduced row-echelon form of this matrix is

$$\begin{bmatrix} 1 & 0 & 0 & 0 & 1 \\ 0 & 1 & 0 & 0 & 1 \\ 0 & 0 & 1 & 0 & -1 \\ 0 & 0 & 0 & 1 & 1 \end{bmatrix}$$

Thus $c_1 = 1$, $c_2 = 1$, $c_3 = -1$, and $c_4 = 1$.

4. If we equate the second vector components in this equation, we have $0c_1 + 0c_2 + 0c_3 = -2$. Hence, there do not exist scalars c_1, c_2, and c_3 which satisfy the given equation.

5. **(c)** $\|\mathbf{v}\| = \left[3^2 + 4^2 + 0^2 + (-12)^2\right]^{1/2} = \sqrt{169} = 13$

6. **(a)** $\|\mathbf{u} + \mathbf{v}\| = \|(4, 4, 10, 1)\| = \left[4^2 + 4^2 + 10^2 + 1^2\right]^{1/2} = \sqrt{133}$

   (c) $\| - 2\mathbf{u}\| + 2\|\mathbf{u}\| = \left[(-8)^2 + (-2)^2 + (-4)^2 + (-6)^2\right]^{1/2} + 2\left[4^2 + 1^2 + 2^2 + 3^2\right]^{1/2}$

   $$= 120^{1/2} + 2[30]^{1/2} = 4\sqrt{30}$$

   (e) $\dfrac{1}{\|\mathbf{w}\|}\mathbf{w} = \dfrac{1}{[3^2 + 1^2 + 2^2 + 2^2]^{1/2}}(3, 1, 2, 2) = \left(\dfrac{1}{\sqrt{2}}, \dfrac{1}{3\sqrt{2}}, \dfrac{\sqrt{2}}{3}, \dfrac{\sqrt{2}}{3}\right)$

8. Since $\|k\mathbf{v}\| = \left[(-2k)^2 + (3k)^2 + 0^2 + (6k)^2\right]^{1/2} = \left[49k^2\right]^{1/2} = 7|k|$, we have $\|k\mathbf{v}\| = 5$ if and only if $k = \pm 5/7$.

9. **(a)** $(2, 5) \cdot (-4, 3) = (2)(-4) + (5)(3) = 7$

   (c) $(3, 1, 4, -5) \cdot (2, 2, -4, -3) = 6 + 2 - 16 + 15 = 7$

10. **(a)** Let $\mathbf{v} = (x, y)$ where $\|\mathbf{v}\| = 1$. We are given that $\mathbf{v} \cdot (3, -1) = 0$. Thus, $3x - y = 0$ or $y = 3x$. But $\|\mathbf{v}\| = 1$ implies that $x^2 + y^2 = x^2 + 9x^2 = 1$ or $x = \pm 1/\sqrt{10}$. Thus, the only possibilities are $\mathbf{v} = (1/\sqrt{10},\ 3/\sqrt{10})$ or $\mathbf{v} = (-1/\sqrt{10},\ -3/\sqrt{10})$. You should graph these two vectors and the vector $(3, -1)$ in an xy-coordinate system.

   (b) Let $\mathbf{v} = (x, y, z)$ be a vector with norm 1 such that

   $$x - 3y + 5z = 0$$

   This equation represents a plane through (0,0,0) which is perpendicular to $(1, -3, 5)$. There are infinitely many vectors $\mathbf{v}$ which lie in this plane and have norm 1 and initial point (0,0,0).

11. **(a)** $d(\mathbf{u}, \mathbf{v}) = \left[(1-2)^2 + (-2-1)^2\right]^{1/2} = \sqrt{10}$

(c) $d(\mathbf{u}, \mathbf{v}) = \left[(0+3)^2 + (-2-2)^2 + (-1-4)^2 + (1-4)^2\right]^{1/2} = \sqrt{59}$

14. **(e)** Since $\mathbf{u} \cdot \mathbf{v} = 0 + 6 + 2 + 0 = 8$, the vectors are not orthogonal.

15. **(a)** We look for values of k such that

$$\mathbf{u} \cdot \mathbf{v} = 2 + 7 + 3k = 0$$

Clearly $k = -3$ is the only possiblity.

16. We must find two vectors $\mathbf{x} = (x_1, x_2, x_3, x_4)$ such that $\mathbf{x} \cdot \mathbf{x} = 1$ and $\mathbf{x} \cdot \mathbf{u} = \mathbf{x} \cdot \mathbf{v} = \mathbf{x} \cdot \mathbf{w} = 0$. Thus x_1, x_2, x_3, and x_4 must satisfy the equations

$$
\begin{aligned}
x_1^2 + x_2^2 + x_3^2 + x_4^2 &= 1 \\
2x_1 + x_2 - 4x_3 &= 0 \\
-x_1 - x_2 + 2x_3 + 2x_4 &= 0 \\
3x_1 + 2x_2 + 5x_3 + 4x_4 &= 0
\end{aligned}
$$

The solution to the three linear equations is $x_1 = -34t$, $x_2 = 44t$, $x_3 = -6t$, and $x_4 = 11t$. If we substitute these values into the quadratic equation, we get

$$\left[(-34)^2 + (44)^2 + (-6)^2 + (11)^2\right] t^2 = 1$$

or

$$t = \pm \frac{1}{\sqrt{3249}} = \pm \frac{1}{57}$$

Therefore, the two vectors are

$$\pm \frac{1}{57}(-34, 44, -6, 11)$$

17. (a) We have $|\mathbf{u} \cdot \mathbf{v}| = |3(4) + 2(-1)| = 10$, while

$$\|\mathbf{u}\| \, \|\mathbf{v}\| = [3^2 + 2^2]^{1/2}[4^2 + (-1)^2]^{1/2} = \sqrt{221}.$$

(d) Here $|\mathbf{u} \cdot \mathbf{v}| = 0 + 2 + 2 + 1 = 5$, while

$$\|\mathbf{u}\| \, \|\mathbf{v}\| = [0^2 + (-2)^2 + 2^2 + 1^2]^{1/2}[(-1)^2 + (-1)^2 + 1^2 + 1^2]^{1/2} = 6.$$

18. (a) In this case

$$A\mathbf{u} \cdot \mathbf{v} = \begin{bmatrix} 5 \\ 13 \end{bmatrix} \cdot \begin{bmatrix} -2 \\ 6 \end{bmatrix} = 68$$

$$\mathbf{u} \cdot A^T\mathbf{v} = \begin{bmatrix} 3 \\ 1 \end{bmatrix} \cdot \begin{bmatrix} 2 & 3 \\ -1 & 4 \end{bmatrix}\begin{bmatrix} -2 \\ 6 \end{bmatrix} = \begin{bmatrix} 3 \\ 1 \end{bmatrix} \cdot \begin{bmatrix} 14 \\ 26 \end{bmatrix} = 68$$

and

$$\mathbf{u} \cdot A\mathbf{v} = \begin{bmatrix} 3 \\ 1 \end{bmatrix} \cdot \begin{bmatrix} -10 \\ 18 \end{bmatrix} = -12$$

$$A^T\mathbf{u} \cdot \mathbf{v} = \begin{bmatrix} 9 \\ 1 \end{bmatrix}\begin{bmatrix} -2 \\ 6 \end{bmatrix} = -12$$

20. By Theorem 4.1.6, we have $\mathbf{u} \cdot \mathbf{v} = \frac{1}{4}\|\mathbf{u} + \mathbf{v}\|^2 - \frac{1}{4}\|\mathbf{u} - \mathbf{v}\|^2 = \frac{1}{4}(1)^2 - \frac{1}{4}(5^2) = -6.$

22. Note that $\mathbf{u} \cdot \mathbf{a} = 4$ and $\|\mathbf{a}\|^2 = 15$. Hence, by Theorem 3.3.3, we have

$$\text{proj}_\mathbf{a}\mathbf{u} = \frac{\mathbf{u} \cdot \mathbf{a}}{\|\mathbf{a}\|^2}\mathbf{a} = \frac{4}{15}(-1, 1, 2, 3)$$

and

$$\mathbf{u} - \text{proj}_\mathbf{a}\mathbf{u} = \mathbf{u} - \frac{\mathbf{u} \cdot \mathbf{a}}{\|\mathbf{a}\|^2}\mathbf{a} = (2, 1, 4, -1) - \left(-\frac{4}{15}, \frac{4}{15}, \frac{8}{15}, \frac{4}{5}\right)$$

$$= \left(\frac{34}{15}, \frac{11}{15}, \frac{52}{15}, -\frac{9}{5}\right)$$

23. The lines will intersect if and only if there are numbers t_1 and t_2 such that

$$(3, 2, 3, -1) + t_1(4, 6, 4, -2) = (0, 3, 5, 4) + t_2(1, -3, -4, -2).$$

That is,

$$
\begin{array}{lcl}
3 + 4t_1 = t_2 & & 4t_1 - t_2 = -3 \\
2 + 6t_1 = 3 - 3t_2 & & 6t_1 + 3t_2 = 1 \\
3 + 4t_1 = 5 - 4t_2 & \text{or} & 2t_1 + 2t_2 = 1 \\
-1 - 2t_1 = 4 - 2t_2 & & -2t_1 + 2t_2 = 5
\end{array}
$$

If we solve the first and second equations for t_1 and t_2, we obtain $t_1 = -4/9$ and $t_2 = 11/9$. But these values of t_1 and t_2 do not satisfy the third equation. Hence, this system of equations has no solution and the lines do not intersect.

24. Note that

$$\|\mathbf{v}_1 + \cdots + \mathbf{v}_r\|^2 = (\mathbf{v}_1 + \cdots + \mathbf{v}_r) \cdot (\mathbf{v}_1 + \cdots + \mathbf{v}_r)$$

$$= \sum_{i=1}^{r} \mathbf{v}_i \cdot \mathbf{v}_i + \sum_{i \neq j} \mathbf{v}_i \cdot \mathbf{v}_j$$

$$= \|\mathbf{v}_1\|^2 + \cdots + \|\mathbf{v}_r\|^2 + \sum_{i \neq j} \mathbf{v}_i \cdot \mathbf{v}_j$$

If $\mathbf{v}_i \cdot \mathbf{v}_j = 0$ whenever $i \neq j$, then the result follows.

25. This is just the Cauchy–Schwarz Inequality applied to the vectors $\mathbf{v}^T A^T$ and $\mathbf{u}^T A^T$ with both sides of the inequality squared. Why?

26. If we apply the Cauchy–Schwarz Inequality to the vectors $\mathbf{u} = (a, b)$ and $\mathbf{v} = (\cos\theta, \sin\theta)$, the result follows directly.

27. Let $\mathbf{u} = (u_1, \ldots, u_n)$, $\mathbf{v} = (\mathbf{v}_1, \ldots, \mathbf{v}_n)$, and $\mathbf{w} = (w_1, \ldots, w_n)$.

(a) $\mathbf{u} \cdot (k\mathbf{v}) = (u_1, \ldots, u_n) \cdot (kv_1, \ldots, kv_n)$

$$= u_1 k v_1 + \cdots + u_n k v_n$$

$$= k(u_1 v_1 + \cdots + u_n v_n)$$

$$= k(\mathbf{u} \cdot \mathbf{v})$$

27. **(b)** $\mathbf{u} \cdot (\mathbf{v} + \mathbf{w}) = (u_1, \ldots, u_n) \cdot (v_1 + w_1, \ldots, v_n + w_n)$

$$= u_1(v_1 + w_1) + \cdots + u_n(v_n + w_n)$$

$$= (u_1 v_1 + \cdots + u_n v_n) + (u_1 w_1 + \cdots + u_n w_n)$$

$$= \mathbf{u} \cdot \mathbf{v} + \mathbf{u} \cdot \mathbf{w}$$

30. Let $\mathbf{u} = (u_1, \ldots, u_n)$ and $\mathbf{v} = (v_1, \ldots, v_n)$.

(a) $\mathbf{u} \cdot \mathbf{v} = u_1 v_1 + \ldots + u_n v_n = v_1 u_1 + \cdots + v_n u_n = \mathbf{v} \cdot \mathbf{u}$

(c) $(k\mathbf{u}) \cdot \mathbf{v} = (ku_1) + \ldots + (ku_n)v_n = k(u_1 v_1 + \cdots + u_n v_n) = k(\mathbf{u} \cdot \mathbf{v})$

32. Let $\mathbf{u} = (u_1, u_2, \ldots, u_n)$ and $\mathbf{v} = (v_1, v_2, \ldots, v_n)$.

(a) Since $d(\mathbf{u}, \mathbf{v})$ is a square root of a sum of squares, it cannot be negative.

(b) Since $d(\mathbf{u}, \mathbf{v}) = [(u_1 - v_1)^2 + (u_2 - v_2)^2 + \cdots + (u_n - v_n)^2]^{1/2} = 0$ if and only if $u_i - v_i = 0$ for $i = 1, 2, \ldots, n$, we have $d(\mathbf{u}, \mathbf{v}) = 0$ if and only if $\mathbf{u} = \mathbf{v}$.

(c) Here $d(\mathbf{u}, \mathbf{v}) = [(u_1 - v_1)^2 + \cdots + (u_n - v_n)^2]^{1/2} = [(v_1 - u_1)^2 + \cdots + (v_n - u_n)^2]^{1/2} = d(\mathbf{v}, \mathbf{u})$.

34. The result follows from the equations used to prove Theorem 4.1.6. It says that the sum of the squares of the lengths of the diagonals of the parallelogram with sides $\mathbf{u}$ and $\mathbf{v}$ is the sum of the squares of the lengths of the 4 sides.

35. **(a)** By theorem 4.1.7, we have $d(\mathbf{u}, \mathbf{v}) = \|\mathbf{u} - \mathbf{v}\| = \sqrt{\|\mathbf{u}\|^2 + \| - \mathbf{v}\|^2} = \sqrt{2}$.

36. Since $\|\mathbf{u} - \mathbf{v}\|^2 = (\mathbf{u} - \mathbf{v}) \cdot (\mathbf{u} - \mathbf{v}) = \|\mathbf{u}\|^2 + \|\mathbf{v}\|^2 - 2(\mathbf{u} \cdot \mathbf{v})$ for all vectors $\mathbf{u}$ and $\mathbf{v}$ in R^n, we need only show that $\cos\theta = \dfrac{\mathbf{u} \cdot \mathbf{v}}{\|\mathbf{u}\| \|\mathbf{v}\|}$ for the formula to hold. Since this result is valid for $\cos\theta$ in R^2 and R^3, it would seem to be the logical way to define $\cos\theta$ and hence θ in R^n for $n > 3$. To show that this makes sense, we must check that $\dfrac{|\mathbf{u} \cdot \mathbf{v}|}{\|\mathbf{u}\| \|\mathbf{v}\|} \leq 1$

so that $\cos\theta$ is well defined. But the above inequality is just the Cauchy-Schwarz inequality in R^n. Thus the formula is valid in R^n.

37. **(a)** True. In general, we know that

$$\|\mathbf{u} + \mathbf{v}\|^2 = \|\mathbf{u}\|^2 + \|\mathbf{v}\|^2 + 2(\mathbf{u} \cdot \mathbf{v})$$

So in this case $\mathbf{u} \cdot \mathbf{v} = 0$ and the vectors are orthogonal.

(b) True. We are given that $\mathbf{u} \cdot \mathbf{v} = \mathbf{u} \cdot \mathbf{w} = 0$. But since $\mathbf{u} \cdot (\mathbf{v} + \mathbf{w}) = \mathbf{u} \cdot \mathbf{v} + \mathbf{u} \cdot \mathbf{w}$, it follows that $\mathbf{u}$ is orthogonal to $\mathbf{v} + \mathbf{w}$.

(c) False. To obtain a counterexample, let $\mathbf{u} = (1, 0, 0)$, $\mathbf{v} = (1, 1, 0)$, and $\mathbf{w} = (-1, 1, 0)$.

EXERCISE SET 4.2

1. **(b)** Since the transformation maps (x_1, x_2) to (w_1, w_2, w_3), the domain is R^2 and the codomain is R^3. The transformation is not linear because of the terms $2x_1x_2$ and $3x_1x_2$.

2. **(a)** Since we can write the transformation as

$$\begin{bmatrix} w_1 \\ w_2 \end{bmatrix} = \begin{bmatrix} 2 & -3 & 0 & 1 \\ 3 & 5 & 0 & -1 \end{bmatrix} \begin{bmatrix} x_1 \\ x_2 \\ x_3 \\ x_4 \end{bmatrix}$$

the standard matrix is $\begin{bmatrix} 2 & -3 & 0 & 1 \\ 3 & 5 & 0 & -1 \end{bmatrix}$.

3. The standard matrix is A, where

$$\mathbf{w} = A\mathbf{x} = \begin{bmatrix} 3 & 5 & -1 \\ 4 & -1 & 1 \\ 3 & 2 & -1 \end{bmatrix} \begin{bmatrix} x_1 \\ x_2 \\ x_3 \end{bmatrix}$$

so that

$$T(-1, 2, 4) = \begin{bmatrix} 3 & 5 & -1 \\ 4 & -1 & 1 \\ 3 & 2 & -1 \end{bmatrix} \begin{bmatrix} -1 \\ 2 \\ 4 \end{bmatrix} = \begin{bmatrix} 3 \\ -2 \\ -3 \end{bmatrix}$$

4. **(a)** The standard matrix is

$$\begin{bmatrix} 2 & -1 \\ 1 & 1 \end{bmatrix}$$

It is interesting to note that $T(1,0) = (2,1)$ and $T(0,1) = (-1,1)$.

5. **(a)** The standard matrix is

$$\begin{bmatrix} 0 & 1 \\ -1 & 0 \\ 1 & 3 \\ 1 & -1 \end{bmatrix}$$

Note that $T(1,0) = (0,-1,1,1)$ and $T(0,1) = (1,0,3,-1)$.

6. **(b)** We have

$$T(\mathbf{x}) = \begin{bmatrix} -1 & 2 & 0 \\ 3 & 1 & 5 \end{bmatrix} \begin{bmatrix} -1 \\ 1 \\ 3 \end{bmatrix} = \begin{bmatrix} 3 \\ 13 \end{bmatrix}$$

7. **(b)** Here

$$T(2,1,-3) = \begin{bmatrix} 2 & -1 & 1 \\ 0 & 1 & 1 \\ 0 & 0 & 0 \end{bmatrix} \begin{bmatrix} 2 \\ 1 \\ -3 \end{bmatrix} = \begin{bmatrix} 0 \\ -2 \\ 0 \end{bmatrix}$$

8. **(c)** Here

$$T(-1,2) = \begin{bmatrix} 0 & 1 \\ 1 & 0 \end{bmatrix} \begin{bmatrix} -1 \\ 2 \end{bmatrix} = \begin{bmatrix} 2 \\ -1 \end{bmatrix}$$

so the reflection of $(-1,2)$ is $(2,-1)$.

9. **(a)** In this case,

$$T(2, -5, 3) = \begin{bmatrix} 1 & 0 & 0 \\ 0 & 1 & 0 \\ 0 & 0 & -1 \end{bmatrix} \begin{bmatrix} 2 \\ -5 \\ 3 \end{bmatrix} = \begin{bmatrix} 2 \\ -5 \\ -3 \end{bmatrix}$$

so the reflection of $(2, -5, 3)$ is $(2, -5, -3)$.

10. **(a)** The desired projection is $\begin{bmatrix} 1 & 0 \\ 0 & 0 \end{bmatrix} \begin{bmatrix} 2 \\ -5 \end{bmatrix} = \begin{bmatrix} 2 \\ 0 \end{bmatrix}$, or $(2, 0)$.

12. **(b)** The image of $(3, -4)$ is

$$\begin{bmatrix} \cos(60°) & \sin(60°) \\ -\sin(60°) & \cos(60°) \end{bmatrix} \begin{bmatrix} 3 \\ -4 \end{bmatrix} = \begin{bmatrix} \dfrac{1}{2} & \dfrac{\sqrt{3}}{2} \\ -\dfrac{\sqrt{3}}{2} & \dfrac{1}{2} \end{bmatrix} \begin{bmatrix} 3 \\ -4 \end{bmatrix} = \begin{bmatrix} \dfrac{3 - 4\sqrt{3}}{2} \\ -\dfrac{4 + 3\sqrt{3}}{2} \end{bmatrix}$$

or $\left(\dfrac{3 - 4\sqrt{3}}{2}, \ -\dfrac{4 + 3\sqrt{3}}{2} \right)$.

(d) The image of $(3, -4)$ is $(4, 3)$, since

$$\begin{bmatrix} \cos(90°) & -\sin(90°) \\ \sin(90°) & \cos(90°) \end{bmatrix} \begin{bmatrix} 3 \\ -4 \end{bmatrix} = \begin{bmatrix} 0 & -1 \\ 1 & 0 \end{bmatrix} \begin{bmatrix} 3 \\ -4 \end{bmatrix} = \begin{bmatrix} 4 \\ 3 \end{bmatrix}$$

13. **(b)** The image of $(-2, 1, 2)$ is $(0, 1, 2\sqrt{2})$, since

$$\begin{bmatrix} \cos(45°) & 0 & \sin(45°) \\ 0 & 1 & 0 \\ -\sin(45°) & 0 & \cos(45°) \end{bmatrix} \begin{bmatrix} -2 \\ 1 \\ 2 \end{bmatrix} = \begin{bmatrix} \dfrac{1}{\sqrt{2}} & 0 & \dfrac{1}{\sqrt{2}} \\ 0 & 1 & 0 \\ -\dfrac{1}{\sqrt{2}} & 0 & \dfrac{1}{\sqrt{2}} \end{bmatrix} \begin{bmatrix} -2 \\ 1 \\ 2 \end{bmatrix} = \begin{bmatrix} 0 \\ 1 \\ 2\sqrt{2} \end{bmatrix}$$

15.   **(b)**   The image of $(-2, 1, 2)$ is $(0, 1, 2\sqrt{2})$, since

$$\begin{bmatrix} \cos(-45°) & 0 & -\sin(-45°) \\ 0 & 1 & 0 \\ \sin(-45°) & 0 & \cos(-45°) \end{bmatrix} \begin{bmatrix} -2 \\ 1 \\ 2 \end{bmatrix}$$

$$= \begin{bmatrix} \dfrac{1}{\sqrt{2}} & 0 & \dfrac{1}{\sqrt{2}} \\ 0 & 1 & 0 \\ -\dfrac{1}{\sqrt{2}} & 0 & \dfrac{1}{\sqrt{2}} \end{bmatrix} \begin{bmatrix} -2 \\ 1 \\ 2 \end{bmatrix} = \begin{bmatrix} 0 \\ 1 \\ 2\sqrt{2} \end{bmatrix}$$

16.   **(b)**   The standard matrix is $\begin{bmatrix} 1/2 & 0 \\ 0 & 1/2 \end{bmatrix} \begin{bmatrix} 0 & 0 \\ 0 & 1 \end{bmatrix} = \begin{bmatrix} 0 & 0 \\ 0 & 1/2 \end{bmatrix}$.

17.   **(a)**   The standard matrix is

$$\begin{bmatrix} 0 & 1 \\ 1 & 0 \end{bmatrix} \begin{bmatrix} 1 & 0 \\ 0 & 0 \end{bmatrix} \begin{bmatrix} \dfrac{1}{2} & -\dfrac{\sqrt{3}}{2} \\ \dfrac{\sqrt{3}}{2} & \dfrac{1}{2} \end{bmatrix} = \begin{bmatrix} 0 & 0 \\ \dfrac{1}{2} & -\dfrac{\sqrt{3}}{2} \end{bmatrix}$$

(c)   The standard matrix for a counterclockwise rotation of $15° + 105° + 60° = 180°$ is

$$\begin{bmatrix} \cos(180°) & -\sin(180°) \\ \sin(180°) & \cos(180°) \end{bmatrix} = \begin{bmatrix} -1 & 0 \\ 0 & -1 \end{bmatrix}$$

18.   **(a)**   The standard matrix is

$$\begin{bmatrix} 1 & 0 & 0 \\ 0 & 0 & 0 \\ 0 & 0 & 1 \end{bmatrix} \begin{bmatrix} -1 & 0 & 0 \\ 0 & 1 & 0 \\ 0 & 0 & 1 \end{bmatrix} = \begin{bmatrix} -1 & 0 & 0 \\ 0 & 0 & 0 \\ 0 & 0 & 1 \end{bmatrix}$$

(b) The standard matrix is

$$\begin{bmatrix} \sqrt{2} & 0 & 0 \\ 0 & \sqrt{2} & 0 \\ 0 & 0 & \sqrt{2} \end{bmatrix} \begin{bmatrix} \dfrac{1}{\sqrt{2}} & 0 & \dfrac{1}{\sqrt{2}} \\ 0 & 1 & 0 \\ -\dfrac{1}{\sqrt{2}} & 0 & \dfrac{1}{\sqrt{2}} \end{bmatrix} = \begin{bmatrix} 1 & 0 & 1 \\ 0 & \sqrt{2} & 0 \\ -1 & 0 & 1 \end{bmatrix}$$

19. **(c)** The standard matrix is

$$\begin{bmatrix} \cos(180°) & -\sin(180°) & 0 \\ \sin(180°) & \cos(180°) & 0 \\ 0 & 0 & 1 \end{bmatrix} \begin{bmatrix} \cos(90°) & 0 & \sin(90°) \\ 0 & 1 & 0 \\ -\sin(90°) & 0 & \cos(90°) \end{bmatrix} \begin{bmatrix} 1 & 0 & 0 \\ 0 & \cos(270°) & -\sin(270°) \\ 0 & \sin(270°) & \cos(270°) \end{bmatrix}$$

$$= \begin{bmatrix} -1 & 0 & 0 \\ 0 & 1 & 0 \\ 0 & 0 & 1 \end{bmatrix} \begin{bmatrix} 0 & 0 & 1 \\ 0 & 1 & 0 \\ -1 & 0 & 0 \end{bmatrix} \begin{bmatrix} 1 & 0 & 0 \\ 0 & 0 & 1 \\ 0 & -1 & 0 \end{bmatrix}$$

$$= \begin{bmatrix} 0 & 1 & 0 \\ 0 & 0 & -1 \\ -1 & 0 & 0 \end{bmatrix}$$

20. **(a)** Projection on either axis followed by projection on the other will yield the zero vector. Therefore the transformations commute. In matrix form, we have

$$[T_1]\,[T_2] = \begin{bmatrix} 1 & 0 \\ 0 & 0 \end{bmatrix} \begin{bmatrix} 0 & 0 \\ 0 & 1 \end{bmatrix} = \begin{bmatrix} 0 & 0 \\ 0 & 0 \end{bmatrix} = \begin{bmatrix} 0 & 0 \\ 0 & 1 \end{bmatrix} \begin{bmatrix} 1 & 0 \\ 0 & 0 \end{bmatrix} = [T_2]\,[T_1]$$

(c) Geometrically, it is clear that the transformations do not commute. In matrix form, we have

$$[T_1]\,[T_2] = \begin{bmatrix} 1 & 0 \\ 0 & 0 \end{bmatrix} \begin{bmatrix} \cos\theta & -\sin\theta \\ \sin\theta & \cos\theta \end{bmatrix} = \begin{bmatrix} \cos\theta & -\sin\theta \\ 0 & 0 \end{bmatrix}$$

and

$$[T_2]\,[T_1] = \begin{bmatrix} \cos\theta & -\sin\theta \\ \sin\theta & \cos\theta \end{bmatrix} \begin{bmatrix} 1 & 0 \\ 0 & 0 \end{bmatrix} = \begin{bmatrix} \cos\theta & 0 \\ \sin\theta & 0 \end{bmatrix}$$

21. **(a)** Geometrically, it doesn't make any difference whether we rotate and then dilate or whether we dilate and then rotate. In matrix terms, a dilation or contraction is represented by a scalar multiple of the identity matrix. Since such a matrix commutes with any square matrix of the appropriate size, the transformations commute.

22. **(a)** We consider only T_1. It is linear because it can be represented by the matrix

$$\begin{bmatrix} 1 & 0 & 0 \\ 0 & 0 & 0 \\ 0 & 0 & 0 \end{bmatrix}$$

(b) Again we consider only T_1. We have $T_1(\mathbf{x}) = (x,0,0)$ and $\mathbf{x} - T_1(\mathbf{x}) = (x,y,z) - (x,0,0) = (0,y,z)$, so that $T_1(\mathbf{x}) \cdot (\mathbf{x} - T_1\mathbf{x}) = 0$ and the vectors are orthogonal.

(c)

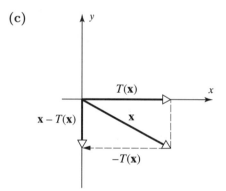

23. Set (a,b,c) equal to $(1,0,0)$, $(0,1,0)$, and $(0,0,1)$ in turn.

24. Let $(a,b,c) = \mathbf{v}/\|\mathbf{v}\| = (1/\sqrt{3},\ 1/\sqrt{3},\ 1/\sqrt{3})$ and $\theta = 90°$ in Formula (17).

25. **(a)** Since $T_2(T_1(x_1,x_2)) = (3(x_1+x_2),\ 2(x_1+x_2)+4(x_1-x_2)) = (3x_1+3x_2, 6x_1-2x_2)$, we have

$$[T_2 \circ T_1] = \begin{bmatrix} 3 & 3 \\ 6 & -2 \end{bmatrix}$$

We also have

$$[T_2][T_1] = \begin{bmatrix} 3 & 0 \\ 2 & 4 \end{bmatrix} \begin{bmatrix} 1 & 1 \\ 1 & -1 \end{bmatrix} = \begin{bmatrix} 3 & 3 \\ 6 & -2 \end{bmatrix}$$

27. Compute the trace of the matrix given in Formula (17) and use the fact that (a, b, c) is a unit vector.

28. **(a)** By inspection, $\mathbf{c}_1 \cdot \mathbf{c}_2 = \mathbf{c}_1 \cdot \mathbf{c}_3 = \mathbf{c}_2 \cdot \mathbf{c}_3 = 0$ and $\|\mathbf{c}_1\| = \|\mathbf{c}_2\| = \|\mathbf{c}_3\| = 1$ where $\mathbf{c}_1$, $\mathbf{c}_2$, and $\mathbf{c}_3$ are the column vectors of A. It is also easy to show that $\det(A) = 1$.

(b) If we let $\mathbf{x} = \begin{bmatrix} 1 \\ 0 \\ 0 \end{bmatrix}$, for instance, an axis of rotation is given by the vector

$$\mathbf{u} = \begin{bmatrix} \frac{1}{9} & -\frac{4}{9} & \frac{8}{9} \\ \frac{8}{9} & \frac{4}{9} & \frac{1}{9} \\ -\frac{4}{9} & \frac{7}{9} & \frac{4}{9} \end{bmatrix} \begin{bmatrix} 1 \\ 0 \\ 0 \end{bmatrix} + \begin{bmatrix} \frac{1}{9} & \frac{8}{9} & -\frac{4}{9} \\ -\frac{4}{9} & \frac{4}{9} & \frac{7}{9} \\ \frac{8}{9} & \frac{1}{9} & \frac{4}{9} \end{bmatrix} \begin{bmatrix} 1 \\ 0 \\ 0 \end{bmatrix} + [1 - 1] \begin{bmatrix} 1 \\ 0 \\ 0 \end{bmatrix}$$

$$= \begin{bmatrix} \frac{1}{9} \\ \frac{8}{9} \\ -\frac{4}{9} \end{bmatrix} + \begin{bmatrix} \frac{1}{9} \\ -\frac{4}{9} \\ \frac{8}{9} \end{bmatrix} = \begin{bmatrix} \frac{2}{9} \\ \frac{4}{9} \\ \frac{4}{9} \end{bmatrix} = \frac{2}{3} \begin{bmatrix} \frac{1}{3} \\ \frac{2}{3} \\ \frac{2}{3} \end{bmatrix}$$

Thus $\left(\frac{1}{3}, \frac{2}{3}, \frac{2}{3}\right)$ is an appropriate vector of length 1.

(c) From Exercise 27, we have $\cos\theta = \dfrac{1 - 1}{2} = 0$, so $\theta = 90° + 180°n$ for $n = 0, 1, 2, \ldots$. Substituting into Formula (17), we find that $\theta = 90° + 360°n$ for $n = 0, 1, 2, \ldots$.

29. **(a)** This is an orthogonal projection on the x-axis and a dilation by a factor of 2.

(b) This is a reflection about the x-axis and a dilation by a factor of 2.

30. **(a)** This transformation dilates the x coordinates by a factor of 2 and the y coordinates by a factor of 3.

(b) This is a rotation through angle of $\pi/6$.

31. Since $\cos(2\theta) = \cos^2\theta - \sin^2\theta$ and $\sin(2\theta) = 2\sin\theta\cos\theta$, this represents a rotation through an angle of 2θ.

32. We have

$$A^T = \begin{bmatrix} \cos\theta & \sin\theta \\ -\sin\theta & \cos\theta \end{bmatrix} = \begin{bmatrix} \cos(-\theta) & -\sin(-\theta) \\ \sin(-\theta) & \cos(-\theta) \end{bmatrix}$$

which represents a rotation through an angle $-\theta$.

EXERCISE SET 4.3

1. **(a)** Projections are not one-to-one since two distinct vectors can have the same image vector.

   (b) Since a reflection is its own inverse, it is a one-to-one mapping of R^2 or R^3 onto itself.

2. **(a)** The standard matrix is $A = \begin{bmatrix} 8 & 4 \\ 2 & 1 \end{bmatrix}$. Since $\det(A) = 0$, A is not invertible and hence the operator is not one-to-one.

   (b) The standard matrix is $A = \begin{bmatrix} 2 & -3 \\ 5 & 1 \end{bmatrix}$. Since $\det(A) = 17 \neq 0$, A is invertible and hence the operator is one-to-one.

3. If we reduce the system of equations to row-echelon form, we find that $w_1 = 2w_2$, so that any vector in the range must be of the form $(2w, w)$. Thus $(3, 1)$, for example, is not in the range.

5. **(a)** Since the determinant of the matrix

$$[T] = \begin{bmatrix} 1 & 2 \\ -1 & 1 \end{bmatrix}$$

   is 3, the transformation T is one-to-one with

$$[T]^{-1} = \begin{bmatrix} \dfrac{1}{3} & -\dfrac{2}{3} \\ \dfrac{1}{3} & \dfrac{1}{3} \end{bmatrix}$$

   Thus $T^{-1}(w_1, w_2) = \left(\dfrac{1}{3}w_1 - \dfrac{2}{3}w_2, \ \dfrac{1}{3}w_1 + \dfrac{1}{3}w_2 \right)$.

5. **(b)** Since the determinant of the matrix

$$[T] = \begin{bmatrix} 4 & -6 \\ -2 & 3 \end{bmatrix}$$

is zero, T is not one-to-one.

6. **(a)** Since the matrix

$$[T] = \begin{bmatrix} 1 & -2 & 2 \\ 2 & 1 & 1 \\ 1 & 1 & 0 \end{bmatrix}$$

is invertible, T is one-to-one. Since

$$[T]^{-1} = \begin{bmatrix} 1 & -2 & 4 \\ -1 & 2 & -3 \\ -1 & 3 & -5 \end{bmatrix}$$

$$T^{-1}(w_1, w_2, w_3) = (w_1 - 2w_2 + 4w_3, \ -w_1 + 2w_2 - 3w_3, \ -w_1 + 3w_2 - 5w_3)$$

(b) Since the matrix

$$[T] = \begin{bmatrix} 1 & -3 & 4 \\ -1 & 1 & 1 \\ 0 & -2 & 5 \end{bmatrix}$$

is not invertible, T is not one-to-one.

8. **(a)** T is linear since

$$T((x_1, y_1) + (x_2, y_2)) = (2(x_1 + x_2), \ y_1 + y_2)$$
$$= (2x_1, y_1) + (2x_2, y_2)$$
$$= T(x_1, y_1) + T(x_2, y_2)$$

and

$$T(k(x,y)) = T(kx, ky) = (2kx, ky)$$
$$= k(2x, y) = kT(x,y)$$

(b) T is not linear since

$$T((x_1, y_1) + (x_2, y_2)) = ((x_1 + x_2)^2, \; y_1 + y_2)$$
$$= (x_1^2, y_1) + (2x_1x_2, 0) + (x_2^2, y_2)$$
$$= T(x_1, y_1) + T(x_2, y_2) + (2x_1x_2, 0)$$
$$\neq T(x_1, y_1) + T(x_2, y_2) \quad \text{if } x_1x_2 \neq 0$$

and

$$T(k(x,y)) = (k^2x^2, ky) = k(kx^2, y)$$
$$\neq kT(x,y) \quad \text{if } k \neq 0, 1 \text{ and } x \neq 0$$

9. (a) T is linear since

$$T((x_1, y_1) + (x_2, y_2)) = (2(x_1 + x_2) + (y_1 + y_2), \; (x_1 + x_2) - (y_1 + y_2))$$
$$= (2x_1 + y_1, \; x_1 - y_1) + (2x_2 + y_2, \; x_2 - y_2)$$
$$= T(x_1, y_1) + T(x_2, y_2)$$

and

$$T(k(x,y)) = (2kx + ky, \; kx - ky)$$
$$= k(2x + y, \; x - y) = kT(x,y)$$

(b) Since

$$T((x_1, y_1) + (x_2, y_2)) = (x_1 + x_2 + 1, \; y_1 + y_2)$$
$$= (x_1 + 1, \; y_1) + (x_2, y_2)$$
$$\neq T(x_1, y_1) + T(x_2, y_2)$$

and $T(k(x,y)) = (kx + 1, ky) \neq kT(x,y)$ unless $k = 1$, T is nonlinear.

10. **(a)** T is linear since

$$T((x_1, y_1, z_1) + (x_2, y_2, z_2)) = (x_1 + x_2, \ x_1 + x_2 + y_1 + y_2 + z_1 + z_2)$$
$$= (x_1, \ x_1 + y_1 + z_1) + (x_2, \ x_2 + y_2 + z_2)$$
$$= T(x_1, y_1, z_1) + T(x_2, y_2, z_2)$$

and

$$T(k(x, y, z)) = (kx, \ kx + ky + kz)$$
$$= kT(x, y, z)$$

(b) Since for all vectors $\mathbf{u}$ and $\mathbf{v}$,

$$T(\mathbf{u} + \mathbf{v}) = (1, 1) \neq T(\mathbf{u}) + T(\mathbf{v}) = (2, 2)$$

and

$$T(k\mathbf{u}) = (1, 1) \neq kT(\mathbf{u}) = (k, k) \ \text{ if } k \neq 1$$

T is nonlinear.

13. **(a)** The projection sends $\mathbf{e}_1$ to itself and the reflection sends $\mathbf{e}_1$ to $-\mathbf{e}_1$, while the projection sends $\mathbf{e}_2$ to the zero vector, which remains fixed under the reflection. Therefore $T(\mathbf{e}_1) = (-1, 0)$ and $T(\mathbf{e}_2) = (0, 0)$, so that $[T] = \begin{bmatrix} -1 & 0 \\ 0 & 0 \end{bmatrix}$.

(b) We have $\mathbf{e}_1 = (1, 0) \rightarrow (0, 1) \rightarrow (0, -1) = 0\mathbf{e}_1 - \mathbf{e}_2$ while $\mathbf{e}_2 = (0, 1) \rightarrow (1, 0) \rightarrow (1, 0) = \mathbf{e}_1 + 0\mathbf{e}_2$. Hence $[T] = \begin{bmatrix} 0 & 1 \\ -1 & 0 \end{bmatrix}$.

(c) Here $\mathbf{e}_1 = (1, 0) \rightarrow (3, 0) \rightarrow (0, 3) = 0\mathbf{e}_1 + 3\mathbf{e}_2$ and $\mathbf{e}_2 = (0, 1) \rightarrow (0, 3) \rightarrow (3, 0) \rightarrow (0, 0) = 0\mathbf{e}_1 + 0\mathbf{e}_2$. Therefore $[T] = \begin{bmatrix} 0 & 0 \\ 3 & 0 \end{bmatrix}$.

14. **(a)** Here $\mathbf{e}_1 = (1, 0, 0) \rightarrow (1, 0, 0) \rightarrow (1/5, 0, 0) = \dfrac{1}{5}\mathbf{e}_1$, $\mathbf{e}_2 = (0, 1, 0) \rightarrow (0, -1, 0) \rightarrow (0, -1/5, 0) = -\dfrac{1}{5}\mathbf{e}_2$, and $\mathbf{e}_3 = (0, 0, 1) \rightarrow (0, 0, 1) \rightarrow (0, 0, 1/5) = \dfrac{1}{5}\mathbf{e}_3$. Therefore

$$[T] = \begin{bmatrix} 1/5 & 0 & 0 \\ 0 & -1/5 & 0 \\ 0 & 0 & 1/5 \end{bmatrix}$$

17. **(a)** By the result of Example 5,

$$T\left(\begin{bmatrix} -1 \\ 2 \end{bmatrix}\right) = \begin{bmatrix} (1/\sqrt{2})^2 & (1/\sqrt{2})(1/\sqrt{2}) \\ (1/\sqrt{2})(1/\sqrt{2}) & (1/\sqrt{2})^2 \end{bmatrix} \begin{bmatrix} -1 \\ 2 \end{bmatrix} = \begin{bmatrix} 1/2 \\ 1/2 \end{bmatrix}$$

or $T(-1, 2) = (1/2, 1/2)$.

18. **(a)** This transformation maps vectors on the x-axis to themselves, vectors on the y-axis into their negatives, and maps no other vectors into scalar multiples of themselves. Thus the eigenvalues are $\lambda = \pm 1$ and the eigenvectors are vectors (x, y) with either $x = 0$ or $y = 0$, but not both. To verify this, we observe that since $\mathbf{e}_1 \to \mathbf{e}_1$ and $\mathbf{e}_2 \to -\mathbf{e}_1$, the standard matrix for the transformation is $\begin{bmatrix} 1 & 0 \\ 0 & -1 \end{bmatrix}$. Hence the characteristic equation is

$$\det\left(\lambda \begin{bmatrix} 1 & 0 \\ 0 & 1 \end{bmatrix} - \begin{bmatrix} 1 & 0 \\ 0 & -1 \end{bmatrix}\right) = 0$$

or $\lambda^2 - 1 = 0$ with solutions $\lambda = \pm 1$. If (x, y) is an eigenvector corresponding to $\lambda = 1$, then

$$\begin{bmatrix} 0 & 0 \\ 0 & 2 \end{bmatrix} \begin{bmatrix} x \\ y \end{bmatrix} = \begin{bmatrix} 0 \\ 2y \end{bmatrix} = \begin{bmatrix} 0 \\ 0 \end{bmatrix}$$

or $y = 0$, so the vector must lie on the x-axis. If (x, y) is an eigenvector corresponding to $\lambda = -1$, then

$$\begin{bmatrix} -2 & 0 \\ 0 & 0 \end{bmatrix} \begin{bmatrix} x \\ y \end{bmatrix} = \begin{bmatrix} -2x \\ 0 \end{bmatrix} = \begin{bmatrix} 0 \\ 0 \end{bmatrix}$$

or $x = 0$, so the vector must lie on the y-axis.

18. **(c)** This transformation leaves vectors on the x-axis fixed and maps vectors on the y-axis to $(0,0)$. No other vectors are mapped into scalar multiples of themselves. Thus the only eigenvalues are $\lambda = 1$ and $\lambda = 0$ with corresponding eigenvectors $(x, 0)$ and $(0, y)$, respectively, where $xy \neq 0$. To verify this, observe that the characteristic equation is

$$\det \left(\lambda \begin{bmatrix} 1 & 0 \\ 0 & 1 \end{bmatrix} - \begin{bmatrix} 1 & 0 \\ 0 & 0 \end{bmatrix} \right) = \begin{vmatrix} \lambda - 1 & 0 \\ 0 & \lambda \end{vmatrix} = 0$$

or $\lambda(\lambda - 1) = 0$. Thus $\lambda = 0$ or $\lambda = 1$. If (x, y) is an eigenvector corresponding to $\lambda = 0$, then

$$\begin{bmatrix} -1 & 0 \\ 0 & 0 \end{bmatrix} \begin{bmatrix} x \\ y \end{bmatrix} = \begin{bmatrix} -x \\ 0 \end{bmatrix} = \begin{bmatrix} 0 \\ 0 \end{bmatrix}$$

so that $x = 0$. If (x, y) is an eigenvector corresponding to $\lambda = 1$, then

$$\begin{bmatrix} 0 & 0 \\ 0 & 1 \end{bmatrix} \begin{bmatrix} x \\ y \end{bmatrix} = \begin{bmatrix} 0 \\ y \end{bmatrix} = \begin{bmatrix} 0 \\ 0 \end{bmatrix}$$

so that $y = 0$. Thus the eigenvectors must indeed lie along the coordinate axes.

19. **(c)** This transformation doubles the length of each vector while leaving its direction unchanged. Therefore $\lambda = 2$ is the only eigenvalue and every nonzero vector in R^3 is a corresponding eigenvector. To verify this, observe that the characteristic equation is

$$\begin{vmatrix} \lambda - 2 & 0 & 0 \\ 0 & \lambda - 2 & 0 \\ 0 & 0 & \lambda - 2 \end{vmatrix} = 0$$

or $(\lambda - 2)^3 = 0$. Thus the only eigenvalue is $\lambda = 2$. If (x, y, z) is a corresponding eigenvector, then

$$\begin{bmatrix} 0 & 0 & 0 \\ 0 & 0 & 0 \\ 0 & 0 & 0 \end{bmatrix} \begin{bmatrix} x \\ y \\ z \end{bmatrix} = \begin{bmatrix} 0 \\ 0 \\ 0 \end{bmatrix}$$

Since the above equation holds for every vector (x, y, z), every nonzero vector is an eigenvector.

(d) Since the transformation leaves all vectors on the z-axis unchanged and alters (but does not reverse) the direction of all other vectors, its only eigenvalue is $\lambda = 1$ with corresponding eigenvectors $(0, 0, z)$ with $z \neq 0$. To verify this, observe that the characteristic equation is

$$\begin{vmatrix} \lambda - 1/\sqrt{2} & 1/\sqrt{2} & 0 \\ -1/\sqrt{2} & \lambda - 1/\sqrt{2} & 0 \\ 0 & 0 & \lambda - 1 \end{vmatrix} = 0$$

or

$$(\lambda - 1)\begin{vmatrix} \lambda - 1/\sqrt{2} & 1/\sqrt{2} \\ -1/\sqrt{2} & \lambda - 1/\sqrt{2} \end{vmatrix} = (\lambda - 1)[(\lambda - 1/\sqrt{2})^2 + (1/\sqrt{2})^2] = 0$$

Since the quadratic $(\lambda - 1/\sqrt{2})^2 + 1/2 = 0$ has no real roots, $\lambda = 1$ is the only eigenvalue. If (x, y, z) is a corresponding eigenvector, then

$$\begin{bmatrix} 1 - 1/\sqrt{2} & 1/\sqrt{2} & 0 \\ -1/\sqrt{2} & 1 - 1/\sqrt{2} & 0 \\ 0 & 0 & 0 \end{bmatrix}\begin{bmatrix} x \\ y \\ z \end{bmatrix} = \begin{bmatrix} (1 - 1/\sqrt{2})x + (1/\sqrt{2})y \\ -(1/\sqrt{2})x + (1 - 1/\sqrt{2})y \\ 0 \end{bmatrix} = \begin{bmatrix} 0 \\ 0 \\ 0 \end{bmatrix}$$

You should verify that the above equation is valid if and only if $x = y = 0$. Therefore the corresponding eigenvectors are all of the form $(0, 0, z)$ with $z \neq 0$.

20. **(a)** Yes. If T_1 and T_2 are both one-to-one (whether or not they are linear), then each maps distinct vectors to distinct vectors and therefore so does $T_2 \circ T_1$.

(b) If T_2 is one-to-one and T_1 is not, then $T_2 \circ T_1$ cannot be one-to-one since there are distinct vectors $\mathbf{u}$ and $\mathbf{v}$ such that $T_1(\mathbf{u}) = T_1(\mathbf{v})$ and therefore such that $T_2 \circ T_1(\mathbf{u}) = T_2 \circ T_1(\mathbf{v})$. This is true whether or not the transformations are linear. However $T_1 \circ T_2$ may indeed be one-to-one. To see this, consider, for instance, the transformations given by

$$[T_1] = \begin{bmatrix} 1 & 0 & 0 \\ 0 & 1 & 0 \end{bmatrix} \quad \text{and} \quad [T_2] = \begin{bmatrix} 1 & 0 \\ 0 & 1 \\ 0 & 0 \end{bmatrix}$$

It is easy to check that T_2 is one-to-one and T_1 is not (since $T_1(\mathbf{e}_3) = \mathbf{0}$). Moreover, $T_1 \circ T_2$ is given by the matrix

$$\begin{bmatrix} 1 & 0 & 0 \\ 0 & 1 & 0 \end{bmatrix} \begin{bmatrix} 1 & 0 \\ 0 & 1 \\ 0 & 0 \end{bmatrix} = \begin{bmatrix} 1 & 0 \\ 0 & 1 \end{bmatrix}$$

which is invertible, so $T_1 \circ T_2$ is one-to-one.

21. Since $T(x, y) = (0, 0)$ has the standard matrix $\begin{bmatrix} 0 & 0 \\ 0 & 0 \end{bmatrix}$, it is linear. If $T(x, y) = (1, 1)$ were linear, then we would have

$$(1, 1) = T(0, 0) = T(0 + 0, 0 + 0) = T(0, 0) + T(0, 0) = (1, 1) + (1, 1) = (2, 2)$$

Since this is a contradiction, T cannot be linear.

22. Since T is linear,

$$T(\mathbf{0}) = T(\mathbf{0} + \mathbf{0}) = T(\mathbf{0}) + T(\mathbf{0}) = 2T(\mathbf{0})$$

so $T(\mathbf{0}) = \mathbf{0}$.

23. From Figure 1, we see that $T(\mathbf{e}_1) = (\cos 2\theta, \sin 2\theta)$ and from Figure 2, that $T(\mathbf{e}_2) = \left(\cos \left(\dfrac{3\pi}{2} + 2\theta \right), \sin \left(\dfrac{3\pi}{2} + 2\theta \right) \right) = (\sin 2\theta, -\cos 2\theta)$.

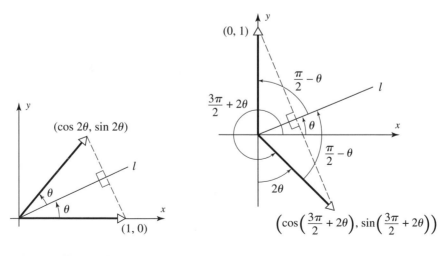

Figure 1   Figure 2

This, of course, should be checked for all possible diagrams, and in particular for the case $\frac{\pi}{2} < \theta < \pi$. The resulting standard matrix is

$$\begin{bmatrix} \cos 2\theta & \sin 2\theta \\ \sin 2\theta & -\cos 2\theta \end{bmatrix}$$

25. **(a)** False. The transformation $T(x_1, x_2) = x_1^2$ from R^2 to R^1 is not linear, but $T(\mathbf{0}) = \mathbf{0}$.

(b) True. If not, $T(\mathbf{u}) = T(\mathbf{v})$ where $\mathbf{u}$ and $\mathbf{v}$ are distinct. Why?

(c) False. One must also demand that $\mathbf{x} \neq \mathbf{0}$.

(d) True. If $c_1 = c_2 = 1$, we obtain equation (a) of Theorem 4.3.2 and if $c_2 = 0$, we obtain equation (b).

26. **(a)** The range of T cannot be all of R^n, since otherwise T would be invertible and $\det(A) \neq 0$. For instance, the matrix $\begin{bmatrix} 1 & 0 \\ 0 & 0 \end{bmatrix}$ sends the entire xy-plane to the x-axis.

(b) Since $\det(A) = 0$, the equation $T(\mathbf{x}) = A\mathbf{x} = \mathbf{0}$ will have a non-trivial solution and hence, T will map infinitely many vectors to $\mathbf{0}$.

EXERCISE SET 5.1

6. The pair $(1, -2)$ is in the set but the pair $(-1)(1, -2) = (-1, 2)$ is not because the first component is negative; hence Axiom 6 fails. Axiom 5 also fails.

8. Axioms 1, 2, 3, 6, 9, and 10 are easily verified. Axiom 4 holds with $\mathbf{0} = (-1, -1)$ and Axiom 5 holds with $-(x, y) = (-x - 2, -y - 2)$. Axiom 7 fails because

$$k\left((x, y) + (x', y')\right) = k(x + x' + 1, \ y + y' + 1)$$
$$= (kx + kx' + k, \ ky + ky' + k)$$

while

$$k(x, y) + k(x', y') = (kx, ky) + (kx', ky')$$
$$= (kx + kx' + 1, \ ky + ky' + 1)$$

Hence, $k(\mathbf{u} + \mathbf{v}) = k\mathbf{u} + k\mathbf{v}$ only if $k = 1$. Axiom 8 also fails, since if $\mathbf{u} = (x, y)$, then

$$(k + \ell)\mathbf{u} = ((k + \ell)x, \ (k + \ell)y)$$

but

$$k\mathbf{u} + \ell\mathbf{u} = (kx, ky) + (\ell x, \ell y)$$
$$= ((k + \ell)x + 1, \ (k + \ell)y + 1)$$

10. This is a vector space. Axioms 2, 3, 7, 8, 9, and 10 follow from properties of matrix addition and scalar multiplication. We verify the remaining axioms.

(1) If we add two matrices of this form, the result will again be a matrix of this form:

$$(*) \qquad \begin{bmatrix} a & 0 \\ 0 & b \end{bmatrix} + \begin{bmatrix} c & 0 \\ 0 & d \end{bmatrix} = \begin{bmatrix} a + c & 0 \\ 0 & b + d \end{bmatrix}$$

(4)   The 2×2 zero matrix is of the appropriate form and has the desired properties.

(5)   If $\mathbf{u}$ is a matrix of the given form, then

$$-\mathbf{u} = \begin{bmatrix} -a & 0 \\ 0 & -b \end{bmatrix}$$

is again of the desired form and $\mathbf{u} + (-\mathbf{u}) = (-\mathbf{u}) + \mathbf{u} = \mathbf{0}$.

(6)   If $\mathbf{u}$ is any matrix of this form, then $k\mathbf{u}$ is

$$(**) \qquad k \begin{bmatrix} a & 0 \\ 0 & b \end{bmatrix} = \begin{bmatrix} ka & 0 \\ 0 & kb \end{bmatrix}$$

and $k\mathbf{u}$ has the desired form.

11.   This is a vector space. We shall check only four of the axioms because the others follow easily from various properties of the real numbers.

(1)   If f and g are real-valued functions defined everywhere, then so is $f + g$. We must also check that if $f(1) = g(1) = 0$, then $(f + g)(1) = 0$. But $(f + g)(1) = f(1) + g(1) = 0 + 0 = 0$.

(4)   The zero vector is the function z which is zero everywhere on the real line. In particular, $z(1) = 0$.

(5)   If f is a function in the set, then $-f$ is also in the set since it is defined for all real numbers and $-f(1) = -0 = 0$. Moreover, $f + (-f) = (-f) + f = z$.

(6)   If f is in the set and k is any real number, then kf is a real valued function defined everywhere. Moreover, $kf(1) = k0 = 0$.

12.   This is a vector space and the proof is almost a direct repeat of that for Problem 10. In fact, we need only modify the two equations $(*)$ and $(**)$ in the following way:

$$\begin{bmatrix} a & a+b \\ a+b & b \end{bmatrix} + \begin{bmatrix} c & c+d \\ c+d & d \end{bmatrix} = \begin{bmatrix} a+c & (a+c)+(b+d) \\ (a+c)+(b+d) & b+d \end{bmatrix}$$

$$k \begin{bmatrix} a & a+b \\ a+b & b \end{bmatrix} = \begin{bmatrix} ka & ka+kb \\ ka+kb & kb \end{bmatrix}$$

Note that if $\mathbf{u}$ is any matrix of this form, then

$$-\mathbf{u} = \begin{bmatrix} -a & -(a+b) \\ -(a+b) & -b \end{bmatrix}$$

13. This is a vector space with $\mathbf{0} = (1,0)$ and $-\mathbf{x} = (1,-x)$. The details are easily checked.

15. We must check all ten properties:

(1) If x and y are positive reals, so is $x + y = xy$.

(2) $x + y = xy = yx = y + x$

(3) $x + (y + z) = x(yz) = (xy)z = (x + y) + z$

(4) There is an object $\mathbf{0}$, the positive real number 1, which is such that

$$1 + x = 1 \cdot x = x = x \cdot 1 = x + 1$$

for all positive real numbers x.

(5) For each positive real x, the positive real $1/x$ acts as a negative:

$$x + (1/x) = x(1/x) = 1 = \mathbf{0} = 1 = (1/x)x = (1/x) + x$$

(6) If k is a real and x is a positive real, then $kx = x^k$ is again a positive real.

(7) $k(x + y) = (xy)^k = x^k y^k = kx + ky$

(8) $(k + \ell)x = x^{k+\ell} = x^k x^\ell = kx + \ell x$

(9) $k(\ell x) = (\ell x)^k = (x^\ell)^k = x^{\ell k} = x^{k\ell} = (k\ell)x$

(10) $1x = x^1 = x$

16. This is not a vector space, since properties (7) and (8) fail. It is easy to show, for instance, that $k\mathbf{u} + k\mathbf{v} = k^2(\mathbf{u} + \mathbf{v})$. Property (4) holds with $\mathbf{0} = (1,1)$, but Property (5) fails because, for instance, $(0,1)$ has no inverse.

18. Properties (2), (3), and (7) – (10) all hold because a line passing through the origin in 3-space is a collection of triples of real numbers and the set of all such triples with the usual operations is a vector space. To verify the remaining four properties, we need only check that

(1) If $\mathbf{u}$ and $\mathbf{v}$ lie on the line, so does $\mathbf{u} + \mathbf{v}$.

(4) The vector $(0,0,0)$ lies on the line (which it does by hypothesis, since the line passes through the origin).

(5) If $\mathbf{u}$ lies on the line, so does $-\mathbf{u}$.

(6) If $\mathbf{u}$ lies on the line, so does any real multiple $k\mathbf{u}$ of $\mathbf{u}$.

We check (1), leaving (5) and (6) to you.

The line passes through the origin and therefore has the parametric equations $x = at$, $y = bt$, $z = ct$ where a, b, and c are fixed real numbers and t is the parameter. Thus, if $\mathbf{u}$ and $\mathbf{v}$ lie on the line, we have $\mathbf{u} = (at_1, bt_1, ct_1)$ and $\mathbf{v} = (at_2, bt_2, ct_2)$. Therefore $\mathbf{u} + \mathbf{v} = (a(t_1 + t_2), b(t_1 + t_2), c(t_1 + t_2))$, which is also on the line.

21. No. Planes which do not pass through the origin do not contain the zero vector.

22. No. The set of polynomials of exactly degree 1 does not contain the zero polynomial, and hence has no zero vector.

23. Since this space has only one element, it would have to be the zero vector. In fact, this is just the zero vector space.

24. If a vector space V had just two distinct vectors, $\mathbf{0}$ and $\mathbf{u}$, then we would have to define vector addition and scalar multiplication on V. Theorem 5.1.1 ensures that $k\mathbf{u} \neq \mathbf{0}$ unless $k = 0$. Thus we would have to define $k\mathbf{u} = \mathbf{u}$ for all $k \neq 0$. But then we would have

$$2\mathbf{u} = (1+1)\mathbf{u} = 1\mathbf{u} + 1\mathbf{u} = \mathbf{u} + \mathbf{u} \neq \mathbf{0},$$

so that $\mathbf{u} + \mathbf{u} = \mathbf{u}$. However,

$$0\mathbf{u} = (1-1)\mathbf{u} = 1\mathbf{u} + (-1)\mathbf{u} = \mathbf{u} + (-1)\mathbf{u} = \mathbf{0}$$

which implies that $(-1)\mathbf{u} \neq \mathbf{u}$. Hence $(-1)\mathbf{u} = \mathbf{0}$, which is contrary to Theorem 5.1.1. Thus V cannot be a vector space.

26. We are given that $k\mathbf{u} = \mathbf{0}$. Suppose that $k \neq 0$. Then

$$\frac{1}{k}(k\mathbf{u}) = \left(\frac{1}{k}k\right)\mathbf{u} = (1)\mathbf{u} = \mathbf{u} \qquad \text{(Axioms 9 and 10)}$$

But

$$\frac{1}{k}(k\mathbf{u}) = \frac{1}{k}\mathbf{0} = \mathbf{0} \qquad \text{(By hypothesis and Part (b))}$$

Thus $\mathbf{u} = \mathbf{0}$. That is, either $k = 0$ or $\mathbf{u} = \mathbf{0}$.

28. Suppose that there are two zero vectors, $\mathbf{0}_1$ and $\mathbf{0}_2$. If we apply Axiom 4 to both of these zero vectors, we have

$$\mathbf{0}_1 = \mathbf{0}_1 + \mathbf{0}_2 = \mathbf{0}_2$$

Hence, the two zero vectors are identical.

29. Suppose that $\mathbf{u}$ has two negatives, $(-\mathbf{u})_1$ and $(-\mathbf{u})_2$. Then

$$(-\mathbf{u})_1 = (-\mathbf{u})_1 + \mathbf{0} = (-\mathbf{u})_1 + (\mathbf{u} + (-\mathbf{u})_2) = ((-\mathbf{u})_1 + \mathbf{u}) + (-\mathbf{u})_2 = \mathbf{0} + (-\mathbf{u})_2 = (-\mathbf{u})_2$$

Axiom 5 guarantees that $\mathbf{u}$ must have at least one negative. We have proved that it has at most one.

30. Following the hint, we have

$$
\begin{aligned}
(\mathbf{u} + \mathbf{v}) - (\mathbf{v} + \mathbf{u}) &= (\mathbf{u} + \mathbf{v}) + (-(\mathbf{v} + \mathbf{u})) && \text{by Theorem 5.1.1} \\
&= (\mathbf{u} + \mathbf{v}) + ((-1)\mathbf{v} + (-1)\mathbf{u}) && \text{by Property (7)} \\
&= (\mathbf{u} + \mathbf{v}) + (-\mathbf{v} + (-\mathbf{u})) && \text{by Theorem 5.1.1} \\
&= ((\mathbf{u} + \mathbf{v}) + (-\mathbf{v})) + (-\mathbf{u}) && \text{by Property (3)} \\
&= (\mathbf{u} + (\mathbf{v} + (-\mathbf{v})) + (-\mathbf{u}) && \text{by Property (3)} \\
&= (\mathbf{u} + \mathbf{0}) + (-\mathbf{u}) && \text{by Property (5)} \\
&= \mathbf{u} + (-\mathbf{u}) && \text{by Property (4)} \\
&= \mathbf{0} && \text{by Property (5)}
\end{aligned}
$$

Thus $(\mathbf{u} + \mathbf{v}) + (-1)(\mathbf{v} + \mathbf{u}) = \mathbf{0}$, from which it follows that

$$(\mathbf{u} + \mathbf{v}) + (-(\mathbf{v} + \mathbf{u})) = \mathbf{0} \qquad \text{by Theorem 5.1.1}$$

so that

$$[(\mathbf{u} + \mathbf{v}) + (-(\mathbf{v} + \mathbf{u}))] + (\mathbf{v} + \mathbf{u}) = \mathbf{0} + (\mathbf{v} + \mathbf{u}) \qquad \text{adding to both sides of the equation}$$

and thus

$$(\mathbf{u} + \mathbf{v}) + (-(\mathbf{v} + \mathbf{u}) + (\mathbf{v} + \mathbf{u})) = \mathbf{0} + (\mathbf{v} + \mathbf{u}) \qquad \text{by Property (3)}$$

or

$$(\mathbf{u} + \mathbf{v}) + \mathbf{0} = \mathbf{0} + (\mathbf{v} + \mathbf{u}) \qquad \text{by Property (5)}$$

so that finally $\mathbf{u} + \mathbf{v} = \mathbf{v} + \mathbf{u}$ by Property (4).

EXERCISE SET 5.2

1. **(a)** The set is closed under vector addition because

$$(a, 0, 0) + (b, 0, 0) = (a + b, 0, 0)$$

   It is closed under scalar multiplication because

$$k(a, 0, 0) = (ka, 0, 0)$$

   Therefore it is a subspace of R^3.

   (b) This set is not closed under either addition or scalar multiplication. For example, $(a, 1, 1) + (b, 1, 1) = (a + b, 2, 2)$ and $(a + b, 2, 2)$ does not belong to the set. Thus it is not a subspace.

2. **(a)** This set is closed under addition since the sum of two integers is again an integer. However, it is not closed under scalar multiplication since the product ku where k is real and a is an integer need not be an integer. Thus, the set is not a subspace.

   (c) If $\det(A) = \det(B) = 0$, it does not necessarily follow that $\det(A + B) = 0$. For instance, let $A = \begin{bmatrix} 1 & 0 \\ 0 & 0 \end{bmatrix}$ and $B = \begin{bmatrix} 0 & 0 \\ 0 & 1 \end{bmatrix}$. Thus the set is not a subspace.

3. **(a)** This is the set of all polynominals with degree ≤ 3 and with a constant term which is equal to zero. Certainly, the sum of any two such polynomials is a polynomial with degree ≤ 3 and with a constant term which is equal to zero. The same is true of a constant multiple of such a polynomial. Hence, this set is a subspace of P_3.

   (c) The sum of two polynomials, each with degree ≤ 3 and each with integral coefficients, is again a polynomial with degree ≤ 3 and with integral coefficients. Hence, the subset is closed under vector addition. However, a constant multiple of such a polynomial will not necessarily have integral coefficients since the constant need not be an integer. Thus, the subset is not closed under scalar multiplication and is therefore not a subspace.

4. **(a)** The function $f(x) = -1$ for all x belongs to the set, but the function $(-1)f(x) = 1$ for all x does not. Hence, the set is not closed under scalar multiplication and is therefore not a subspace.

(c) Suppose that f and g are in the set. Then

$$(f + g)(0) = f(0) + g(0) = 2 + 2 = 4$$

and

$$-2f(0) = (-2)(2) = -4$$

This set is therefore not closed under either operation.

(e) Let $f(x) = a + b \sin x$ and $g(x) = c + d \sin x$ be two functions in this set. Then

$$(f + g)(x) = (a + c) + (b + d) \sin x$$

and

$$k(f(x)) = ka + kb \sin x$$

Thus, both closure properties are satisfied and the set is a subspace.

5. **(b)** If A and B are in the set, then $a_{ij} = -a_{ji}$ and $b_{ij} = -b_{ji}$ for all i and j. Thus $a_{ij} + b_{ij} = -(a_{ji} + b_{ji})$ so that $A + B$ is also in the set. Also $a_{ij} = -a_{ji}$ implies that $ka_{ij} = -(ka_{ji})$, so that kA is in the set for all real k. Thus the set is a subspace.

(c) For A and B to be in the set it is necessary and sufficient for both to be invertible, but the sum of 2 invertible matrices need not be invertible. (For instance, let $B = -A$.) Thus $A + B$ need not be in the set, so the set is not a subspace.

6. **(b)** The matrix reduces to

$$\begin{bmatrix} 1 & -2 & 0 \\ 0 & 0 & 1 \\ 0 & 0 & 0 \end{bmatrix}$$

so the solution space is the line $x = 2t, y = t, z = 0$.

(d) The matrix reduces to the identity matrix, so the solution space is the origin.

(f) The matrix reduces to

$$\begin{bmatrix} 1 & -3 & 1 \\ 0 & 0 & 0 \\ 0 & 0 & 0 \end{bmatrix}$$

so the solution space is the plane $x - 3y + z = 0$.

7. **(a)** We look for constants a and b such that $a\mathbf{u} + b\mathbf{v} = (2, 2, 2)$, or

$$a(0, -2, 2) + b(1, 3, -1) = (2, 2, 2)$$

Equating corresponding vector components gives the following system of equations:

$$b = 2$$
$$-2a + 3b = 2$$
$$2a - b = 2$$

From the first equation, we see that $b = 2$. Substituting this value into the remaining equations yields $a = 2$. Thus $(2,2,2)$ is a linear combination of $\mathbf{u}$ and $\mathbf{v}$.

(c) We look for constants a and b such that $a\mathbf{u} + b\mathbf{v} = (0, 4, 5)$, or

$$a(0, -2, 2) + b(1, 3, -1) = (0, 4, 5)$$

Equating corresponding components gives the following system of equations:

$$b = 0$$
$$-2a + 3b = 4$$
$$2a - b = 5$$

From the first equation, we see that $b = 0$. If we substitute this value into the remaining equations, we find that $a = -2$ and $a = 5/2$. Thus, the system of equations is inconsistent and therefore $(0,4,5)$ is not a linear combination of $\mathbf{u}$ and $\mathbf{v}$.

8. **(a)** We look for constants a, b, and c such that $a\mathbf{u} + b\mathbf{v} + c\mathbf{w} = (-9, -7, -15)$; that is, such that

$$a(2, 1, 4) + b(1, -1, 3) + c(3, 2, 5) = (-9, -7, -15)$$

If we equate corresponding components, we obtain the system

$$2a + b + 3c = -9$$
$$a - b + 2c = -7$$
$$4a + 3b + 5c = -15$$

The augmented matrix for this system is

$$\begin{bmatrix} 2 & 1 & 3 & -9 \\ 1 & -1 & 2 & -7 \\ 4 & 3 & 5 & -15 \end{bmatrix}$$

The reduced row-echelon form of this matrix is

$$\begin{bmatrix} 1 & 0 & 0 & -2 \\ 0 & 1 & 0 & 1 \\ 0 & 0 & 1 & -2 \end{bmatrix}$$

Thus $a = -2$, $b = 1$, and $c = -2$ and $(-9, -7, -15)$ is therefore a linear combination of $\mathbf{u}$, $\mathbf{v}$, and $\mathbf{w}$.

8. **(c)** This time we look for constants a, b, and c such that

$$a\mathbf{u} + b\mathbf{v} + c\mathbf{w} = (0, 0, 0)$$

If we choose $a = b = c = 0$, then it is obvious that $a\mathbf{u} + b\mathbf{v} + c\mathbf{w} = (0, 0, 0)$. We now proceed to show that $a = b = c = 0$ is the only choice. To this end, we equate components to obtain a system of equations whose augmented matrix is

$$\begin{bmatrix} 2 & 1 & 3 & 0 \\ 1 & -1 & 2 & 0 \\ 4 & 3 & 5 & 0 \end{bmatrix}$$

From Part (a), we know that this matrix can be reduced to

$$\begin{bmatrix} 1 & 0 & 0 & 0 \\ 0 & 1 & 0 & 0 \\ 0 & 0 & 1 & 0 \end{bmatrix}$$

Thus, $a = b = c = 0$ is the only solution.

9. (a) We look for constants a, b, and c such that

$$a\mathbf{p}_1 + b\mathbf{p}_2 + c\mathbf{p}_3 = -9 - 7x - 15x^2$$

If we substitute the expressions for $\mathbf{p}_1$, $\mathbf{p}_2$, and $\mathbf{p}_3$ into the above equation and equate corresponding coefficients, we find that we have exactly the same system of equations that we had in Problem 8(a), above. Thus, we know that $a = -2$, $b = 1$, and $c = -2$ and thus $-2\mathbf{p}_1 + 1\mathbf{p}_2 - 2\mathbf{p}_3 = -9 - 7x - 15x^2$.

(c) Just as Problem 9(a) was Problem 8(a) in disguise, Problem 9(c) is Problem 8(c) in different dress. The constants are the same, so that $0 = 0\mathbf{p}_1 + 0\mathbf{p}_2 + 0\mathbf{p}_3$.

10. (a) We ask if there are constants a, b, and c such that

$$a\begin{bmatrix} 4 & 0 \\ -2 & -2 \end{bmatrix} + b\begin{bmatrix} 1 & -1 \\ 2 & 3 \end{bmatrix} + c\begin{bmatrix} 0 & 2 \\ 1 & 4 \end{bmatrix} = \begin{bmatrix} 6 & -8 \\ -1 & -8 \end{bmatrix}$$

If we multiply, add, and equate corresponding matrix entries, we obtain the following system of equations:

$$
\begin{aligned}
4a + b &= 6 \\
-b + 2c &= -8 \\
-2a + 2b + c &= -1 \\
-2a + 3b + 4c &= -8
\end{aligned}
$$

This system has the solution $a = 1$, $b = 2$, $c = -3$; thus, the matrix *is* a linear combination of the three given matrices.

(b) Clearly the zero matrix is a linear combination of any set of matrices since we can always choose the scalars to be zero.

11. (a) Given any vector (x, y, z) in R^3, we must determine whether or not there are constants a, b, and c such that

$$
\begin{aligned}
(x, y, z) &= a\mathbf{v}_1 + b\mathbf{v}_2 + c\mathbf{v}_3 \\
&= a(2, 2, 2) + b(0, 0, 3) + c(0, 1, 1) \\
&= (2a,\ 2a + c,\ 2a + 3b + c)
\end{aligned}
$$

or

$$x = 2a$$

$$y = 2a \qquad + c$$

$$z = 2a + 3b + c$$

This is a system of equations for a, b, and c. Since the determinant of the system is nonzero, the system of equations must have a solution for any values of x, y, and z, whatsoever. Therefore, $\mathbf{v}_1$, $\mathbf{v}_2$, and $\mathbf{v}_3$ do indeed span R^3.

Note that we can also show that the system of equations has a solution by solving for a, b, and c explicitly.

11. **(c)** We follow the same procedure that we used in Part (a). This time we obtain the system of equations

$$3a + 2b + 5c + \quad d = x$$

$$a - 3b - 2c + 4d = y$$

$$4a + 5b + 9c - \quad d = z$$

The augmented matrix of this system is

$$\begin{bmatrix} 3 & 2 & 5 & 1 & x \\ 1 & -3 & -2 & 4 & y \\ 4 & 5 & 9 & -1 & z \end{bmatrix}$$

which reduces to

$$\begin{bmatrix} 1 & -3 & -2 & 4 & y \\ 0 & 1 & 1 & -1 & \dfrac{x - 3y}{11} \\ 0 & 0 & 0 & 0 & \dfrac{z - 4y}{17} - \dfrac{x - 3y}{11} \end{bmatrix}$$

Thus the system is inconsistent unless the last entry in the last row of the above matrix is zero. Since this is not the case for all values of x, y, and z, the given vectors do not span R^3.

12. **(a)** Since $\cos(2x) = (1)\cos^2 x + (-1)\sin^2 x$ for all x, it follows that $\cos(2x)$ lies in the space spanned by $\cos^2 x$ and $\sin^2 x$.

(b) Suppose that $3 + x^2$ is in the space spanned by $\cos^2 x$ and $\sin^2 x$; that is, $3 + x^2 = a\cos^2 x + b\sin^2 x$ for some constants a and b. This equation must hold for all x. If we set $x = 0$, we find that $a = 3$. However, if we set $x = \pi$, we find $a = 3 + \pi^2$. Thus we have a contradiction, so $3 + x^2$ is not in the space spanned by $\cos^2 x$ and $\sin^2 x$.

13. Given an arbitrary polynomial $a_0 + a_1 x + a_2 x^2$ in P_2, we ask whether there are numbers a, b, c and d such that

$$a_0 + a_1 x + a_2 x^2 = a\mathbf{p}_1 + b\mathbf{p}_2 + c\mathbf{p}_3 + d\mathbf{p}_4$$

If we equate coefficients, we obtain the system of equations:

$$a_0 = a + 3b + 5c - 2d$$
$$a_1 = -a + b - c - 2d$$
$$a_2 = 2a + 4c + 2d$$

A row-echelon form of the augmented matrix of this system is

$$\begin{bmatrix} 1 & 3 & 5 & -2 & a_0 \\ 0 & 1 & 1 & -1 & \dfrac{a_0 + a_1}{4} \\ 0 & 0 & 0 & 0 & -a_0 + 3a_1 + 2a_2 \end{bmatrix}$$

Thus the system is inconsistent whenever $a_0 + 3a_1 + 2a_2 \neq 0$ (for example, when $a_0 = 0$, $a_1 = 0$, and $a_2 = 1$). Hence the given polynomials do not span P_2.

14. **(a)** As before, we look for constants a, b, and c such that

$$(2, 3, -7, 3) = a\mathbf{v}_1 + b\mathbf{v}_2 + c\mathbf{v}_3$$

If we equate components, we obtain the following system of equations:

$$2a + 3b - c = 2$$
$$a - b = 3$$
$$5b + 2c = -7$$
$$3a + 2b + c = 3$$

The augmented matrix of this system is

$$\begin{bmatrix} 2 & 3 & -1 & 2 \\ 1 & -1 & 0 & 3 \\ 0 & 5 & 2 & -7 \\ 3 & 2 & 1 & 3 \end{bmatrix}$$

This reduces to

$$\begin{bmatrix} 1 & 0 & 0 & 2 \\ 0 & 1 & 0 & -1 \\ 0 & 0 & 1 & -1 \\ 0 & 0 & 0 & 0 \end{bmatrix}$$

Thus $a = 2$, $b = -1$, and $c = -1$, and the existence of a solution guarantees that the given vector is in span$\{\mathbf{v}_1, \mathbf{v}_2, \mathbf{v}_3\}$.

14. **(c)** Proceeding as in Part (a), we obtain the matrix

$$\begin{bmatrix} 2 & 3 & -1 & 1 \\ 1 & -1 & 0 & 1 \\ 0 & 5 & 2 & 1 \\ 3 & 2 & 1 & 1 \end{bmatrix}$$

This reduces to a matrix whose last row is $[0\ 0\ 0\ 1]$. Thus the system is inconsistent and hence the given vector is not in span$\{\mathbf{v}_1, \mathbf{v}_2, \mathbf{v}_3\}$.

15. The plane has the vector $\mathbf{u} \times \mathbf{v} = (0, 7, -7)$ as a normal and passes through the point $(0,0,0)$. Thus its equation is $y - z = 0$.

Alternatively, we look for conditions on a vector (x, y, z) which will insure that it lies in span$\{\mathbf{u}, \mathbf{v}\}$. That is, we look for numbers a and b such that

$$(x, y, z) = a\mathbf{u} + b\mathbf{v}$$
$$= a(-1, 1, 1) + b(3, 4, 4)$$

If we expand and equate components, we obtain a system whose augmented matrix is

$$\begin{bmatrix} -1 & 3 & x \\ 1 & 4 & y \\ 1 & 4 & z \end{bmatrix}$$

This reduces to the matrix

$$\begin{bmatrix} 1 & -3 & -x \\ 0 & 1 & \dfrac{x+y}{7} \\ 0 & 0 & \dfrac{-y+z}{7} \end{bmatrix}$$

Thus the system is consistent if and only if $\dfrac{-y+z}{7} = 0$ or $y = z$.

17. The set of solution vectors of such a system does not contain the zero vector. Hence it cannot be a subspace of R^n.

18. Suppose that span$\{v_1, v_2, \ldots, v_r\}$ = span$\{w_1, w_2, \ldots, w_k\}$. Since each vector v_i in S belongs to span$\{v_1, v_2, \ldots, v_r\}$, it must, by the definition of span$\{w_1, w_2, \ldots, w_k\}$, be a linear combination of the vectors in S'. The converse must also hold.

Now suppose that each vector in S is a linear combination of those in S' and conversely. Then we can express each vector v_i as a linear combination of the vectors $w_1, w_2, \ldots, w_k$, so span$\{v_1, v_2, \ldots, v_r\} \subseteq$ span$\{w_1, w_2, \ldots, w_k\}$. But conversely we have span$\{w_1, w_2, \ldots, w_k\} \subseteq$ span$\{v_1, v_2, \ldots, v_r\}$, so the two sets are equal.

19. Note that if we solve the system $v_1 = aw_1 + bw_2$, we find that $v_1 = w_1 + w_2$. Similarly, $v_2 = 2w_1 + w_2, v_3 = -w_1 + 0w_2, w_1 = 0v_1 + 0v_2 - v_3$, and $w_2 = v_1 + 0v_2 + v_3$.

21. **(a)** We simply note that the sum of two continuous functions is a continuous function and that a constant times a continuous function is a continuous function.

(b) We recall that the sum of two differentiable functions is a differentiable function and that a constant times a differentiable function is a differentiable function.

23. **(a)** False. The system has the form $A\mathbf{x} = \mathbf{b}$ where $\mathbf{b}$ has at least one nonzero entry. Suppose that $\mathbf{x}_1$ and $\mathbf{x}_2$ are two solutions of this system; that is, $A\mathbf{x}_1 = \mathbf{b}$ and $A\mathbf{x}_2 = \mathbf{b}$. Then

$$A(\mathbf{x}_1 + \mathbf{x}_2) = A\mathbf{x}_1 + A\mathbf{x}_2 = \mathbf{b} + \mathbf{b} \neq \mathbf{b}$$

Thus the solution set is not closed under addition and so cannot form a subspace of R^n. Alternatively, we could show that it is not closed under scalar multiplication.

(b) True. Let $\mathbf{u}$ and $\mathbf{v}$ be vectors in W. Then we are given that $k\mathbf{u} + \mathbf{v}$ is in W for all scalars k. If $k = 1$, this shows that W is closed under addition. If $k = -1$ and $\mathbf{u} = \mathbf{v}$, then the zero vector of V must be in W. Thus, we can let $\mathbf{v} = \mathbf{0}$ to show that W is closed under scalar multiplication.

(d) True. Let W_1 and W_2 be subspaces of V. Then if $\mathbf{u}$ and $\mathbf{v}$ are in $W_1 \cap W_2$, we know that $\mathbf{u} + \mathbf{v}$ must be in both W_1 and W_2, as must $k\mathbf{u}$ for every scalar k. This follows from the closure of both W_1 and W_2 under addition and scalar multiplication.

(e) False. Span$\{\mathbf{v}\} = $ span$\{2\mathbf{v}\}$, but $\mathbf{v} \neq 2\mathbf{v}$ in general.

24. **(a)** Two vectors in R^3 will span a plane if and only if one is not a constant multiple of the other. They will span a line if and only if one is a constant multiple of the other.

(b) Span$\{\mathbf{u}\} = $ span$\{\mathbf{v}\}$ if and only if $\mathbf{u}$ is a nonzero multiple of $\mathbf{v}$. Why?

(c) The solution set will be a subspace of R^n if and only if $\mathbf{b} = \mathbf{0}$. See Exercise 23(a).

25. No. For instance, $(1, 1)$ is in W_1 and $(1, -1)$ is in W_2, but $(1, 1) + (1, -1) = (2, 0)$ is in neither W_1 nor W_2.

26. **(a)** The matrices $\begin{bmatrix} 1 & 0 \\ 0 & 0 \end{bmatrix}$, $\begin{bmatrix} 0 & 1 \\ 0 & 0 \end{bmatrix}$, $\begin{bmatrix} 0 & 0 \\ 1 & 0 \end{bmatrix}$, and $\begin{bmatrix} 0 & 0 \\ 0 & 1 \end{bmatrix}$ span M_{22}.

27. They cannot all lie in the same plane.

EXERCISE SET 5.3

2. (a) Clearly neither of these vectors is a multiple of the other. Thus they are linearly independent.

(b) Following the technique used in Example 4, we consider the system of linear equations

$$-3k_1 + 5k_2 + k_3 = 0$$
$$- k_2 + k_3 = 0$$
$$4k_1 + 2k_2 + 3k_3 = 0$$

Since the determinant of the coefficient matrix is nonzero, the system has *only* the trivial solution. Therefore, the three vectors are linearly independent.

(d) By Theorem 5.3.3, any four vectors in R^3 are linearly dependent.

3. (a) Following the technique used in Example 4, we obtain the system of equations

$$3k_1 + k_2 + 2k_3 + k_4 = 0$$
$$8k_1 + 5k_2 - k_3 + 4k_4 = 0$$
$$7k_1 + 3k_2 + 2k_3 = 0$$
$$-3k_1 - k_2 + 6k_3 + 3k_4 = 0$$

Since the determinant of the coefficient matrix is nonzero, the system has *only* the trivial solution. Hence, the four vectors are linearly independent.

3. **(b)** Again following the technique of Example 4, we obtain the system of equations

$$3k_2 + k_3 = 0$$
$$3k_2 + k_3 = 0$$
$$2k_1 \qquad\quad = 0$$
$$2k_1 \quad - k_3 = 0$$

The third equation, above, implies that $k_1 = 0$. This implies that k_3 and hence k_2 must also equal zero. Thus the three vectors are linearly independent.

4. **(a)** We ask whether there exist constants a, b, and c such that

$$a(2 - x + 4x^2) + b(3 + 6x + 2x^2) + c(2 + 10x - 4x^2) = 0$$

If we equate the coefficients of x^0, x, and x^2 in the above polynomial to zero, we obtain the following system of equations:

$$2a + 3b + \ 2c = 0$$
$$-a + 6b + 10c = 0$$
$$4a + 2b - \ 4c = 0$$

Since the coefficient matrix of this system is invertible, the trivial solution is the only solution. Hence, the polynomials are linearly independent.

(d) If we set up this problem in the same way we set up Part (a), above, we obtain three equations in four unknowns. Since this is equivalent to having four vectors in R^3, the vectors are linearly dependent by Theorem 5.3.3.

5. **(a)** The vectors lie in the same plane through the origin if and only if they are linearly dependent. Since the determinant of the matrix

$$\begin{bmatrix} 2 & 6 & 2 \\ -2 & 1 & 0 \\ 0 & 4 & -4 \end{bmatrix}$$

is not zero, the matrix is invertible and the vectors are linearly independent. Thus they do not lie in the same plane.

6. (a) Since $\mathbf{v}_2 = -2\mathbf{v}_1$, the vectors $\mathbf{v}_1$ and $\mathbf{v}_2$ lie on the same line. But since $\mathbf{v}_3$ is not a multiple of $\mathbf{v}_1$ or $\mathbf{v}_2$, the three vectors do not lie on the same line through the origin.

    (c) Since $\mathbf{v}_1 = 2\mathbf{v}_2 = -2\mathbf{v}_3$, these vectors all lie on the same line through the origin.

7. (a) Note that $7\mathbf{v}_1 - 2\mathbf{v}_2 + 3\mathbf{v}_3 = \mathbf{0}$.

8. If there are constants a, b, and c such that

$$a(\lambda, \ -1/2, \ -1/2) + b(-1/2, \ \lambda, \ -1/2) + c(-1/2, \ -1/2, \ \lambda) = (0,0,0)$$

then

$$\begin{bmatrix} \lambda & -1/2 & -1/2 \\ -1/2 & \lambda & -1/2 \\ -1/2 & -1/2 & \lambda \end{bmatrix} \begin{bmatrix} a \\ b \\ c \end{bmatrix} = \begin{bmatrix} 0 \\ 0 \\ 0 \end{bmatrix}$$

The determinant of the coefficient matrix is

$$\lambda^3 - \frac{3}{4}\lambda - \frac{1}{4} = (\lambda - 1)\left(\lambda + \frac{1}{2}\right)^2$$

This equals zero if and only if $\lambda = 1$ or $\lambda = -1/2$. Thus the vectors are linearly dependent for these two values of λ and linearly independent for all other values.

10. Suppose that S has a linearly dependent subset T. Denote its vectors by $\mathbf{w}_1, \ldots, \mathbf{w}_m$. Then there exist constants k_i, not all zero, such that

$$k_1\mathbf{w}_1 + \cdots + k_m\mathbf{w}_m = \mathbf{0}$$

But if we let $\mathbf{u}_1, \ldots, \mathbf{u}_{n-m}$ denote the vectors which are in S but not in T, then

$$k_1\mathbf{w}_1 + \cdots + k_m\mathbf{w}_m + 0\mathbf{u}_1 + \cdots + 0\mathbf{u}_{n-m} = \mathbf{0}$$

Thus we have a linear combination of the vectors $\mathbf{v}_1, \ldots, \mathbf{v}_n$ which equals $\mathbf{0}$. Since not all of the constants are zero, it follows that S is not a linearly independent set of vectors, contrary to the hypothesis. That is, if S is a linearly independent set, then so is every non-empty subset T.

12. This is similar to Problem 10. Since $\{\mathbf{v}_1, \mathbf{v}_2, \ldots, \mathbf{v}_r\}$ is a linearly dependent set of vectors, there exist constants $c_1, c_2, \ldots, c_r$ not all zero such that

$$c_1\mathbf{v}_1 + c_2\mathbf{v}_2 + \cdots + c_r\mathbf{v}_r = \mathbf{0}$$

But then

$$c_1\mathbf{v}_1 + c_2\mathbf{v}_2 + \cdots + c_r\mathbf{v}_r + 0\mathbf{v}_{r+1} + \cdots + 0\mathbf{v}_n = \mathbf{0}$$

The above equation implies that the vectors $\mathbf{v}_1, \ldots, \mathbf{v}_n$ are linearly dependent.

14. Suppose that $\{\mathbf{v}_1, \mathbf{v}_2, \mathbf{v}_3\}$ is linearly dependent. Then there exist constants a, b, and c not all zero such that

$$(*) \qquad\qquad a\mathbf{v}_1 + b\mathbf{v}_2 + c\mathbf{v}_3 = \mathbf{0}$$

<u>Case 1</u>: $c = 0$. Then $(*)$ becomes

$$a\mathbf{v}_1 + b\mathbf{v}_2 = \mathbf{0}$$

where not both a and b are zero. But then $\{\mathbf{v}_1, \mathbf{v}_2\}$ is linearly dependent, contrary to hypothesis.

<u>Case 2</u>: $c \neq 0$. Then solving $(*)$ for $\mathbf{v}_3$ yields

$$\mathbf{v}_3 = -\frac{a}{c}\mathbf{v}_1 - \frac{b}{c}\mathbf{v}_2$$

This equation implies that $\mathbf{v}_3$ is in span$\{\mathbf{v}_1, \mathbf{v}_2\}$, contrary to hypothesis. Thus, $\{\mathbf{v}_1, \mathbf{v}_2, \mathbf{v}_3\}$ is linearly independent.

15. Note that $(\mathbf{u} - \mathbf{v}) + (\mathbf{v} - \mathbf{w}) + (\mathbf{w} - \mathbf{u}) = \mathbf{0}$.

17. Any nonzero vector forms a linearly independent set. The only scalar multiple of a nonzero vector which can equal the zero vector is the zero scalar times the vector.

19. **(a)** Since $\sin^2 x + \cos^2 x = 1$, we observe that

$$2(3\sin^2 x) + 3(2\cos^2 x) + (-1)(6) = 0$$

Hence the vectors are linearly dependent.

(c) Suppose that there are constants a, b, and c such that

$$a(1) + b\sin x + c\sin 2x = 0$$

Setting $x = 0$ yields $a = 0$. Setting $x = \pi/2$ yields $b = 0$, and thus, since $\sin 2x \not\equiv 0$, we must also have $c = 0$. Therefore the vectors are linearly independent.

(e) Suppose that there are constants a, b, and c such that

$$a(3 - x)^2 + b(x^2 - 6x) + c(5) = 0$$

or

$$(9a + 5c) + (-6a - 6b)x + (a + b)x^2 = 0$$

Clearly $a = -b = -(5/9)c$. Thus $a = 5$, $b = -5$, $c = -9$ is one solution and the vectors are linearly dependent.

This conclusion may also be reached by noting that the determinant of the system of equations

$$
\begin{aligned}
9a \qquad\ \ + 5c &= 0 \\
-6a - 6b \qquad\quad &= 0 \\
a + \ b \qquad\quad &= 0
\end{aligned}
$$

is zero.

20. **(a)** The Wronskian is

$$
\begin{vmatrix}
1 & x & e^x \\
0 & 1 & e^x \\
0 & 0 & e^x
\end{vmatrix}
= e^x \not\equiv 0
$$

Thus the vectors are linearly independent.

20. **(b)** The Wronskian is

$$\begin{vmatrix} \sin x & \cos x & x \sin x \\ \cos x & -\sin x & \sin x + x \cos x \\ -\sin x & -\cos x & 2\cos x - x \sin x \end{vmatrix} = \begin{vmatrix} \sin x & \cos x & x \sin x \\ \cos x & -\sin x & \sin x + x \cos x \\ 0 & 0 & 2\cos x \end{vmatrix}$$

$$= 2\cos x (-\sin^2 x - \cos^2 x) = -2\cos x \neq 0$$

Thus the vectors are linearly independent.

22. Use Theorem 5.3.1, Part (a).

23. **(a)** False. There are 6 such matrices, namely

$$A_1 = \begin{bmatrix} 1 & 1 \\ 0 & 0 \end{bmatrix} \quad A_2 = \begin{bmatrix} 1 & 0 \\ 1 & 0 \end{bmatrix} \quad A_3 = \begin{bmatrix} 1 & 0 \\ 0 & 1 \end{bmatrix}$$

$$A_4 = \begin{bmatrix} 0 & 1 \\ 1 & 0 \end{bmatrix} \quad A_5 = \begin{bmatrix} 0 & 1 \\ 0 & 1 \end{bmatrix} \quad A_6 = \begin{bmatrix} 0 & 0 \\ 1 & 1 \end{bmatrix}$$

Since M_{22} has dimension 4, at least two of these matrices must be linear combinations of the other four. In fact, it is easy to show that A_1, A_2, A_3, and A_4 are linearly independent and that $A_5 = -A_2 + A_3 + A_4$ and $A_6 = -A_1 + A_3 + A_4$.

(b) False. One of the vectors might be the zero vector. Otherwise it would be true.

(d) False. A finite set of vectors can be linearly dependent without containing the zero vector.

25. We could think of any 3 linearly independent vectors in R^3 as spanning R^3. That is, they could determine directions for 3 (not necessarily orthogonal) coordinate axes. Then any fourth vector would represent a point in this coordinate system and hence be a linear combination of the other 3 vectors.

EXERCISE SET 5.4

2. **(a)** This set is a basis. It has two vectors and neither is a multiple of the other.

   (c) This set is not a basis since one vector is a multiple of the other.

3. **(a)** This set has the correct number of vectors and they are linearly independent because

$$\begin{vmatrix} 1 & 2 & 3 \\ 0 & 2 & 3 \\ 0 & 0 & 3 \end{vmatrix} = 6 \neq 0$$

   Hence, the set is a basis.

   (c) The vectors in this set are linearly dependent because

$$\begin{vmatrix} 2 & 4 & 0 \\ -3 & 1 & -7 \\ 1 & 1 & 1 \end{vmatrix} = 0$$

   Hence, the set is not a basis.

4. **(a)** The vectors in this set are linearly dependent because

$$\begin{vmatrix} 1 & 1 & 1 \\ -3 & 1 & -7 \\ 2 & 4 & 0 \end{vmatrix} = 0$$

   Thus, the set is not a basis. (Compare with Problem 3(c), above.)

4. (c) This set has the correct number of vectors and

$$\begin{vmatrix} 1 & 0 & 0 \\ 1 & 1 & 0 \\ 1 & 1 & 1 \end{vmatrix} = 1 \neq 0$$

Hence, the vectors are linearly independent and therefore are a basis.

5. The set has the correct number of vectors. To show that they are linearly independent, we consider the equation

$$a\begin{bmatrix} 3 & 6 \\ 3 & -6 \end{bmatrix} + b\begin{bmatrix} 0 & -1 \\ -1 & 0 \end{bmatrix} + c\begin{bmatrix} 0 & -8 \\ -12 & -4 \end{bmatrix} + d\begin{bmatrix} 1 & 0 \\ -1 & 2 \end{bmatrix} = \begin{bmatrix} 0 & 0 \\ 0 & 0 \end{bmatrix}$$

If we add matrices and equate corresponding entries, we obtain the following system of equations:

$$3a \qquad\qquad + d = 0$$
$$6a - b - 8c \qquad = 0$$
$$3a - b - 12c - d = 0$$
$$-6a \qquad - 4c + 2d = 0$$

Since the determinant of the coefficient matrix is nonzero, the system of equations has only the trivial solution; hence, the vectors are linearly independent.

6. (a) Recall that $\cos 2x = \cos^2 x - \sin^2 x$; that is,

$$1\mathbf{v}_1 + (-1)\mathbf{v}_2 + (-1)\mathbf{v}_3 = \mathbf{0}$$

Hence, S is not a linearly independent set of vectors.

(b) We can use the above identity to write any one of the vectors $\mathbf{v}_i$ as a linear combination of the other two. Since no one of these vectors is a multiple of any other, they are pairwise linearly independent. Thus any two of these vectors form a basis for V.

7. (a) Clearly $\mathbf{w} = 3\mathbf{u}_1 - 7\mathbf{u}_2$, so the coordinate vector relative to $\{\mathbf{u}_1, \mathbf{u}_2\}$ is $(3, -7)$

(b) If $\mathbf{w} = a\mathbf{u}_1 + b\mathbf{u}_2$, then equating coordinates yields the system of equations

$$2a + 3b = 1$$

$$-4a + 8b = 1$$

This system has the solution $a = 5/28$, $b = 3/14$. Thus the desired coordinate vector is $(5/28, 3/14)$.

8. (a) If $\mathbf{v} = a\mathbf{v}_1 + b\mathbf{v}_2 + c\mathbf{v}_3$, then

$$a + 2b + 3c = 2$$

$$2b + 3c = -1$$

$$3c = 3$$

From the third equation, $c = 1$. Plugging this value into the second equation yields $b = -2$, and finally, the first equation yields $a = 3$. Thus the desired coordinate vector is $(3, -2, 1)$.

9. (b) If $\mathbf{p} = a\mathbf{p}_1 + b\mathbf{p}_2 + c\mathbf{p}_3$, then equating coefficients yields the system of equations

$$a + b = 2$$

$$a + c = -1$$

$$b + c = 1$$

This system has the solution $a = 0$, $b = 2$, and $c = -1$. Thus the desired coordinate vector is $(0, 2, -1)$.

11. The augmented matrix of the system reduces to

$$\begin{bmatrix} 1 & 0 & -1 & 0 \\ 0 & 1 & 0 & 0 \\ 0 & 0 & 0 & 0 \end{bmatrix}$$

Hence, $x_1 = s$, $x_2 = 0$, and $x_3 = s$. Thus the solution space is spanned by $(1,0,1)$ and has dimension 1.

14. If we reduce the augmented matrix to row-echelon form, we obtain

$$\begin{bmatrix} 1 & -3 & 1 & 0 \\ 0 & 0 & 0 & 0 \\ 0 & 0 & 0 & 0 \end{bmatrix}$$

Thus $x_1 = 3r - s$, $x_2 = r$, and $x_3 = s$, and the solution vector is

$$\begin{bmatrix} x_1 \\ x_2 \\ x_3 \end{bmatrix} = \begin{bmatrix} 3r - s \\ r \\ s \end{bmatrix} = \begin{bmatrix} 3 \\ 1 \\ 0 \end{bmatrix} r + \begin{bmatrix} -1 \\ 0 \\ 1 \end{bmatrix} s$$

Since $(3, 1, 0)$ and $(-1, 0, 1)$ are linearly independent, they form a basis for the solution space and the dimension of the solution space is 2.

15. Since the determinant of the system is not zero, the only solution is $x_1 = x_2 = x_3 = 0$. Hence there is no basis for the solution space and its dimension is zero.

17. **(a)** Any two linearly independent vectors in the plane form a basis. For instance, $(1, -1, -1)$ and $(0, 5, 2)$ are a basis because they satisfy the plane equation and neither is a multiple of the other.

(c) Any nonzero vector which lies on the line forms a basis. For instance, $(2, -1, 4)$ will work, as will any nonzero multiple of this vector.

(d) The vectors $(1, 1, 0)$ and $(0, 1, 1)$ form a basis because they are linearly independent and

$$a(1, 1, 0) + c(0, 1, 1) = (a, a + c, c)$$

19. This space is spanned by the vectors 0, x, x^2 and x^3. Only the last three vectors form a linearly independent triple. Thus the space has dimension 3.

20. **(a)** We consider the three linear systems

$$\begin{array}{rcrcccc} -k_1 & + & k_2 & = & 1 & 0 & 0 \\ 2k_1 & - & 2k_2 & = & 0 & 1 & 0 \\ 3k_1 & - & 2k_2 & = & 0 & 0 & 1 \end{array}$$

which give rise to the matrix

$$\begin{bmatrix} -1 & 1 & 1 & 0 & 0 \\ 2 & -2 & 0 & 1 & 0 \\ 3 & -2 & 0 & 0 & 1 \end{bmatrix}$$

A row-echelon form of the matrix is

$$\begin{bmatrix} 1 & -1 & -1 & 0 & 0 \\ 0 & 1 & 3 & 0 & 1 \\ 0 & 0 & 1 & 1/2 & 0 \end{bmatrix}$$

from which we conclude that e_3 is in the span of $\{v_1, v_2\}$, but e_1 and e_2 are not. Thus $\{v_1, v_2, e_1\}$ and $\{v_1, v_2, e_2\}$ are both bases for R^3.

21. We consider the linear system of equations $k_1v_1 + k_2v_2 + k_3e_1 + k_4e_2 + k_5e_3 + k_6e_4 = \mathbf{0}$ which give rise to the coefficient matrix

$$\begin{bmatrix} 1 & -3 & 1 & 0 & 0 & 0 \\ -4 & 8 & 0 & 1 & 0 & 0 \\ 2 & -4 & 0 & 0 & 1 & 0 \\ -3 & 6 & 0 & 0 & 0 & 1 \end{bmatrix}$$

A row-echelon form of the matrix is

$$\begin{bmatrix} 1 & -3 & 1 & 0 & 0 & 0 \\ 0 & 1 & -1 & -1/4 & 0 & 0 \\ 0 & 0 & 0 & 1 & 2 & 0 \\ 0 & 0 & 0 & 0 & 1 & 2/3 \end{bmatrix}$$

If we eliminate any two of the final four columns of this matrix, we obtain a 4×4 matrix. If the determinant of this matrix is zero, then the system of equations has a nontrivial solution, so the corresponding vectors are linearly dependent. Otherwise, they are linearly independent. If we include column 3 (which corresponds to e_1) in the determinant, its value is zero. Otherwise, it is not. Thus we may add any two of the vectors e_2, e_3, and e_4 to the set $\{v_1, v_2\}$ to obtain a basis for R^4.

22. Since $\{\mathbf{u}_1, \mathbf{u}_2, \mathbf{u}_3\}$ has the correct number of elements, we need only show that they are linearly independent. Let

$$a\mathbf{u}_1 + b\mathbf{u}_2 + c\mathbf{u}_3 = \mathbf{0}$$

Thus

$$a\mathbf{v}_1 + b(\mathbf{v}_1 + \mathbf{v}_2) + c(\mathbf{v}_1 + \mathbf{v}_2 + \mathbf{v}_3) = \mathbf{0}$$

or

$$(a + b + c)\mathbf{v}_1 + (b + c)\mathbf{v}_2 + c\mathbf{v}_3 = \mathbf{0}$$

Since $\{\mathbf{v}_1, \mathbf{v}_2, \mathbf{v}_3\}$ is a linearly independent set, the above equation implies that $a + b + c = b + c = c = 0$. Thus, $a = b = c = 0$ and $\{\mathbf{u}_1, \mathbf{u}_2, \mathbf{u}_3\}$ is also linearly independent.

23. **(a)** Note that the polynomials $1, x, x^2, \ldots, x^n$ form a set of $n + 1$ linearly independent functions in $F(-\infty, \infty)$.

(b) From Part (a), dim $F(-\infty, \infty) > n$ for any positive integer n. Thus $F(-\infty, \infty)$ is infinite-dimensional.

24. First notice that if $\mathbf{v}$ and $\mathbf{w}$ are vectors in V and a and b are scalars, then $(a\mathbf{v} + b\mathbf{w})_S = a(\mathbf{v})_S + b(\mathbf{w})_S$. This follows from the definition of coordinate vectors. Clearly, this result applies to any finite sum of vectors. Also notice that if $(\mathbf{v})_S = (\mathbf{0})_S$, then $\mathbf{v} = \mathbf{0}$. Why?

Now suppose that $k_1\mathbf{v}_1 + \cdots + k_r\mathbf{v}_r = \mathbf{0}$. Then

$$(k_1\mathbf{v}_1 + \cdots + k_r\mathbf{v}_r)_S = k_1(\mathbf{v}_1)_S + \cdots + k_r(\mathbf{v}_r)_S$$

$$= (\mathbf{0})_S$$

Conversely, if $k_1(\mathbf{v}_1)_S + \cdots + k_r(\mathbf{v}_r)_S = (\mathbf{0})_S$, then

$$(k_1\mathbf{v}_1 + \cdots + k_r\mathbf{v}_r)_S = (\mathbf{0})_S, \quad \text{or} \quad k_1\mathbf{v}_1 + \cdots + k_r\mathbf{v}_r = \mathbf{0}$$

Thus the vectors $\mathbf{v}_1, \ldots, \mathbf{v}_r$ are linearly independent in V if and only if the coordinate vectors $(\mathbf{v}_1)_S, \ldots, (\mathbf{v}_r)_S$ are linearly independent in R^n.

25. If every vector $\mathbf{v}$ in V can be written as a linear combination $\mathbf{v} = a_1\mathbf{v}_1 + \cdots + a_r\mathbf{v}_r$ of $\mathbf{v}_1, \ldots, \mathbf{v}_r$, then, as in Exercise 24, we have $(\mathbf{v})_S = a_1(\mathbf{v}_1)_S + \cdots + a_r(\mathbf{v}_r)_S$. Hence, the vectors $(\mathbf{v}_1)_S, \ldots, (\mathbf{v}_r)_S$ span a subspace of R^n. But since V is an n-dimensional space with, say, the basis $S = \{\mathbf{u}_1, \ldots, \mathbf{u}_n\}$, then if $\mathbf{u} = b_1\mathbf{u}_1 + \cdots + b_n\mathbf{u}_n$, we have $(\mathbf{u})_S = (b_1, \ldots, b_n)$; that is, every vector in R^n represents a vector in V. Hence $\{(\mathbf{v}_1)_S, \ldots, (\mathbf{v}_r)_S\}$ spans R^n.

Conversely, if $\{(\mathbf{v}_1)_S, \ldots, (\mathbf{v}_r)_S\}$ spans R^n, then for every vector $(b_1, \ldots, b_n)$ in R^n, there is an r-tuple $(a_1, \ldots, a_r)$ such that

$$(b_1, \ldots, b_n) = a_1(\mathbf{v}_1)_S + \cdots + a_r(\mathbf{v}_r)_S$$

$$= (a_1\mathbf{v}_1 + \cdots + a_r\mathbf{v}_r)_S$$

Thus $a_1\mathbf{v}_1 + \cdots + a_r\mathbf{v}_r = b_1\mathbf{u}_1 + \cdots + b_n\mathbf{u}_n$, so that every vector in V can be represented as a linear combination of $\mathbf{v}_1, \ldots, \mathbf{v}_r$.

26. **(a)** Let $\mathbf{v}_1$, $\mathbf{v}_2$, and $\mathbf{v}_3$ denote the vectors. Since $S = \{1, x, x^2\}$ is the standard basis for P_2, we have $(\mathbf{v}_1)_S = (-1, 1, -2)$, $(\mathbf{v}_2)_S = (3, 3, 6)$, and $(\mathbf{v}_3)_S = (9, 0, 0)$. Since $\{(-1, 1, -2), (3, 3, 6), (9, 0, 0)\}$ is a linearly independent set of three vectors in R^3, then it spans R^3. Thus, by Exercises 24 and 25, $\{\mathbf{v}_1, \mathbf{v}_2, \mathbf{v}_3\}$ is linearly independent and spans P_2. Hence it is a basis for P_2.

27. **(a)** It is clear from the picture that the x'-y' coordinates of $(1,1)$ are $(0, \sqrt{2})$.

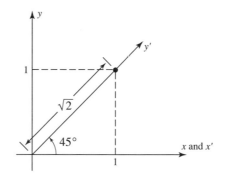

(d) Let (a, b) and (a', b') denote the coordinates of a point with respect to the x-y and x'-y' coordinate systems, respectively. If (a, b) is positioned as in Figure (1), then we have

$$a' = a - b \quad \text{and} \quad b' = \sqrt{2}b$$

27. (d)

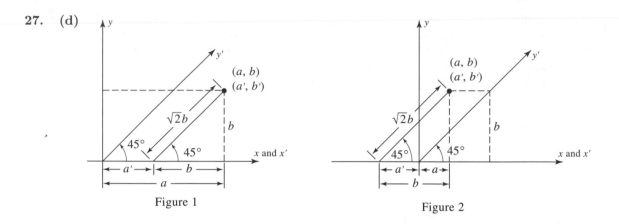

Figure 1   Figure 2

These formulas hold no matter where (a, b) lies in relation to the coordinate axes. Figure (2) shows another configuration, and you should draw similar pictures for all of the remaining cases.

29. See Theorem 5.4.7.

30. There is. Consider, for instance, the set of matrices

$$A = \begin{bmatrix} 0 & 1 \\ 1 & 1 \end{bmatrix} \qquad B = \begin{bmatrix} 1 & 0 \\ 1 & 1 \end{bmatrix}$$

$$C = \begin{bmatrix} 1 & 1 \\ 0 & 1 \end{bmatrix} \qquad \text{and} \qquad D = \begin{bmatrix} 1 & 1 \\ 1 & 0 \end{bmatrix}$$

Each of these matrices is clearly invertible. To show that they are linearly independent, consider the equation

$$aA + bB + cC + dD = \begin{bmatrix} 0 & 0 \\ 0 & 0 \end{bmatrix}$$

This implies that

$$\begin{bmatrix} 0 & 1 & 1 & 1 \\ 1 & 0 & 1 & 1 \\ 1 & 1 & 0 & 1 \\ 1 & 1 & 1 & 0 \end{bmatrix} \begin{bmatrix} a \\ b \\ c \\ d \end{bmatrix} = \begin{bmatrix} 0 \\ 0 \\ 0 \\ 0 \end{bmatrix}$$

The above 4×4 matrix is invertible, and hence $a = b = c = d = 0$ is the only solution. And since the set, A, B, C, and D consists of 4 linearly independent vectors, it forms a basis for M_{22}.

31. (b) The most general 2×2 symmetric matrix has the form $\begin{bmatrix} a & c \\ c & b \end{bmatrix}$. Hence, if $n = 2$, the matrices

$$\begin{bmatrix} 1 & 0 \\ 0 & 0 \end{bmatrix}, \begin{bmatrix} 0 & 0 \\ 0 & 1 \end{bmatrix}, \quad \text{and} \quad \begin{bmatrix} 0 & 1 \\ 1 & 0 \end{bmatrix}$$

form a basis and the dimension is therefore 3.

If $n = 3$, the matrices

$$\begin{bmatrix} 1 & 0 & 0 \\ 0 & 0 & 0 \\ 0 & 0 & 0 \end{bmatrix}, \begin{bmatrix} 0 & 0 & 0 \\ 0 & 1 & 0 \\ 0 & 0 & 0 \end{bmatrix}, \begin{bmatrix} 0 & 0 & 0 \\ 0 & 0 & 0 \\ 0 & 0 & 1 \end{bmatrix}$$

$$\begin{bmatrix} 0 & 1 & 0 \\ 1 & 0 & 0 \\ 0 & 0 & 0 \end{bmatrix}, \begin{bmatrix} 0 & 0 & 1 \\ 0 & 0 & 0 \\ 1 & 0 & 0 \end{bmatrix}, \quad \text{and} \quad \begin{bmatrix} 0 & 0 & 0 \\ 0 & 0 & 1 \\ 0 & 1 & 0 \end{bmatrix}$$

form a basis and the dimension is therefore 6.

In general, there are n^2 elements in an $n \times n$ matrix with $\dfrac{n^2 + n}{2} = \dfrac{n(n+1)}{2}$ on or above the main diagonal. Therefore the dimension of the subspace of $n \times n$ symmetric matrices is $n(n+1)/2$.

(c) Since there are n^2 elements in each $n \times n$ matrix, with n elements on the main diagonal, there are $(n^2 - n)/2$ elements above the diagonal. Thus, any triangular matrix can have at most $(n^2 - n)/2 + n = (n^2 + n)/2$ nonzero elements. The set of $n \times n$ matrices consisting of all zeros except for a 1 in each of these spots will form a basis for the space. Consequently, the space will have dimension $n(n+1)/2$.

32. (a) The set has 10 elements in a 9 dimensional space.

34. **(b)** The equation $x_1 + x_2 + \cdots + x_n = 0$ can be written as $x_1 = -x_2 - x_3 - \cdots - x_n$ where $x_2, x_3, \ldots, x_n$ can all be assigned arbitrary values. Thus, its solution space should have dimension $n - 1$. To see this, we can write

$$
\begin{bmatrix} x_1 \\ x_2 \\ x_3 \\ \vdots \\ x_n \end{bmatrix} = \begin{bmatrix} -x_2 - x_3 - \cdots - x_n \\ x_2 \\ x_3 \\ \vdots \\ x_n \end{bmatrix}
$$

$$
= x_2 \begin{bmatrix} -1 \\ 1 \\ 0 \\ 0 \\ \vdots \\ 0 \end{bmatrix} + x_3 \begin{bmatrix} -1 \\ 0 \\ 1 \\ 0 \\ \vdots \\ 0 \end{bmatrix} + \cdots + x_n \begin{bmatrix} -1 \\ 0 \\ 0 \\ 0 \\ \vdots \\ 1 \end{bmatrix}
$$

The $n - 1$ vectors in the above equation are linearly independent, so the vectors do form a basis for the solution space.

35. **(a)** If $p_1(1) = p_2(1) = 0$, then $(p_1 + p_2)(1) = p_1(1) + p_2(1) = 0$. Also, $kp_1(1) = 0$. Hence W is closed under both vector addition and scalar multiplication.

(b) We are looking at the set W of polynomials $ax^2 + bx + c$ in P_2 where $a + b + c = 0$ or $c = -(a + b)$. Thus it would appear that the dimension of W is 2.

(c) The polynomial $p(x) = ax^2 + bx + c$ will be in the subspace W if and only if $p(1) = a + b + c = 0$, or $c = -a - b$. Thus,

$$
\begin{bmatrix} a \\ b \\ c \end{bmatrix} = \begin{bmatrix} a \\ b \\ -a - b \end{bmatrix} = a \begin{bmatrix} 1 \\ 0 \\ -1 \end{bmatrix} + b \begin{bmatrix} 0 \\ 1 \\ -1 \end{bmatrix}
$$

where a and b are arbitrary. The vectors $\begin{bmatrix} 1 \\ 0 \\ -1 \end{bmatrix}$ and $\begin{bmatrix} 0 \\ 1 \\ -1 \end{bmatrix}$ are linearly independent and hence the polynomials $x^2 - 1$ and $x - 1$ form a basis for W.

EXERCISE SET 5.5

3. **(b)** Since the equation $A\mathbf{x} = \mathbf{b}$ has no solution, $\mathbf{b}$ is not in the column space of A.

(c) Since $A \begin{bmatrix} 1 \\ -3 \\ 1 \end{bmatrix} = \mathbf{b}$, we have $\mathbf{b} = \mathbf{c}_1 - 3\mathbf{c}_2 + \mathbf{c}_3$.

(d) Since $A \begin{bmatrix} 1 \\ t-1 \\ t \end{bmatrix} = \mathbf{b}$, we have $\mathbf{b} = \mathbf{c}_1 + (t-1)\mathbf{c}_2 + t\mathbf{c}_3$ for all real numbers t.

5. **(a)** The general solution is $x_1 = 1 + 3t$, $x_2 = t$. Its vector form is

$$\begin{bmatrix} 1 \\ 0 \end{bmatrix} + t \begin{bmatrix} 3 \\ 1 \end{bmatrix}$$

Thus the vector form of the general solution to $A\mathbf{x} = \mathbf{0}$ is

$$t \begin{bmatrix} 3 \\ 1 \end{bmatrix}$$

(c) The general solution is $x_1 = -1 + 2r - s - 2t$, $x_2 = r$, $x_3 = s$, $x_4 = t$. Its vector form is

$$\begin{bmatrix} -1 \\ 0 \\ 0 \\ 0 \end{bmatrix} + r \begin{bmatrix} 2 \\ 1 \\ 0 \\ 0 \end{bmatrix} + s \begin{bmatrix} -1 \\ 0 \\ 1 \\ 0 \end{bmatrix} + t \begin{bmatrix} -2 \\ 0 \\ 0 \\ 1 \end{bmatrix}$$

Thus the vector form of the general solution to $A\mathbf{x} = \mathbf{0}$ is

$$r \begin{bmatrix} 2 \\ 1 \\ 0 \\ 0 \end{bmatrix} + s \begin{bmatrix} -1 \\ 0 \\ 1 \\ 0 \end{bmatrix} + t \begin{bmatrix} -2 \\ 0 \\ 0 \\ 1 \end{bmatrix}$$

6. **(a)** Since the reduced row-echelon form of A is

$$\begin{bmatrix} 1 & 0 & -16 \\ 0 & 1 & -19 \\ 0 & 0 & 0 \end{bmatrix}$$

the solution to the equation $A\mathbf{x} = \mathbf{0}$ is $x_1 = 16t$, $x_2 = 19t$, $x_3 = t$. Thus

$$\begin{bmatrix} 16 \\ 19 \\ 1 \end{bmatrix}$$

is a basis for the nullspace.

(c) Since the reduced row-echelon form of A is

$$\begin{bmatrix} 1 & 0 & 1 & -2/7 \\ 0 & 1 & 1 & 4/7 \\ 0 & 0 & 0 & 0 \end{bmatrix}$$

one solution to the equation $A\mathbf{x} = \mathbf{0}$ is $x_1 = -s + 2t$, $x_2 = -s - 4t$, $x_3 = s$, $x_4 = 7t$. Thus the set of vectors

$$\begin{bmatrix} -1 \\ -1 \\ 1 \\ 0 \end{bmatrix} \quad \text{and} \quad \begin{bmatrix} 2 \\ -4 \\ 0 \\ 7 \end{bmatrix}$$

is a basis for the nullspace.

(e) The reduced row-echelon form of A is

$$\begin{bmatrix} 1 & 0 & 0 & 2 & 4/3 \\ 0 & 1 & 0 & 0 & -1/6 \\ 0 & 0 & 1 & 0 & -5/12 \\ 0 & 0 & 0 & 0 & 0 \\ 0 & 0 & 0 & 0 & 0 \end{bmatrix}$$

One solution to the equation $A\mathbf{x} = \mathbf{0}$ is $x_1 = -2s - 16t$, $x_2 = 2t$, $x_3 = 5t$, $x_4 = s$, and $x_5 = 12t$. Hence the set of vectors

$$\begin{bmatrix} -2 \\ 0 \\ 0 \\ 1 \\ 0 \end{bmatrix} \quad \text{and} \quad \begin{bmatrix} -16 \\ 2 \\ 5 \\ 0 \\ 12 \end{bmatrix}$$

is a basis for the nullspace of A.

8. (a) From a row-echelon form of A, we have that the vectors $(1, -1, 3)$ and $(0, 1, -19)$ are a basis for the row space of A.

(c) From a row-echelon form of A, we have that the vectors $(1, 4, 5, 2)$ and $(0, 1, 1, 4/7)$ are a basis for the row space of A.

9. (a) One row-echelon form of A^T is

$$\begin{bmatrix} 1 & 5 & 7 \\ 0 & 1 & 1 \\ 0 & 0 & 0 \end{bmatrix}$$

Thus a basis for the column space of A is

$$\begin{bmatrix} 1 \\ 5 \\ 7 \end{bmatrix} \quad \text{and} \quad \begin{bmatrix} 0 \\ 1 \\ 1 \end{bmatrix}$$

9. **(c)** One row-echelon form of A^T is

$$\begin{bmatrix} 1 & 2 & -1 \\ 0 & 1 & -1 \\ 0 & 0 & 0 \\ 0 & 0 & 0 \end{bmatrix}$$

Thus a basis for the column space of A is

$$\begin{bmatrix} 1 \\ 2 \\ -1 \end{bmatrix} \quad \text{and} \quad \begin{bmatrix} 0 \\ 1 \\ -1 \end{bmatrix}$$

10. **(a)** Since, by 8(a), the row space of A has dimension 2, any two linearly independent row vectors of A will form a basis for the row space. Because no row of A is a multiple of another, any two rows will do. In particular, the first two rows form a basis and Row 3 = Row 2 + 2(Row 1).

(c) Refer to 8(c) and use the solution to 10(a), above. In particular, the first two rows form a basis and Row 3 = Row 1 − Row 2.

(e) Let $\mathbf{r}_1, \mathbf{r}_2, \cdots, \mathbf{r}_5$ denote the rows of A. If we observe, for instance, that $\mathbf{r}_1 = -\mathbf{r}_3 + \mathbf{r}_4$ and that $\mathbf{r}_2 = 2\mathbf{r}_1 + \mathbf{r}_5$, then we see that $\{\mathbf{r}_3, \mathbf{r}_4, \mathbf{r}_5\}$ spans the row space. Since the dimension of this space is 3 (see the solution to Exercise 6(e)), the set forms a basis.

For those who don't wish to rely on insight, set $a\mathbf{r}_1 + b\mathbf{r}_2 + c\mathbf{r}_3 + d\mathbf{r}_4 + e\mathbf{r}_5 = \mathbf{0}$ and solve the resulting homogeneous system of equations by finding the reduced row-echelon form for A^T. This yields

$$a = -s + 2t$$
$$b = -t$$
$$c = -s$$
$$d = s$$
$$e = t$$

so that $(-s + 2t)\mathbf{r}_1 - t\mathbf{r}_2 - s\mathbf{r}_3 + s\mathbf{r}_4 + t\mathbf{r}_5 = \mathbf{0}$, or

$$s(-\mathbf{r}_1 - \mathbf{r}_3 + \mathbf{r}_4) + t(2\mathbf{r}_1 - \mathbf{r}_2 + \mathbf{r}_5) = \mathbf{0}$$

Since this equation must hold for all values of s and t, we have

$$\mathbf{r}_1 = -\mathbf{r}_3 + \mathbf{r}_4 \quad \text{and} \quad \mathbf{r}_2 = 2\mathbf{r}_1 + \mathbf{r}_5$$

which is the result obtained above.

11. **(a)** The space spanned by these vectors is the row space of the matrix

$$\begin{bmatrix} 1 & 1 & -4 & -3 \\ 2 & 0 & 2 & -2 \\ 2 & -1 & 3 & 2 \end{bmatrix}$$

One row-echelon form of the above matrix is

$$\begin{bmatrix} 1 & 1 & -4 & -3 \\ 0 & 1 & -5 & -2 \\ 0 & 0 & 1 & -1/2 \end{bmatrix}$$

and the reduced row-echelon form is

$$\begin{bmatrix} 1 & 0 & 0 & -1/2 \\ 0 & 1 & 0 & -9/2 \\ 0 & 0 & 1 & -1/2 \end{bmatrix}$$

Thus $\{(1, 1, -4, -3), (0, 1, -5, -2), (0, 0, 1, -1/2)\}$ is one basis. Another basis is $\{(1, 0, 0, -1/2), (0, 1, 0, -9/2), (0, 0, 1, -1/2)\}$.

12. **(a)** If we solve the vector equation

$$(*) \qquad\qquad a\mathbf{v}_1 + b\mathbf{v}_2 + c\mathbf{v}_3 + d\mathbf{v}_4 = 0$$

we obtain the homogeneous system

$$a - 3b - c - 5d = 0$$
$$3b + 3c + 3d = 0$$
$$a + 7b + 9c + 5d = 0$$
$$a + b + 3c - d = 0$$

The reduced row-echelon form of the augmented matrix is

$$\begin{bmatrix} 1 & 0 & 2 & -2 & 0 \\ 0 & 1 & 1 & 1 & 0 \\ 0 & 0 & 0 & 0 & 0 \\ 0 & 0 & 0 & 0 & 0 \end{bmatrix}$$

Thus $\{\mathbf{v}_1, \mathbf{v}_2\}$ forms a basis for the space. The solution is $a = -2s + 2t$, $b = -s - t$, $c = s$, $d = t$. This yields

$$(-2s + 2t)\mathbf{v}_1 + (-s - t)\mathbf{v}_2 + s\mathbf{v}_3 + t\mathbf{v}_4 = \mathbf{0}$$

or

$$s(-2\mathbf{v}_1 - \mathbf{v}_2 + \mathbf{v}_3) + t(2\mathbf{v}_1 - \mathbf{v}_2 + \mathbf{v}_4) = \mathbf{0}$$

Since s and t are arbitrary, set $s = 1$, $t = 0$ and then $s = 0$, $t = 1$ to obtain the dependency equations

$$-2\mathbf{v}_1 - \mathbf{v}_2 + \mathbf{v}_3 = \mathbf{0}$$

$$2\mathbf{v}_1 - \mathbf{v}_2 + \mathbf{v}_4 = \mathbf{0}$$

Thus

$$\mathbf{v}_3 = 2\mathbf{v}_1 + \mathbf{v}_2$$

and

$$\mathbf{v}_4 = -2\mathbf{v}_1 + \mathbf{v}_2$$

12. (c) If we solve the vector equation

$$(*) \qquad\qquad a\mathbf{v}_1 + b\mathbf{v}_2 + c\mathbf{v}_3 + d\mathbf{v}_4 + e\mathbf{v}_5 = \mathbf{0}$$

we obtain the homogeneous system

$$a - 2b + 4c \qquad - 7e = 0$$

$$-a + 3b - 5c + 4d + 18e = 0$$

$$5a + b + 9c + 2d + 2e = 0$$

$$2a \qquad + 4c - 3d - 8e = 0$$

The reduced row-echelon form of the augmented matrix is

$$\begin{bmatrix} 1 & 0 & 2 & 0 & -1 & 0 \\ 0 & 1 & -1 & 0 & 3 & 0 \\ 0 & 0 & 0 & 1 & 2 & 0 \\ 0 & 0 & 0 & 0 & 0 & 0 \end{bmatrix}$$

This tells us that $\{\mathbf{v}_1, \mathbf{v}_2, \mathbf{v}_4\}$ is the desired basis. The solution is $a = -2s + t$, $b = s - 3t$, $c = s$, $d = -2t$, and $e = t$. This yields

$$(-2s + t)\mathbf{v}_1 + (s - 3t)\mathbf{v}_2 + s\mathbf{v}_3 - 2t\mathbf{v}_4 + t\mathbf{v}_5 = \mathbf{0}$$

or

$$s(-2\mathbf{v}_1 + \mathbf{v}_2 + \mathbf{v}_3) + t(\mathbf{v}_1 - 3\mathbf{v}_2 - 2\mathbf{v}_4 + \mathbf{v}_5) = \mathbf{0}$$

Since s and t are arbitrary, set $s = 1$, $t = 0$ and then $s = 0$, $t = 1$ to obtain the dependency equations

$$-2\mathbf{v}_1 + \mathbf{v}_2 + \mathbf{v}_3 = \mathbf{0}$$

$$\mathbf{v}_1 - 3\mathbf{v}_2 - 2\mathbf{v}_4 + \mathbf{v}_5 = \mathbf{0}$$

Thus

$$\mathbf{v}_3 = 2\mathbf{v}_1 - \mathbf{v}_2$$

and

$$\mathbf{v}_5 = -\mathbf{v}_1 + 3\mathbf{v}_2 + 2\mathbf{v}_4$$

13. Let A be an $n \times n$ invertible matrix. Since A^T is also invertible, it is row equivalent to I_n. It is clear that the column vectors of I_n are linearly independent. Hence, by virtue of Theorem 5.5.5, the column vectors of A^T, which are just the row vectors of A, are also linearly independent. Therefore the rows of A form a set of n linearly independent vectors in R^n, and consequently form a basis for R^n.

15. (a) True. Since premultiplication by an elementary matrix is equivalent to an elementary row operation, the result follows from Theorem 5.5.3.

15. **(c)** False. For instance, let $A = \begin{bmatrix} 1 & 1 \\ 2 & 2 \end{bmatrix}$ and $EA = \begin{bmatrix} 1 & 1 \\ 0 & 0 \end{bmatrix}$. Then the column space of A is $\text{span}\left(\begin{bmatrix} 1 \\ 2 \end{bmatrix} \right)$ and the column space of EA is $\text{span}\left(\begin{bmatrix} 1 \\ 0 \end{bmatrix} \right)$. These are not the same spaces.

(d) True by Theorem 5.5.1

(e) False. The row space of an invertible $n \times n$ matrix is the same as the row space of I_n, which is R^n. The nullspace of an invertible matrix is just the zero vector.

16. **(a)** The matrices will all have the form $\begin{bmatrix} 3s - 5s \\ 3t - 5t \end{bmatrix} = s \begin{bmatrix} 3 & -5 \\ 0 & 0 \end{bmatrix} + t \begin{bmatrix} 0 & 0 \\ 3 & -5 \end{bmatrix}$ where s and t are any real numbers.

(b) Since A and B are invertible, their nullspaces are the origin. The nullspace of C is the line $3x + y = 0$. The nullspace of D is the entire xy-plane.

17. Let $A = [1 \ 1 \ 1]$. Then $A\mathbf{x} = [1]$ has the particular solution $[1 \ 0 \ 0]^T$ and $A\mathbf{x} = [0]$ has the general solution $[s \ t \ -s - t]^T$. Thus the general solution can be written as

$$\begin{bmatrix} x_1 \\ x_2 \\ x_3 \end{bmatrix} = \begin{bmatrix} 1 \\ 0 \\ 0 \end{bmatrix} + s \begin{bmatrix} 1 \\ 0 \\ -1 \end{bmatrix} + t \begin{bmatrix} 0 \\ 1 \\ -1 \end{bmatrix}$$

18. Theorem: If A and B are $n \times n$ matrices and A is invertible, then the row space of AB is the row space of B.

Proof: If A is invertible, then there exist elementary matrices $E_1, E_2, \ldots, E_k$ such that

$$A = E_1 E_2 \ldots E_k I_n$$

or

$$AB = E_1 E_2 \ldots E_k B$$

Thus, Theorem 5.5.4 guarantees that AB and B will have the same row spaces.

EXERCISE SET 5.6

2. **(a)** The reduced row-echelon form for A is

$$\begin{bmatrix} 1 & 0 & -16 \\ 0 & 1 & -19 \\ 0 & 0 & 0 \end{bmatrix}$$

Thus rank$(A) = 2$. The solution to $A\mathbf{x} = 0$ is $x = 16t$, $y = 19t$, $z = t$, so that the nullity is one. There are three columns, so we have $2 + 1 = 3$.

(c) The reduced row-echelon form for A is

$$\begin{bmatrix} 1 & 0 & 1 & -2/7 \\ 0 & 1 & 1 & 4/7 \\ 0 & 0 & 0 & 0 \end{bmatrix}$$

Thus rank$(A) = 2$. The null space will have dimension two since the solution to $A\mathbf{x} = \mathbf{0}$ has two parameters. There are four columns, so we have $2 + 2 = 4$.

4. Recall that rank(A) is the dimension of both the row and column spaces of A. Use the Dimension Theorem to find the dimensions of the nullspace of A and of A^T, recalling that if A is $m \times n$, then A^T is $n \times m$, or just refer to the chart in the text.

7. Use Theorems 5.6.5 and 5.6.7.

(a) The system is consistent because the two ranks are equal. Since $n = r = 3$, $n - r = 0$ and therefore the number of parameters is 0.

(b) The system is inconsistent because the two ranks are not equal.

7. **(d)** The system is consistent because the two ranks are equal. Here $n = 9$ and $r = 2$, so that $n - r = 7$ parameters will appear in the solution.

(f) Since the ranks are equal, the system is consistent. However A must be the zero matrix, so the system gives no information at all about its solution. This is reflected in the fact that $n - r = 4 - 0 = 4$, so that there will be 4 parameters in the solution for the 4 variables.

9. The system is of the form $A\mathbf{x} = \mathbf{b}$ where $\text{rank}(A) = 2$. Therefore it will be consistent if and only if $\text{rank}([A|\mathbf{b}]) = 2$. Since $[A|\mathbf{b}]$ reduces to

$$\begin{bmatrix} 1 & -3 & b_1 \\ 0 & 1 & b_2 - b_1 \\ 0 & 0 & b_3 - 4b_2 + 3b_1 \\ 0 & 0 & b_4 + b_2 - 2b_1 \\ 0 & 0 & b_5 - 8b_2 + 7b_1 \end{bmatrix}$$

the system will be consistent if and only if $b_3 = 4b_2 - 3b_1$, $b_4 = -b_2 + 2b_1$, and $b_5 = 8b_2 - 7b_1$, where b_1 and b_2 can assume any values.

10. Suppose that A has rank 2. Then two of its column vectors are linearly independent. Thus, by Theorem 5.6.9, at least one of the 2×2 submatrices has nonzero determinant.

Conversely, if at least one of the determinants of the 2×2 submatrices is nonzero, then, by Theorem 5.6.9, at least two of the column vectors must be linearly independent. Thus the rank of A must be at least 2. But since the dimension of the row space of A is at most 2, A has rank at most 2. Thus, the rank of A is exactly 2.

11. If the nullspace of A is a line through the origin, then it has the form $x = at$, $y = bt$, $z = ct$ where t is the only parameter. Thus nullity$(A) = 3- \text{rank}(A) = 1$. That is, the row and column spaces of A have dimension 2, so neither space can be a line. Why?

12. **(a)** If we attempt to reduce A to row-echelon form, we find that

$$A \to \begin{bmatrix} 1 & 1 & t \\ 0 & t-1 & 1-t \\ t & 1 & 1 \end{bmatrix} \to \begin{bmatrix} 1 & 1 & t \\ 0 & t-1 & 1-t \\ 0 & 1-t & 1-t^2 \end{bmatrix} \quad \text{if } t \neq 0$$

$$\to \begin{bmatrix} 1 & 1 & t \\ 0 & 1 & -1 \\ 0 & -1 & -(1+t) \end{bmatrix} \quad \text{if } t \neq 1$$

$$\to \begin{bmatrix} 1 & 1 & t \\ 0 & 1 & -1 \\ 0 & 0 & t+2 \end{bmatrix} \to \begin{bmatrix} 1 & 1 & t \\ 0 & 1 & -1 \\ 0 & 0 & 1 \end{bmatrix} \quad \text{if } t \neq -2$$

Thus $\operatorname{rank}(A) = 3$ if $t \neq 0, 1, -2$. If $t = 0$, $\operatorname{rank}(A) = 3$ by direct computation. If $t = 1$, $\operatorname{rank}(A) = 1$ by inspection, and if $t = -2$, $\operatorname{rank}(A) = 2$ by the above reduction.

13. Call the matrix A. If $r = 2$ and $s = 1$, then clearly $\operatorname{rank}(A) = 2$. Otherwise, either $r - 2$ or $s - 1 \neq 0$ and $\operatorname{rank}(A) = 3$. Rank$(A)$ can never be 1.

14. Call the matrix A and note that $\operatorname{rank}(A)$ is either 1 or 2. Why? By Exercise 10, $\operatorname{rank}(A) = 1$ if and only if

$$\begin{vmatrix} x & y \\ 1 & x \end{vmatrix} = 0, \quad \begin{vmatrix} x & z \\ 1 & y \end{vmatrix} = 0, \quad \text{and} \quad \begin{vmatrix} y & z \\ x & y \end{vmatrix} = 0$$

Thus we must have $x^2 - y = xy - z = y^2 - xz = 0$. If we let $x = t$, the result follows.

16. **(a)** The column space of the matrix

$$\begin{bmatrix} 1 & 0 & 1 \\ 0 & 1 & 0 \\ 0 & 0 & 0 \end{bmatrix}$$

is the xy-plane.

16. **(b)** The nullspace is the line $x = t$, $y = 0$, $z = -t$.

(c) The row space is the plane $x - z = 0$.

(d) In general, if the column space of a 3×3 matrix is a plane through the origin then the nullspace will be a line through the origin and the row space will be a plane through the origin. This follows from the fact that if the column space is a plane, the matrix must have rank 2 and therefore nullity 1. The one dimensional subspaces of R^3 are lines through the origin and the two dimensional subspaces are planes through the origin.

Similarly, if the column space represents a line through the origin, then so does the row space. In this case, the nullspace will represent a plane through the origin.

17. **(a)** False. Let $A = \begin{bmatrix} 1 & 0 & 0 \\ 0 & 1 & 0 \end{bmatrix}$.

(c) True. If A were an $m \times n$ matrix where, say, $m > n$, then it would have m rows, each of which would be a vector in R_n. Thus, by Theorem 5.4.2, they would form a linearly dependent set.

18. **(a)** Since the row rank equals the column rank and since A has 3 rows and more than 3 columns, its maximum rank is 3. Hence, the number of leading 1's in its reduced row-echelon form is less than or equal to 3.

(b) Since nullity $(A) \leq 5$, there could be as many as 5 parameters in the general solution of $A\mathbf{x} = \mathbf{0}$. The maximum of 5 would occur if A were the zero matrix.

(c) Since A has 3 columns and more than 3 rows, its maximum rank is 3. Hence the maximum number of leading 1's in the reduced row-echelon form of A is 3.

(d) Since nullity $(A) \leq 3$, there could be as many as 3 parameters in the general solution to $A\mathbf{x} = \mathbf{0}$.

SUPPLEMENTARY EXERCISES 5

1. **(b)** The augmented matrix of this system reduces to

$$\begin{bmatrix} 2 & -3 & 1 & 0 \\ 0 & 0 & 0 & 0 \\ 0 & 0 & 0 & 0 \end{bmatrix}$$

Therefore, the solution space is a plane with equation $2x - 3y + z = 0$

(c) The solution is $x = 2t$, $y = t$, $z = 0$, which is a line.

2. Let A be the coefficient matrix. Since $\det(A) = -(1-s)^2(2+s)$, the solution space is the origin unless $s = 1$ or $s = -2$. If $s = 1$, the solution space is the plane $x_1 + x_2 + x_3 = 0$. If $s = -2$, the solution space is the line $x_1 = t, x_2 = t, x_3 = t$.

   Alternative Solution: Let A be the coefficient matrix. We can use the Dimension Theorem and the result of Exercise 12(a) of Section 5.6 to solve this problem. Recall that rank$(A) = 1$ if $s = 1$, rank$(A) = 2$ if $s = -2$, and rank$(A) = 3$ for all other values of s. Hence nullity$(A) = 2$ if $s = 1$, nullity$(A) = 1$ if $s = -2$, and nullity$(A) = 0$ for all other values of s. Thus, if $s = 1$, then the solution space is a two-dimensional subspace of R^3, i.e., a plane through the origin. If $s = -2$, then the solution space is a one-dimensional subspace of R^3, i.e., a line through the origin. If $s \neq 1$ and $s \neq -2$, then the solution space is the zero-dimensional subspace of R^3, i.e., $\{\mathbf{0}\}$.

4. **(a)** The identities

$$\sin(x + \theta) = \cos\theta\sin x + \sin\theta\cos x$$

$$\cos(x + \theta) = \cos\theta\cos x - \sin\theta\sin x$$

hold for all values of x and θ. Hence

$$\mathbf{f}_1 = (\cos \theta)\mathbf{f} + (\sin \theta)\mathbf{g}$$
$$(*)$$
$$\mathbf{g}_1 = (-\sin \theta)\mathbf{f} + (\cos \theta)\mathbf{g}$$

That is, $\mathbf{f}_1$ and $\mathbf{g}_1$ are linear combinations of $\mathbf{f}$ and $\mathbf{g}$ and therefore belong to W.

4.   **(b)** If we solve the system $(*)$ for $\mathbf{f}$ and $\mathbf{g}$, we obtain

$$\mathbf{f} = (\cos \theta)\mathbf{f}_1 + (-\sin \theta)\mathbf{g}_1$$
$$\mathbf{g} = (\sin \theta)\mathbf{f}_1 + (\cos \theta)\mathbf{g}_1$$

Hence, any linear combination of $\mathbf{f}$ and $\mathbf{g}$ is also a linear combination of $\mathbf{f}_1$ and $\mathbf{g}_1$ and thus $\mathbf{f}_1$ and $\mathbf{g}_1$ span W. Since the dimension of W is 2 (it is spanned by 2 linearly independent vectors), Theorem 5.4.6(b) implies that $\mathbf{f}_1$ and $\mathbf{g}_1$ form a basis for W.

5.   **(a)** We look for constants a, b, and c such that $\mathbf{v} = a\mathbf{v}_1 + b\mathbf{v}_2 + c\mathbf{v}_3$, or

$$a + 3b + 2c = 1$$
$$-a \quad\ \ + \ c = 1$$

This system has the solution

$$a = t - 1 \qquad b = \frac{2}{3} - t \qquad c = t$$

where t is arbitrary. If we set $t = 0$ and $t = 1$, we obtain $\mathbf{v} = (-1)\mathbf{v}_1 + (2/3)\mathbf{v}_2$ and $\mathbf{v} = (-1/3)\mathbf{v}_2 + \mathbf{v}_3$, respectively. There are infinitely many other possibilities.

   (b) Since $\mathbf{v}_1$, $\mathbf{v}_2$, and $\mathbf{v}_3$ all belong to R^2 and $\dim(R^2) = 2$, it follows from Theorem 5.4.2 that these three vectors do not form a basis for R^2. Hence, Theorem 5.4.1 does not apply.

6.   Suppose that there are constants $c_1, \ldots, c_n$, not all zero, such that $c_1 A\mathbf{v}_1 + \cdots + c_n A\mathbf{v}_n = \mathbf{0}$. Then

$$A(c_1\mathbf{v}_1 + \cdots + c_n\mathbf{v}_n) = \mathbf{0}$$

Since the vectors $\mathbf{v}_1, \ldots, \mathbf{v}_n$ are linearly independent, the $n \times 1$ matrix $c_1\mathbf{v}_1 + \cdots + c_n\mathbf{v}_n$ cannot equal $\mathbf{0}$. Thus the equation $A\mathbf{x} = \mathbf{0}$ has a non-trivial solution, and so A is not invertible. Therefore, by Theorem 1.5.3, A is invertible if and only if $A\mathbf{v}_1, A\mathbf{v}_2, \ldots, A\mathbf{v}_n$ are linearly independent.

7.   Consider the polynomials x and $x + 1$ in P_1. Verify that these polynomials form a basis for P_1.

8.   **(c)**   Since the odd numbered rows are all repeats of Row 1 and the even numbered rows are all repeats of Row 2, while Rows 1 and 2 are linearly independent, an $n \times n$ checker board matrix has rank 2 whenever $n \geq 2$. Since the nullity is n minus the rank, we have nullity $= n - 2$.

10.   **(a)**   If p belongs to the set, then it contains only even powers of x. Since this set is closed under polynomial addition and scalar multiplication (Why?), it is a subspace of P_n. One basis is the set $\{1, x^2, x^4, \ldots, x^{2m}\}$ where $2m = n$ if n is even and $2m = n - 1$ if n is odd.

   (b)   If p belongs to this set, then its constant term must be zero. Since this set is closed under polynomial addition and scalar multiplication (Why?), it is a subspace of P_n. One basis is the set $\{x, x^2, \ldots, x^n\}$.

12.   **(a)**   Since $\begin{vmatrix} 1 & 0 \\ 2 & -1 \end{vmatrix} = -1 \neq 0$, the rank is 2.

   (b)   Since all three 2×2 subdeterminants are zero, the rank is 1.

   (c)   Since the determinant of the matrix is zero, its rank is less than 3. Since $\begin{vmatrix} 1 & 0 \\ 2 & -1 \end{vmatrix} = -1 \neq 0$, the rank is 2.

   (d)   Since the determinant of the 3×3 submatrix obtained by deleting the last column is $30 \neq 0$, the rank of the matrix is 3.

13.   Call the matrix A. Since the determinant of every 5×5, 4×4, and 3×3 submatrix is zero, $\text{rank}(A) \leq 2$. Since

$$\det\left(\begin{bmatrix} 0 & a_{i6} \\ a_{5j} & a_{56} \end{bmatrix}\right) = -a_{5j}a_{i6}$$

for $i = 1, 2, \ldots, 4$ and $j = 1, \ldots, 5$, then $\text{rank}(A) = 2$ if any of these determinants is nonzero. Otherwise, if any of the numbers $a_{ij} \neq 0$, then $\text{rank}(A) = 1$ and if $a_{ij} = 0$ for all i and j, then $\text{rank}(A) = 0$.

14. **(b)** Let $S = \{\mathbf{v}_1, \ldots, \mathbf{v}_n\}$ and let $\mathbf{u} = u_1\mathbf{v}_1 + \cdots + u_n\mathbf{v}_n$. Thus $(\mathbf{u})_S = (u_1, \ldots, u_n)$. We have

$$k\mathbf{u} = ku_1\mathbf{v}_1 + \cdots + ku_n\mathbf{v}_n$$

so that $(k\mathbf{u})_S = (ku_1, \ldots, ku_n) = k(u_1, \ldots, u_n)$. Therefore $(k\mathbf{u})_S = k(\mathbf{u})_S$.

EXERCISE SET 6.1

1. **(c)** Since $\mathbf{v} + \mathbf{w} = (3, 11)$, we have

$$\langle \mathbf{u}, \mathbf{v} + \mathbf{w} \rangle = 3(3) + (-2)(11) = -13$$

On the other hand,

$$\langle \mathbf{u}, \mathbf{v} \rangle = 3(4) + (-2)(5) = 2$$

and

$$\langle \mathbf{u}, \mathbf{w} \rangle = 3(-1) + (-2)(6) = -15$$

(d) Since $k\mathbf{u} = (-12, 8)$ and $k\mathbf{v} = (-16, -20)$, we have

$$\langle k\mathbf{u}, \mathbf{v} \rangle = (-12)(4) + (8)(5) = -8$$

and

$$\langle \mathbf{u}, k\mathbf{v} \rangle = 3(-16) + (-2)(-20) = -8$$

Since $\langle \mathbf{u}, \mathbf{v} \rangle = 2$, $k\langle \mathbf{u}, \mathbf{v} \rangle = -8$.

2. **(c)** Since $\mathbf{v} + \mathbf{w} = (3, 11)$, we have

$$\langle \mathbf{u}, \mathbf{v} + \mathbf{w} \rangle = 4(3)(3) + 5(-2)(11) = -74$$

On the other hand,

$$\langle \mathbf{u}, \mathbf{v} \rangle = 4(3)(4) + 5(-2)(5) = -2$$

and

$$\langle \mathbf{u}, \mathbf{w} \rangle = 4(3)(-1) + 5(-2)(6) = -72$$

3. **(a)** $\langle \mathbf{u}, \mathbf{v} \rangle = 3(-1) - 2(3) + 4(1) + 8(1) = 3$

4. **(a)** $\langle \mathbf{p}, \mathbf{q} \rangle = (-2)(4) + 1(0) + 3(-7) = -29$

5. **(a)** By Formula (4),

$$\langle \mathbf{u}, \mathbf{v} \rangle = \begin{bmatrix} v_1 & v_2 \end{bmatrix} \begin{bmatrix} 3 & 0 \\ 0 & 2 \end{bmatrix} \begin{bmatrix} 3 & 0 \\ 0 & 2 \end{bmatrix} \begin{bmatrix} u_1 \\ u_2 \end{bmatrix}$$

$$= \begin{bmatrix} v_1 & v_2 \end{bmatrix} \begin{bmatrix} 9 & 0 \\ 0 & 4 \end{bmatrix} \begin{bmatrix} u_1 \\ u_2 \end{bmatrix}$$

$$= \begin{bmatrix} 9v_1 & 4v_2 \end{bmatrix} \begin{bmatrix} u_1 \\ u_2 \end{bmatrix}$$

$$= 9u_1 v_1 + 4u_2 v_2$$

(b) We have $\langle \mathbf{u}, \mathbf{v} \rangle = 9(-3)(1) + 4(2)(7) = 29$.

6. **(a)** Since

$$A^T A = \begin{bmatrix} 5 & -1 \\ -1 & 10 \end{bmatrix}$$

then

$$\begin{bmatrix} v_1 & v_2 \end{bmatrix} \begin{bmatrix} 5 & -1 \\ -1 & 10 \end{bmatrix} \begin{bmatrix} u_1 \\ u_2 \end{bmatrix} = \begin{bmatrix} 5v_1 - v_2 & -v_1 + 10v_2 \end{bmatrix} \begin{bmatrix} u_1 \\ u_2 \end{bmatrix}$$

$$= 5u_1 v_1 - u_1 v_2 - u_2 v_1 + 10u_2 v_2$$

7. **(a)** By Formula (4), we have $\langle \mathbf{u}, \mathbf{v} \rangle = \mathbf{v}^T A^T A \mathbf{u}$ where

$$A = \begin{bmatrix} \sqrt{3} & 0 \\ 0 & \sqrt{5} \end{bmatrix}$$

8. **(a)** (1) $\langle \mathbf{u}, \mathbf{v} \rangle = 3u_1v_1 + 5u_2v_2 = 3v_1u_1 + 5v_2u_2 = \langle \mathbf{v}, \mathbf{u} \rangle$

     (2) If $\mathbf{w} = (w_1, w_2)$, then

$$\langle \mathbf{u} + \mathbf{v}, \mathbf{w} \rangle = 3(u_1 + v_1)w_1 + 5(u_2 + v_2)w_2$$
$$= (3u_1w_1 + 5u_2w_2) + (3v_1w_1 + 5v_2w_2)$$
$$= \langle \mathbf{u}, \mathbf{w} \rangle + \langle \mathbf{v}, \mathbf{w} \rangle$$

     (3) $\langle k\mathbf{u}, \mathbf{v} \rangle = 3(ku_1)v_1 + 5(ku_2)v_2$

$$= k(3u_1v_1 + 5u_2v_2)$$
$$= k\langle \mathbf{u}, \mathbf{v} \rangle$$

     (4) $\langle \mathbf{v}, \mathbf{v} \rangle = 3v_1^2 + 5v_2^2 \geq 0$

       Moreover, $\langle \mathbf{v}, \mathbf{v} \rangle = 0$ if and only if $v_1 = v_2 = 0$, or $\mathbf{v} = \mathbf{0}$.

9. **(b)** Axioms 1 and 4 are easily checked. However, if $\mathbf{w} = (w_1, w_2, w_3)$, then

$$\langle \mathbf{u} + \mathbf{v}, \mathbf{w} \rangle = (u_1 + v_1)^2 w_1^2 + (u_2 + v_2)^2 w_2^2 + (u_3 + v_3)^2 w_3^2$$
$$= \langle \mathbf{u}, \mathbf{w} \rangle + \langle \mathbf{v}, \mathbf{w} \rangle + 2u_1v_1w_1^2 + 2u_2v_2w_2^2 + 2u_3v_3w_3^2$$

If, for instance, $\mathbf{u} = \mathbf{v} = \mathbf{w} = (1, 0, 0)$, then Axiom 2 fails.

     To check Axiom 3, we note that $\langle k\mathbf{u}, \mathbf{v} \rangle = k^2 \langle \mathbf{u}, \mathbf{v} \rangle$. Thus $\langle k\mathbf{u}, \mathbf{v} \rangle \neq k\langle \mathbf{u}, \mathbf{v} \rangle$ unless $k = 0$ or $k = 1$, so Axiom 3 fails.

(c) (1) Axiom 1 follows from the commutativity of multiplication in R.

     (2) If $\mathbf{w} = (w_1, w_2, w_3)$, then

$$\langle \mathbf{u} + \mathbf{v}, \mathbf{w} \rangle = 2(u_1 + v_1)w_1 + (u_2 + v_2)w_2 + 4(u_3 + v_3)w_3$$
$$= 2u_1w_1 + u_2w_2 + 4u_3w_3 + 2v_1w_1 + v_2w_2 + 4v_3w_3$$
$$= \langle \mathbf{u}, \mathbf{w} \rangle + \langle \mathbf{v}, \mathbf{w} \rangle$$

     (3) $\langle k\mathbf{u}, \mathbf{v} \rangle = 2(ku_1)v_1 + (ku_2)v_2 + 4(ku_3)v_3 = k\langle \mathbf{u}, \mathbf{v} \rangle$

     (4) $\langle \mathbf{v}, \mathbf{v} \rangle = 2v_1^2 + v_2^2 + 4v_3^2 \geq 0$

$$= 0 \text{ if and only if } v_1 = v_2 = v_3 = 0, \text{ or } \mathbf{v} = \mathbf{0}$$

Thus this is an inner product for R^3.

10.   **(c)**   Since $A^T A = \begin{bmatrix} 2 & -1 \\ -1 & 13 \end{bmatrix}$, we have

$$\|\mathbf{w}\|^2 = \left(\begin{bmatrix} -1 & 3 \end{bmatrix} \begin{bmatrix} 2 & -1 \\ -1 & 13 \end{bmatrix} \right) \begin{bmatrix} -1 \\ 3 \end{bmatrix}$$

$$= \begin{bmatrix} -5 & 40 \end{bmatrix} \begin{bmatrix} -1 \\ 3 \end{bmatrix} = 125$$

Thus $\|\mathbf{w}\| = \sqrt{125} = 5\sqrt{5}$.

11.   We have $\mathbf{u} - \mathbf{v} = (-3, -3)$.

(b) $d(\mathbf{u}, \mathbf{v}) = \|(-3, -3)\| = [3(9) + 2(9)]^{1/2} = \sqrt{45} = 3\sqrt{5}$

(c)   From Problem 10(**c**), we have

$$[d(\mathbf{u}, \mathbf{v})]^2 = \begin{bmatrix} -3 & -3 \end{bmatrix} \begin{bmatrix} 2 & -1 \\ -1 & 13 \end{bmatrix} \begin{bmatrix} -3 \\ -3 \end{bmatrix} = 117$$

Thus

$$d(\mathbf{u}, \mathbf{v}) = \sqrt{117} = 3\sqrt{13}$$

12.   **(a)** $\|\mathbf{p}\| = \left[(-2)^2 + (3)^2 + (2)^2 \right]^{1/2} = \sqrt{17}$

13.   **(a)** $\|A\| = \left[(-2)^2 + (5)^2 + (3)^2 + (6)^2 \right]^{1/2} = \sqrt{74}$

14.   Since $\mathbf{p} - \mathbf{q} = 1 - x - 4x^2$, we have

$$d(\mathbf{p}, \mathbf{q}) = \langle \mathbf{p} - \mathbf{q}, \ \mathbf{p} - \mathbf{q} \rangle^{1/2} = \left[1^2 + (-1)^2 + (-4)^2 \right]^{1/2} = 3\sqrt{2}$$

15. **(a)** Since $A - B = \begin{bmatrix} 6 & -1 \\ 8 & -2 \end{bmatrix}$, we have

$$d(A, B) = \langle A - B, \ A - B \rangle^{1/2} = \left[6^2 + (-1)^2 + 8^2 + (-2)^2\right]^{1/2} = \sqrt{105}$$

16. **(a)** We have $\langle \mathbf{u} + \mathbf{v}, \mathbf{v} + \mathbf{w} \rangle = \langle \mathbf{u}, \mathbf{v} \rangle + \langle \mathbf{u}, \mathbf{w} \rangle + \langle \mathbf{v}, \mathbf{v} \rangle + \langle \mathbf{v}, \mathbf{w} \rangle = 2 + 5 + (2)^2 - 3 = 8$.

(d) We have $\|\mathbf{u} + \mathbf{v}\|^2 = \langle \mathbf{u} + \mathbf{v}, \mathbf{u} + \mathbf{v} \rangle = \langle \mathbf{u}, \mathbf{u} \rangle + 2\langle \mathbf{u}, \mathbf{v} \rangle + \langle \mathbf{v}, \mathbf{v} \rangle = (1)^2 + 2(2) + (2)^2 = 9$. Hence $\|\mathbf{u} + \mathbf{v}\| = 3$.

17. **(a)** For instance, $\|x\| = \left(\displaystyle\int_{-1}^{1} x^2 \, dx\right)^{1/2} = \left(\dfrac{x^3}{3}\Big]_{-1}^{1}\right)^{1/2} = \left(\dfrac{2}{3}\right)^{1/2}$.

(b) We have

$$d(\mathbf{p}, \mathbf{q}) = \|\mathbf{p} - \mathbf{q}\|$$

$$= \|1 - x\|$$

$$= \left(\int_{-1}^{1} (1 - x)^2 \, dx\right)^{1/2}$$

$$= \left(\int_{-1}^{1} (1 - 2x + x^2) \, dx\right)^{1/2}$$

$$= \left(\left(x - x^2 + \frac{x^3}{3}\right)\Big]_{-1}^{1}\right)^{1/2}$$

$$= 2\left(\frac{2}{3}\right)^{1/2} = \frac{2}{3}\sqrt{6}$$

18. **(a)** Since we are looking for points (x, y) with $\|(x, y)\| = 1$, we have $\left(\frac{1}{4}x^2 + \frac{1}{16}y^2\right)^{1/2} = 1$

or $\frac{1}{4}x^2 + \frac{1}{16}y^2 = 1$. This is the ellipse shown in the figure.

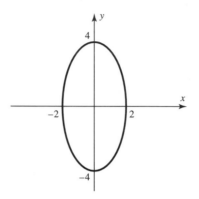

20. Observe that

$$(*) \qquad \|\mathbf{u} + \mathbf{v}\|^2 = \langle \mathbf{u} + \mathbf{v}, \mathbf{u} + \mathbf{v} \rangle = \langle \mathbf{u}, \mathbf{u} \rangle + 2\langle \mathbf{u}, \mathbf{v} \rangle + \langle \mathbf{v}, \mathbf{v} \rangle$$

and

$$(**) \qquad \|\mathbf{u} - \mathbf{v}\|^2 = \langle \mathbf{u} - \mathbf{v}, \mathbf{u} - \mathbf{v} \rangle = \langle \mathbf{u}, \mathbf{u} \rangle - 2\langle \mathbf{u}, \mathbf{v} \rangle + \langle \mathbf{v}, \mathbf{v} \rangle$$

Thus if we add $(*)$ and $(**)$, we obtain the equation

$$\|\mathbf{u} + \mathbf{v}\|^2 + \|\mathbf{u} - \mathbf{v}\|^2 = 2\|\mathbf{u}\|^2 + 2\|\mathbf{v}\|^2$$

21. If, in the solution to Exercise 20, we subtract $(**)$ from $(*)$ and divide by 4, we obtain the desired result.

22. If, for instance, $V = \begin{bmatrix} 0 & 1 \\ -1 & 0 \end{bmatrix}$, then $\langle V, V \rangle = -2 < 0$. Thus Axiom 4 fails.

23. Axioms 1 and 3 are easily verified. So is Axiom 2, as shown: Let $\mathbf{r} = r(x)$ be a polynomial in P_2. Then

$$\langle \mathbf{p} + \mathbf{q}, \mathbf{r} \rangle = [(p+q)(0)]r(0) + [(p+q)(1/2)]r(1/2) + [(p+q)(1)]r(1)$$
$$= p(0)r(0) + p(1/2)r(1/2) + p(1)r(1) + q(0)r(0) + q(1/2)r(1/2) + q(1)r(1)$$
$$= \langle \mathbf{p}, \mathbf{r} \rangle + \langle \mathbf{q}, \mathbf{r} \rangle$$

It remains to verify Axiom 4:

$$\langle \mathbf{p}, \mathbf{p} \rangle = [p(0)]^2 + [p(1/2)]^2 + [p(1)]^2 \geq 0$$

and

$$\langle \mathbf{p}, \mathbf{p} \rangle = 0 \quad \text{if and only if} \quad p(0) = p(1/2) = p(1) = 0$$

But a quadratic polynomial can have at most two zeros unless it is identically zero. Thus $\langle \mathbf{p}, \mathbf{p} \rangle - 0$ if and only if $\mathbf{p}$ is identically zero, or $\mathbf{p} = \mathbf{0}$.

24. Using the hint and properties of the transpose, we have

$$\langle \mathbf{u}, A\mathbf{v} \rangle = (A\mathbf{v})^T \mathbf{u} = (\mathbf{v}^T A^T)\mathbf{u} = \mathbf{v}^T(A^T\mathbf{u}) = \langle A^T\mathbf{u}, \mathbf{v} \rangle$$

26. (1) Axiom 1 follows immediately from the commutativity of multiplication in the real numbers.

(2) If $\mathbf{z} = (z_1, \ldots, z_n)$, then

$$\langle \mathbf{u} + \mathbf{v}, \mathbf{z} \rangle = w_1(u_1 + v_1)z_1 + \cdots + w_n(u_n + v_n)z_n$$
$$= (w_1 u_1 z_1 + \cdots + w_n u_n z_n) + (w_1 v_1 z_1 + \cdots + w_n v_n z_n)$$
$$= \langle \mathbf{u}, \mathbf{z} \rangle + \langle \mathbf{v}, \mathbf{z} \rangle$$

(3) If k is any constant, then

$$\langle k\mathbf{u}, \mathbf{v} \rangle = w_1(ku_1)v_1 + \cdots + w_n(ku_n)v_n$$
$$= k(w_1 u_1 v_1 + \cdots + w_n u_n v_n)$$
$$= k\langle \mathbf{u}, \mathbf{v} \rangle$$

(4)   Since $\langle \mathbf{v}, \mathbf{v} \rangle = w_1 v_1^2 + \cdots + w_n v_n^2$ where $w_1, \ldots, w_n$ are positive reals, then $\langle \mathbf{v}, \mathbf{v} \rangle \geq 0$ and $\langle \mathbf{v}, \mathbf{v} \rangle = 0$ if and only if $v_1 = \cdots = v_n = 0$ or $\mathbf{v} = \mathbf{0}$. If one of the constants, w_i, were negative or zero, then Axiom 4 would fail to hold.

27.   **(b)** $\displaystyle \langle \mathbf{p}, \mathbf{q} \rangle = \int_{-1}^{1} (x - 5x^3)(2 + 8x^2)\, dx = \int_{-1}^{1} (2x - 2x^3 - 40x^5)\, dx$

$$= x^2 - x^4/2 - 20x^6/3 \Big]_{-1}^{1} = 0$$

28.   **(a)** $\displaystyle \langle \mathbf{f}, \mathbf{g} \rangle = \int_{0}^{1} \cos 2\pi x \, \sin 2\pi x \, dx = (1/4\pi)\, \sin^2 2\pi x \Big]_{0}^{1} = 0$

29.   We have $\langle U, V \rangle = u_1 v_1 + u_2 v_2 + u_3 v_3 + u_4 v_4$ and

$$\mathrm{tr}(U^T V) = \mathrm{tr}\left(\begin{bmatrix} u_1 & u_3 \\ u_2 & u_4 \end{bmatrix} \begin{bmatrix} v_1 & v_2 \\ v_3 & v_4 \end{bmatrix} \right)$$

$$= \mathrm{tr}\left(\begin{bmatrix} u_1 v_1 + u_3 v_3 & u_1 v_2 + u_3 v_4 \\ u_2 v_1 + u_4 v_3 & u_2 v_2 + u_4 v_4 \end{bmatrix} \right)$$

$$= u_1 v_1 + u_3 v_3 + u_2 v_2 + u_4 v_4$$

which does, indeed, equal $\langle U, V \rangle$.

30.   We use the formula $\langle \mathbf{u}, \mathbf{v} \rangle = \mathbf{v}^T A^T A \mathbf{u}$ as suggested in the hint.

(1)   We must show that $\langle \mathbf{u}, \mathbf{v} \rangle = \langle \mathbf{v}, \mathbf{u} \rangle$. Since $\langle \mathbf{v}, \mathbf{u} \rangle$ is a real number which can be thought of as a 1×1 matrix, it will suffice to show that $\langle \mathbf{u}, \mathbf{v} \rangle = \langle \mathbf{v}, \mathbf{u} \rangle^T$. But

$$\langle \mathbf{v}, \mathbf{u} \rangle^T = (\mathbf{u}^T A^T A \mathbf{v})^T = \mathbf{v}^T A^T A \mathbf{u} = \langle \mathbf{u}, \mathbf{v} \rangle$$

(2) We have

$$\langle \mathbf{u} + \mathbf{v}, \mathbf{w} \rangle = \mathbf{w}^T A^T A (\mathbf{u} + \mathbf{v})$$

$$= \mathbf{w}^T A^T A \mathbf{u} + \mathbf{w}^T A^T A \mathbf{v}$$

$$= \langle \mathbf{u}, \mathbf{w} \rangle + \langle \mathbf{v}, \mathbf{w} \rangle$$

(3) Here $\langle k\mathbf{u}, \mathbf{v} \rangle = \mathbf{v}^T A^T A (k\mathbf{u}) = k(\mathbf{v}^T A^T A \mathbf{u}) = k \langle \mathbf{u}, \mathbf{v} \rangle$.

(4) Since $\langle \mathbf{v}, \mathbf{v} \rangle = \mathbf{v}^T A^T A \mathbf{v} = A\mathbf{v} \cdot A\mathbf{v} \geq 0$, we need only show that $\langle \mathbf{v}, \mathbf{v} \rangle = 0$ if and only if $\mathbf{v} = \mathbf{0}$. But $\langle \mathbf{v}, \mathbf{v} \rangle = 0$ if and only if $A\mathbf{v} = \mathbf{0}$, or $\mathbf{v} = A^{-1}\mathbf{0} = \mathbf{0}$, since A is invertible.

Therefore $\langle \mathbf{v}, \mathbf{v} \rangle \geq 0$ for every vector $\mathbf{v}$. It is clear that if $\mathbf{v} = \mathbf{0}$, then $\langle \mathbf{v}, \mathbf{v} \rangle = 0$. It remains to show that if $\mathbf{v} \neq \mathbf{0}$, then $\langle \mathbf{v}, \mathbf{v} \rangle \neq 0$.

To see this, we notice that the only way that $\langle \mathbf{v}, \mathbf{v} \rangle$ can equal zero is for $\mathbf{v} \cdot \mathbf{r}_i$ to equal zero for every row $\mathbf{r}_i$ of A. But this implies that a linear combination of the columns of A is the $n \times 1$ zero matrix. That is, if we let $\mathbf{c}_i$ be the ith column of A, then

$$v_1 \mathbf{c}_1 + v_2 \mathbf{c}_2 + \cdots + v_n \mathbf{c}_n = \mathbf{0}$$

Therefore the column vectors of A are linearly dependent, and hence A is not invertible. This is why the matrix A has to be invertible to generate an inner product. If A is invertible, then $\langle \mathbf{v}, \mathbf{v} \rangle$ can only $= 0$ when $\mathbf{v} = \mathbf{0}$.

31. Calling the matrix A, we have

$$\langle \mathbf{u}, \mathbf{v} \rangle = \mathbf{v}^T A^T A \mathbf{u} = \mathbf{v}^T A^2 \mathbf{u} = w_1 u_1 v_1 + \cdots + w_n u_n v_n$$

33. To prove Part (a) of Theorem 6.1.1 first observe that $\langle \mathbf{0}, \mathbf{v} \rangle = \langle \mathbf{v}, \mathbf{0} \rangle$ by the symmetry axiom. Moreover,

$$\langle \mathbf{0}, \mathbf{v} \rangle = \langle 0\mathbf{0}, \mathbf{v} \rangle \quad \text{by Theorem 5.1.1}$$
$$= 0\langle \mathbf{0}, \mathbf{v} \rangle \quad \text{by the homogeneity axiom}$$
$$= 0$$

Alternatively,

$$\langle \mathbf{0}, \mathbf{v} \rangle + \langle \mathbf{0}, \mathbf{v} \rangle = \langle \mathbf{0} + \mathbf{0}, \mathbf{v} \rangle \quad \text{by additivity}$$
$$= \langle \mathbf{0}, \mathbf{v} \rangle \qquad \text{by definition of the zero vector}$$

But $\langle \mathbf{0}, \mathbf{v} \rangle = 2\langle \mathbf{0}, \mathbf{v} \rangle$ only if $\langle \mathbf{0}, \mathbf{v} \rangle = 0$.

To prove Part (d), observe that, by Theorem 5.1.1, $-\mathbf{v}$ (the inverse of $\mathbf{v}$) and $(-1)\mathbf{v}$ are the same vector. Thus,

$$\langle \mathbf{u} - \mathbf{v}, \mathbf{w} \rangle = \langle \mathbf{u} + (-\mathbf{v}), \mathbf{w} \rangle$$
$$= \langle \mathbf{u}, \mathbf{w} \rangle + \langle -\mathbf{v}, \mathbf{w} \rangle \quad \text{by additivity}$$
$$= \langle \mathbf{u}, \mathbf{w} \rangle - \langle \mathbf{v}, \mathbf{w} \rangle \qquad \text{by homogeneity}$$

34. The set of points $\langle (x, y), (x, y) \rangle = x^2/5^2 + y^2/4^2 = 1$ is the unit circle with respect to the weighted Euclidean inner product $\langle \mathbf{u}, \mathbf{v} \rangle = (1/25)u_1 v_1 + (1/16)u_2 v_2$.

EXERCISE SET 6.2

1. **(e)** Since $\mathbf{u} \cdot \mathbf{v} = 0 + 6 + 2 + 0 = 8$, the vectors are not orthogonal.

2. We have $\langle \mathbf{u}, \mathbf{w}_1 \rangle = \langle \mathbf{u}, \mathbf{w}_3 \rangle = 0$ but $\langle \mathbf{u}, \mathbf{w}_2 \rangle = -2 \neq 0$. Hence, $\mathbf{u}$ is not orthogonal to W.

3. **(a)** $\cos \theta = \dfrac{\langle (1, -3),\ (2, 4) \rangle}{\|(1, -3)\|\, \|(2, 4)\|} = \dfrac{2 - 12}{\sqrt{10}\,\sqrt{20}} = \dfrac{-1}{\sqrt{2}}$

   (c) $\cos \theta = \dfrac{\langle (-1, 5, 2),\ (2, 4, -9) \rangle}{\|(-1, 5, 2)\|\, \|(2, 4, -9)\|} = \dfrac{-2 + 20 - 18}{\sqrt{30}\,\sqrt{101}} = 0$

   (e) $\cos \theta = \dfrac{\langle (1, 0, 1, 0),\ (-3, -3, -3, -3) \rangle}{\|(1, 0, 1, 0)\|\, \|(-3, -3, -3, -3)\|} = \dfrac{-3 - 3}{\sqrt{2}\,\sqrt{36}} = \dfrac{-1}{\sqrt{2}}$

4. **(a)** $\cos \theta = \dfrac{\langle -1 + 5x + 2x^2,\ 2 + 4x - 9x^2 \rangle}{\| -1 + 5x + 2x^2 \|\, \|2 + 4x - 9x^2\|} = \dfrac{-2 + 20 - 18}{\sqrt{30}\,\sqrt{101}} = 0$

   (Compare with 3(c), above.)

5. $\langle \mathbf{p}, \mathbf{q} \rangle = (1)(0) + (-1)(2) + (2)(1) = 0$

6. **(a)** $\cos \theta = \dfrac{\langle A, B \rangle}{\|A\|\, \|B\|} = \dfrac{6 + 12 + 1 + 0}{\sqrt{50}\,\sqrt{14}} = \dfrac{19}{10\sqrt{7}}$

7. **(b)** $\left\langle \begin{bmatrix} 2 & 1 \\ -1 & 3 \end{bmatrix}, \begin{bmatrix} 1 & 1 \\ 0 & -1 \end{bmatrix} \right\rangle = (2)(1) + (1)(1) + (-1)(0) + (3)(-1) = 0$

   Thus the matrices are orthogonal.

7. (d) $\left\langle \begin{bmatrix} 2 & 1 \\ -1 & 3 \end{bmatrix}, \begin{bmatrix} 2 & 1 \\ 5 & 2 \end{bmatrix} \right\rangle = 4 + 1 - 5 + 6 = 6 \neq 0$

Thus the matrices are not orthogonal.

8. (a) We look for values of k such that

$$\langle \mathbf{u}, \mathbf{v} \rangle = 2 + 7 + 3k = 0$$

Clearly $k = -3$ is the only possible value.

9. We must find two vectors $\mathbf{x} = (x_1, x_2, x_3, x_4)$ such that $\langle \mathbf{x}, \mathbf{x} \rangle = 1$ and $\langle \mathbf{x}, \mathbf{u} \rangle = \langle \mathbf{x}, \mathbf{v} \rangle = \langle \mathbf{x}, \mathbf{w} \rangle = 0$. Thus x_1, x_2, x_3, and x_4 must satisfy the equations

$$
\begin{aligned}
x_1^2 + \ x_2^2 + \ x_3^2 + \ x_4^2 &= 1 \\
2x_1 + \ x_2 - 4x_3 \quad\quad &= 0 \\
-x_1 - \ x_2 + 2x_3 + 2x_4 &= 0 \\
3x_1 + 2x_2 + 5x_3 + 4x_4 &= 0
\end{aligned}
$$

The solution to the three linear equations is $x_1 = -34t$, $x_2 = 44t$, $x_3 = -6t$, and $x_4 = 11t$. If we substitute these values into the quadratic equation, we get

$$\left[(-34)^2 + (44)^2 + (-6)^2 + (11)^2 \right] t^2 = 1$$

or

$$t = \pm \frac{1}{\sqrt{57}}$$

Therefore, the two vectors are

$$\pm \frac{1}{\sqrt{57}} (-34, 44, -6, 11)$$

10. (a) Here $\langle \mathbf{u}, \mathbf{v} \rangle^2 = (3(4) + 2(-1))^2 = 100$, while $\langle \mathbf{u}, \mathbf{u} \rangle \langle \mathbf{v}, \mathbf{v} \rangle = (3^2 + 2^2)(4^2 + (-1)^2) = 221$.

11. **(a)** Here $\langle \mathbf{u}, \mathbf{v} \rangle^2 = (3(-2)(1) + 2(1)(0))^2 = 36$, while, on the other hand, $\langle \mathbf{u}, \mathbf{u} \rangle \langle \mathbf{v}, \mathbf{v} \rangle = (3(-2)^2 + 2(1)^2)(3(1)^2 + 2(0)^2) = 42$.

12. Since W is the line $y = 2x$, then the space $W^\perp$ of all vectors which are perpendicular to every vector in W is just the line $y = -\frac{1}{2}x$. Note that $(x, 2x) \cdot \left(x, -\frac{1}{2}x\right) = 0$.

13. **(a)** Here $W^\perp$ is the line which is normal to the plane and which passes through the origin. By inspection, a normal vector to the plane is $(1, -2, -3)$. Hence this line has parametric equations $x = t$, $y = -2t$, $z = -3t$.

14. **(a)** A row-echelon form of A is

$$\begin{bmatrix} 1 & 2 & -1 & 2 \\ 0 & 1 & -3 & 2 \\ 0 & 0 & 0 & 0 \end{bmatrix}$$

Therefore the vectors $\mathbf{v}_1 = (1, 2, -1, 2)$ and $\mathbf{v}_2 = (0, 1, -3, 2)$ form a basis for the row space of A. From the above matrix, we see that the nullspace of A consists of all vectors of the form $(-5s + 2t, 3s - 2t, s, t)$ where s and t are parameters. If we set $s = 1$, $t = 0$ and $s = 0$, $t = 1$, we obtain a basis $\mathbf{w}_1 = (-5, 3, 1, 0)$, $\mathbf{w}_2 = (2, -2, 0, 1)$ for the nullspace of A. It is easy to check that $\mathbf{v}_1 \cdot \mathbf{w}_1 = \mathbf{v}_1 \cdot \mathbf{w}_2 = \mathbf{v}_2 \cdot \mathbf{w}_1 = \mathbf{v}_2 \cdot \mathbf{w}_2 = 0$, which gives the desired result.

16. **(a)** The subspace of R^3 spanned by the given vectors is the row space of the matrix

$$\begin{bmatrix} 1 & -1 & 3 \\ 5 & -4 & -4 \\ 7 & -6 & 2 \end{bmatrix} \quad \text{which reduces to} \quad \begin{bmatrix} 1 & -1 & 3 \\ 0 & 1 & -19 \\ 0 & 0 & 0 \end{bmatrix}$$

The space we are looking for is the nullspace of this matrix. From the reduced form, we see that the nullspace consists of all vectors of the form $(16, 19, 1)t$, so that the vector $(16, 19, 1)$ is a basis for this space.

Alternatively the vectors $\mathbf{w}_1 = (1, -1, 3)$ and $\mathbf{w}_2 = (0, 1, -19)$ form a basis for the row space of the matrix. They also span a plane, and the orthogonal complement of this plane is the line spanned by the normal vector $\mathbf{w}_1 \times \mathbf{w}_2 = (16, 19, 1)$.

16. **(b)** Since $\mathbf{v}_2 = 2\mathbf{v}_1$, we are looking for a basis for the orthogonal complement of the space spanned by $\mathbf{v}_1$. (This space is the line containing $\mathbf{v}_1$ and its orthogonal complement is the plane $2x - z = 0$.) The solution space for the equation $2x_1 + 0x_2 - x_3 = 0$ is the set of vectors $(\frac{1}{2}t, s, t) = (0, 1, 0)s + (\frac{1}{2}, 0, 1)t$. Thus the vectors $(0, 1, 0)$ and $(1, 0, 2)$ form a basis for the orthogonal complement. (They also span the plane $2x - z = 0$.)

Alternatively, the subspace W of R^3 spanned by the given vectors $\mathbf{v}_1$ and $\mathbf{v}_2$ is the row space of the matrix A, where

$$A = \begin{bmatrix} 2 & 0 & -1 \\ 4 & 0 & -2 \end{bmatrix} \rightarrow \begin{bmatrix} 2 & 0 & -1 \\ 0 & 0 & 0 \end{bmatrix}$$

Since $(2, 0, -1)$ is a basis for the row space, then $(2, 0, -1)$ is a basis for W. To find a basis for $W^{\perp}$, note that $(1, 0, 2)$ and $(0, 1, 0)$ are a basis for the null-space of A and hence for $W^{\perp}$. It is obvious that $W^{\perp}$ is the plane $2x - z = 0$.

(c) The orthogonal complement of the subspace of R^4 spanned by the given vectors is the nullspace of the matrix

$$\begin{bmatrix} 1 & 4 & 5 & 2 \\ 2 & 1 & 3 & 0 \\ -1 & 3 & 2 & 2 \end{bmatrix} \quad \text{which reduces to} \quad \begin{bmatrix} 1 & 4 & 5 & 2 \\ 0 & 1 & 1 & 4/7 \\ 0 & 0 & 0 & 0 \end{bmatrix}$$

Thus the nullspace consists of all vectors of the form $(-s + 2t, -s - 4t, s, 7t)$ and the vectors $(-1, -1, 1, 0)$ and $(2, -4, 0, 7)$ form a basis for this space.

17. If $\mathbf{u}$ and $\mathbf{v}$ are orthogonal vectors with norm 1, then

$$\|\mathbf{u} - \mathbf{v}\| = \langle \mathbf{u} - \mathbf{v}, \mathbf{u} - \mathbf{v} \rangle^{1/2}$$

$$= [\langle \mathbf{u}, \mathbf{u} \rangle - 2\langle \mathbf{u}, \mathbf{v} \rangle + \langle \mathbf{v}, \mathbf{v} \rangle]^{1/2}$$

$$= [1 - 2(0) + 1]^{1/2}$$

$$= \sqrt{2}$$

18. If $\langle \mathbf{w}, \mathbf{u}_1 \rangle = \langle \mathbf{w}, \mathbf{u}_2 \rangle = 0$, then

$$\langle \mathbf{w}, k_1\mathbf{u}_1 + k_2\mathbf{u}_2 \rangle = \langle \mathbf{w}, k_1\mathbf{u}_1 \rangle + \langle \mathbf{w}, k_2\mathbf{u}_2 \rangle$$

$$= k_1\langle \mathbf{w}, \mathbf{u}_1 \rangle + k_2\langle \mathbf{w}, \mathbf{u}_2 \rangle$$

$$= 0$$

Thus $\mathbf{w}$ is orthogonal to $k_1\mathbf{u}_1 + k_2\mathbf{u}_2$.

Now consider R^3 with the Euclidean inner product. If $\mathbf{w}$ is perpendicular to both $\mathbf{u}_1$ and $\mathbf{u}_2$ and they determine a plane, then $\mathbf{w}$ is perpendicular to every vector in that plane. If $\mathbf{u}_1$ and $\mathbf{u}_2$ determine a line, then $\mathbf{w}$ must be perpendicular to every vector in that line. If $\mathbf{u}_1$ and $\mathbf{u}_2$ determine neither a plane nor a line, then $\mathbf{u}_1 = \mathbf{u}_2 = \mathbf{0}$ and the result is not of interest.

19. By definition, $\mathbf{u}$ is in span$\{\mathbf{u}_1, \mathbf{u}_2, \ldots, \mathbf{u}_r\}$ if and only if there exist constants $c_1, c_2, \ldots, c_r$ such that

$$\mathbf{u} = c_1 \mathbf{u}_1 + c_2 \mathbf{u}_2 + \cdots + c_r \mathbf{u}_r$$

But if $\langle \mathbf{w}, \mathbf{u}_1 \rangle = \langle \mathbf{w}, \mathbf{u}_2 \rangle = \cdots = \langle \mathbf{w}, \mathbf{u}_r \rangle = 0$, then $\langle \mathbf{w}, \mathbf{u} \rangle = 0$.

20. If a vector $\mathbf{z}$ is orthogonal to all of the basis vectors, then, by the result of Exercise 19, it is orthogonal to every linear combination of the basis vectors, and hence to every vector in V. Thus, in particular, $\langle \mathbf{z}, \mathbf{z} \rangle = 0$, so, by Axiom 4, $\mathbf{z} = \mathbf{0}$.

21. We have that $W = \text{span}\{\mathbf{w}_1, \mathbf{w}_2, \ldots, \mathbf{w}_k\}$

Suppose that $\mathbf{w}$ is in $W^\perp$. Then, by definition, $\langle \mathbf{w}, \mathbf{w}_i \rangle = 0$ for each basis vector $\mathbf{w}_i$ of W.

Conversely, if a vector $\mathbf{w}$ of V is orthogonal to each basis vector of W, then, by Problem 20, it is orthogonal to every vector in W.

22. Note that

$$\|\mathbf{v}_1 + \cdots + \mathbf{v}_r\|^2 = \langle \mathbf{v}_1 + \cdots + \mathbf{v}_r,\ \mathbf{v}_1 + \cdots + \mathbf{v}_r \rangle$$

$$= \sum_{i=1}^{r} \langle \mathbf{v}_i, \mathbf{v}_i \rangle + \sum_{i \neq j} \langle \mathbf{v}_i, \mathbf{v}_j \rangle$$

$$= \|\mathbf{v}_1\|^2 + \cdots + \|\mathbf{v}_r\|^2 + \sum_{i \neq j} \langle \mathbf{v}_i, \mathbf{v}_j \rangle$$

If $\langle \mathbf{v}_i, \mathbf{v}_j \rangle = 0$ whenever $i \neq j$, then the result follows.

23. **(c)** By Property (3) in the definition of inner product, we have

$$\|k\mathbf{u}\|^2 = \langle k\mathbf{u}, k\mathbf{u} \rangle = k^2 \langle \mathbf{u}, \mathbf{u} \rangle = k^2 \|\mathbf{u}\|^2$$

Therefore $\|k\mathbf{u}\| = |k|\, \|\mathbf{u}\|$.

24.   (d) We must show that for all $\mathbf{u}, \mathbf{v}$, and $\mathbf{w}$

$$\|\mathbf{u} - \mathbf{v}\| \le \|\mathbf{u} - \mathbf{w}\| + \|\mathbf{w} - \mathbf{v}\|$$

If we apply the triangle inequality to the vectors $\mathbf{u} - \mathbf{w}$ and $\mathbf{w} - \mathbf{v}$, then we have

$$\|(\mathbf{u} - \mathbf{w}) + (\mathbf{w} - \mathbf{v})\| \le \|\mathbf{u} - \mathbf{w}\| + \|\mathbf{w} - \mathbf{v}\|$$

and the result follows.

25. This is just the Cauchy-Schwarz Inequality using the inner product on R^n generated by A (see Formula (4) of Section 6.1).

26. If we let R^2 have the Euclidean inner product and apply the Cauchy-Schwarz Inequality to the vectors $\mathbf{u} = (a, b)$ and $\mathbf{v} = (\cos \theta, \sin \theta)$, the result follows directly.

28. Suppose that

$$(*) \qquad\qquad\qquad \langle \mathbf{u}, \mathbf{v} \rangle^2 = \langle \mathbf{u}, \mathbf{u} \rangle \langle \mathbf{v}, \mathbf{v} \rangle$$

Following the proof of Theorem 6.2.1, we see that $(*)$ can hold if and only if either $\mathbf{u} = \mathbf{0}$ or the quadratic equation $\langle t\mathbf{u} + \mathbf{v},\, t\mathbf{u} + \mathbf{v} \rangle = 0$ has just one real root. If $\mathbf{u} = \mathbf{0}$, then $\mathbf{u}$ and $\mathbf{v}$ are linearly dependent. If the equation has one real root, t_0, then $\langle t_0\mathbf{u} + \mathbf{v},\, t_0\mathbf{u} + \mathbf{v} \rangle = 0$. Thus, by Axiom 4, $t_0\mathbf{u} + \mathbf{v} = \mathbf{0}$, and again $\mathbf{u}$ and $\mathbf{v}$ are linearly dependent.

Conversely, suppose that $\mathbf{u}$ and $\mathbf{v}$ are linearly dependent. Then either $\mathbf{u} = \mathbf{0}$ or $\mathbf{v}$ can be uniquely expressed as a multiple, $-t_0$, of $\mathbf{u}$; that is, $t_0\mathbf{u} + \mathbf{v} = \mathbf{0}$. Thus if we work backward through the above argument, we can verify that $\langle \mathbf{u}, \mathbf{v} \rangle^2 = \langle \mathbf{u}, \mathbf{u} \rangle \langle \mathbf{v}, \mathbf{v} \rangle$.

29. We wish to show that $\angle ABC$ is a right angle, or that $\overrightarrow{AB}$ and $\overrightarrow{BC}$ are orthogonal. Observe that $\overrightarrow{AB} = \mathbf{u} - (-\mathbf{v})$ and $\overrightarrow{BC} = \mathbf{v} - \mathbf{u}$ where $\mathbf{u}$ and $\mathbf{v}$ are radii of the circle, as shown in the figure. Thus $\|\mathbf{u}\| = \|\mathbf{v}\|$. Hence

$$
\begin{aligned}
\langle \overrightarrow{AB}, \overrightarrow{BC} \rangle &= \langle \mathbf{u} + \mathbf{v},\ \mathbf{v} - \mathbf{u} \rangle \\
&= \langle \mathbf{u}, \mathbf{v} \rangle + \langle \mathbf{v}, \mathbf{v} \rangle + \langle \mathbf{u}, -\mathbf{u} \rangle + \langle \mathbf{v}, -\mathbf{u} \rangle \\
&= \langle \mathbf{v}, \mathbf{u} \rangle + \langle \mathbf{v}, \mathbf{v} \rangle - \langle \mathbf{u}, \mathbf{u} \rangle - \langle \mathbf{v}, \mathbf{u} \rangle \\
&= \|\mathbf{v}\|^2 - \|\mathbf{u}\|^2 \\
&= 0
\end{aligned}
$$

30. Suppose that $\langle \mathbf{u}, \mathbf{v} \rangle = au_1 v_1 + bu_2 v_2$ is a weighted Euclidean inner product on R^2. For $\mathbf{u}$ and $\mathbf{v}$ to be orthogonal unit vectors with respect to this inner product, we must have

$$\langle \mathbf{u}, \mathbf{u} \rangle = \quad a + 3b = 1$$

$$\langle \mathbf{v}, \mathbf{v} \rangle = \quad a + 3b = 1$$

$$\langle \mathbf{u}, \mathbf{v} \rangle = -a + 3b = 0$$

which yields $a = 1/2$ and $b = 1/6$.

31. **(a)** As noted in Example 9 of Section 6.1, $\displaystyle\int_0^1 f(x)g(x)dx$ is an inner product on $C[0,1]$. Thus the Cauchy-Schwarz inequality must hold, and that is exactly what we're asked to prove.

(b) In the inner product notation, we must show that

$$\langle \mathbf{f} + \mathbf{g}, \ \mathbf{f} + \mathbf{g} \rangle^{1/2} \leq \langle \mathbf{f}, \mathbf{f} \rangle^{1/2} + \langle \mathbf{g}, \mathbf{g} \rangle^{1/2}$$

or, squaring both sides, that

$$\langle \mathbf{f} + \mathbf{g}, \ \mathbf{f} + \mathbf{g} \rangle \leq \langle \mathbf{f}, \mathbf{f} \rangle + 2\langle \mathbf{f}, \mathbf{f} \rangle^{1/2}\langle \mathbf{g}, \mathbf{g} \rangle^{1/2} + \langle \mathbf{g}, \mathbf{g} \rangle$$

For any inner product, we know that

$$\langle \mathbf{f} + \mathbf{g}, \ \mathbf{f} + \mathbf{g} \rangle = \langle \mathbf{f}, \mathbf{f} \rangle + 2\langle \mathbf{f}, \mathbf{g} \rangle + \langle \mathbf{g}, \mathbf{g} \rangle$$

By the Cauchy-Schwarz inequality

$$\langle \mathbf{f}, \mathbf{g} \rangle^2 \leq \langle \mathbf{f}, \mathbf{f} \rangle \langle \mathbf{g}, \mathbf{g} \rangle$$

or

$$\langle \mathbf{f}, \mathbf{g} \rangle \leq \langle \mathbf{f}, \mathbf{f} \rangle^{1/2}\langle \mathbf{g}, \mathbf{g} \rangle^{1/2}$$

If we substitute the above inequality into the equation for $\langle \mathbf{f} + \mathbf{g}, \ \mathbf{f} + \mathbf{g} \rangle$, we obtain

$$\langle \mathbf{f} + \mathbf{g}, \ \mathbf{f} + \mathbf{g} \rangle \leq \langle \mathbf{f}, \mathbf{f} \rangle + 2\langle \mathbf{f}, \mathbf{f} \rangle^{1/2}\langle \mathbf{g}, \mathbf{g} \rangle^{1/2} + \langle \mathbf{g}, \mathbf{g} \rangle$$

as required.

32. Recall that $\cos(\alpha + \beta) = \cos \alpha \cos \beta - \sin \alpha \sin \beta$. Thus

$$\cos(\alpha - \beta) = \cos \alpha \cos(-\beta) - \sin \alpha \sin(-\beta)$$
$$= \cos \alpha \cos \beta + \sin \alpha \sin \beta$$

Adding these two equations gives the identity

$$\cos(\alpha + \beta) + \cos(\alpha - \beta) = 2 \cos \alpha \cos \beta$$

Thus if $k \neq \ell$, then $k + \ell$ and $k - \ell$ are nonzero and

$$\langle \mathbf{f}_k, \mathbf{f}_\ell \rangle = \int_0^\pi \cos(kx) \cos(\ell x) \, dx$$

$$= \int_0^\pi \frac{1}{2} [\cos(k + \ell)x + \cos(k - \ell)x] \, dx$$

$$= \frac{1}{2} \left[\frac{\sin(k + \ell)x}{k + \ell} + \frac{\sin(k - \ell)x}{k - \ell} \right]_0^\pi$$

$$= 0$$

33. **(a)** $W^\perp$ is the line $y = -x$.

(b) $W^\perp$ is the xz-plane.

(c) $W^\perp$ is the x-axis.

34. **(a)** If the solution space is a line in R^3, then, by Theorem 6.2.6, the row space is the orthogonal complement to that line, or the plane through the origin orthogonal to the line.

(c) If the homogeneous system $A^T \mathbf{x} = \mathbf{0}$ has a unique solution, then that solution must be the zero vector. Therefore the nullspace of A^T is the vector space consisting of just the zero vector. Its orthogonal complement, which is all of R^3, is the column space and hence also, because they have the same dimension, the row space of A. This also follows from the fact that A^T and hence also A must be invertible.

35. **(a)** False. Let $n = 3$, let V be the xy-plane, and let W be the x-axis. Then $V^\perp$ is the z-axis and $W^\perp$ is the yz-plane. In fact $V^\perp$ is a subspace of $W^\perp$.

(c) True. The two spaces are orthogonal complements and the only vector orthogonal to itself is the zero vector.

(d) False. For instance, if A is invertible, then both its row space and its column space are all of R^n.

36. **(a)** Let W be the subspace of all diagonal matrices in M_{22}. Thus, a vector $D = \begin{bmatrix} a & 0 \\ 0 & b \end{bmatrix}$ in W is fully described by its diagonal elements a and b. A vector $V = \begin{bmatrix} x & y \\ z & w \end{bmatrix}$ in M_{22} is in $W^\perp$ if and only if

$$\langle D, V \rangle = \operatorname{tr}(D^T V) = ax + bw = 0$$

for all possible values of a and b. This can only happen if $x = w = 0$. Thus, $W^\perp$ is the subspace of all matrices in M_{22} with only zeros on the diagonal.

(b) Let W be the subspace of all symmetric matrices in M_{22}. Thus, the elements of W are matrices of the form $S = \begin{bmatrix} a & c \\ c & b \end{bmatrix}$. A matrix $V = \begin{bmatrix} x & y \\ z & w \end{bmatrix}$ in M_{22} is in $W^\perp$ if and only if

$$\langle S, V \rangle = \operatorname{tr}(S^T V) = \operatorname{tr}(SV)$$
$$= ax + cz + cy + bw = 0$$

for all values of a, b, and c. This implies that $x = w = 0$ and $y = -z$. Thus, $W^\perp$ is the subspace of all skew-symmetric matrices in M_{22}.

EXERCISE SET 6.3

2. **(a)** The vectors are orthogonal, but since $\|(2,0)\| \neq 1$, they are not orthonormal.

   (c) The vectors are not orthogonal and therefore not orthonormal.

4. **(a)** The set is not orthonormal because the last two vectors are not orthogonal.

   (c) The set is not orthonormal because the last two vectors are not orthogonal.

5. See Exercise 3, Parts (b) and (c).

6. **(a)** Denote the vectors by $\mathbf{v}_1$, $\mathbf{v}_2$, $\mathbf{v}_3$, and $\mathbf{v}_4$. Then

$$\langle \mathbf{v}_1, \mathbf{v}_i \rangle = 0 \text{ for } i = 2, 3, 4$$

$$\langle \mathbf{v}_2, \mathbf{v}_3 \rangle = \frac{4}{9} - \frac{2}{9} - \frac{2}{9} = 0$$

$$\langle \mathbf{v}_2, \mathbf{v}_4 \rangle = \frac{2}{9} + \frac{2}{9} - \frac{4}{9} = 0$$

$$\langle \mathbf{v}_3, \mathbf{v}_4 \rangle = \frac{2}{9} - \frac{4}{9} + \frac{2}{9} = 0$$

Thus, the vectors are orthogonal. Moreover, $\|\mathbf{v}_1\| = 1$ and $\|\mathbf{v}_2\| = \|\mathbf{v}_3\| = \|\mathbf{v}_4\| = \frac{4}{9} + \frac{1}{9} + \frac{4}{9} = 1$. The vectors are therefore orthonormal.

7. **(b)** Call the vectors $\mathbf{u}_1, \mathbf{u}_2$ and $\mathbf{u}_3$. Then $\langle \mathbf{u}_1, \mathbf{u}_2 \rangle = 2 - 2 = 0$ and $\langle \mathbf{u}_1, \mathbf{u}_3 \rangle = \langle \mathbf{u}_2, \mathbf{u}_3 \rangle = 0$. The set is therefore orthogonal. Moreover, $\|\mathbf{u}_1\| = \sqrt{2}$, $\|\mathbf{u}_2\| = \sqrt{8} = 2\sqrt{2}$, and $\|\mathbf{u}_3\| = \sqrt{25} = 5$. Thus $\left\{ \dfrac{1}{\sqrt{2}}\mathbf{u}_1, \ \dfrac{1}{2\sqrt{2}}\mathbf{u}_2, \ \dfrac{1}{5}\mathbf{u}_3 \right\}$ is an orthonormal set.

9. It is easy to verify that $\mathbf{v}_1 \cdot \mathbf{v}_2 = \mathbf{v}_1 \cdot \mathbf{v}_3 = \mathbf{v}_2 \cdot \mathbf{v}_3 = 0$ and that $\|\mathbf{v}_3\| = 1$. Moreover, $\|\mathbf{v}_1\|^2 = (-3/5)^2 + (4/5)^2 = 1$ and $\|\mathbf{v}_2\| = (4/5)^2 + (3/5)^2 = 1$. Thus $\{\mathbf{v}_1, \mathbf{v}_2, \mathbf{v}_3\}$ is an orthonormal set in R^3. It will be an orthonormal basis provided that the three vectors are linearly independent, which is guaranteed by Theorem 6.3.3.

(b) By Theorem 6.3.1, we have

$$(3, -7, 4) = \left(-\frac{9}{5} - \frac{28}{5} + 0 \right) \mathbf{v}_1 + \left(\frac{12}{5} - \frac{21}{5} + 0 \right) \mathbf{v}_2 + 4\mathbf{v}_3$$

$$= (-37/5)\mathbf{v}_1 + (-9/5)\mathbf{v}_2 + 4\mathbf{v}_3$$

10. It is easy to check that $\mathbf{v}_1 \cdot \mathbf{v}_2 = \mathbf{v}_1 \cdot \mathbf{v}_3 = \mathbf{v}_1 \cdot \mathbf{v}_4 = \mathbf{v}_2 \cdot \mathbf{v}_3 = \mathbf{v}_2 \cdot \mathbf{v}_4 = \mathbf{v}_3 \cdot \mathbf{v}_4 = 0$, so that we have 4 orthogonal vectors in R^4. Thus, by Theorem 6.3.3, they form an orthogonal basis.

(b) By Formula (1), we have

$$(\sqrt{2}, \ -3\sqrt{2}, \ 5\sqrt{2}, \ -\sqrt{2}) = \frac{\sqrt{2} + 3\sqrt{2} + 10\sqrt{2} + \sqrt{2}}{1 + 1 + 4 + 1}\mathbf{v}_1 + \frac{-2\sqrt{2} - 6\sqrt{2} + 15\sqrt{2} - 2\sqrt{2}}{4 + 4 + 9 + 4}\mathbf{v}_2$$

$$+ \frac{\sqrt{2} - 6\sqrt{2} + 0 + \sqrt{2}}{1 + 4 + 0 + 1}\mathbf{v}_3 + \frac{\sqrt{2} + 0 + 0 - \sqrt{2}}{1 + 0 + 0 + 1}\mathbf{v}_4$$

$$= \frac{15\sqrt{2}}{7}\mathbf{v}_1 + \frac{5\sqrt{2}}{21}\mathbf{v}_2 - \frac{2\sqrt{2}}{3}\mathbf{v}_3 + 0\mathbf{v}_4$$

11. **(a)** We have $(\mathbf{w})_S = (\langle \mathbf{w}, \mathbf{u}_1 \rangle, \langle \mathbf{w}, \mathbf{u}_2 \rangle) = \left(\dfrac{-4}{\sqrt{2}}, \ \dfrac{10}{\sqrt{2}} \right) = (-2\sqrt{2}, \ 5\sqrt{2})$.

12. **(a)** We have $\mathbf{u} = \mathbf{w}_1 + \mathbf{w}_2 = \left(\dfrac{7}{5}, -\dfrac{1}{5}\right)$ and $\mathbf{v} = -\mathbf{w}_1 + 4\mathbf{w}_2 = \left(\dfrac{13}{5}, \dfrac{16}{5}\right)$.

(b) By Theorem 6.3.2, these numbers can be computed directly using the Euclidean inner product without reference to the basis vectors. We find that $\|\mathbf{u}\| = \sqrt{2}$, $d(\mathbf{u}, \mathbf{v}) = \sqrt{13}$, and $\langle \mathbf{u}, \mathbf{v} \rangle - 3$. Similar computations with the versions of $\mathbf{u}$ and $\mathbf{v}$ obtained in Part (a) fortunately yield the same results.

14. **(a)** $\|\mathbf{u}\| = (1 + 4 + 1 + 9)^{1/2} = \sqrt{15}$

$\|\mathbf{v} - \mathbf{w}\| = \|(2, 1, -2, 4)\| = (4 + 1 + 4 + 16)^{1/2} = 5$

$\|\mathbf{v} + \mathbf{w}\| = \|(-2, -7, 4, 6)\| = (4 + 49 + 16 + 36)^{1/2} = \sqrt{105}$

$\langle \mathbf{v}, \mathbf{w} \rangle = 0 + 12 + 3 + 5 = 20$

16. **(a)** Let

$$\mathbf{v}_1 = \dfrac{\mathbf{u}_1}{\|\mathbf{u}_1\|} = \left(\dfrac{1}{\sqrt{10}}, -\dfrac{3}{\sqrt{10}}\right)$$

Since $\langle \mathbf{u}_2, \mathbf{v}_1 \rangle = \dfrac{2}{\sqrt{10}} - \dfrac{6}{\sqrt{10}} = -\dfrac{4}{\sqrt{10}}$, we have

$$\mathbf{u}_2 - \langle \mathbf{u}_2, \mathbf{v}_1 \rangle \mathbf{v}_1 = (2, \ 2) + \dfrac{4}{\sqrt{10}}\left(\dfrac{1}{\sqrt{10}}, -\dfrac{3}{\sqrt{10}}\right) = \left(\dfrac{12}{5}, \dfrac{4}{5}\right)$$

This vector has norm $\left\|\left(\dfrac{12}{5}, \dfrac{4}{5}\right)\right\| = \dfrac{4\sqrt{10}}{5}$. Thus

$$\mathbf{v}_2 = \dfrac{5}{4\sqrt{10}}\left(\dfrac{12}{5}, \dfrac{4}{5}\right) = \left(\dfrac{3}{\sqrt{10}}, \dfrac{1}{\sqrt{10}}\right)$$

and $\{\mathbf{v}_1, \mathbf{v}_2\}$ is the desired orthonormal basis.

17. **(a)** Let

$$\mathbf{v}_1 = \dfrac{\mathbf{u}_1}{\|\mathbf{u}_1\|} = \left(\dfrac{1}{\sqrt{3}}, \dfrac{1}{\sqrt{3}}, \dfrac{1}{\sqrt{3}}\right)$$

Since $\langle \mathbf{u}_2, \mathbf{v}_1 \rangle = 0$, we have

$$\mathbf{v}_2 = \dfrac{\mathbf{u}_2}{\|\mathbf{u}_2\|} = \left(-\dfrac{1}{\sqrt{2}}, \dfrac{1}{\sqrt{2}}, 0\right)$$

Since $\langle \mathbf{u}_3, \mathbf{v}_1 \rangle = \dfrac{4}{\sqrt{3}}$ and $\langle \mathbf{u}_3, \mathbf{v}_2 \rangle = \dfrac{1}{\sqrt{2}}$, we have

$$\mathbf{u}_3 - \langle \mathbf{u}_3, \mathbf{v}_1 \rangle \mathbf{v}_1 - \langle \mathbf{u}_3, \mathbf{v}_2 \rangle \mathbf{v}_2$$

$$= (1,\ 2,\ 1) - \frac{4}{\sqrt{3}} \left(\frac{1}{\sqrt{3}},\ \frac{1}{\sqrt{3}},\ \frac{1}{\sqrt{3}} \right) - \frac{1}{\sqrt{2}} \left(-\frac{1}{\sqrt{2}},\ \frac{1}{\sqrt{2}},\ 0 \right)$$

$$= \left(\frac{1}{6},\ \frac{1}{6},\ -\frac{1}{3} \right)$$

This vector has norm $\left\| \left(\dfrac{1}{6},\ \dfrac{1}{6},\ -\dfrac{1}{3} \right) \right\| = \dfrac{1}{\sqrt{6}}$. Thus

$$\mathbf{v}_3 = \left(\frac{1}{\sqrt{6}},\ \frac{1}{\sqrt{6}},\ -\frac{2}{\sqrt{6}} \right)$$

and $\{\mathbf{v}_1, \mathbf{v}_2, \mathbf{v}_3\}$ is the desired orthonormal basis.

19. Since the third vector is the sum of the first two, we ignore it. Let $\mathbf{u}_1 = (0,1,2)$ and $\mathbf{u}_2 = (-1,0,1)$. Then

$$\mathbf{v}_1 = \frac{\mathbf{u}_1}{\|\mathbf{u}_1\|} = \left(0,\ \frac{1}{\sqrt{5}},\ \frac{2}{\sqrt{5}} \right)$$

Since $\langle \mathbf{u}_2, \mathbf{v}_1 \rangle = \dfrac{2}{\sqrt{5}}$, then

$$\mathbf{u}_2 - \langle \mathbf{u}_2, \mathbf{v}_1 \rangle \mathbf{v}_1 = \left(-1,\ -\frac{2}{5},\ \frac{1}{5} \right)$$

where $\left\| \left(-1,\ -\dfrac{2}{5},\ \dfrac{1}{5} \right) \right\| = \dfrac{\sqrt{30}}{5}$. Hence

$$\mathbf{v}_2 = \left(-\frac{5}{\sqrt{30}},\ \frac{-2}{\sqrt{30}},\ \frac{1}{\sqrt{30}} \right)$$

Thus $\{\mathbf{v}_1, \mathbf{v}_2\}$ is an orthonormal basis.

21. Note that $\mathbf{u}_1$ and $\mathbf{u}_2$ are orthonormal. Thus we apply Theorem 6.3.5 to obtain

$$\mathbf{w}_1 = \langle \mathbf{w}, \mathbf{u}_1 \rangle \mathbf{u}_1 + \langle \mathbf{w}, \mathbf{u}_2 \rangle \mathbf{u}_2$$

$$= -\left(\frac{4}{5}, \ 0, \ -\frac{3}{5} \right) + 2(0, \ 1, \ 0)$$

$$= \left(-\frac{4}{5}, \ 2, \ \frac{3}{5} \right)$$

and

$$\mathbf{w}_2 = \mathbf{w} - \mathbf{w}_1$$

$$= \left(\frac{9}{5}, \ 0, \ \frac{12}{5} \right)$$

22. Since $\mathbf{u}_1$ and $\mathbf{u}_2$ are not orthogonal, we must apply the Gram-Schmidt process before we can invoke Theorem 6.3.5. Thus we let

$$\mathbf{v}_1 = \frac{\mathbf{u}_1}{\|\mathbf{u}_1\|} = \left(\frac{1}{\sqrt{3}}, \ \frac{1}{\sqrt{3}}, \ \frac{1}{\sqrt{3}} \right)$$

Since $\langle \mathbf{u}_2, \mathbf{v}_1 \rangle = \dfrac{1}{\sqrt{3}}$, we have

$$\mathbf{u}_2 - \langle \mathbf{u}_2, \mathbf{v}_1 \rangle \mathbf{v}_1 = \left(\frac{5}{3}, \ -\frac{1}{3}, \ -\frac{4}{3} \right)$$

where $\left\| \left(\dfrac{5}{3}, \ -\dfrac{1}{3}, \ -\dfrac{4}{3} \right) \right\| = \dfrac{\sqrt{42}}{3}$. Thus

$$\mathbf{v}_2 = \left(\frac{5}{\sqrt{42}}, \ -\frac{1}{\sqrt{42}}, \ -\frac{4}{\sqrt{42}} \right)$$

Hence $\{\mathbf{v}_1, \mathbf{v}_2\}$ is the desired orthonormal set. Next, we compute $\langle \mathbf{w}, \mathbf{v}_1 \rangle = 2\sqrt{3}$ and $\langle \mathbf{w}, \mathbf{v}_2 \rangle = -\dfrac{7\sqrt{42}}{6}$. Theorem 6.3.5 then gives

$$\mathbf{w}_1 = \langle \mathbf{w}, \mathbf{v}_1 \rangle \mathbf{v}_1 + \langle \mathbf{w}, \mathbf{v}_2 \rangle \mathbf{v}_2 = \left(\frac{13}{14}, \ \frac{31}{14}, \ \frac{40}{14} \right)$$

and

$$\mathbf{w}_2 = \mathbf{w} - \mathbf{w}_1 = \left(\frac{1}{14}, \; -\frac{3}{14}, \; \frac{2}{14} \right)$$

24. **(a)** If we apply the Gram-Schmidt process to the column vectors

$$\mathbf{u}_1 = \begin{bmatrix} 1 \\ 2 \end{bmatrix}, \mathbf{u}_2 = \begin{bmatrix} -1 \\ 3 \end{bmatrix} \quad \text{we obtain} \quad \mathbf{q}_1 = \begin{bmatrix} 1/\sqrt{5} \\ 2/\sqrt{5} \end{bmatrix}, \mathbf{q}_2 = \begin{bmatrix} -2/\sqrt{5} \\ 1/\sqrt{5} \end{bmatrix}$$

Thus

$$R = \begin{bmatrix} \langle \mathbf{u}_1, \mathbf{q}_1 \rangle & \langle \mathbf{u}_2, \mathbf{q}_1 \rangle \\ 0 & \langle \mathbf{u}_2, \mathbf{q}_2 \rangle \end{bmatrix} = \begin{bmatrix} \sqrt{5} & \sqrt{5} \\ 0 & \sqrt{5} \end{bmatrix}$$

and

$$A = \begin{bmatrix} 1/\sqrt{5} & -2/\sqrt{5} \\ 2/\sqrt{5} & 1/\sqrt{5} \end{bmatrix} \begin{bmatrix} \sqrt{5} & \sqrt{5} \\ 0 & \sqrt{5} \end{bmatrix}$$

(c) If we apply the Gram-Schmidt process to the column vectors

$$\mathbf{u}_1 = \begin{bmatrix} 1 \\ -2 \\ 2 \end{bmatrix}, \mathbf{u}_2 = \begin{bmatrix} 1 \\ 1 \\ 1 \end{bmatrix} \quad \text{we obtain} \quad \mathbf{q}_1 = \begin{bmatrix} 1/3 \\ -2/3 \\ 2/3 \end{bmatrix}, \mathbf{q}_2 = \begin{bmatrix} 8/3\sqrt{26} \\ 11/3\sqrt{26} \\ 7/3\sqrt{26} \end{bmatrix}$$

Thus

$$R = \begin{bmatrix} \langle \mathbf{u}_1, \mathbf{q}_1 \rangle & \langle \mathbf{u}_2, \mathbf{q}_1 \rangle \\ 0 & \langle \mathbf{u}_2, \mathbf{q}_2 \rangle \end{bmatrix} = \begin{bmatrix} 3 & 1/3 \\ 0 & \sqrt{26}/3 \end{bmatrix}$$

and

$$A = \begin{bmatrix} 1/3 & 8/3\sqrt{26} \\ -2/3 & 11/3\sqrt{26} \\ 2/3 & 7/3\sqrt{26} \end{bmatrix} \begin{bmatrix} 3 & 1/3 \\ 0 & \sqrt{26}/3 \end{bmatrix}$$

25. By Theorem 6.3.1, we know that

$$\mathbf{w} = a_1\mathbf{v}_1 + a_2\mathbf{v}_2 + a_3\mathbf{v}_3$$

where $a_i = \langle \mathbf{w}, \mathbf{v}_i \rangle$. Thus

$$\|\mathbf{w}\|^2 = \langle \mathbf{w}, \mathbf{w} \rangle$$

$$= \sum_{i=1}^{3} a_i^2 \langle \mathbf{v}_i, \mathbf{v}_i \rangle + \sum_{i \neq j} a_i a_j \langle \mathbf{v}_i, \mathbf{v}_j \rangle$$

But $\langle \mathbf{v}_i, \mathbf{v}_j \rangle = 0$ if $i \neq j$ and $\langle \mathbf{v}_i, \mathbf{v}_i \rangle = 1$ because the set $\{\mathbf{v}_1, \mathbf{v}_2, \mathbf{v}_3\}$ is orthonormal. Hence

$$\|\mathbf{w}\|^2 = a_1^2 + a_2^2 + a_3^2$$

$$= \langle \mathbf{w}, \mathbf{v}_1 \rangle^2 + \langle \mathbf{w}, \mathbf{v}_2 \rangle^2 + \langle \mathbf{w}, \mathbf{v}_3 \rangle^2$$

26. Generalize the solution of Exercise 25.

27. Suppose the contrary; that is, suppose that

$$(*) \qquad \mathbf{u}_3 - \langle \mathbf{u}_3, \mathbf{v}_1 \rangle \mathbf{v}_1 - \langle \mathbf{u}_3, \mathbf{v}_2 \rangle \mathbf{v}_2 = \mathbf{0}$$

Then $(*)$ implies that $\mathbf{u}_3$ is a linear combination of $\mathbf{v}_1$ and $\mathbf{v}_2$. But $\mathbf{v}_1$ is a multiple of $\mathbf{u}_1$ while $\mathbf{v}_2$ is a linear combination of $\mathbf{u}_1$ and $\mathbf{u}_2$. Hence, $(*)$ implies that $\mathbf{u}_3$ is a linear combination of $\mathbf{u}_1$ and $\mathbf{u}_2$ and therefore that $\{\mathbf{u}_1, \mathbf{u}_2, \mathbf{u}_3\}$ is linearly dependent, contrary to the hypothesis that $\{\mathbf{u}_1, \ldots, \mathbf{u}_n\}$ is linearly independent. Thus, the assumption that $(*)$ holds leads to a contradiction.

28. The diagonal entries of R are $\langle \mathbf{u}_i, \mathbf{q}_i \rangle$ for $i = 1, \ldots, n$, where $\mathbf{q}_i = \mathbf{v}_i/\|\mathbf{v}_i\|$ is a vector resulting from an application of the Gram-Schmidt process. That process first generates an orthogonal set of vectors $\mathbf{v}_1, \ldots, \mathbf{v}_n$ where each vector $\mathbf{v}_i$ is $\mathbf{u}_i$ minus a linear combination of the vectors $\mathbf{v}_1, \ldots, \mathbf{v}_{i-1}$. Therefore $\langle \mathbf{u}_i, \mathbf{v}_i \rangle = \langle \mathbf{v}_i, \mathbf{v}_i \rangle$ for $i = 1, \ldots, n$. But since $\langle \mathbf{u}_i, \mathbf{q}_i \rangle = \langle \mathbf{u}_i, \mathbf{v}_i \rangle/\|\mathbf{v}_i\| = \langle \mathbf{v}_i, \mathbf{v}_i \rangle/\|\mathbf{v}_i\| = \|\mathbf{v}_i\|$ and each of the vectors $\mathbf{v}_i$ is nonzero, we have that each diagonal entry of R is nonzero.

29. We have $\mathbf{u}_1 = 1$, $\mathbf{u}_2 = x$, and $\mathbf{u}_3 = x^2$. Since

$$\|\mathbf{u}_1\|^2 = \langle \mathbf{u}_1, \mathbf{u}_1 \rangle = \int_{-1}^{1} 1 \, dx = 2$$

we let

$$\mathbf{v}_1 = \frac{1}{\sqrt{2}}$$

Then

$$\langle \mathbf{u}_2, \mathbf{v}_1 \rangle = \frac{1}{\sqrt{2}} \int_{-1}^{1} x \, dx = 0$$

and thus $\mathbf{v}_2 = \mathbf{u}_2 / \|\mathbf{u}_2\|$ where

$$\|\mathbf{u}_2\|^2 = \int_{-1}^{1} x^2 \, dx = \frac{2}{3}$$

Hence

$$\mathbf{v}_2 = \sqrt{\frac{3}{2}} \, x$$

In order to compute $\mathbf{v}_3$, we note that

$$\langle \mathbf{u}_3, \mathbf{v}_1 \rangle = \frac{1}{\sqrt{2}} \int_{-1}^{1} x^2 dx = \frac{\sqrt{2}}{3}$$

and

$$\langle \mathbf{u}_3, \mathbf{v}_2 \rangle = \sqrt{\frac{3}{2}} \int_{-1}^{1} x^3 dx = 0$$

Thus

$$\mathbf{u}_3 - \langle \mathbf{u}_3, \mathbf{v}_1 \rangle \mathbf{v}_1 - \langle \mathbf{u}_3, \mathbf{v}_2 \rangle \mathbf{v}_2 = x^2 - \frac{1}{3}$$

and

$$\left\| x^2 - \frac{1}{3} \right\|^2 = \int_{-1}^{1} \left[x^2 - \frac{1}{3} \right]^2 dx = \frac{8}{45}$$

Hence,

$$\mathbf{v}_3 = \sqrt{\frac{45}{8}} \left(x^2 - \frac{1}{3} \right) \quad \text{or} \quad \mathbf{v}_3 = \frac{\sqrt{5}}{2\sqrt{2}} \left(3x^2 - 1 \right)$$

30. **(b)** If we call the three polynomials $\mathbf{v}_1$, $\mathbf{v}_2$, and $\mathbf{v}_3$, and let $\mathbf{u} = 2 - 7x^2$, then, by virtue of Theorem 6.3.1,

$$\mathbf{u} = \langle \mathbf{u}, \mathbf{v}_1 \rangle \mathbf{v}_1 + \langle \mathbf{u}, \mathbf{v}_2 \rangle \mathbf{v}_2 + \langle \mathbf{u}, \mathbf{v}_3 \rangle \mathbf{v}_3$$

Here

$$\langle \mathbf{u}, \mathbf{v}_1 \rangle = \frac{1}{\sqrt{2}} \int_{-1}^{1} \left(2 - 7x^2 \right) dx = -\frac{\sqrt{2}}{3}$$

$$\langle \mathbf{u}, \mathbf{v}_2 \rangle = \sqrt{\frac{3}{2}} \int_{-1}^{1} \left(2x - 7x^3 \right) dx = 0$$

$$\langle \mathbf{u}, \mathbf{v}_3 \rangle = \frac{1}{2} \sqrt{\frac{5}{2}} \int_{-1}^{1} \left(2 - 7x^2 \right) \left(3x^2 - 1 \right) dx = -\frac{28}{15} \sqrt{\frac{5}{2}}$$

It is easy to check that $\mathbf{u} = -\dfrac{\sqrt{2}}{3} \mathbf{v}_1 - \dfrac{28}{15} \sqrt{\dfrac{5}{2}} \mathbf{v}_3$.

31. This is similar to Exercise 29 except that the lower limit of integration is changed from -1 to 0. If we again set $\mathbf{u}_1 = 1$, $\mathbf{u}_2 = x$, and $\mathbf{u}_3 = x^2$, then $\| \mathbf{u}_1 \| = 1$ and thus

$$\mathbf{v}_1 = 1$$

Then $\langle \mathbf{u}_2, \mathbf{v}_1 \rangle = \displaystyle\int_0^1 x \, dx = \frac{1}{2}$ and thus

$$\mathbf{v}_2 = \frac{x - 1/2}{\| x - 1/2 \|} = \sqrt{12}(x - 1/2)$$

or

$$\mathbf{v}_2 = \sqrt{3}(2x - 1)$$

Finally,

$$\langle \mathbf{u}_3, \mathbf{v}_1 \rangle = \int_0^1 x^2 dx = \frac{1}{3}$$

and

$$\langle \mathbf{u}_3, \mathbf{v}_2 \rangle = \sqrt{3} \int_0^1 (2x^3 - x^2) dx = \frac{\sqrt{3}}{6}$$

Thus

$$\mathbf{v}_3 = \frac{x^2 - \dfrac{1}{3} - \dfrac{1}{2}(2x - 1)}{\left\| x^2 - \dfrac{1}{3} - \dfrac{1}{2}(2x - 1) \right\|} = 6\sqrt{5} \left(x^2 - x + \frac{1}{6} \right)$$

or

$$\mathbf{v}_3 = \sqrt{5}(6x^2 - 6x + 1)$$

32. **(a)** Let $S = \{\mathbf{w}_1, \mathbf{w}_2, \ldots, \mathbf{w}_n\}$ be the orthonormal basis. Then $(\mathbf{u})_S = (u_1, u_2, \ldots, u_n)$ means that $\mathbf{u} = u_1\mathbf{w}_1 + u_2\mathbf{w}_2 + \cdots + u_n\mathbf{w}_n$. Thus

$$\|\mathbf{u}\|^2 = \langle u_1\mathbf{w}_1 + \cdots + u_n\mathbf{w}_n, u_1\mathbf{w}_1 + \cdots + u_n\mathbf{w}_n \rangle$$

$$= \sum_{i=1}^{n} u_i u_i \langle \mathbf{w}_i, \mathbf{w}_i \rangle + \sum_{i \neq j} u_i u_j \langle \mathbf{w}_i, \mathbf{w}_j \rangle$$

$$= \sum_{i=1}^{n} u_i^2 \qquad \text{(because } S \text{ is orthonormal)}$$

Taking positive square roots yields Theorem 6.3.2(a).

Note that this result also follows from Theorem 6.3.1 and Exercise 26.

(b) Since $d(\mathbf{u}, \mathbf{v}) = \|\mathbf{u} - \mathbf{v}\|$, we can apply Theorem 6.3.2(a) to $\mathbf{u} - \mathbf{v}$ to obtain Theorem 6.3.2(b).

(c) Let $S = \{\mathbf{w}_1, \mathbf{w}_2, \ldots, \mathbf{w}_n\}$ be the orthonormal basis. Then

$$\mathbf{u} = \sum_{i=1}^{n} u_i\mathbf{w}_i \qquad \text{and} \qquad \mathbf{v} = \sum_{i=1}^{n} v_i\mathbf{w}_i$$

Hence

$$\langle \mathbf{u}, \mathbf{v} \rangle = \langle u_1 \mathbf{w}_1 + \cdots + u_n \mathbf{w}_n, \; v_1 \mathbf{w}_1 + \cdots + v_n \mathbf{w}_n \rangle$$

$$= \sum_{i=1}^{n} u_i v_i \langle \mathbf{w}_i, \mathbf{w}_i \rangle + \sum_{i \neq j} u_i v_j \langle \mathbf{w}_i, \mathbf{w}_j \rangle$$

$$= \sum_{i=1}^{n} u_i v_i \qquad \text{(because } S \text{ is orthonormal)}$$

33. Let W be a finite dimensional subspace of the inner product space V and let $\{\mathbf{v}_1, \mathbf{v}_2, \ldots, \mathbf{v}_r\}$ be an orthonormal basis for W. Then if $\mathbf{u}$ is any vector in V, we know from Theorem 6.3.4 that $\mathbf{u} = \mathbf{w}_1 + \mathbf{w}_2$ where $\mathbf{w}_1$ is in W and $\mathbf{w}_2$ is in $W^\perp$. Moreover, this decomposition of $\mathbf{u}$ is unique. Theorem 6.3.5 gives us a candidate for $\mathbf{w}_1$. To prove the theorem, we must show that if $\mathbf{w}_1 = \langle \mathbf{u}, \mathbf{v}_1 \rangle \mathbf{v}_1 + \cdots + \langle \mathbf{u}, \mathbf{v}_r \rangle \mathbf{v}_r$ and, therefore, that $\mathbf{w}_2 = \mathbf{u} - \mathbf{w}_1$ then

$$\text{(i)} \qquad \mathbf{w}_1 \text{ is in } W$$

and

$$\text{(ii)} \qquad \mathbf{w}_2 \text{ is orthogonal to } W.$$

That is, we must show that this candidate "works." Then, since $\mathbf{w}_1$ is unique, it will be $\text{proj}_W \mathbf{u}$.

Part (i) follows immediately because $\mathbf{w}_1$ is, by definition, a linear combination of the vectors $\mathbf{v}_1, \mathbf{v}_2, \ldots, \mathbf{v}_r$.

To prove Part (ii), we use the facts that $\langle \mathbf{v}_i, \mathbf{v}_j \rangle = 0$ if $i \neq j$ and $\langle \mathbf{v}_i, \mathbf{v}_i \rangle = 1$, together with the definition of $\mathbf{w}_1$, to show that for $i = 1, \ldots, r$

$$\langle \mathbf{w}_2, \mathbf{v}_i \rangle = \langle \mathbf{u} - \mathbf{w}_1, \mathbf{v}_i \rangle$$

$$= \langle \mathbf{u}, \mathbf{v}_i \rangle - \langle \mathbf{w}_1, \mathbf{v}_i \rangle$$

$$= \langle \mathbf{u}, \mathbf{v}_i \rangle - \langle \mathbf{u}, \mathbf{v}_i \rangle \langle \mathbf{v}_i, \mathbf{v}_i \rangle$$

$$= \langle \mathbf{u}, \mathbf{v}_i \rangle - \langle \mathbf{u}, \mathbf{v}_i \rangle$$

$$= 0$$

Thus, $\mathbf{w}_2$ is orthogonal to each of the vectors $\mathbf{v}_1, \mathbf{v}_2, \ldots, \mathbf{v}_r$ and hence $\mathbf{w}_2$ is in $W^\perp$.

If the vectors $\mathbf{v}_i$ form an orthogonal set, not necessarily orthonormal, then we must normalize them to obtain Part (b) of the theorem.

34. (a) Call the plane Γ and note that $\mathbf{n} = (a, b, c)$ is the normal vector to Γ. Choose an arbitrary vector $\mathbf{u}$ in Γ and note that $\mathbf{u} \cdot \mathbf{n} = 0$. Hence $\mathbf{u} \times \mathbf{n}$ (and $\mathbf{n} \times \mathbf{u}$) must lie in Γ and be perpendicular to both $\mathbf{u}$ and $\mathbf{n}$. Therefore $\mathbf{u}$ and $\mathbf{u} \times \mathbf{n}$ (or $\mathbf{n} \times \mathbf{u}$) form an orthonogonal basis for Γ. Normalize each of these vectors to obtain the required orthonormal basis.

(b) In this case, $\mathbf{n} = (1, 2, -1)$. The vector $\mathbf{u} = (1, 1, 3)$ lies in the plane Γ given by $x + 2y - z = 0$. Note that $\mathbf{n} \times \mathbf{u} = (7, -4, -1)$ lies in the plane and that $\mathbf{u} \cdot (\mathbf{n} \times \mathbf{u}) = 0$. So an orthonormal basis for Γ consists of the vectors

$$\frac{\mathbf{u}}{\|\mathbf{u}\|} = \left(\frac{1}{\sqrt{11}}, \frac{1}{\sqrt{11}}, \frac{3}{\sqrt{11}} \right)$$

and

$$\frac{\mathbf{n} \times \mathbf{u}}{\|\mathbf{n} \times \mathbf{u}\|} = \left(\frac{7}{\sqrt{66}}, \frac{-4}{\sqrt{66}}, \frac{-1}{\sqrt{66}} \right)$$

35. The vectors $\mathbf{x} = (1/\sqrt{3}, 0)$ and $\mathbf{y} = (0, 1/\sqrt{2})$ are orthonormal with respect to the given inner product. However, although they are orthogonal with respect to the Euclidean inner product, they are not orthonormal.

The vectors $\mathbf{x} = (2/\sqrt{30}, 3/\sqrt{30})$ and $\mathbf{y} = (1/\sqrt{5}, -1/\sqrt{5})$ are orthonormal with respect to the given inner product. However, they are neither orthogonal nor of unit length with respect to the Euclidean inner product.

37. (a) True. Suppose that $\mathbf{v}_1, \mathbf{v}_2, \ldots, \mathbf{v}_n$ is an orthonormal set of vectors. If they were linearly dependent, then there would be a linear combination

$$c_1 \mathbf{v}_1 + c_2 \mathbf{v}_2 + \cdots + c_n \mathbf{v}_n = 0$$

where at least one of the numbers $c_i \neq 0$. But

$$c_i = \langle \mathbf{v}_i, c_1 \mathbf{v}_1 + c_2 \mathbf{v}_2 + \cdots + c_n \mathbf{v}_n \rangle = \langle \mathbf{v}_i, 0 \rangle = 0$$

for $i = 1, \ldots, n$. Thus, the orthonormal set of vectors cannot be linearly dependent.

(b) False. The zero vector space has basis $\mathbf{0}$, which cannot be converted to a unit vector.

(c) True, since $\text{proj}_W\,\mathbf{u}$ is in W and $\text{proj}_{W^\perp}\mathbf{u}$ is in $W^\perp$.

(d) True. If A is a (necessarily square) matrix with a nonzero determinant, then A has linearly independent column vectors. Thus, by Theorem 6.3.7, A has a QR decomposition.

EXERCISE SET 6.4

1. **(a)** If we call the system $A\mathbf{x} = \mathbf{b}$, then the associated normal system is $A^T A\mathbf{x} = A^T\mathbf{b}$, or

$$\begin{bmatrix} 1 & 2 & 4 \\ -1 & 3 & 5 \end{bmatrix} \begin{bmatrix} 1 & -1 \\ 2 & 3 \\ 4 & 5 \end{bmatrix} \begin{bmatrix} x_1 \\ x_2 \end{bmatrix} = \begin{bmatrix} 1 & 2 & 4 \\ -1 & 3 & 5 \end{bmatrix} \begin{bmatrix} 2 \\ -1 \\ 5 \end{bmatrix}$$

which simplifies to

$$\begin{bmatrix} 21 & 25 \\ 25 & 35 \end{bmatrix} \begin{bmatrix} x_1 \\ x_2 \end{bmatrix} = \begin{bmatrix} 20 \\ 20 \end{bmatrix}$$

2. **(a)** We have

$$A^T A = \begin{bmatrix} -1 & 2 & 0 \\ 3 & 1 & 1 \\ 2 & 3 & 1 \end{bmatrix} \begin{bmatrix} -1 & 3 & 2 \\ 2 & 1 & 3 \\ 0 & 1 & 1 \end{bmatrix} = \begin{bmatrix} 5 & -1 & 4 \\ -1 & 11 & 10 \\ 4 & 10 & 14 \end{bmatrix}$$

so that $\det(A^T A) = 0$. Hence A does not have linearly independent column vectors.

3. **(a)** The associated normal system is $A^T A\mathbf{x} = A^T\mathbf{b}$, or

$$\begin{bmatrix} 1 & -1 & -1 \\ 1 & 1 & 2 \end{bmatrix} \begin{bmatrix} 1 & 1 \\ -1 & 1 \\ -1 & 2 \end{bmatrix} \begin{bmatrix} x_1 \\ x_2 \end{bmatrix} = \begin{bmatrix} 1 & -1 & -1 \\ 1 & 1 & 2 \end{bmatrix} \begin{bmatrix} 7 \\ 0 \\ -7 \end{bmatrix}$$

or

$$\begin{bmatrix} 3 & -2 \\ -2 & 6 \end{bmatrix} \begin{bmatrix} x_1 \\ x_2 \end{bmatrix} = \begin{bmatrix} 14 \\ -7 \end{bmatrix}$$

This system has solution $x_1 = 5, x_2 = 1/2$, which is the least squares solution of $A\mathbf{x} = \mathbf{b}$.

The orthogonal projection of $\mathbf{b}$ on the column space of A is $A\mathbf{x}$, or

$$\begin{bmatrix} 1 & 1 \\ -1 & 1 \\ -1 & 2 \end{bmatrix} \begin{bmatrix} 5 \\ 1/2 \end{bmatrix} = \begin{bmatrix} 11/2 \\ -9/2 \\ -4 \end{bmatrix}$$

3. (c) The associated normal system is

$$\begin{bmatrix} 1 & 2 & 1 & 1 \\ 0 & 1 & 1 & 1 \\ -1 & -2 & 0 & -1 \end{bmatrix} \begin{bmatrix} 1 & 0 & -1 \\ 2 & 1 & -2 \\ 1 & 1 & 0 \\ 1 & 1 & -1 \end{bmatrix} \begin{bmatrix} x_1 \\ x_2 \\ x_3 \end{bmatrix}$$

$$= \begin{bmatrix} 1 & 2 & 1 & 1 \\ 0 & 1 & 1 & 1 \\ -1 & -2 & 0 & -1 \end{bmatrix} \begin{bmatrix} 6 \\ 0 \\ 9 \\ 3 \end{bmatrix}$$

or

$$\begin{bmatrix} 7 & 4 & -6 \\ 4 & 3 & -3 \\ -6 & -3 & 6 \end{bmatrix} \begin{bmatrix} x_1 \\ x_2 \\ x_3 \end{bmatrix} = \begin{bmatrix} 18 \\ 12 \\ -9 \end{bmatrix}$$

This system has solution $x_1 = 12, x_2 = -3, x_3 = 9$, which is the least squares solution of $A\mathbf{x} = \mathbf{b}$.

The orthogonal projection of $\mathbf{b}$ on the column space of A is $A\mathbf{x}$, or

$$\begin{bmatrix} 1 & 0 & -1 \\ 2 & 1 & -2 \\ 1 & 1 & 0 \\ 1 & 1 & -1 \end{bmatrix} \begin{bmatrix} 12 \\ -3 \\ 9 \end{bmatrix} = \begin{bmatrix} 3 \\ 3 \\ 9 \\ 0 \end{bmatrix}$$

which can be written as $(3, 3, 9, 0)$.

4. **(a)** First we find a least squares solution of $A\mathbf{x} = \mathbf{u}$ where $A = [\mathbf{u}_1^T | \mathbf{v}_2^T]$. The associated normal system is

$$\begin{bmatrix} 1 & 1 & 0 \\ 1 & 2 & 1 \end{bmatrix} \begin{bmatrix} 1 & 1 \\ 1 & 2 \\ 0 & 1 \end{bmatrix} \begin{bmatrix} x_1 \\ x_2 \end{bmatrix} = \begin{bmatrix} 1 & 1 & 0 \\ 1 & 2 & 1 \end{bmatrix} \begin{bmatrix} 2 \\ 1 \\ 3 \end{bmatrix}$$

or

$$\begin{bmatrix} 2 & 3 \\ 3 & 6 \end{bmatrix} \begin{bmatrix} x_1 \\ x_2 \end{bmatrix} = \begin{bmatrix} 3 \\ 7 \end{bmatrix}$$

This system has solution $x_1 = -1$, $x_2 = 5/3$, which is the least squares solution. The desired orthogonal projection is $A\mathbf{x}$, or

$$\begin{bmatrix} 1 & 1 \\ 1 & 2 \\ 0 & 1 \end{bmatrix} \begin{bmatrix} -1 \\ 5/3 \end{bmatrix} = \begin{bmatrix} 2/3 \\ 7/3 \\ 5/3 \end{bmatrix}$$

or $(2/3, 7/3, 5/3)$.

5. **(a)** First we find a least squares solution of $A\mathbf{x} = \mathbf{u}$ where $A = [\mathbf{v}_1^T | \mathbf{v}_2^T | \mathbf{v}_3^T]$. The associated normal system is

$$\begin{bmatrix} 2 & 1 & 1 & 1 \\ 1 & 0 & 1 & 1 \\ -2 & -1 & 0 & -1 \end{bmatrix} \begin{bmatrix} 2 & 1 & -2 \\ 1 & 0 & -1 \\ 1 & 1 & 0 \\ 1 & 1 & -1 \end{bmatrix} \begin{bmatrix} x_1 \\ x_2 \\ x_3 \end{bmatrix}$$

$$= \begin{bmatrix} 2 & 1 & 1 & 1 \\ 1 & 0 & 1 & 1 \\ -2 & -1 & 0 & -1 \end{bmatrix} \begin{bmatrix} 6 \\ 3 \\ 9 \\ 6 \end{bmatrix}$$

or

$$\begin{bmatrix} 7 & 4 & -6 \\ 4 & 3 & -3 \\ -6 & -3 & 6 \end{bmatrix} \begin{bmatrix} x_1 \\ x_2 \\ x_3 \end{bmatrix} = \begin{bmatrix} 30 \\ 21 \\ -21 \end{bmatrix}$$

This system has solution $x_1 = 6$, $x_2 = 3$, $x_3 = 4$, which is the least squares solution. The desired orthogonal projection is $A\mathbf{x}$, or

$$
\begin{bmatrix} 2 & 1 & -2 \\ 1 & 0 & -1 \\ 1 & 1 & 0 \\ 1 & 1 & -1 \end{bmatrix}
\begin{bmatrix} 6 \\ 3 \\ 4 \end{bmatrix}
=
\begin{bmatrix} 7 \\ 2 \\ 9 \\ 5 \end{bmatrix}
$$

or $(7, 2, 9, 5)$.

6. The solution of the given system can be expressed as $x_1 = -s + t$, $x_2 = -s - t$, $x_3 = 2s$, $x_4 = 2t$. Thus the solution space is spanned by the vectors

$$
\mathbf{v}_1 = \begin{bmatrix} -1 \\ -1 \\ 2 \\ 0 \end{bmatrix}
\quad \text{and} \quad
\mathbf{v}_2 = \begin{bmatrix} 1 \\ -1 \\ 0 \\ 2 \end{bmatrix}
$$

Let $A = [\mathbf{v}_1 | \mathbf{v}_2]$. We next find a least squares solution of $A\mathbf{x} = \mathbf{u}$. The associated normal system is

$$
\begin{bmatrix} -1 & -1 & 2 & 0 \\ 1 & -1 & 0 & 2 \end{bmatrix}
\begin{bmatrix} -1 & 1 \\ -1 & -1 \\ 2 & 0 \\ 0 & 2 \end{bmatrix}
\begin{bmatrix} x_1 \\ x_2 \end{bmatrix}
=
\begin{bmatrix} -1 & -1 & 2 & 0 \\ 1 & -1 & 0 & 2 \end{bmatrix}
\begin{bmatrix} 5 \\ 6 \\ 7 \\ 2 \end{bmatrix}
$$

or

$$
\begin{bmatrix} 6 & 0 \\ 0 & 6 \end{bmatrix}
\begin{bmatrix} x_1 \\ x_2 \end{bmatrix}
=
\begin{bmatrix} 3 \\ 3 \end{bmatrix}
$$

This system has solution $x_1 = 1/2$, $x_2 = 1/2$, which is the least squares solution. The desired orthogonal projection is $A\mathbf{x}$, or

$$
\begin{bmatrix} -1 & 1 \\ -1 & -1 \\ 2 & 0 \\ 0 & 2 \end{bmatrix}
\begin{bmatrix} 1/2 \\ 1/2 \end{bmatrix}
=
\begin{bmatrix} 0 \\ -1 \\ 1 \\ 1 \end{bmatrix}
$$

or $(0, -1, 1, 1)$.

7. **(a)** If we use the vector $(1,0)$ as a basis for the x-axis and let $A = \begin{bmatrix} 1 \\ 0 \end{bmatrix}$, then we have

$$[P] = A(A^T A)^{-1} A^T = \begin{bmatrix} 1 \\ 0 \end{bmatrix} [\,1\,] \begin{bmatrix} 1 & 0 \end{bmatrix} = \begin{bmatrix} 1 & 0 \\ 0 & 0 \end{bmatrix}$$

8. **(a)** If we use the vectors $(1,0,0)$ and $(0,0,1)$ as a basis for the xz-plane and let

$$A = \begin{bmatrix} 1 & 0 \\ 0 & 0 \\ 0 & 1 \end{bmatrix}$$

then we have

$$[P] = \begin{bmatrix} 1 & 0 \\ 0 & 0 \\ 0 & 1 \end{bmatrix} \left(\begin{bmatrix} 1 & 0 & 0 \\ 0 & 0 & 1 \end{bmatrix} \begin{bmatrix} 1 & 0 \\ 0 & 0 \\ 0 & 1 \end{bmatrix} \right)^{-1} \begin{bmatrix} 1 & 0 & 0 \\ 0 & 0 & 1 \end{bmatrix}$$

$$= \begin{bmatrix} 1 & 0 \\ 0 & 0 \\ 0 & 1 \end{bmatrix} \begin{bmatrix} 1 & 0 \\ 0 & 1 \end{bmatrix} \begin{bmatrix} 1 & 0 & 0 \\ 0 & 0 & 1 \end{bmatrix}$$

$$= \begin{bmatrix} 1 & 0 & 0 \\ 0 & 0 & 0 \\ 0 & 0 & 1 \end{bmatrix}$$

9. **(a)** By inspection, the vectors $(1, 0, -5)$ and $(0, 1, 3)$ lie in the plane W and they form a linearly independent set with the appropriate number of vectors. Hence they are a basis for W.

(b) Using the basis for W found in Part (a), we let

$$A = \begin{bmatrix} 1 & 0 \\ 0 & 1 \\ -5 & 3 \end{bmatrix}$$

Thus the standard matrix for the orthogonal projection on W is

$$[P] = \begin{bmatrix} 1 & 0 \\ 0 & 1 \\ -5 & 3 \end{bmatrix} \left(\begin{bmatrix} 1 & 0 & -5 \\ 0 & 1 & 3 \end{bmatrix} \begin{bmatrix} 1 & 0 \\ 0 & 1 \\ -5 & 3 \end{bmatrix} \right)^{-1} \begin{bmatrix} 1 & 0 & -5 \\ 0 & 1 & 3 \end{bmatrix}$$

$$= \begin{bmatrix} 1 & 0 \\ 0 & 1 \\ -5 & 3 \end{bmatrix} \begin{bmatrix} 26 & -15 \\ -15 & 10 \end{bmatrix}^{-1} \begin{bmatrix} 1 & 0 & -5 \\ 0 & 1 & 3 \end{bmatrix}$$

$$= \begin{bmatrix} 1 & 0 \\ 0 & 1 \\ -5 & 3 \end{bmatrix} \begin{bmatrix} 2/7 & 3/7 \\ 3/7 & 26/35 \end{bmatrix} \begin{bmatrix} 1 & 0 & -5 \\ 0 & 1 & 3 \end{bmatrix}$$

$$= \begin{bmatrix} 2/7 & 3/7 & -1/7 \\ 3/7 & 26/35 & 3/35 \\ -1/7 & 3/35 & 34/35 \end{bmatrix} = \frac{1}{35} \begin{bmatrix} 10 & 15 & -5 \\ 15 & 26 & 3 \\ -5 & 3 & 34 \end{bmatrix}$$

9. (c) By Part (b), the point (x_0, y_0, z_0) projects to the point

$$\frac{1}{35} \begin{bmatrix} 10 & 15 & -5 \\ 15 & 26 & 3 \\ -5 & 3 & 34 \end{bmatrix} \begin{bmatrix} x_0 \\ y_0 \\ z_0 \end{bmatrix} = \begin{bmatrix} (2x_0 + 3y_0 - z_0)/7 \\ (15x_0 + 26y_0 + 3z_0)/35 \\ (-5x_0 + 3y_0 + 34z_0)/35 \end{bmatrix}$$

on the plane W.

(d) By Part (c), the point $(1, -2, 4)$ projects to the point $(-8/7, -5/7, 25/7)$ on the plane. The distance between these two points is $3\sqrt{35}/7$.

10. (a) The vector $\mathbf{v} = (2, -1, 4)$ forms a basis for the line W.

(b) If we let $A = [\mathbf{v}^T]$, then the standard matrix for the orthogonal projection on W is

$$[P] = A(A^T A)^{-1} A^T = \begin{bmatrix} 2 \\ -1 \\ 4 \end{bmatrix} \left(\begin{bmatrix} 2 & -1 & 4 \end{bmatrix} \begin{bmatrix} 2 \\ -1 \\ 4 \end{bmatrix} \right)^{-1} \begin{bmatrix} 2 & -1 & 4 \end{bmatrix}$$

$$= \begin{bmatrix} 2 \\ -1 \\ 4 \end{bmatrix} \begin{bmatrix} \dfrac{1}{21} \end{bmatrix} \begin{bmatrix} 2 & -1 & 4 \end{bmatrix}$$

$$= \frac{1}{21} \begin{bmatrix} 4 & -2 & 8 \\ -2 & 1 & -4 \\ 8 & -4 & 16 \end{bmatrix}$$

(c) By Part (b), the point $P_0(x_0, y_0, z_0)$ projects to the point on the line W given by

$$\frac{1}{21} \begin{bmatrix} 4 & -2 & 8 \\ -2 & 1 & -4 \\ 8 & -4 & 16 \end{bmatrix} \begin{bmatrix} x_0 \\ y_0 \\ z_0 \end{bmatrix} = \begin{bmatrix} (4x_0 - 2y_0 + 8z_0)/21 \\ (-2x_0 + y_0 - 4z_0)/21 \\ (8x_0 - 4y_0 + 16z_0)/21 \end{bmatrix}$$

(d) By the result in Part (c), the point $(2, 1, -3)$ projects to the point $(-6/7, 3/7, -12/7)$. The distance between these two points is $\sqrt{497}/7$.

11. **(a)** Using horizontal vector notation, we have $\mathbf{b} = (7, 0, -7)$ and $A\bar{\mathbf{x}} = (11/2, -9/2, -4)$. Therefore $A\bar{\mathbf{x}} - \mathbf{b} = (-3/2, -9/2, 3)$, which is orthogonal to both of the vectors $(1, -1, -1)$ and $(1, 1, 2)$ which span the column space of A. Hence the error vector is orthogonal to the column space of A.

(c) In horizontal vector notation, $\mathbf{b} = (6, 0, 9, 3)$ and $A\bar{\mathbf{x}} = (3, 3, 9, 0)$. Hence $A\bar{\mathbf{x}} - \mathbf{b} = (-3, 3, 0, -3)$, which is orthogonal to the three vectors $(1, 2, 1, 1), (0, 1, 1, 1)$, and $(-1, -2, 0, -1)$ which span the column space of A. Therefore $A\bar{\mathbf{x}} - \mathbf{b}$ is orthogonal to the column space of A.

12. Since A has linearly independent column vectors, Theorem 6.4.4 guarantees that there is a unique least squares solution $\bar{\mathbf{x}}$ of the system $A\mathbf{x} = \mathbf{b}$. Since $\bar{\mathbf{x}}$ is a least squares solution, it minimizes $\|A\mathbf{x} - \mathbf{b}\|$. But if there is a vector $\bar{\bar{\mathbf{x}}}$ for which $A\bar{\bar{\mathbf{x}}} = \mathbf{b}$, then this minimum is zero and occurs when $\mathbf{x} = \bar{\bar{\mathbf{x}}}$. Hence $\bar{\mathbf{x}} = \bar{\bar{\mathbf{x}}}$ by the uniqueness of the least squares solution.

13.   Recall that if $\mathbf{b}$ is orthogonal to the column space of A, then $\text{proj}_W \mathbf{b} = \mathbf{0}$.

14.   **(a)**   We have that $[P] = A(A^T A)^{-1} A^T$ where A is any matrix formed using a set of basis vectors for W as its column vectors. Therefore

$$[P]^2 = [A(A^T A)^{-1} A^T][A(A^T A)^{-1} A^T]$$

$$= A[(A^T A)^{-1}(A^T A)](A^T A)^{-1} A^T$$

$$= A(A^T A)^{-1} A^T$$

$$= [P]$$

(c)   We need to show that $[P]^T = [P]$. Define A as in Part (a). Then

$$[P]^T = [A(A^T A)^{-1} A^T]^T$$

$$= (A^T)^T[(A^T A)^{-1}]^T A^T \quad \text{because } (BC)^T = C^T B^T$$

$$= A[(A^T A)^{-1}]^T A^T \quad \text{because } (A^T)^T = A$$

We need only show that $[(A^T A)^{-1}]^T = (A^T A)^{-1}$ to have the desired result. This follows from the fact that if B is any invertible matrix, then $(B^{-1})^T = (B^T)^{-1}$, so that $[(A^T A)^{-1}]^T = [(A^T A)^T]^{-1} = (A^T A)^{-1}$. To see the above result, note that if B is invertible, then

$$BB^{-1} = (BB^{-1})^T = I$$

so that

$$(B^{-1})^T B^T = I$$

or

$$(B^{-1})^T = (B^T)^{-1}$$

15.   If A is an $m \times n$ matrix with linearly independent row vectors, then A^T is an $n \times m$ matrix with linearly independent column vectors which span the row space of A. Therefore, by Formula (6) and the fact that $(A^T)^T = A$, the standard matrix for the orthogonal projection, S, of R^n on the row space of A is $[S] = A^T(AA^T)^{-1}A$.

17. **(a)** Call the vector $A\mathbf{x}$. Then by Theorem 6.4.4, $A\mathbf{x} = A(A^T A)^{-1} A^T \mathbf{b}$.

   (c) The error is $\|A\mathbf{x} - \mathbf{b}\|$.

   (d) The standard matrix is $A(A^T A)^{-1} A^T$.

EXERCISE SET 6.5

2. **(b)** $T(\mathbf{x}) = (14/3, -5/3, 11/3)$ and $\|T(\mathbf{x})\| = \|\mathbf{x}\| = \sqrt{38}$.

3. **(b)** Since the row vectors form an orthonormal set, the matrix is orthogonal. Therefore its inverse is its transpose,

$$
\begin{bmatrix}
1/\sqrt{2} & 1/\sqrt{2} \\
-1/\sqrt{2} & 1/\sqrt{2}
\end{bmatrix}
$$

(c) Since the Euclidean inner product of Column 2 and Column 3 is not zero, the column vectors do not form an orthonormal set and the matrix is not orthogonal.

(f) Since the norm of Column 3 is not 1, the matrix is not orthogonal.

5. **(b)** We have $(\mathbf{w})_S = (a, b)$ where $\mathbf{w} = a\mathbf{u}_1 + b\mathbf{u}_2$. Thus

$$2a + 3b = 1$$
$$-4a + 8b = 1$$

or $a = \dfrac{5}{28}$ and $b = \dfrac{3}{14}$. Hence $(\mathbf{w})_S = \left(\dfrac{5}{28}, \dfrac{3}{14}\right)$ and

$$
[\mathbf{w}]_S =
\begin{bmatrix}
\dfrac{5}{28} \\
\dfrac{3}{14}
\end{bmatrix}
$$

6.   (a)   Let $\mathbf{v} = a\mathbf{v}_1 + b\mathbf{v}_2 + c\mathbf{v}_3$. Then

$$a + 2b + 3c = 2$$
$$2b + 3c = -1$$
$$3c = 3$$

so that $a = 3$, $b = -2$, and $c = 1$. Thus $(\mathbf{v})_S = (3, \ -2, \ 1)$ and

$$[\mathbf{v}]_S = \begin{bmatrix} 3 \\ -2 \\ 1 \end{bmatrix}$$

7.   (b)   Let $\mathbf{p} = a\mathbf{p}_1 + b\mathbf{p}_2 + c\mathbf{p}_3$. Then

$$a + b = 2$$
$$a + c = -1$$
$$b + c = 1$$

or $a = 0$, $b = 2$, and $c = -1$. Thus $(\mathbf{v})_S = (0, 2, -1)$ and

$$[\mathbf{v}]_S = \begin{bmatrix} 0 \\ 2 \\ -1 \end{bmatrix}$$

8.   Let $A = aA_1 + bA_2 + cA_3 + dA_4$. Then

$$-a + b = 2$$
$$a + b = 0$$
$$c = -1$$
$$d = 3$$

so that $a = -1$, $b = 1$, $c = -1$, and $d = 3$. Thus $(A)_S = (-1, \ 1, \ -1, \ 3)$ and

$$[A]_S = \begin{bmatrix} -1 \\ 1 \\ -1 \\ 3 \end{bmatrix}$$

9. **(a)** We have $\mathbf{w} = 6\mathbf{v}_1 - \mathbf{v}_2 + 4\mathbf{v}_3 = (16, 10, 12)$.

(c) We have $B = -8A_1 + 7A_2 + 6A_3 + 3A_4 = \begin{bmatrix} 15 & -1 \\ 6 & 3 \end{bmatrix}$.

11. **(a)** Since $\mathbf{v}_1 = \dfrac{13}{10}\mathbf{u}_1 - \dfrac{2}{5}\mathbf{u}_2$ and $\mathbf{v}_2 = -\dfrac{1}{2}\mathbf{u}_1 + 0\mathbf{u}_2$, the transition matrix is

$$Q = \begin{bmatrix} \dfrac{13}{10} & -\dfrac{1}{2} \\ -\dfrac{2}{5} & 0 \end{bmatrix}$$

(b) Since $\mathbf{u}_1 = 0\mathbf{v}_1 - 2\mathbf{v}_2$ and $\mathbf{u}_2 = -\dfrac{5}{2}\mathbf{v}_1 - \dfrac{13}{2}\mathbf{v}_2$, the transition matrix is

$$P = \begin{bmatrix} 0 & -\dfrac{5}{2} \\ -2 & -\dfrac{13}{2} \end{bmatrix}$$

Note that $P = Q^{-1}$.

(c) We find that $\mathbf{w} = -\dfrac{17}{10}\mathbf{u}_1 + \dfrac{8}{5}\mathbf{u}_2$; that is

$$[\mathbf{w}]_B = \begin{bmatrix} -\dfrac{17}{10} \\ \dfrac{8}{5} \end{bmatrix}$$

and hence

$$[\mathbf{w}]_{B'} = \begin{bmatrix} 0 & -\dfrac{5}{2} \\ -2 & -\dfrac{13}{2} \end{bmatrix} \begin{bmatrix} -\dfrac{17}{10} \\ \dfrac{8}{5} \end{bmatrix} = \begin{bmatrix} -4 \\ -7 \end{bmatrix}$$

11. **(d)** Verify that $\mathbf{w} = (-4)\mathbf{v}_1 + (-7)\mathbf{v}_2$.

12. **(a)** We must find constants c_{ij} such that

$$\mathbf{u}_j = c_{1j}\mathbf{v}_1 + c_{2j}\mathbf{v}_2 + c_{3j}\mathbf{v}_3$$

for $j = 1, 2, 3$. For $j = 1$, this reduces to the system

$$-6c_{11} - 2c_{21} - 2c_{31} = -3$$
$$-6c_{11} - 6c_{21} - 3c_{31} = 0$$
$$4c_{21} + 7c_{31} = -3$$

which has the solution $c_{11} = 3/4$, $c_{21} = -3/4$, $c_{31} = 0$. For $j = 2$ and $j = 3$, we have systems with the same coefficient matrix but with different right-hand sides. These have solutions $c_{12} = 3/4$, $c_{22} = -17/12$, $c_{32} = 2/3$, $c_{13} = 1/12$, $c_{23} = -17/12$, and $c_{33} = 2/3$. Thus the transition matrix is

$$P = \begin{bmatrix} 3/4 & 3/4 & 1/12 \\ -3/4 & -17/12 & -17/12 \\ 0 & 2/3 & 2/3 \end{bmatrix}$$

(b) We must find constants a, b, c such that $\mathbf{w} = a\mathbf{u}_1 + b\mathbf{u}_2 + c\mathbf{u}_3$. This means solving the system of equations

$$-3a - 3b + c = -5$$
$$2b + 6c = 8$$
$$-3a - b - c = -5$$

The solution is $a = 1$, $b = 1$, $c = 1$. Thus

$$[\mathbf{w}]_B = \begin{bmatrix} 1 \\ 1 \\ 1 \end{bmatrix}$$

and hence

$$[\mathbf{w}]_{B'} = \begin{bmatrix} 3/4 & 3/4 & 1/12 \\ -3/4 & -17/12 & -17/12 \\ 0 & 2/3 & 2/3 \end{bmatrix} \begin{bmatrix} 1 \\ 1 \\ 1 \end{bmatrix} = \begin{bmatrix} 19/12 \\ -43/12 \\ 4/3 \end{bmatrix}$$

(c) Let $\mathbf{w} = a\mathbf{v}_1 + b\mathbf{v}_2 + c\mathbf{v}_3$ and solve the equations

$$-6a - 2b - 2c = -5$$
$$-6a - 6b - 3c = 8$$
$$4b + 7c = -5$$

for a, b, and c.

14. **(a)** The transition matrix from B' to B is

$$\begin{bmatrix} 3/4 & 7/2 \\ 3/2 & 1 \end{bmatrix}^{-1} = \begin{bmatrix} -2/9 & 7/9 \\ 1/3 & -1/6 \end{bmatrix}$$

(b) Since $\mathbf{p}_1 = (3/4)\mathbf{q}_1 + (3/2)\mathbf{q}_2$ and $\mathbf{p}_2 = (7/2)\mathbf{q}_1 + \mathbf{q}_2$, the transition matrix is

$$\begin{bmatrix} 3/4 & 7/2 \\ 3/2 & 1 \end{bmatrix}$$

(c) Let $\mathbf{p} = a\mathbf{p}_1 + b\mathbf{p}_2$. Thus

$$6a + 10b = -4$$
$$3a + 2b = 1$$

so that $a = 1$ and $b = -1$. Therefore $[\mathbf{p}]_B = \begin{bmatrix} 1 \\ -1 \end{bmatrix}$ and hence

$$[\mathbf{p}]_{B'} = \begin{bmatrix} 3/4 & 7/2 \\ 3/2 & 1 \end{bmatrix} \begin{bmatrix} 1 \\ -1 \end{bmatrix} = \begin{bmatrix} -11/4 \\ 1/2 \end{bmatrix}$$

(d) Let $\mathbf{p} = a\mathbf{q}_1 + b\mathbf{q}_2$ and solve for a and b.

15. **(a)** By hypothesis, $\mathbf{f}_1$ and $\mathbf{f}_2$ span V. Since neither is a multiple of the other, then $\{\mathbf{f}_1, \mathbf{f}_2\}$ is a linearly independent set and hence is a basis for V. Now by inspection, $\mathbf{f}_1 = \frac{1}{2}\mathbf{g}_1 + \left(-\frac{1}{6}\right)\mathbf{g}_2$ and $\mathbf{f}_2 = \frac{1}{3}\mathbf{g}_2$. Therefore, $\{\mathbf{g}_1, \mathbf{g}_2\}$ must also be a basis for V because it is a spanning set which contains the correct number of vectors.

15. (b) The transition matrix is

$$
\begin{bmatrix} \dfrac{1}{2} & 0 \\[2mm] -\dfrac{1}{6} & \dfrac{1}{3} \end{bmatrix}^{-1} = \begin{bmatrix} 2 & 0 \\ 1 & 3 \end{bmatrix}
$$

(c) From the observations in Part (a), we have

$$
P = \begin{bmatrix} \dfrac{1}{2} & 0 \\[2mm] -\dfrac{1}{6} & \dfrac{1}{3} \end{bmatrix}
$$

(d) Since $\mathbf{h} = 2\mathbf{f}_1 + (-5)\mathbf{f}_2$, we have $[\mathbf{h}]_B = \begin{bmatrix} 2 \\ -5 \end{bmatrix}$; thus

$$
[\mathbf{h}]_{B'} = \begin{bmatrix} \dfrac{1}{2} & 0 \\[2mm] -\dfrac{1}{6} & \dfrac{1}{3} \end{bmatrix} \begin{bmatrix} 2 \\ -5 \end{bmatrix} = \begin{bmatrix} 1 \\ -2 \end{bmatrix}
$$

16. (a) Equation (9) yields

$$
\begin{bmatrix} x' \\ y' \end{bmatrix} = \begin{bmatrix} \cos(3\pi/4) & \sin(3\pi/4) \\ -\sin(3\pi/4) & \cos(3\pi/4) \end{bmatrix} \begin{bmatrix} x \\ y \end{bmatrix}
$$

$$
= \begin{bmatrix} -1/\sqrt{2} & 1/\sqrt{2} \\ -1/\sqrt{2} & -1/\sqrt{2} \end{bmatrix} \begin{bmatrix} -2 \\ 6 \end{bmatrix} = \begin{bmatrix} 4\sqrt{2} \\ -2\sqrt{2} \end{bmatrix}
$$

(b) If we compute the inverse of the transition matrix in (a), we have

$$
\begin{bmatrix} x \\ y \end{bmatrix} \begin{bmatrix} -1/\sqrt{2} & -1/\sqrt{2} \\ 1/\sqrt{2} & -1/\sqrt{2} \end{bmatrix} \begin{bmatrix} 5 \\ 2 \end{bmatrix} = \begin{bmatrix} -7/\sqrt{2} \\ 3/\sqrt{2} \end{bmatrix}
$$

18. (a)
$$
\begin{bmatrix} x' \\ y' \\ z' \end{bmatrix} = \begin{bmatrix} \cos(\pi/4) & \sin(\pi/4) & 0 \\ -\sin(\pi/4) & \cos(\pi/4) & 0 \\ 0 & 0 & 1 \end{bmatrix} \begin{bmatrix} -1 \\ 2 \\ 5 \end{bmatrix} = \begin{bmatrix} 1/\sqrt{2} \\ 3/\sqrt{2} \\ 5 \end{bmatrix}
$$

$$
\textbf{(b)} \quad
\begin{bmatrix} x \\ y \\ z \end{bmatrix}
=
\begin{bmatrix}
1/\sqrt{2} & -1/\sqrt{2} & 0 \\
1/\sqrt{2} & 1/\sqrt{2} & 0 \\
0 & 0 & 1
\end{bmatrix}
\begin{bmatrix} 1 \\ 6 \\ -3 \end{bmatrix}
=
\begin{bmatrix} -5/\sqrt{2} \\ 7/\sqrt{2} \\ -3 \end{bmatrix}
$$

19. The general transition matrix will be

$$
\begin{bmatrix}
\cos\theta & 0 & -\sin\theta \\
0 & 1 & 0 \\
\sin\theta & 0 & \cos\theta
\end{bmatrix}
$$

In particular, if we rotate through $\theta = \dfrac{\pi}{3}$, then the transition matrix is

$$
\begin{bmatrix}
\dfrac{1}{2} & 0 & -\dfrac{\sqrt{3}}{2} \\
0 & 1 & 0 \\
\dfrac{\sqrt{3}}{2} & 0 & \dfrac{1}{2}
\end{bmatrix}
$$

21. **(a)** See Exercise 19, above.

22. Use the result of Example 7 to find P_1 where $\mathbf{x}' = P_1\mathbf{x}$. Use the result of Exercise 21(a), above, to find P_2 where $\mathbf{x}'' = P_2\mathbf{x}'$. Then $\mathbf{x}'' = P_2 P_1 \mathbf{x}$, and hence $A = P_2 P_1$.

23. Since the row vectors (and the column vectors) of the given matrix are orthogonal, the matrix will be orthogonal provided these vectors have norm 1. A necessary and sufficient condition for this is that $a^2 + b^2 = 1/2$. Why?

24. Suppose that $A = [a_{ij}]$ is a 2×2 orthogonal matrix. Since the column vectors form an orthonormal set, we have the equations $a_{11}^2 + a_{21}^2 = a_{12}^2 + a_{22}^2 = 1$ and $a_{11}a_{12} + a_{21}a_{22} = 0$. Thus $|a_{ij}| \leq 1$ for all a_{ij}. Since $\cos\alpha$ takes on all values between -1 and $+1$ for $0 \leq \alpha \leq \pi$, we can let $a_{11} = \cos\theta$ for some value of θ where $0 \leq \theta \leq \pi$. Alternatively, we could let $a_{11} = \cos(2\pi - \theta)$. Since $a_{11}^2 + a_{21}^2 = 1$, we have $a_{21}^2 = 1 - \cos^2\theta = \sin^2\theta$, so that $a_{21} = \pm\sin\theta$, or alternatively, $a_{21} = \mp\sin(2\pi - \theta)$. Thus the orthogonality equation becomes $a_{12}\cos\theta \pm a_{22}\sin\theta = 0$, so that $a_{12} = k\sin\theta$ and $a_{22} = \mp k\cos\theta$. Since $a_{12}^2 + a_{22}^2 = 1$, $k = \pm 1$. Thus we have four possible matrices:

$$
\begin{bmatrix} \cos\theta & -\sin\theta \\ \sin\theta & \cos\theta \end{bmatrix},
\begin{bmatrix} \cos\theta & -\sin\theta \\ -\sin\theta & -\cos\theta \end{bmatrix},
\begin{bmatrix} \cos\theta & \sin\theta \\ -\sin\theta & \cos\theta \end{bmatrix},
\begin{bmatrix} \cos\theta & \sin\theta \\ \sin\theta & -\cos\theta \end{bmatrix}
$$

If we substitute $\theta' = 2\pi - \theta$ for θ in the third and fourth matrices, they become the first and second matrices, respectively. Thus for some value of θ where $0 \le \theta \le 2\pi$, the first two matrices represent all possible forms of a 2×2 orthogonal matrix. Since $\theta = 2\pi$ gives the same value as $\theta = 0$, we can restrict θ to the range $0 \le \theta < 2\pi$.

25. Multiplication by the first matrix A in Exercise 24 represents a rotation and $\det(A) = 1$. The second matrix has determinant -1 and can be written as

$$
\begin{bmatrix} \cos\theta & -\sin\theta \\ -\sin\theta & -\cos\theta \end{bmatrix} =
\begin{bmatrix} 1 & 0 \\ 0 & -1 \end{bmatrix}
\begin{bmatrix} \cos\theta & -\sin\theta \\ \sin\theta & \cos\theta \end{bmatrix}
$$

Thus it represents a rotation followed by a reflection about the x-axis.

28. Since a composition of rotations can be represented as a product of orthogonal matrices with determinant 1 and, by Theorem 6.5.2, such a product is also orthogonal, it must represent either a single rotation or a rotation followed by a reflection. However, if it is the product of matrices with determinant 1, its determinant is also 1, and hence it represents a rotation.

29. Note that A is orthogonal if and only if A^T is orthogonal. Since the rows of A^T are the columns of A, we need only apply the equivalence of Parts (a) and (b) to A^T to obtain the equivalence of Parts (a) and (c).

30. **(a)** Rotations about the origin and reflections about any line through the origin are rigid. So is any combination of these transformations.

(b) Rotations about the origin, dilations and contractions, reflections about lines through the origin, and combinations of these are all angle preserving.

(c) Suppose that T is a linear operator in R^2 and let A be the matrix associated with T. If T is rigid, then $\|A\mathbf{x}\| = \|\mathbf{x}\|$ for all $\mathbf{x}$ in R^2. Then Theorem 6.5.3 guarantees that $A\mathbf{x} \cdot A\mathbf{y} = \mathbf{x} \cdot \mathbf{y}$ or, since $\|A\mathbf{x}\| = \|\mathbf{x}\|$, that

$$
\frac{A\mathbf{x} \cdot A\mathbf{y}}{\|A\mathbf{x}\|\,\|A\mathbf{y}\|} = \frac{\mathbf{x} \cdot \mathbf{y}}{\|\mathbf{x}\|\,\|\mathbf{y}\|}
$$

Therefore, the cosine of the angle between $A\mathbf{x}$ and $A\mathbf{y}$ is equal to the cosine of the angle between $\mathbf{x}$ and $\mathbf{y}$ and thus the angles themselves are the same. Hence, if T is rigid, then T is angle preserving.

The converse is not true. Simply note that dilations in R^2 by a factor k where $k \neq 1$ preserve the angle between two vectors but not their lengths.

31. If A is the standard matrix associated with a rigid transformation, then Theorem 6.5.3 guarantees that A must be orthogonal. But if A is orthogonal, then Theorem 6.5.2 guarantees that $\det(A) = \pm 1$.

32. If A is orthogonal, then its row vectors must form an orthonormal set $\{\mathbf{r}_1, \mathbf{r}_2, \mathbf{r}_3\}$. Since $\|\mathbf{r}_1\| = a^2 + 1$, then $a = 0$. This guarantees that $\mathbf{r}_1 \cdot \mathbf{r}_2 = \mathbf{r}_1 \cdot \mathbf{r}_3 = 0$. It remains to find b and c such that $b^2 + 1/3 = 1$, $c^2 + 2/3 = 1$, and $bc + 2/\sqrt{18} = 0$. The first two equations imply that $b = \pm\sqrt{2}/\sqrt{3}$ and $c = \pm 1/\sqrt{3}$, while the third equation tells us that b and c must have different signs. Therefore $a = 0$, $b = \sqrt{2}/\sqrt{3}$, $c = -1/\sqrt{3}$ and $a = 0$, $b = -\sqrt{2}/\sqrt{3}$, $c = 1/\sqrt{3}$ are the only two possibilities.

SUPPLEMENTARY EXERCISES 6

1. **(a)** We must find a vector $\mathbf{x} = (x_1, x_2, x_3, x_4)$ such that

$$\mathbf{x} \cdot \mathbf{u}_1 = 0, \quad \mathbf{x} \cdot \mathbf{u}_4 = 0, \quad \text{and} \quad \frac{\mathbf{x} \cdot \mathbf{u}_2}{\|\mathbf{x}\| \, \|\mathbf{u}_2\|} = \frac{\mathbf{x} \cdot \mathbf{u}_3}{\|\mathbf{x}\| \, \|\mathbf{u}_3\|}$$

The first two conditions guarantee that $x_1 = x_4 = 0$. The third condition implies that $x_2 = x_3$. Thus any vector of the form $(0, a, a, 0)$ will satisfy the given conditions provided $a \neq 0$.

(b) We must find a vector $\mathbf{x} = (x_1, x_2, x_3, x_4)$ such that $\mathbf{x} \cdot \mathbf{u}_1 = \mathbf{x} \cdot \mathbf{u}_4 = 0$. This implies that $x_1 = x_4 = 0$. Moreover, since $\|\mathbf{x}\| = \|\mathbf{u}_2\| = \|\mathbf{u}_3\| = 1$, the cosine of the angle between $\mathbf{x}$ and $\mathbf{u}_2$ is $\mathbf{x} \cdot \mathbf{u}_2$ and the cosine of the angle between $\mathbf{x}$ and $\mathbf{u}_3$ is $\mathbf{x} \cdot \mathbf{u}_3$. Thus we are looking for a vector $\mathbf{x}$ such that $\mathbf{x} \cdot \mathbf{u}_2 = 2\mathbf{x} \cdot \mathbf{u}_3$, or $x_2 = 2x_3$. Since $\|\mathbf{x}\| = 1$, we have $\mathbf{x} = (0, 2x_3, x_3, 0)$ where $4x_3^2 + x_3^2 = 1$ or $x_3 = \pm 1/\sqrt{5}$. Therefore

$$\mathbf{x} = \pm \left(0, \ \frac{2}{\sqrt{5}}, \ \frac{1}{\sqrt{5}}, \ 0 \right)$$

2. We first show that A is symmetric, or that $A = A^T$. Observe that

$$A^T = \left(I_n - \frac{2}{\|\mathbf{x}\|^2} \mathbf{x}\mathbf{x}^T \right)^T = I_n^T - \frac{2}{\|\mathbf{x}\|^2} (\mathbf{x}\mathbf{x}^T)^T$$

$$= I_n - \frac{2}{\|\mathbf{x}\|^2} (\mathbf{x}^T)^T \mathbf{x}^T = I_n - \frac{2}{\|\mathbf{x}\|^2} \mathbf{x}\mathbf{x}^T = A$$

To show that A is orthogonal, it will suffice to show that $A^{-1} = A^T$, or that $AA^T = I_n$, or, in this case, that $A^2 = I_n$, or that $A^2 - I_n = O_n$. To this end, observe that

$$A^2 = I_n - \frac{4}{\|\mathbf{x}\|^2} \mathbf{x}\mathbf{x}^T + \frac{4}{\|\mathbf{x}\|^4} (\mathbf{x}\mathbf{x}^T)^2$$

309

or that

$$A^2 - I_n = -\frac{4}{\|\mathbf{x}\|^2}\left(\mathbf{x}\mathbf{x}^T - \frac{1}{\|\mathbf{x}\|^2}(\mathbf{x}\mathbf{x}^T)^2\right)$$

Thus it will suffice to show that $\mathbf{x}\mathbf{x}^T = \dfrac{1}{\|\mathbf{x}\|^2}(\mathbf{x}\mathbf{x}^T)^2$. To this end, let $\mathbf{x} = [x_1 x_2 \cdots x_n]^T$.
Then

$$\mathbf{x}\mathbf{x}^T = \begin{bmatrix} x_1\mathbf{x}^T \\ x_2\mathbf{x}^T \\ \vdots \\ x_n\mathbf{x}^T \end{bmatrix} = \begin{bmatrix} x_1\mathbf{x} & x_2\mathbf{x} & \cdots & x_n\mathbf{x} \end{bmatrix}$$

so that

$$(\mathbf{x}\mathbf{x}^T)^2 = \begin{bmatrix} x_1\mathbf{x} & x_2\mathbf{x} & \cdots & x_n\mathbf{x} \end{bmatrix} \begin{bmatrix} x_1\mathbf{x}^T \\ x_2\mathbf{x}^T \\ \vdots \\ x_n\mathbf{x}^T \end{bmatrix}$$

$$= x_1^2\mathbf{x}\mathbf{x}^T + x_2^2\mathbf{x}\mathbf{x}^T + \cdots + x_n^2\mathbf{x}\mathbf{x}^T$$

$$= (x_1^2 + x_2^2 + \cdots + x_n^2)\mathbf{x}\mathbf{x}^T = \|\mathbf{x}\|\mathbf{x}\mathbf{x}^T$$

This completes the proof.

4. Let $\mathbf{u} = (\sqrt{a_1}, \ldots, \sqrt{a_n})$ and $\mathbf{v} = (1/\sqrt{a_1}, \ldots, 1/\sqrt{a_n})$. By the Cauchy-Schwarz Inequality,

$$(\mathbf{u} \cdot \mathbf{v})^2 = \underbrace{(1 + \cdots + 1)}_{n \text{ terms}}^2 \le \|\mathbf{u}\|^2\|\mathbf{v}\|^2$$

or

$$n^2 \le (a_1 + \cdots + a_n)\left(\frac{1}{a_1} + \cdots + \frac{1}{a_n}\right)$$

6. If $\mathbf{v} = (a, b, c)$ where $\|\mathbf{v}\| = 1$ and $\mathbf{v} \cdot \mathbf{u}_1 = \mathbf{v} \cdot \mathbf{u}_2 = \mathbf{v} \cdot \mathbf{u}_3 = 0$, then

$$a + b - c = 0$$

$$-2a - b + 2c = 0$$

$$-a + c = 0$$

where $a^2 + b^2 + c^2 = 1^2$. The above system has the solution $a = t$, $b = 0$, and $c = t$ where t is arbitrary. But $a^2 + b^2 + c^2 = 1^2$ implies that $2t^2 = 1^2$ or $t = \pm 1/\sqrt{2}$. The two vectors are therefore $\mathbf{v} = \pm(1/\sqrt{2},\, 0,\, 1/\sqrt{2})$.

7. Let

$$(*) \qquad\qquad \langle \mathbf{u}, \mathbf{v} \rangle = w_1 u_1 v_1 + w_2 u_2 v_2 + \cdots + w_n u_n v_n$$

be the weighted Euclidean inner product. Since $\langle \mathbf{v}_i, \mathbf{v}_j \rangle = 0$ whenever $i \neq j$, the vectors $\{\mathbf{v}_1, \mathbf{v}_2, \ldots, \mathbf{v}_n\}$ form an orthogonal set with respect to $(*)$ for *any* choice of the constants $w_1, w_2, \ldots, w_n$. We must now choose the positive constants $w_1, w_2, \ldots, w_n$ so that $\|\mathbf{v}_k\| = 1$ for all k. But $\|\mathbf{v}_k\|^2 = k w_k$. If we let $w_k = 1/k$ for $k = 1, 2, \ldots, n$, the given vectors will then form an orthonormal set with respect to $(*)$.

8. Suppose that $\langle \mathbf{u}, \mathbf{v} \rangle = w_1 u_1 v_1 + w_2 u_2 v_2$ were such an inner product. Then

$$\langle (1, 2),\ (3, -1) \rangle = 0 \quad \Rightarrow \quad 3w_1 - 2w_2 = 0$$

$$\|(1, 2)\| = 1 \quad \Rightarrow \quad w_1 + 4w_2 = 1$$

$$\|(3, -1)\| = 1 \quad \Rightarrow \quad 9w_1 + w_2 = 1$$

But since the three equations for w_1 and w_2 above are inconsistent, there is no such inner product.

9. Let $Q = [a_{ij}]$ be orthogonal. Then $Q^{-1} = Q^T$ and $\det(Q) = \pm 1$. If C_{ij} is the cofactor of a_{ij}, then

$$Q = [(a_{ij})] = (Q^{-1})^T = \left(\frac{1}{\det(Q)} (C_{ij})^T \right)^T = \det(Q)(C_{ij})$$

so that $a_{ij} = \det(Q) C_{ij}$.

10.   From the definition of the norm, we have

$$\|\mathbf{u} - \mathbf{v}\|^2 = \langle \mathbf{u} - \mathbf{v}, \ \mathbf{u} - \mathbf{v} \rangle$$

$$= \langle \mathbf{u}, \mathbf{u} \rangle + \langle \mathbf{v}, \mathbf{v} \rangle - 2\langle \mathbf{u}, \mathbf{v} \rangle$$

$$= \|\mathbf{u}\|^2 + \|\mathbf{v}\|^2 - 2\|\mathbf{u}\|\|\mathbf{v}\| \cos \theta$$

11.   **(a)**   The length of each "side" of this "cube" is $|k|$. The length of the "diagonal" is $\sqrt{n}|k|$. The inner product of any "side" with the "diagonal" is k^2. Therefore,

$$\cos \theta = \frac{k^2}{|k| \ \sqrt{n} \ |k|} = \frac{1}{\sqrt{n}}$$

   (b)   As $n \to +\infty$, $\cos \theta \to 0$, so that $\theta \to \pi/2$.

12.   **(a)**   We have that $\mathbf{u} + \mathbf{v}$ and $\mathbf{u} - \mathbf{v}$ are orthogonal if and only if

$$\langle \mathbf{u} + \mathbf{v}, \ \mathbf{u} - \mathbf{v} \rangle = 0$$

But

$$\langle \mathbf{u} + \mathbf{v}, \ \mathbf{u} - \mathbf{v} \rangle = \|\mathbf{u}\| - \|\mathbf{v}\|$$

by the properties of an inner product (Why?). Thus $\mathbf{u} + \mathbf{v}$ and $\mathbf{u} - \mathbf{v}$ are orthogonal if and only if $\|\mathbf{u}\| = \|\mathbf{v}\|$.

   (b)   Since $\mathbf{u} + \mathbf{v}$ and $\mathbf{u} - \mathbf{v}$ can be thought of as diagonals of a parallelogram with sides $\mathbf{u}$ and $\mathbf{v}$, the above result says that the two diagonals are perpendicular if and only if the parallelogram has equal sides.

13.   Recall that $\mathbf{u}$ can be expressed as the linear combination

$$\mathbf{u} = a_1 \mathbf{v}_1 + \cdots + a_n \mathbf{v}_n$$

where $a_i = \langle \mathbf{u}, \mathbf{v}_i \rangle$ for $i = 1, \ldots, n$. Thus

$$\cos^2 \alpha_i = \left(\frac{\langle \mathbf{u}, \mathbf{v}_i \rangle}{\|\mathbf{u}\| \, \|\mathbf{v}_i\|} \right)^2$$

$$= \left(\frac{a_i}{\|\mathbf{u}\|} \right)^2 \qquad (\|\mathbf{v}_i\| = 1)$$

$$= \frac{a_i^2}{a_1^2 + a_2^2 + \cdots + a_n^2} \qquad \text{(Why?)}$$

Therefore

$$\cos^2 \alpha_1 + \cdots + \cos^2 \alpha_n = \frac{a_1^2 + a_2^2 + \cdots + a_n^2}{a_1^2 + a_2^2 + \cdots + a_n^2} = 1$$

14. We must show that $\langle \mathbf{u}, \mathbf{v} \rangle$ satisfies the four axioms which define an inner product. We show (2) and (4), and leave the others to you.

(2) Using Axiom 3 for $\langle \mathbf{u}, \mathbf{v} \rangle_1$ and $\langle \mathbf{u}, \mathbf{v} \rangle_2$, we have

$$\langle \mathbf{u} + \mathbf{v}, \mathbf{w} \rangle = \langle \mathbf{u} + \mathbf{v}, \mathbf{w} \rangle_1 + \langle \mathbf{u} + \mathbf{v}, \mathbf{w} \rangle_2$$

$$= \langle \mathbf{u}, \mathbf{w} \rangle_1 + \langle \mathbf{v}, \mathbf{w} \rangle_1 + \langle \mathbf{u}, \mathbf{w} \rangle_2 + \langle \mathbf{v}, \mathbf{w} \rangle_2$$

$$= \langle \mathbf{u}, \mathbf{w} \rangle_1 + \langle \mathbf{u}, \mathbf{w} \rangle_2 + \langle \mathbf{v}, \mathbf{w} \rangle_1 + \langle \mathbf{v}, \mathbf{w} \rangle_2$$

$$= \langle \mathbf{u}, \mathbf{w} \rangle + \langle \mathbf{v}, \mathbf{w} \rangle$$

(4) Using Axiom 4 for $\langle \mathbf{u}, \mathbf{v} \rangle_1$ and $\langle \mathbf{u}, \mathbf{v} \rangle_2$, we have

$$\langle \mathbf{v}, \mathbf{v} \rangle = \langle \mathbf{v}, \mathbf{v} \rangle_1 + \langle \mathbf{v}, \mathbf{v} \rangle_2 \geq 0$$

since both $\langle \mathbf{v}, \mathbf{v} \rangle_1$ and $\langle \mathbf{v}, \mathbf{v} \rangle_2$ are nonnegative. Moreover, $\langle \mathbf{v}, \mathbf{v} \rangle = 0$ if and only if $\langle \mathbf{v}, \mathbf{v} \rangle_1 = \langle \mathbf{v}, \mathbf{v} \rangle_2 = 0$, which is true if and only if $\mathbf{v} = \mathbf{0}$.

15. Recall that A is orthogonal provided $A^{-1} = A^T$. Hence

$$\langle \mathbf{u}, \mathbf{v} \rangle = \mathbf{v}^T A^T A \mathbf{u}$$

$$= \mathbf{v}^T A^{-1} A \mathbf{u} = \mathbf{v}^T \mathbf{u}$$

which is the Euclidean inner product.

16. To show that $(W^\perp)^\perp = W$, we first show that $W \subseteq (W^\perp)^\perp$. If $\mathbf{w}$ is in W, then $\mathbf{w}$ is orthogonal to every vector in $W^\perp$, so that $\mathbf{w}$ is in $(W^\perp)^\perp$. Thus $W \subseteq (W^\perp)^\perp$.

To show that $(W^\perp)^\perp \subseteq W$, let $\mathbf{v}$ be in $(W^\perp)^\perp$. Since $\mathbf{v}$ is in V, we have, by the Projection Theorem, that $\mathbf{v} = \mathbf{w}_1 + \mathbf{w}_2$ where $\mathbf{w}_1$ is in W and $\mathbf{w}_2$ is in $W^\perp$. By definition, $\langle \mathbf{v}, \mathbf{w}_2 \rangle = \langle \mathbf{w}_1, \mathbf{w}_2 \rangle = 0$. But

$$\langle \mathbf{v}, \mathbf{w}_2 \rangle = \langle \mathbf{w}_1 + \mathbf{w}_2, \mathbf{w}_2 \rangle$$
$$= \langle \mathbf{w}_1, \mathbf{w}_2 \rangle + \langle \mathbf{w}_2, \mathbf{w}_2 \rangle$$
$$= \langle \mathbf{w}_2, \mathbf{w}_2 \rangle$$

so that $\langle \mathbf{w}_2, \mathbf{w}_2 \rangle = 0$. Hence $\mathbf{w}_2 = \mathbf{0}$ and therefore $\mathbf{v} = \mathbf{w}_1$, so that $\mathbf{v}$ is in W. Thus $(W^\perp)^\perp \subseteq W$.

EXERCISE SET 7.1

1. **(a)** Since

$$\det(\lambda I - A) = \det \begin{bmatrix} \lambda - 3 & 0 \\ -8 & \lambda + 1 \end{bmatrix} = (\lambda - 3)(\lambda + 1)$$

the characteristic equation is $\lambda^2 - 2\lambda - 3 = 0$.

(e) Since

$$\det(\lambda I - A) = \det \begin{bmatrix} \lambda & 0 \\ 0 & \lambda \end{bmatrix} = \lambda^2$$

the characteristic equation is $\lambda^2 = 0$.

3. **(a)** The equation $(\lambda I - A)\mathbf{x} = \mathbf{0}$ becomes

$$\begin{bmatrix} \lambda - 3 & 0 \\ -8 & \lambda + 1 \end{bmatrix} \begin{bmatrix} x_1 \\ x_2 \end{bmatrix} = \begin{bmatrix} 0 \\ 0 \end{bmatrix}$$

The eigenvalues are $\lambda = 3$ and $\lambda = -1$. Substituting $\lambda = 3$ into $(\lambda I - A)\mathbf{x} = \mathbf{0}$ yields.

$$\begin{bmatrix} 0 & 0 \\ -8 & 4 \end{bmatrix} \begin{bmatrix} x_1 \\ x_2 \end{bmatrix} = \begin{bmatrix} 0 \\ 0 \end{bmatrix}$$

or

$$-8x_1 + 4x_2 = 0$$

Thus $x_1 = \dfrac{1}{2}s$ and $x_2 = s$ where s is arbitrary, so that a basis for the eigenspace corresponding to $\lambda = 3$ is $\begin{bmatrix} 1/2 \\ 1 \end{bmatrix}$. Of course, $\begin{bmatrix} 1 \\ 2 \end{bmatrix}$ and $\begin{bmatrix} \pi \\ 2\pi \end{bmatrix}$ are also bases.

Substituting $\lambda = -1$ into $(\lambda I - A)\mathbf{x} = \mathbf{0}$ yields

$$\begin{bmatrix} -4 & 0 \\ -8 & 0 \end{bmatrix} \begin{bmatrix} x_1 \\ x_2 \end{bmatrix} = \begin{bmatrix} 0 \\ 0 \end{bmatrix}$$

or

$$-4x_1 = 0$$
$$-8x_1 = 0$$

Hence, $x_1 = 0$ and $x_2 = s$ where s is arbitrary. In particular, if $s = 1$, then a basis for the eigenspace corresponding to $\lambda = -1$ is $\begin{bmatrix} 0 \\ 1 \end{bmatrix}$.

3. (e) The equation $(\lambda I - A)\mathbf{x} = \mathbf{0}$ becomes

$$\begin{bmatrix} \lambda & 0 \\ 0 & \lambda \end{bmatrix} \begin{bmatrix} x_1 \\ x_2 \end{bmatrix} = \begin{bmatrix} 0 \\ 0 \end{bmatrix}$$

Clearly, $\lambda = 0$ is the only eigenvalue. Substituting $\lambda = 0$ into the above equation yields $x_1 = s$ and $x_2 = t$ where s and t are arbitrary. In particular, if $s = t = 1$, then we find that $\begin{bmatrix} 1 \\ 0 \end{bmatrix}$ and $\begin{bmatrix} 0 \\ 1 \end{bmatrix}$ form a basis for the eigenspace associated with $\lambda = 0$.

4. (c) Since

$$\det[\lambda I - A] = \det \begin{bmatrix} \lambda + 2 & 0 & -1 \\ 6 & \lambda + 2 & 0 \\ -19 & -5 & \lambda + 4 \end{bmatrix}$$

$$= \lambda^3 + 8\lambda^2 + \lambda + 8$$

the characteristic equation of A is

$$\lambda^3 + 8\lambda^2 + \lambda + 8 = 0$$

5. **(c)** From the solution to 4(c), we have

$$\lambda^3 + 8\lambda^2 + \lambda + 8 = (\lambda + 8)(\lambda^2 + 1)$$

Since $\lambda^2 + 1 = 0$ has no real solutions, then $\lambda = -8$ is the only (real) eigenvalue.

6. **(c)** The only eigenvalue is $\lambda = -8$. If we substitute the value $\lambda = -8$ into the equation $(\lambda I - A)\mathbf{x} = \mathbf{0}$, we obtain

$$\begin{bmatrix} -6 & 0 & -1 \\ 6 & -6 & 0 \\ -19 & -5 & -4 \end{bmatrix} \begin{bmatrix} x_1 \\ x_2 \\ x_3 \end{bmatrix} = \begin{bmatrix} 0 \\ 0 \\ 0 \end{bmatrix}$$

The augmented matrix of the above system can be reduced to

$$\begin{bmatrix} 1 & 0 & 1/6 & 0 \\ 0 & 1 & 1/6 & 0 \\ 0 & 0 & 0 & 0 \end{bmatrix}$$

Thus, $x_1 = -\dfrac{1}{6}s$, $x_2 = -\dfrac{1}{6}s$, $x_3 = s$ is a solution where s is arbitrary. Therefore

$$\begin{bmatrix} -1/6 \\ -1/6 \\ 1 \end{bmatrix}$$ forms a basis for the eigenspace.

(f) The eigenvalues are $\lambda - -4$ and $\lambda - 3$. If we substitute $\lambda - -4$ into the equation $(\lambda I - A)\mathbf{x} = \mathbf{0}$, we obtain

$$\begin{bmatrix} -9 & -6 & -2 \\ 0 & -3 & 8 \\ -1 & 0 & -2 \end{bmatrix} \begin{bmatrix} x_1 \\ x_2 \\ x_3 \end{bmatrix} = \begin{bmatrix} 0 \\ 0 \\ 0 \end{bmatrix}$$

If we reduce the augmented matrix to row-echelon form, we obtain

$$\begin{bmatrix} 1 & 0 & 2 & 0 \\ 0 & 1 & -8/3 & 0 \\ 0 & 0 & 0 & 0 \end{bmatrix}$$

This implies that $x_1 = -2s$, $x_2 = \dfrac{8}{3}s$, $x_3 = s$. If we set $s = 1$, we find that

$$\begin{bmatrix} -2 \\ 8/3 \\ 1 \end{bmatrix}$$

is a basis for the eigenspace associated with $\lambda = -4$.

If we substitute $\lambda = 3$ into the equation $(\lambda I - A)\mathbf{x} = \mathbf{0}$, we obtain

$$\begin{bmatrix} -2 & -6 & -2 \\ 0 & 4 & 8 \\ -1 & 0 & 5 \end{bmatrix} \begin{bmatrix} x_1 \\ x_2 \\ x_3 \end{bmatrix} = \begin{bmatrix} 0 \\ 0 \\ 0 \end{bmatrix}$$

If we reduce the augmented matrix to row-echelon form, we obtain

$$\begin{bmatrix} 1 & 0 & -5 & 0 \\ 0 & 1 & 2 & 0 \\ 0 & 0 & 0 & 0 \end{bmatrix}$$

It then follows that

$$\begin{bmatrix} 5 \\ -2 \\ 1 \end{bmatrix}$$

is a basis for the eigenspace associated with $\lambda = 3$.

7. (a) Since

$$\det[\lambda I - A] = \det \begin{bmatrix} \lambda & 0 & -2 & 0 \\ -1 & \lambda & -1 & 0 \\ 0 & -1 & \lambda + 2 & 0 \\ 0 & 0 & 0 & \lambda - 1 \end{bmatrix}$$

$$= \lambda^4 + \lambda^3 - 3\lambda^2 - \lambda + 2$$

$$= (\lambda - 1)^2(\lambda + 2)(\lambda + 1)$$

the characteristic equation is

$$(\lambda - 1)^2(\lambda + 2)(\lambda + 1) = 0$$

9. **(a)** The eigenvalues are $\lambda = 1$, $\lambda = -2$, and $\lambda = -1$. If we set $\lambda = 1$, then $(\lambda I - A)\mathbf{x} = \mathbf{0}$ becomes

$$\begin{bmatrix} 1 & 0 & -2 & 0 \\ -1 & 1 & -1 & 0 \\ 0 & -1 & 3 & 0 \\ 0 & 0 & 0 & 0 \end{bmatrix} \begin{bmatrix} x_1 \\ x_2 \\ x_3 \\ x_4 \end{bmatrix} = \begin{bmatrix} 0 \\ 0 \\ 0 \\ 0 \end{bmatrix}$$

The augmented matrix can be reduced to

$$\begin{bmatrix} 1 & 0 & -2 & 0 & 0 \\ 0 & 1 & -3 & 0 & 0 \\ 0 & 0 & 0 & 0 & 0 \\ 0 & 0 & 0 & 0 & 0 \end{bmatrix}$$

Thus, $x_1 = 2s$, $x_2 = 3s$, $x_3 = s$, and $x_4 = t$ is a solution for all s and t. In particular, if we let $s = t = 1$, we see that

$$\begin{bmatrix} 2 \\ 3 \\ 1 \\ 0 \end{bmatrix} \quad \text{and} \quad \begin{bmatrix} 0 \\ 0 \\ 0 \\ 1 \end{bmatrix}$$

form a basis for the eigenspace associated with $\lambda = 1$.

   If we set $\lambda = -2$, then $(\lambda I - A)\mathbf{x} = \mathbf{0}$ becomes

$$\begin{bmatrix} -2 & 0 & -2 & 0 \\ -1 & -2 & -1 & 0 \\ 0 & -1 & 0 & 0 \\ 0 & 0 & 0 & -3 \end{bmatrix} \begin{bmatrix} x_1 \\ x_2 \\ x_3 \\ x_4 \end{bmatrix} = \begin{bmatrix} 0 \\ 0 \\ 0 \\ 0 \end{bmatrix}$$

The augmented matrix can be reduced to

$$\begin{bmatrix} 1 & 0 & 1 & 0 & 0 \\ 0 & 1 & 0 & 0 & 0 \\ 0 & 0 & 0 & 1 & 0 \\ 0 & 0 & 0 & 0 & 0 \end{bmatrix}$$

This implies that $x_1 = -s$, $x_2 = x_4 = 0$, and $x_3 = s$. Therefore the vector

$$\begin{bmatrix} -1 \\ 0 \\ 1 \\ 0 \end{bmatrix}$$

forms a basis for the eigenspace associated with $\lambda = -2$.

Finally, if we set $\lambda = -1$, then $(\lambda I - A)\mathbf{x} = \mathbf{0}$ becomes

$$\begin{bmatrix} -1 & 0 & -2 & 0 \\ -1 & -1 & -1 & 0 \\ 0 & -1 & 1 & 0 \\ 0 & 0 & 0 & -2 \end{bmatrix} \begin{bmatrix} x_1 \\ x_2 \\ x_3 \\ x_4 \end{bmatrix} = \begin{bmatrix} 0 \\ 0 \\ 0 \\ 0 \end{bmatrix}$$

The augmented matrix can be reduced to

$$\begin{bmatrix} 1 & 0 & 2 & 0 & 0 \\ 0 & 1 & -1 & 0 & 0 \\ 0 & 0 & 0 & 1 & 0 \\ 0 & 0 & 0 & 0 & 0 \end{bmatrix}$$

Thus, $x_1 = -2s$, $x_2 = s$, $x_3 = s$, and $x_4 = 0$ is a solution. Therefore the vector

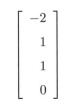

forms a basis for the eigenspace associated with $\lambda = -1$.

11. By Theorem 7.1.1, the eigenvalues of A are 1, $1/2, 0$, and 2. Thus by Theorem 7.1.3, the eigenvalues of A^9 are $1^9 = 1$, $(1/2)^9 = 1/512$, $0^9 = 0$, and $2^9 = 512$.

12. Notice that $A^2 = I$. Therefore $A^{25} = A$.

13. The vectors $A\mathbf{x}$ and $\mathbf{x}$ will lie on the same line through the origin if and only if there exists a real number λ such that $A\mathbf{x} = \lambda\mathbf{x}$, that is, if and only if λ is a real eigenvalue for A and $\mathbf{x}$ is the associated eigenvector.

  (a) In this case, the eigenvalues are $\lambda = 3$ and $\lambda = 2$, while associated eigenvectors are

  $$\begin{bmatrix} 1 \\ 1 \end{bmatrix} \quad \text{and} \quad \begin{bmatrix} 1 \\ 2 \end{bmatrix}$$

  respectively. Hence the lines $y = x$ and $y = 2x$ are the only lines which are invariant under A.

  (b) In this case, the characteristic equation for A is $\lambda^2 + 1 = 0$. Since A has no real eigenvalues, there are no lines which are invariant under A.

14. **(a)** From the proof of Theorem 7.1.4, $p(0) = \det(-A) = 5$. Since each signed elementary product in $\det(-A)$ is just $(-1)^3$ times the corresponding signed elementary product in $\det(A)$, we have that $\det(A) = -\det(-A) = -5$.

15. Let a_{ij} denote the ijth entry of A. Then the characteristic polynomial of A is $\det(\lambda I - A)$ or

$$\det \begin{bmatrix} \lambda - a_{11} & -a_{12} & \cdots & -a_{1n} \\ -a_{21} & \lambda - a_{22} & \cdots & -a_{2n} \\ \vdots & \vdots & & \vdots \\ -a_{n1} & -a_{n2} & \cdots & \lambda - a_{nn} \end{bmatrix}$$

This determinant is a sum each of whose terms is the product of n entries from the given matrix. Each of these entries is either a constant or is of the form $\lambda - a_{ij}$. The only term with a λ in each factor of the product is

$$(\lambda - a_{11})(\lambda - a_{22}) \cdots (\lambda - a_{nn})$$

Therefore, this term must produce the highest power of λ in the characteristic polynomial. This power is clearly n and the coefficient of λ^n is 1.

17. The characteristic equation of A is

$$\lambda^2 - (a + d)\lambda + ad - bc = 0$$

This is a quadratic equation whose discriminant is

$$(a + d)^2 - 4ad + 4bc = a^2 - 2ad + d^2 + 4bc$$
$$= (a - d)^2 + 4bc$$

The roots are

$$\lambda = \frac{1}{2}\left[(a + d) \pm \sqrt{(a - d)^2 + 4bc}\right]$$

If the discriminant is positive, then the equation has two distinct real roots; if it is zero, then the equation has one real root (repeated); if it is negative, then the equation has no real roots. Since the eigenvalues are assumed to be real numbers, the result follows.

18. We are given the two distinct eigenvalues λ_1 and λ_2. To find the corresponding eigenvectors, we set

$$\begin{bmatrix} \lambda_i - a & -b \\ -c & \lambda_i - d \end{bmatrix} \begin{bmatrix} x_1 \\ x_2 \end{bmatrix} = \begin{bmatrix} 0 \\ 0 \end{bmatrix}$$

This yields the equations

$$(\lambda_i - a)x_1 - \qquad bx_2 = 0$$
$$-cx_1 + (\lambda_i - d)x_2 = 0$$

This system of equations is guaranteed to have a nontrivial solution. Since b and $a - \lambda_i$ are not both zero, then any nonzero multiple of $\begin{bmatrix} -b \\ a - \lambda_i \end{bmatrix}$ will be an eigenvector corresponding to λ_i. Therefore

$$\begin{bmatrix} -b \\ a - \lambda_1 \end{bmatrix} \quad \text{and} \quad \begin{bmatrix} -b \\ a - \lambda_2 \end{bmatrix}$$

are eigenvectors corresponding to λ_1 and λ_2, respectively.

19. As in Exercise 17, we have

$$\lambda = \frac{a + d \pm \sqrt{(a-d)^2 + 4bc}}{2}$$

$$= \frac{a + d \pm \sqrt{(c-b)^2 + 4bc}}{2} \qquad \text{because } a - d = c - b$$

$$= \frac{a + d \pm \sqrt{(c+b)^2}}{2}$$

$$= \frac{a + b + c + d}{2} \quad \text{or} \quad \frac{a - b - c + d}{2}$$

$$= a + b \quad \text{or} \quad a - c$$

Alternate Solution: Recall that if r_1 and r_2 are roots of the quadratic equation $x^2 + Bx + C = 0$, then $B = -(r_1 + r_2)$ and $C = r_1 r_2$. The converse of this result is also true. Thus the result will follow if we can show that the system of equations

$$\lambda_1 + \lambda_2 = a + d$$

$$\lambda_1 \lambda_2 = ad - bc$$

is satisfied by $\lambda_1 = a + b$ and $\lambda_2 = a - c$. This is a straightforward computation and we leave it to you.

20. Suppose that $A\mathbf{x} = \lambda\mathbf{x}$ where A is invertible. Then

$$\mathbf{x} = A^{-1}A\mathbf{x} = A^{-1}\lambda\mathbf{x} = \lambda A^{-1}\mathbf{x}$$

Since A is invertible, we know that $\lambda \neq 0$. Thus

$$A^{-1}\mathbf{x} = \frac{1}{\lambda}\mathbf{x}$$

That is, $1/\lambda$ is an eigenvalue of A^{-1} and $\mathbf{x}$ is a corresponding eigenvector.

21. Suppose that $A\mathbf{x} = \lambda\mathbf{x}$. Then

$$(A - sI)\mathbf{x} = A\mathbf{x} - sI\mathbf{x} = \lambda\mathbf{x} - s\mathbf{x} = (\lambda - s)\mathbf{x}$$

That is, $\lambda - s$ is an eigenvalue of $A - sI$ and $\mathbf{x}$ is a corresponding eigenvector.

22. The characteristic equation of A is $\lambda^3 - 6\lambda^2 + 11\lambda - 6 = 0$. The roots of this equation are $\lambda = 1, 2, 3$, so that these are the eigenvalues of A. If $\lambda = 1$, then $\lambda I - A$ reduces to

$$\begin{bmatrix} 1 & 0 & -1 \\ 0 & 1 & 0 \\ 0 & 0 & 0 \end{bmatrix}$$

Thus if $\mathbf{x} = [x_1 \quad x_2 \quad x_3]^T$ is an eigenvector of A corresponding to $\lambda = 1$, then $x_1 = x_3$ and $x_2 = 0$, so that

$$\mathbf{x} = \begin{bmatrix} r \\ 0 \\ r \end{bmatrix} = r \begin{bmatrix} 1 \\ 0 \\ 1 \end{bmatrix}$$

If $\lambda = 2$, then $\lambda I - A$ reduces to

$$\begin{bmatrix} 1 & -1/2 & 0 \\ 0 & 0 & 1 \\ 0 & 0 & 0 \end{bmatrix}$$

Thus any eigenvector $\mathbf{x}$ corresponding to $\lambda = 2$ is of the form

$$\mathbf{x} = \begin{bmatrix} s/2 \\ s \\ 0 \end{bmatrix} = s \begin{bmatrix} 1/2 \\ 1 \\ 0 \end{bmatrix}$$

If $\lambda = 3$, then $\lambda I - A$ reduces to

$$\begin{bmatrix} 1 & 0 & -1 \\ 0 & 1 & -1 \\ 0 & 0 & 0 \end{bmatrix}$$

Thus any eigenvector $\mathbf{x}$ corresponding to $\lambda = 2$ is of the form

$$\mathbf{x} = \begin{bmatrix} t \\ t \\ t \end{bmatrix} = t \begin{bmatrix} 1 \\ 1 \\ 1 \end{bmatrix}$$

(a) By Exercise 20, the eigenvalues of A^{-1} are 1, 1/2, and 1/3. The corresponding eigenspaces are spanned by the vectors

$$\begin{bmatrix} 1 \\ 0 \\ 1 \end{bmatrix} \qquad \begin{bmatrix} 1/2 \\ 1 \\ 0 \end{bmatrix} \qquad \begin{bmatrix} 1 \\ 1 \\ 1 \end{bmatrix}$$

respectively.

23. (a) For any square matrix B, we know that $\det(B) = \det(B^T)$. Thus

$$\det(\lambda I - A) = \det(\lambda I - A)^T$$
$$= \det(\lambda I^T - A^T)$$
$$= \det(\lambda I - A^T)$$

from which it follows that A and A^T have the same eigenvalues because they have the same characteristic equation.

(b) Consider, for instance, the matrix $\begin{bmatrix} 2 & 1 \\ -1 & 0 \end{bmatrix}$ which has $\lambda = 1$ as a (repeated) eigenvalue. Its eigenspace is spanned by the vector $\begin{bmatrix} -1 \\ 1 \end{bmatrix}$, while the eigenspace of its transpose is spanned by the vector $\begin{bmatrix} 1 \\ 1 \end{bmatrix}$.

24. (a) False, since $\mathbf{x}$ must be nonzero.

(b) True. The eigenvalues of A are the *only* values of λ for which the matrix $\lambda I - A$ is singular. So if λ is not an eigenvalue, then $\lambda I - A$ is invertible and the homogeneous system $(\lambda I - A)\mathbf{x} = \mathbf{0}$ has only the trivial solution.

(c) True. By hypothesis, $\lambda = 0$ is an eigenvalue of A. Theorem 7.1.3 then guarantees that λ^2, which is also zero, is an eigenvalue for A^2. Theorem 7.1.4 then guarantees that A^2 is singular.

Alternatively, observe that if $\mathbf{x}$ is the (necessarily nonzero) eigenvector corresponding to λ, then $A\mathbf{x} = \mathbf{0}$. Thus, $AA\mathbf{x} = \mathbf{0}$ for some $\mathbf{x} \neq \mathbf{0}$, so A^2 is singular.

(d) True. Since $\lambda = 0$ is not a root of $\lambda^n + 1$, Theorem 7.1.4 guarantees that A is invertible.

25. (a) Since $p(\lambda)$ has degree 6, A is 6×6.

(b) Yes, A is invertible because $\lambda = 0$ is not an eigenvalue.

(c) A will have 3 eigenspaces corresponding to the 3 eigenvalues.

EXERCISE SET 7.2

1. The eigenspace corresponding to $\lambda = 0$ can have dimension 1 or 2. The eigenspace corresponding to $\lambda = 1$ must have dimension 1. The eigenspace corresponding to $\lambda = 2$ can have dimension 1, 2, or 3.

2. **(a)** The characteristic equation is

$$\begin{vmatrix} \lambda - 4 & 0 & -1 \\ -2 & \lambda - 3 & -2 \\ -1 & 0 & \lambda - 4 \end{vmatrix} = (\lambda - 3)^2(\lambda - 5) = 0$$

Hence the eigenvalues are $\lambda = 3$ and $\lambda = 5$.

(b) If $\lambda = 3$,

$$\lambda I - A = \begin{bmatrix} -1 & 0 & -1 \\ -2 & 0 & -2 \\ -1 & 0 & -1 \end{bmatrix} \rightarrow \begin{bmatrix} 1 & 0 & 1 \\ 0 & 0 & 0 \\ 0 & 0 & 0 \end{bmatrix}$$

which has rank 1. If $\lambda = 5$,

$$\lambda I - A = \begin{bmatrix} 1 & 0 & -1 \\ -2 & 2 & -2 \\ -1 & 0 & 1 \end{bmatrix} \rightarrow \begin{bmatrix} 1 & 0 & -1 \\ 0 & 1 & 0 \\ 0 & 0 & 0 \end{bmatrix}$$

which has rank 2.

(c) The nullities of the above matrices are 2 and 1, so the corresponding eigenspaces will have dimensions 2 and 1. Therefore, the 3×3 matrix A has 3 linearly independent eigenvectors, and is diagonalizable.

5.   Call the matrix A. Since A is triangular, the eigenvalues are $\lambda = 3$ and $\lambda = 2$. The matrices $3I - A$ and $2I - A$ both have rank 2 and hence nullity 1. Thus A has only 2 linearly independent eigenvectors, so it is not diagonalizable.

6.   Call the matrix A. The characteristic equation of A is $(\lambda - 2)(\lambda^2 + \lambda + 1) = 0$, so $\lambda = 2$ is the only (real) eigenvalue. Thus A has only one linearly independent eigenvector, so it is not diagonalizable.

8.   The characteristic equation is $\lambda^2 - 3\lambda + 2 = 0$, the eigenvalues are $\lambda = 2$ and $\lambda = 1$, and the eigenspaces are spanned by the vectors

$$\begin{bmatrix} 3/4 \\ 1 \end{bmatrix} \quad \text{and} \quad \begin{bmatrix} 4/5 \\ 1 \end{bmatrix}$$

respectively. Thus, if we let

$$P = \begin{bmatrix} 3 & 4 \\ 4 & 5 \end{bmatrix}$$

then

$$P^{-1}AP = \begin{bmatrix} 2 & 0 \\ 0 & 1 \end{bmatrix}$$

Clearly, there are other possibilities for P.

10.   The characteristic equation is $\lambda(\lambda - 1)(\lambda - 2) = 0$, the eigenvalues are $\lambda = 0$, $\lambda = 1$, and $\lambda = 2$, and the eigenspaces are spanned by the vectors

$$\begin{bmatrix} 0 \\ -1 \\ 1 \end{bmatrix} \quad \begin{bmatrix} 1 \\ 0 \\ 0 \end{bmatrix} \quad \begin{bmatrix} 0 \\ 1 \\ 1 \end{bmatrix}$$

Thus, if we let

$$P = \begin{bmatrix} 0 & 1 & 0 \\ -1 & 0 & 1 \\ 1 & 0 & 1 \end{bmatrix}$$

then

$$P^{-1}AP = \begin{bmatrix} 0 & 0 & 0 \\ 0 & 1 & 0 \\ 0 & 0 & 2 \end{bmatrix}$$

12. The characteristic equation is $\lambda^3 - 4\lambda^2 + 5\lambda - 2 = 0$, the eigenvalues are $\lambda = 1$ and $\lambda = 2$, and the eigenspaces are spanned by the vectors

$$\begin{bmatrix} 1 \\ 4/3 \\ 1 \end{bmatrix} \quad \text{and} \quad \begin{bmatrix} 3/4 \\ 3/4 \\ 1 \end{bmatrix}$$

respectively. Since A is a 3×3 matrix with only 2 linearly independent eigenvectors, it is not diagonalizable.

13. The characteristic equation is $\lambda^3 - 6\lambda^2 + 11\lambda - 6 = 0$, the eigenvalues are $\lambda = 1$, $\lambda = 2$, and $\lambda = 3$, and the eigenspaces are spanned by the vectors

$$\begin{bmatrix} 1 \\ 1 \\ 1 \end{bmatrix} \quad \begin{bmatrix} 2/3 \\ 1 \\ 1 \end{bmatrix} \quad \begin{bmatrix} 1/4 \\ 3/4 \\ 1 \end{bmatrix}$$

Thus, one possibility is

$$P = \begin{bmatrix} 1 & 2 & 1 \\ 1 & 3 & 3 \\ 1 & 3 & 4 \end{bmatrix}$$

and

$$P^{-1}AP = \begin{bmatrix} 1 & 0 & 0 \\ 0 & 2 & 0 \\ 0 & 0 & 3 \end{bmatrix}$$

15. The characteristic equation is $\lambda^2(\lambda - 1) = 0$; thus $\lambda = 0$ and $\lambda = 1$ are the only eigenvalues.

The eigenspace associated with $\lambda = 0$ is spanned by the vectors $\begin{bmatrix} 1 \\ 0 \\ -3 \end{bmatrix}$ and $\begin{bmatrix} 0 \\ 1 \\ 0 \end{bmatrix}$; the

eigenspace associated with $\lambda = 1$ is spanned by $\begin{bmatrix} 0 \\ 0 \\ 1 \end{bmatrix}$. Thus, one possibility is

$$P = \begin{bmatrix} 1 & 0 & 0 \\ 0 & 1 & 0 \\ -3 & 0 & 1 \end{bmatrix}$$

and hence

$$P^{-1}AP = \begin{bmatrix} 0 & 0 & 0 \\ 0 & 0 & 0 \\ 0 & 0 & 1 \end{bmatrix}$$

18. The eigenvalues of A are $\lambda = 1$ and $\lambda = 2$ and the eigenspaces are spanned by the vectors

$$\begin{bmatrix} 1 \\ 1 \end{bmatrix} \quad \text{and} \quad \begin{bmatrix} 0 \\ 1 \end{bmatrix}$$

Thus if we let

$$P = \begin{bmatrix} 1 & 0 \\ 1 & 1 \end{bmatrix}$$

then

$$P^{-1}AP = \begin{bmatrix} 1 & 0 \\ 0 & 2 \end{bmatrix}$$

Now by Formula (10), we have

$$A^{10} = P \begin{bmatrix} 1 & 0 \\ 0 & 2^{10} \end{bmatrix} P^{-1}$$

$$= \begin{bmatrix} 1 & 0 \\ 1 & 1 \end{bmatrix} \begin{bmatrix} 1 & 0 \\ 0 & 2^{10} \end{bmatrix} \begin{bmatrix} 1 & 0 \\ -1 & 1 \end{bmatrix}$$

$$= \begin{bmatrix} 1 & 0 \\ 1 - 2^{10} & 2^{10} \end{bmatrix}$$

20. By Theorem 7.1.1, the eigenvalues of A are $\lambda_1 = 1$ and $\lambda_2 = \lambda_3 = -1$. Corresponding eigenvectors are

$$\begin{bmatrix} 1 \\ 0 \\ 0 \end{bmatrix} \qquad \begin{bmatrix} 1 \\ 1 \\ 0 \end{bmatrix} \qquad \begin{bmatrix} -4 \\ 0 \\ 1 \end{bmatrix}$$

Since A has three linearly independent eigenvectors, it is diagonalizable. Let

$$P = \begin{bmatrix} 1 & 1 & -4 \\ 0 & 1 & 0 \\ 0 & 0 & 1 \end{bmatrix}$$

Then

$$P^{-1}A\,P = D = \begin{bmatrix} 1 & 0 & 0 \\ 0 & -1 & 0 \\ 0 & 0 & -1 \end{bmatrix}$$

Thus, $A^k = PD^kP^{-1}$ for any positive integer k, or

$$A^k = \begin{cases} PIP^{-1} = PP^{-1} = I & \text{if } k \text{ is even} \\ PDP^{-1} = A & \text{if } k \text{ is odd} \end{cases}$$

Finally, since D is its own inverse, we have

$$A^{-1} = (PDP^{-1})^{-1} = PD^{-1}P^{-1} = A$$

so that

$$A^{-k} = (A^{-1})^k = A^k$$

Of course, all of the above could have been greatly simplified if we had been clever enough to notice at the beginning that $A^2 = I$.

21. The characteristic equation of A is $(\lambda - 1)(\lambda - 3)(\lambda - 4) = 0$ so that the eigenvalues are $\lambda = 1, 3$, and 4. Corresponding eigenvectors are $[1 \quad 2 \quad 1]^T$, $[1 \quad 0 \quad -1]^T$, and $[1 \quad -1 \quad 1]^T$, respectively, so we let

$$P = \begin{bmatrix} 1 & 1 & 1 \\ 2 & 0 & -1 \\ 1 & -1 & 1 \end{bmatrix}$$

Hence

$$P^{-1} = \begin{bmatrix} 1/6 & 1/3 & 1/6 \\ 1/2 & 0 & -1/2 \\ 1/3 & -1/3 & 1/3 \end{bmatrix}$$

and therefore

$$A^n = \begin{bmatrix} 1 & 1 & 1 \\ 2 & 0 & -1 \\ 1 & -1 & 1 \end{bmatrix} \begin{bmatrix} 1^n & 0 & 0 \\ 0 & 3^n & 0 \\ 0 & 0 & 4^n \end{bmatrix} \begin{bmatrix} 1/6 & 1/3 & 1/6 \\ 1/2 & 0 & -1/2 \\ 1/3 & -1/3 & 1/3 \end{bmatrix}$$

22. (a) By Theorem 7.2.3, we can diagonalize A if it has two distinct eigenvalues. According to Exercise 17 of Section 7.1, this will occur if and only if $(a - d)^2 + 4bc > 0$.

(b) See Part (a). If the discriminant, $(a - d)^2 + 4bc$, of the characteristic equation is negative, then there are no real roots and hence no real eigenvalues or eigenvectors. Thus, by Theorem 7.2.1, A is not diagonalizable.

24. Since A is diagonalizable, we have $A = PDP^{-1}$ where D is a diagonal matrix with the eigenvalues of A as its diagonal elements. Thus the rank of D is the number of nonzero eigenvalues of A. Since P and P^{-1} are invertible, they are products of elementary matrices. But multiplication on the left or on the right by an elementary matrix does not change the rank of the product. Therefore $\operatorname{rank}(A) = \operatorname{rank}(D)$, and we are done.

25. **(a)** False. For instance the matrix $\begin{bmatrix} 0 & 1 \\ -1 & 2 \end{bmatrix}$, which has linearly independent column vectors, has characteristic polynomial $(\lambda - 1)^2$. Thus $\lambda = 1$ is the only eigenvalue. The corresponding eigenvectors all have the form $t \begin{bmatrix} 1 \\ 1 \end{bmatrix}$. Thus this 2×2 matrix has only 1 linearly independent eigenvector, and hence is not diagonalizable.

(b) False. Any matrix Q which is obtained from P by multiplying each entry by a nonzero number k will also work. Why?

(c) True by Theorem 7.2.2.

(d) True. Suppose that A is invertible and diagonalizable. Then there is an invertible matrix P such that $P^{-1}AP = D$ where D is diagonal. Since D is the product of invertible matrices, it is invertible, which means that each of its diagonal elements d_i is nonzero and D^{-1} is the diagonal matrix with diagonal elements $1/d_i$. Thus we have

$$(P^{-1}AP)^{-1} = D^{-1}$$

or

$$P^{-1}A^{-1}P = D^{-1}$$

That is, the same matrix P will diagonalize both A and A^{-1}.

26. **(a)** The dimensions of the eigenspaces corresponding to $\lambda = 1$, $\lambda = 3$, and $\lambda = 4$ are at most 1, 2, and 3 respectively. The dimension of the eigenspace corresponding to $\lambda = 1$ must be exactly 1.

(b) If A is diagonalizable, then it follows from Theorem 7.2.1 that the dimensions of the 3 spaces will be exactly 1, 2, and 3.

(c) The eigenvalue must be $\lambda = 4$.

27. **(a)** Since A is diagonalizable, there exists an invertible matrix P such that $P^{-1}AP = D$ where D is a diagonal matrix containing the eigenvalues of A along its diagonal. Moreover, it easily follows that $P^{-1}A^kP = D^k$ for k a positive integer. In addition, Theorem 7.1.3 guarantees that if λ is an eigenvalue for A, then λ^k is an eigenvalue for A^k. In other words, D^k displays the eigenvalues of A^k along its diagonal.

Therefore, the sequence

$$P^{-1}AP = D$$

$$P^{-1}A^2P = D^2$$

$$\vdots$$

$$P^{-1}A^kP = D^k$$

$$\vdots$$

will converge if and only if the sequence $A, A^2, \ldots, A^k, \ldots$ converges. Moreover, this will occur if and only if the sequences $\lambda_i, \lambda_i^2, \ldots, \lambda_i^k, \ldots$ converges for each of the n eigenvalues λ_i of A

(b) In general, a given sequence of real numbers $a, a^2, a^3, \ldots$ will converge to 0 if and only if $-1 < a < 1$ and to 1 if $a = 1$. The sequence diverges for all other values of a.

Recall that $P^{-1}A^kP = D^k$ where D^k is a diagonal matrix containing the eigenvalues $\lambda_1^k, \lambda_2^k, \ldots, \lambda_n^k$ on its diagonal. If $|\lambda_i| < 1$ for all $i = 1, 2, \ldots, n$, then $\lim_{k \to \infty} D^k = O$ and hence $\lim_{k \to \infty} A^k = O$.

If $\lambda_i = 1$ is an eigenvalue of A for one or more values of i and if all of the other eigenvalues satisfy the inequality $|\lambda_j| < 1$, then $\lim_{k \to \infty} A^k$ exists and equals PD_LP^{-1} where D_L is a diagonal matrix with only 1's and 0's on the diagonal.

If A possesses one or more eigenvalues λ which do not satisfy the inequality $-1 < \lambda \le 1$, then $\lim_{k \to \infty} A^k$ does not exist.

28. (a) If $S_n = 1 + x + x^2 + \cdots + x^n$, then $S_n - xS_n = (1 - x)S_n = 1 - x^{n+1}$ or $S_n = (1 - x^{n+1})/(1 - x)$. If $|x| < 1$, then $\lim_{n \to \infty} S_n = 1/(1 - x)$. If $|x| \ge 1$, then the limit doesn't exist.

(b) and (c) If A is diagonalizable and $\lambda_1, \lambda_2, \ldots, \lambda_n$ are the eigenvalues of A, then

$$P^{-1}(I + A + A^2 + \cdots + A^k)P = I + P^{-1}AP + P^{-1}A^2P + \cdots + P^{-1}A^kP$$

is a diagonal matrix with

$$d_i = 1 + \lambda_i + \lambda_i^2 + \cdots + \lambda_i^k$$

as its ith entry. If $|\lambda_i| < 1$ for all $i = 1, 2, \ldots, n$, then the ith entry of that diagonal matrix will converge to $1/(1 - \lambda_i)$.

If A possesses one or more eigenvalues λ such that $|\lambda| \ge 1$, then the limit of the diagonal matrix will not exist.

EXERCISE SET 7.3

1. **(a)** The characteristic equation is $\lambda(\lambda - 5) = 0$. Thus each eigenvalue is repeated once and hence each eigenspace is 1-dimensional.

   (c) The characteristic equation is $\lambda^2(\lambda - 3) = 0$. Thus the eigenspace corresponding to $\lambda = 0$ is 2-dimensional and that corresponding to $\lambda = 3$ is 1-dimensional.

   (e) The characteristic equation is $\lambda^3(\lambda - 8) = 0$. Thus the eigenspace corresponding to $\lambda = 0$ is 3-dimensional and that corresponding to $\lambda = 8$ is 1-dimensional.

2. The eigenvalues of A are $\lambda = 4$ and $\lambda = 2$. The eigenspaces are spanned by the orthogonal vectors

$$\begin{bmatrix} 1 \\ 1 \end{bmatrix} \quad \text{and} \quad \begin{bmatrix} -1 \\ 1 \end{bmatrix}$$

respectively. If we normalize these vectors, we obtain

$$\begin{bmatrix} \dfrac{1}{\sqrt{2}} \\ \dfrac{1}{\sqrt{2}} \end{bmatrix} \quad \text{and} \quad \begin{bmatrix} \dfrac{-1}{\sqrt{2}} \\ \dfrac{1}{\sqrt{2}} \end{bmatrix}$$

Thus

$$P = \begin{bmatrix} \dfrac{1}{\sqrt{2}} & -\dfrac{1}{\sqrt{2}} \\ \dfrac{1}{\sqrt{2}} & \dfrac{1}{\sqrt{2}} \end{bmatrix}$$

and

$$P^{-1}AP = \begin{bmatrix} 4 & 0 \\ 0 & 2 \end{bmatrix}$$

If we interchange the order of the vectors which span the eigenspaces, we obtain

$$P = \begin{bmatrix} -\dfrac{1}{\sqrt{2}} & \dfrac{1}{\sqrt{2}} \\ \dfrac{1}{\sqrt{2}} & \dfrac{1}{\sqrt{2}} \end{bmatrix}$$

and

$$P^{-1}AP = \begin{bmatrix} 2 & 0 \\ 0 & 4 \end{bmatrix}$$

6. The eigenvalues of A are $\lambda = 2$ and $\lambda = 0$. The eigenspace associated with $\lambda = 2$ is spanned by $\begin{bmatrix} 1 \\ 1 \\ 0 \end{bmatrix}$; the eigenspace associated with $\lambda = 0$ is spanned by $\begin{bmatrix} 1 \\ -1 \\ 0 \end{bmatrix}$ and $\begin{bmatrix} 0 \\ 0 \\ 1 \end{bmatrix}$. If we normalize these three orthogonal vectors, we obtain

$$\begin{bmatrix} \dfrac{1}{\sqrt{2}} \\ \dfrac{1}{\sqrt{2}} \\ 0 \end{bmatrix} \quad \begin{bmatrix} \dfrac{1}{\sqrt{2}} \\ -\dfrac{1}{\sqrt{2}} \\ 0 \end{bmatrix} \quad \begin{bmatrix} 0 \\ 0 \\ 1 \end{bmatrix}$$

Thus

$$P = \begin{bmatrix} \dfrac{1}{\sqrt{2}} & \dfrac{1}{\sqrt{2}} & 0 \\ \dfrac{1}{\sqrt{2}} & -\dfrac{1}{\sqrt{2}} & 0 \\ 0 & 0 & 1 \end{bmatrix}$$

and

$$P^{-1}AP = \begin{bmatrix} 2 & 0 & 0 \\ 0 & 0 & 0 \\ 0 & 0 & 0 \end{bmatrix}$$

Another possibility (which results from changing the order of the spanning vectors) is

$$P = \begin{bmatrix} \dfrac{1}{\sqrt{2}} & 0 & \dfrac{1}{\sqrt{2}} \\ -\dfrac{1}{\sqrt{2}} & 0 & \dfrac{1}{\sqrt{2}} \\ 0 & 1 & 0 \end{bmatrix}$$

so that

$$P^{-1}AP = \begin{bmatrix} 0 & 0 & 0 \\ 0 & 0 & 0 \\ 0 & 0 & 2 \end{bmatrix}$$

8. The eigenvalues of A are $\lambda = 0$, $\lambda = 4$, and $\lambda = 2$. The eigenspace associated with $\lambda = 0$

is spanned by $\begin{bmatrix} 0 \\ 0 \\ 1 \\ 0 \end{bmatrix}$ and $\begin{bmatrix} 0 \\ 0 \\ 0 \\ 1 \end{bmatrix}$; the eigenspaces associated with $\lambda = 4$ and $\lambda = 2$ are

spanned by $\begin{bmatrix} 1 \\ 1 \\ 0 \\ 0 \end{bmatrix}$ and $\begin{bmatrix} 1 \\ -1 \\ 0 \\ 0 \end{bmatrix}$ respectively. Hence, the eigenspaces of A are spanned by

the orthogonal vectors

$$\begin{bmatrix} 0 \\ 0 \\ 1 \\ 0 \end{bmatrix} \quad \begin{bmatrix} 0 \\ 0 \\ 0 \\ 1 \end{bmatrix} \quad \begin{bmatrix} 1 \\ 1 \\ 0 \\ 0 \end{bmatrix} \quad \begin{bmatrix} 1 \\ -1 \\ 0 \\ 0 \end{bmatrix}$$

Normalizing each of these vectors yields

$$
\begin{bmatrix} 0 \\ 0 \\ 1 \\ 0 \end{bmatrix}
\quad
\begin{bmatrix} 0 \\ 0 \\ 0 \\ 1 \end{bmatrix}
\quad
\begin{bmatrix} 1/\sqrt{2} \\ 1/\sqrt{2} \\ 0 \\ 0 \end{bmatrix}
\quad
\begin{bmatrix} 1/\sqrt{2} \\ -1/\sqrt{2} \\ 0 \\ 0 \end{bmatrix}
$$

Thus

$$
P = \begin{bmatrix}
0 & 0 & 1/\sqrt{2} & 1/\sqrt{2} \\
0 & 0 & 1/\sqrt{2} & -1/\sqrt{2} \\
1 & 0 & 0 & 0 \\
0 & 1 & 0 & 0
\end{bmatrix}
$$

$$
P^{-1} = \begin{bmatrix}
0 & 0 & 1 & 0 \\
0 & 0 & 0 & 1 \\
1/\sqrt{2} & 1/\sqrt{2} & 0 & 0 \\
1/\sqrt{2} & -1/\sqrt{2} & 0 & 0
\end{bmatrix}
$$

and

$$
P^{-1}AP = \begin{bmatrix}
0 & 0 & 0 & 0 \\
0 & 0 & 0 & 0 \\
0 & 0 & 4 & 0 \\
0 & 0 & 0 & 2
\end{bmatrix}
$$

Changing the order of the spanning vectors may change the position of the eigenvalues on the diagonal of $P^{-1}AP$.

10. The characteristic equation is $\lambda^2 - 2a\lambda + a^2 - b^2 = 0$. By the quadratic formula, $\lambda = a \pm |b|$, or $\lambda = a \pm b$. Since $b \neq 0$, there are 2 distinct eigenvalues with eigenspaces spanned by the orthogonal vectors

$$
\begin{bmatrix} 1 \\ 1 \end{bmatrix}
\quad \text{and} \quad
\begin{bmatrix} -1 \\ 1 \end{bmatrix}
$$

Thus

$$
\begin{bmatrix}
\dfrac{1}{\sqrt{2}} & -\dfrac{1}{\sqrt{2}} \\[2ex]
\dfrac{1}{\sqrt{2}} & \dfrac{1}{\sqrt{2}}
\end{bmatrix}
$$

will orthogonally diagonalize the given matrix.

12. **(a)** By Theorem 7.3.1, it will suffice to show that $I - \mathbf{v}\,\mathbf{v}^T$ is symmetric. This follows from properties of the transpose.

   (b) In this case,

$$
I - \mathbf{v}\,\mathbf{v}^T =
\begin{bmatrix}
1 & 0 & 0 \\
0 & 1 & 0 \\
0 & 0 & 1
\end{bmatrix}
-
\begin{bmatrix}
1 \\ 0 \\ 1
\end{bmatrix}
\begin{bmatrix} 1 & 0 & 1 \end{bmatrix}
$$

$$
=
\begin{bmatrix}
0 & 0 & -1 \\
0 & 1 & 0 \\
-1 & 0 & 0
\end{bmatrix}
$$

The characteristic equation is $(\lambda-1)\,(\lambda-1)(\lambda+1) = 0$, so the eigenvalues are $\lambda = \pm 1$.

Two linearly independent eigenvectors corresponding to $\lambda = 1$ are $\begin{bmatrix} 1 \\ 0 \\ -1 \end{bmatrix}$ and $\begin{bmatrix} 0 \\ 1 \\ 0 \end{bmatrix}$.

An eigenvector corresponding to $\lambda = -1$ is $\begin{bmatrix} 1 \\ 0 \\ 1 \end{bmatrix}$. Thus the matrix

$$
P =
\begin{bmatrix}
1/\sqrt{2} & 0 & 1/\sqrt{2} \\
0 & 1 & 0 \\
-1/\sqrt{2} & 0 & 1/\sqrt{2}
\end{bmatrix}
$$

will orthogonally diagonalize $I - \mathbf{v}\,\mathbf{v}^T$.

13. By the result of Exercise 17, Section 7.1, the eigenvalues of the symmetric 2×2 matrix

$$\begin{bmatrix} a & b \\ b & a \end{bmatrix} \text{ are } \lambda = \frac{1}{2}\left[(a+d) \pm \sqrt{(a-d)^2 + 4b^2}\right]. \text{ Since } (a-d)^2 + 4b^2 \text{ cannot be negative,}$$

the eigenvalues are real.

14. **(a)** True, since both are symmetric.

(b) True, since $\mathbf{v}_1$ and $\mathbf{v}_2$ must be orthogonal.

(c) False. For instance, let $A = \begin{bmatrix} 1/2 & -\sqrt{3}/2 \\ \sqrt{3}/2 & 1/2 \end{bmatrix}$. Then A is orthogonal, but not symmetric and hence not orthogonally diagonalizable.

(d) Since A is orthogonally diagonalizable, it follows from Theorem 7.3.1 that it must be symmetric, that is, $A = A^T$. And since A is also invertible, we have

$$I = I^T = (A^{-1}A)^T = A^T(A^{-1})^T = A(A^{-1})^T$$

This implies that $A^{-1} = (A^{-1})^T$, so that A^{-1} is symmetric and hence is orthogonally diagonalizable.

15. Yes. Notice that the given vectors are pairwise orthogonal, so we consider the equation

$$P^{-1}AP = D$$

or

$$A = PDP^{-1}$$

where the columns of P consist of the given vectors each divided by its norm and where D is the diagonal matrix with the eigenvalues of A along its diagonal. That is,

$$P = \begin{bmatrix} 0 & 1 & 0 \\ 1/\sqrt{2} & 0 & 1/\sqrt{2} \\ -1/\sqrt{2} & 0 & 1/\sqrt{2} \end{bmatrix} \quad \text{and} \quad D = \begin{bmatrix} -1 & 0 & 0 \\ 0 & 3 & 0 \\ 0 & 0 & 7 \end{bmatrix}$$

From this, it follows that

$$A = PDP^{-1} = \begin{bmatrix} 3 & 0 & 0 \\ 0 & 3 & 4 \\ 0 & 4 & 3 \end{bmatrix}$$

Alternatively, we could just substitute the appropriate values for λ and $\mathbf{x}$ in the equation $A\mathbf{x} = \lambda\mathbf{x}$ and solve for the matrix A.

SUPPLEMENTARY EXERCISES 7

1. **(a)** The characteristic equation of A is $\lambda^2 - 2\cos\theta + 1 = 0$. The discriminant of this equation is $4(\cos^2\theta - 1)$, which is negative unless $\cos^2\theta = 1$. Thus A can have no real eigenvalues or eigenvectors in case $0 < \theta < \pi$.

3. **(a)** If

$$D = \begin{bmatrix} a_1 & 0 & \cdots & 0 \\ 0 & a_2 & \cdots & 0 \\ \vdots & \vdots & & \vdots \\ 0 & 0 & \cdots & a_n \end{bmatrix}$$

then $D = S^2$, where

$$S = \begin{bmatrix} \sqrt{a_1} & 0 & \cdots & 0 \\ 0 & \sqrt{a_2} & \cdots & 0 \\ \vdots & \vdots & & \vdots \\ 0 & 0 & \cdots & \sqrt{a_n} \end{bmatrix}$$

Of course, this makes sense only if $a_1 \geq 0, \ldots, a_n \geq 0$.

(b) If A is diagonalizable, then there are matrices P and D such that D is diagonal and $D = P^{-1}AP$. Moreover, if A has nonnegative eigenvalues, then the diagonal entries of D are nonnegative since they are all eigenvalues. Thus there is a matrix T, by virtue of Part (a), such that $D = T^2$. Therefore,

$$A = PDP^{-1} = PT^2P^{-1} = PTP^{-1}PTP^{-1} = (PTP^{-1})^2$$

That is, if we let $S = PTP^{-1}$, then $A = S^2$.

3.   **(c)**   The eigenvalues of A are $\lambda = 9$, $\lambda = 1$, and $\lambda = 4$. The eigenspaces are spanned by the vectors

$$\begin{bmatrix} 1 \\ 2 \\ 2 \end{bmatrix} \qquad \begin{bmatrix} 1 \\ 0 \\ 0 \end{bmatrix} \qquad \begin{bmatrix} 1 \\ 1 \\ 0 \end{bmatrix}$$

Thus, we have

$$P = \begin{bmatrix} 1 & 1 & 1 \\ 2 & 0 & 1 \\ 2 & 0 & 0 \end{bmatrix} \quad \text{and} \quad P^{-1} = \begin{bmatrix} 0 & 0 & 1/2 \\ 1 & -1 & 1/2 \\ 0 & 1 & -1 \end{bmatrix}$$

while

$$D = \begin{bmatrix} 9 & 0 & 0 \\ 0 & 1 & 0 \\ 0 & 0 & 4 \end{bmatrix} \quad \text{and} \quad T = \begin{bmatrix} 3 & 0 & 0 \\ 0 & 1 & 0 \\ 0 & 0 & 2 \end{bmatrix}$$

Therefore

$$S = PTP^{-1} = \begin{bmatrix} 1 & 1 & 0 \\ 0 & 2 & 1 \\ 0 & 0 & 3 \end{bmatrix}$$

4.   **(a)**   Since $(\lambda I - A)^T = \lambda I - A^T$, we have

$$\det(\lambda I - A) = \det((\lambda I - A)^T)$$
$$= \det(\lambda I - A^T)$$

Thus A and A^T have the same characteristic polynomials, and hence the same eigenvalues.

(b)   By the result of Exercise 18, Section 7.1, if A has distinct eigenvalues λ_1 and λ_2, and if $bc \neq 0$, then the eigenvectors of A are

$$\begin{bmatrix} -b \\ a - \lambda_1 \end{bmatrix} \quad \text{and} \quad \begin{bmatrix} -b \\ a - \lambda_2 \end{bmatrix}$$

Corresponding eigenvectors of A^T are

$$\begin{bmatrix} -c \\ a - \lambda_1 \end{bmatrix} \text{ and } \begin{bmatrix} -c \\ a - \lambda_2 \end{bmatrix}$$

Thus we need only look for a matrix with two distinct eigenvalues, with $bc \neq 0$, and with $b \neq c$. For instance, if we let

$$A = \begin{bmatrix} 1 & 1 \\ 3 & -1 \end{bmatrix}$$

then the resulting pairs of eigenvectors will be different.

5. Since $\det(\lambda I - A)$ is a sum of signed elementary products, we ask which terms involve λ^{n-1}. Obviously the signed elementary product

$$q = (\lambda - a_{11})(\lambda - a_{22}) \cdots (\lambda - a_{nn})$$

$$= \lambda^n - (a_{11} + a_{22} + \cdots + a_{nn})\lambda^{n-1}$$

$$+ \text{ terms involving } \lambda^r \text{ where } r < n - 1$$

has a term $- (\text{trace of } A)\lambda^{n-1}$. Any elementary product containing fewer than $n-1$ factors of the form $\lambda - a_{ii}$ cannot have a term which contains λ^{n-1}. But there is no elementary product which contains exactly $n-1$ of these factors (Why?). Thus the coefficient of λ^{n-1} is minus the trace of A.

6. The only eigenvalue is $\lambda = a$ and if $b \neq 0$, then the eigenspace is spanned by the vector $\begin{bmatrix} 1 \\ 0 \end{bmatrix}$. Thus, by Theorem 7.2.1, A is not diagonalizable.

7. **(b)** The characteristic equation is

$$p(\lambda) = -1 + 3\lambda - 3\lambda^2 + \lambda^3 = 0$$

Moreover,

$$A^2 = \begin{bmatrix} 0 & 0 & 1 \\ 1 & -3 & 3 \\ 3 & -8 & 6 \end{bmatrix}$$

and

$$A^3 = \begin{bmatrix} 1 & -3 & 3 \\ 3 & -8 & 6 \\ 6 & -15 & 10 \end{bmatrix}$$

It then follows that

$$p(A) = -I + 3A - 3A^2 + A^3 = 0$$

9. Since $c_0 = 0$ and $c_1 = -5$, we have $A^2 = 5A$, and, in general, $A^n = 5^{n-1}A$.

11. Call the matrix A and show that $A^2 = (c_1 + c_2 + \cdots + c_n)A = [\text{tr}(A)]A$. Thus $A^k = [\text{tr}(A)]^{k-1}A$ for $k = 2, 3, \ldots$. Now if λ is an eigenvalue of A, then λ^k is an eigenvalue of A^k, so that in case $\text{tr}(A) \neq 0$, we have that $\lambda^k/[\text{tr}(A)]^{k-1} = [\lambda/\text{tr}(A)]^{k-1}\lambda$ is an eigenvalue of A for $k = 2, 3, \ldots$. Why? We know that A has at most n eigenvalues, so that this expression can take on only finitely many values. This means that either $\lambda = 0$ or $\lambda = \text{tr}(A)$. Why? In case $\text{tr}(A) = 0$, then all of the eigenvalues of A are 0. Why? Thus the only possible eigenvalues of A are zero and $\text{tr}(A)$. It is easy to check that each of these is, in fact, an eigenvalue of A.

Alternatively, we could evaluate $\det(I\lambda - A)$ by brute force. If we add Column 1 to Column 2, the new Column 2 to Column 3, the new Column 3 to Column 4, and so on, we obtain the equation

$$\det(I\lambda - A) = \det \begin{bmatrix} \lambda - c_1 & \lambda - c_1 - c_2 & \lambda - c_1 - c_2 - c_3 & \cdots & \lambda - c_1 - c_2 - \cdots - c_n \\ -c_1 & \lambda - c_1 - c_2 & \lambda - c_1 - c_2 - c_3 & \cdots & \lambda - c_1 - c_2 - \cdots - c_n \\ -c_1 & -c_1 - c_2 & \lambda - c_1 - c_2 - c_3 & \cdots & \lambda - c_1 - c_2 - \cdots - c_n \\ \vdots & \vdots & \vdots & & \vdots \\ -c_1 & -c_1 - c_2 & -c_1 - c_2 - c_3 & \cdots & \lambda - c_1 - c_2 - \cdots - c_n \end{bmatrix}$$

If we subtract Row 1 from each of the other rows and then expand by cofactors along the nth column, we have

$$\det(I\lambda - A) = \det \begin{bmatrix} \lambda - c_1 & \lambda - c_1 - c_2 & \lambda - c_1 - c_2 - c_3 & \cdots & \lambda - \text{tr}(A) \\ -\lambda & 0 & 0 & \cdots & 0 \\ -\lambda & -\lambda & 0 & \cdots & 0 \\ \vdots & \vdots & \vdots & & \vdots \\ -\lambda & -\lambda & -\lambda & \cdots & 0 \end{bmatrix}$$

$$= (-1)^{n+1}(\lambda - \text{tr}(A)) \det \begin{bmatrix} -\lambda & 0 & 0 & \cdots & 0 \\ -\lambda & -\lambda & 0 & \cdots & 0 \\ -\lambda & -\lambda & -\lambda & \cdots & 0 \\ \vdots & \vdots & \vdots & & \vdots \\ -\lambda & -\lambda & -\lambda & \cdots & -\lambda \end{bmatrix}$$

$$= (-1)^{n+1}(\lambda - \text{tr}(A))(-\lambda)^{n-1} \qquad \text{because the above matrix is triangular}$$

$$= (-1)^{2n}(\lambda - \text{tr}(A))\lambda^{n-1}$$

$$= \lambda^{n-1}(\lambda - \text{tr}(A))$$

Thus $\lambda = \text{tr}(A)$ and $\lambda = 0$ are the eigenvalues, with $\lambda = 0$ repeated $n - 1$ times.

12. **(a)** The characteristic polynomial of the given matrix is

$$p(\lambda) = \det \begin{bmatrix} \lambda & 0 & 0 & \cdots & 0 & c_1 \\ -1 & \lambda & 0 & \cdots & 0 & c_2 \\ 0 & -1 & \lambda & \cdots & 0 & c_3 \\ \vdots & \vdots & \vdots & & \vdots & \vdots \\ 0 & 0 & 0 & \cdots & -1 & \lambda + c_{n-1} \end{bmatrix}$$

If we add λ times the second row to the first row, this becomes

$$p(\lambda) = \det \begin{bmatrix} 0 & \lambda^2 & 0 & \cdots & 0 & c_1 + c_2\lambda \\ -1 & \lambda & 0 & \cdots & 0 & c_2 \\ 0 & -1 & \lambda & \cdots & 0 & c_3 \\ \vdots & \vdots & \vdots & & \vdots & \vdots \\ 0 & 0 & 0 & \cdots & -1 & \lambda + c_{n-1} \end{bmatrix}$$

Expanding by cofactors along the first column gives

$$p(\lambda) = \det \begin{bmatrix} \lambda^2 & 0 & 0 & \cdots & 0 & c_1 + c_2\lambda \\ -1 & \lambda & 0 & \cdots & 0 & c_3 \\ 0 & -1 & \lambda & \cdots & 0 & c_4 \\ \vdots & \vdots & \vdots & & \vdots & \vdots \\ 0 & 0 & 0 & \cdots & -1 & \lambda + c_{n-1} \end{bmatrix}$$

Adding λ^2 times the second row to the first row yields

$$p(\lambda) = \det \begin{bmatrix} 0 & \lambda^3 & 0 & \cdots & 0 & c_1 + c_2\lambda + c_3\lambda^2 \\ -1 & \lambda & 0 & \cdots & 0 & c_3 \\ 0 & -1 & \lambda & \cdots & 0 & c_4 \\ \vdots & \vdots & \vdots & & \vdots & \vdots \\ 0 & 0 & 0 & \cdots & -1 & \lambda + c_{n-1} \end{bmatrix}$$

Expanding by cofactors along the first column gives

$$p(\lambda) = \det \begin{bmatrix} \lambda^3 & 0 & 0 & \cdots & 0 & c_1 + c_2\lambda + c_3\lambda^2 \\ -1 & \lambda & 0 & \cdots & 0 & c_4 \\ 0 & -1 & \lambda & \cdots & 0 & c_5 \\ \vdots & \vdots & \vdots & & \vdots & \vdots \\ 0 & 0 & 0 & \cdots & -1 & \lambda + c_{n-1} \end{bmatrix}$$

Eventually this procedure will yield

$$p(\lambda) = \det \begin{bmatrix} \lambda^{n-2} & c_1 + c_2\lambda + \cdots + c_{n-2}\lambda^{n-3} \\ -1 & \lambda + c_{n-1} \end{bmatrix}$$

Adding λ^{n-2} times the second row to the first row gives

$$p(\lambda) = \det \begin{bmatrix} 0 & c_1 + c_2\lambda + \cdots + c_{n-1}\lambda^{n-2} + \lambda^{n-1} \\ -1 & \lambda + c_{n-1} \end{bmatrix}$$

or

$$p(\lambda) = c_1 + c_2\lambda + \cdots + c_{n-1}\lambda^{n-2} + \lambda^{n-1}$$

which is the desired result.

14. The characteristic equation of an $n \times n$ matrix has degree n and therefore n roots. Since the complex roots of a polynomial equation with real coefficients occur in conjugate pairs, any such equation of odd degree must have at least one real root. Hence A has at least one real eigenvalue.

16. The characteristic equation of the matrix is

$$(\lambda - 1)(\lambda + 2)(\lambda - 3)(\lambda + 3) = \lambda^4 + \lambda^3 - 11\lambda^2 - 9\lambda + 18 = 0$$

  (a) By Exercise 14 of Section 7.1, $\det(A) = 18$.

  (b) By Exercise 5, above, $\operatorname{tr}(A) = -1$.

17. Since every odd power of A is again A, we have that every odd power of an eigenvalue of A is again an eigenvalue of A. Thus the only possible eigenvalues of A are $\lambda = 0, \pm 1$.

EXERCISE SET 8.1

2. Using the usual notation, we have

$$T(\mathbf{u} + \mathbf{v}) = (2(u_1 + v_1) - (u_2 + v_2) + (u_3 + v_3), \ (u_2 + v_2) - 4(u_3 + v_3))$$

$$= (2u_1 - u_2 + u_3, \ u_2 - 4u_3) + (2v_1 - v_2 + v_3, \ v_2 - 4v_3)$$

$$= T(\mathbf{u}) + T(\mathbf{v})$$

and

$$T(c\mathbf{u}) = (2cu_1 - cu_2 + cu_3, \ cu_2 - 4cu_3) = cT(\mathbf{u})$$

3. Since $T(-\mathbf{u}) = \|-\mathbf{u}\| = \|\mathbf{u}\| = T(\mathbf{u}) \neq -T(\mathbf{u})$ unless $\mathbf{u} = \mathbf{0}$, the function is not linear.

5. We observe that

$$T(A_1 + A_2) = (A_1 + A_2)B = A_1 B + A_2 B = T(A_1) + T(A_2)$$

and

$$T(cA) = (cA)B = c(AB) = cT(A)$$

Hence, T is linear.

6. Since $T(A + B) = \operatorname{tr}(A + B) = \operatorname{tr}(A) + \operatorname{tr}(B) = T(A) + T(B)$ and $T(cA) = \operatorname{tr}(cA) = c(\operatorname{tr}(A)) = cT(A)$, T is linear.

8. **(b)** Since both properties fail, T is nonlinear. For instance,

$$T\left(k\begin{bmatrix} a & b \\ c & d \end{bmatrix}\right) = k^2 a^2 + k^2 b^2$$

$$\neq kT\left(\begin{bmatrix} a & b \\ c & d \end{bmatrix}\right) \text{ in general.}$$

10. **(a)** T is not linear because it sends the zero function to $f(x) = 1$.

(b) Since

$$T(f(x) + g(x)) = T((f+g)(x)) = (f+g)(x+1)$$

$$= f(x+1) + g(x+1) = T(f(x)) + T(g(x))$$

and

$$T(cf(x)) = cf(x+1) = cT(f(x))$$

T is linear.

12. Let $\mathbf{v} = (x_1, x_2) = a\mathbf{v}_1 + b\mathbf{v}_2 = a(1,1) + b(1,0)$. This yields the equations $x_1 = a + b$ and $x_2 = a$, so that $\mathbf{v} = x_2\mathbf{v}_1 + (x_1 - x_2)\mathbf{v}_2$. Thus

$$T(\mathbf{v}) = T(x_1, x_2) = x_2 T(\mathbf{v}_1) + (x_1 - x_2)T(\mathbf{v}_2)$$

$$= x_2(1, -2) + (x_1 - x_2)(-4, 1)$$

$$= (-4x_1 + 5x_2, \ x_1 - 3x_2)$$

so that

$$T(2, -3) = (-8 - 15, \ 2 + 9) = (-23, 11)$$

14. Let $\mathbf{v} = (x_1, x_2, x_3) = a\mathbf{v}_1 + b\mathbf{v}_2 + c\mathbf{v}_3 = a(1,1,1) + b(1,1,0) + c(1,0,0)$. Thus $x_1 = a + b + c$, $x_2 = a + b$, and $x_3 = a$, so that $\mathbf{v} = x_3\mathbf{v}_1 + (x_2 - x_3)\mathbf{v}_2 + (x_1 - x_2)\mathbf{v}_3$. Therefore

$$T(\mathbf{v}) = T(x_1, x_2, x_3) = x_3 T(\mathbf{v}_1) + (x_2 - x_3)T(\mathbf{v}_2) + (x_1 - x_2)T(\mathbf{v}_3)$$

$$= x_3(2, -1, 4) + (x_2 - x_3)(3, 0, 1) + (x_1 - x_2)(-1, 5, 1)$$

$$= (-x_1 + 4x_2 - x_3, \ 5x_1 - 5x_2 - x_3, \ x_1 + 3x_3)$$

and

$$T(2, 4, -1) = (-2 + 16 + 1, \ 10 - 20 + 1, \ 2 - 3) = (15, -9, -1)$$

16. We have

$$T(2\mathbf{v}_1 - 3\mathbf{v}_2 + 4\mathbf{v}_3) = 2T(\mathbf{v}_1) - 3T(\mathbf{v}_2) + 4T(\mathbf{v}_3)$$

$$= (2, -2, 4) + (0, -9, -6) + (-12, 4, 8)$$

$$= (-10, -7, 6)$$

17. **(a)** Since T_1 is defined on all of R^2, the domain of $T_2 \circ T_1$ is R^2. We have $T_2 \circ T_1(x, y) = T_2(T_1(x, y)) = T_2(2x, 3y) = (2x - 3y, 2x + 3y)$. Since the system of equations

$$2x - 3y = a$$

$$2x + 3y = b$$

can be solved for all values of a and b, the codomain is also all of R^2.

(d) Since T_1 is defined on all of R^2, the domain of $T_2 \circ T_1$ is R^2. We have

$$T_2(T_1(x, y)) = T_2(x - y, \ y + z, \ x - z) = (0, 2x)$$

Thus the codomain of $T_2 \circ T_1$ is the y-axis.

18. **(a)** Since T_1 is defined on all of R^2, the domain of $T_3 \circ T_2 \circ T_1$ is R^2. We have

$$T_3 \circ T_2 \circ T_1(x, y) = T_3 \circ T_2(-2y, \ 3x, \ x - 2y)$$

$$= T_3(3x, \ x - 2y, \ -2y)$$

$$= (3x - 2y, \ x)$$

This vector can assume all possible values in R^2, so the codomain is R^2.

19. **(a)** We have

$$(T_1 \circ T_2)(A) = \operatorname{tr}(A^T) = \operatorname{tr} \begin{bmatrix} a & c \\ b & d \end{bmatrix} = a + d$$

(b) Since the range of T_1 is not contained in the domain of T_2, $T_2 \circ T_1$ is not well defined.

20. We have

$$(T_1 \circ T_2)(p(x)) = T_1(p(x+1)) = p((x-1)+1) = p(x)$$

and

$$(T_2 \circ T_1)(p(x)) = T_2(p(x-1)) = p((x+1)-1) = p(x)$$

22. We have

$$T_2 \circ T_1 \left(a_0 + a_1 x + a_2 x^2\right) = T_2 \left(a_0 + a_1(x+1) + a_2(x+1)^2\right)$$

$$= T_2 \left((a_0 + a_1 + a_2) + (a_1 + 2a_2)x + a_2 x^2\right)$$

$$= (a_0 + a_1 + a_2)x + (a_1 + 2a_2)x^2 + a_2 x^3$$

24. **(b)** We have

$$(T_3 \circ T_2) \circ T_1(\mathbf{v}) = (T_3 \circ T_2)(T_1(\mathbf{v}))$$

$$= T_3(T_2(T_1(\mathbf{v})))$$

$$= T_3 \circ T_2 \circ T_1(\mathbf{v})$$

25. Since $(1, 0, 0)$ and $(0, 1, 0)$ form an orthonormal basis for the xy-plane, we have $T(x, y, z) = (x, 0, 0) + (0, y, 0) = (x, y, 0)$, which can also be arrived at by inspection. Then $T(T(x, y, z)) = T(x, y, 0) = (x, y, 0) = T(x, y, z)$. This says that T leaves every point in the x-y plane unchanged.

26.   **(a)**   From the linearity of T and the definition of kT, we have

$$(kT)(\mathbf{u} + \mathbf{v}) = k(T(\mathbf{u} + \mathbf{v})) = k(T(\mathbf{u}) + T(\mathbf{v}))$$
$$= kT(\mathbf{u}) + kT(\mathbf{v}) = (kT)(\mathbf{u}) + (kT)(\mathbf{v})$$

and

$$(kT)(c\mathbf{u}) = k(T(c\mathbf{u})) = kcT(\mathbf{u}) = c(kT)(\mathbf{u})$$

Therefore, kT is linear.

28.   **(a)** F is linear since

$$F((x_1, y_1) + (x_2, y_2)) = F(x_1 + x_2, \ y_1 + y_2)$$
$$= (a_1(x_1 + x_2) + b_1(y_1 + y_2), \ a_2(x_1 + x_2) + b_2(y_1 + y_2))$$
$$= (a_1 x_1 + b_1 y_1, \ a_2 x_1 + b_2 y_1) + (a_1 x_2 + b_1 y_2, \ a_2 x_2 + b_2 y_2)$$
$$= F(x_1, y_1) + F(x_2, y_2)$$

and

$$F(c(x, y)) = F(cx, cy)$$
$$= (a_1 cx + b_1 cy, \ a_2 cx + b_2 cy)$$
$$= c(a_1 x + b_1 y, \ a_2 x + b_2 y)$$
$$= cF(x, y)$$

(b)   Since $F(c(x, y)) = F(cx, cy) = c^2 F(x, y)$, F is not a linear operator.

30.   Let $\mathbf{v} = a_1 \mathbf{v}_1 + \cdots + a_n \mathbf{v}_n$ be any vector in V. Then

$$T(\mathbf{v}) = a_1 T(\mathbf{v}_1) + \cdots + a_n T(\mathbf{v}_n)$$
$$= a_1 \mathbf{v}_1 + \cdots + a_n \mathbf{v}_n$$
$$= \mathbf{v}$$

Hence T is the identity transformation on V.

31. **(b)** We have

$$(J \circ D)(\sin x) = \int_0^x (\sin t)' \, dt = \sin(x) - \sin(0) = \sin(x)$$

(c) We have

$$(J \circ D)(e^x + 3) = \int_0^x (e^t + 3)' \, dt = e^x - 1$$

33. **(a)** True. Let $c_1 = c_2 = 1$ to establish Part (a) of the definition and let $c_2 = 0$ to establish Part (b).

(b) False. All linear transformations have this property, and, for instance there is more than one linear transformation from R^2 to R^2.

(c) True. If we let $\mathbf{u} = \mathbf{0}$, then we have $T(\mathbf{v}) = T(-\mathbf{v}) = -T(\mathbf{v})$. That is, $T(\mathbf{v}) = \mathbf{0}$ for all vectors $\mathbf{v}$ in V. But there is only one linear transformation which maps every vector to the zero vector.

(d) False. For this operator T, we have

$$T(\mathbf{v} + \mathbf{v}) = T(2\mathbf{v}) = \mathbf{v}_0 + 2\mathbf{v}$$

But

$$T(\mathbf{v}) + T(\mathbf{v}) = 2T(\mathbf{v}) = 2\mathbf{v}_0 + 2\mathbf{v}$$

Since $\mathbf{v}_0 \neq \mathbf{0}$, these two expressions cannot be equal.

34. Since B is a basis for the vector space V, we can write any vector $\mathbf{v}$ in V in one and only one way as a linear combination $\mathbf{v} = c_1\mathbf{v}_1 + c_2\mathbf{v}_2 + \cdots + c_n\mathbf{v}_n$ of vectors from B. Thus for any linear operator T, we have

$$T(\mathbf{v}) = c_1 T(\mathbf{v}_1) + c_2 T(\mathbf{v}_2) + \cdots + c_n T(\mathbf{v}_n)$$

That is, to define T, we need only specify where it maps each of the basis vectors in B. If T is to map B into itself, then we have n choices for each of the n vectors $T(\mathbf{v}_i)$, and hence a total of n^n possible operators T.

EXERCISE SET 8.2

1. (a) If $(1, -4)$ is in $R(T)$, then there must be a vector (x, y) such that $T(x, y) = (2x - y, -8x + 4y) = (1, -4)$. If we equate components, we find that $2x - y = 1$ or $y = t$ and $x = (1 + t)/2$. Thus T maps infinitely many vectors into $(1, -4)$.

    (b) Proceeding as above, we obtain the system of equations

    $$2x - y = 5$$
    $$-8x + 4y = 0$$

    Since $2x - y = 5$ implies that $-8x + 4y = -20$, this system has no solution. Hence $(5, 0)$ is not in $R(T)$.

2. (a) The vector $(5, 10)$ is in $\ker(T)$ since

    $$T(5, 10) = (2(5) - 10, \ -8(5) + 4(10)) = (0, 0)$$

    (b) The vector $(3, 2)$ is not in $\ker(T)$ since $T(3, 2) = (4, -16) \neq (0, 0)$.

3. (b) The vector $(1, 3, 0)$ is in $R(T)$ if and only if the following system of equations has a solution:

    $$4x + y - 2z - 3w = 1$$
    $$2x + y + z - 4w = 3$$
    $$6x - 9z + 9w = 0$$

    This system has infinitely many solutions $x = (3/2)(t - 1)$, $y = 10 - 4t$, $z = t$, $w = 1$ where t is arbitrary. Thus $(1, 3, 0)$ is in $R(T)$.

4. **(b)** Since $T(0,0,0,1) = (-3,-4,9) \neq (0,0,0)$, the vector $(0,0,0,1)$ is not in $\ker(T)$.

5. **(a)** Since $T(x^2) = x^3 \neq 0$, the polynomial x^2 is not in $\ker(T)$.

6. **(a)** Since $T(1+x) = x + x^2$, the polynomial $x + x^2$ is in $R(T)$.

(c) Since there doesn't exist a polynomial $p(x)$ such that $xp(x) = 3 - x^2$, then $3 - x^2$ is not in $R(T)$.

7. **(a)** We look for conditions on x and y such that $2x - y = 0$ and $-8x + 4y = 0$. Since these equations are satisfied if and only if $y = 2x$, the kernel will be spanned by the vector $(1, 2)$, which is then a basis.

(c) Since the only vector which is mapped to zero is the zero vector, the kernel is $\{\mathbf{0}\}$ and has dimension zero so the basis is the empty set.

8. **(a)** Since $-8x + 4y = -4(2x - y)$ and $2x - y$ can assume any real value, the range of T is the set of vectors $(x, -4x)$ or the line $y = -4x$. The vector $(1, -4)$ is a basis for this space.

(c) The range of T is just the set of all polynomials in P_3 with constant term zero. The set $\{x, x^2, x^3\}$ is a basis for this space.

9. **(a)** Here $n = \dim(R^2) = 2$, $\text{rank}(T) = 1$ by the result of Exercise 8(a), and $\text{nullity}(T) = 1$ by Exercise 7(a). Recall that $1 + 1 = 2$.

(c) Here $n = \dim(P_2) = 3$, $\text{rank}(T) = 3$ by virtue of Exercise 8(c), and $\text{nullity}(T) = 0$ by Exercise 7(c). Thus we have $3 = 3 + 0$.

10. **(a)** We know from Example 1 that the range of T is the column space of A. Using elementary column operations, we reduce A to the matrix

$$\begin{bmatrix} 1 & 0 & 0 \\ 5 & 1 & 0 \\ 7 & 1 & 0 \end{bmatrix} \quad \text{or} \quad \begin{bmatrix} 1 & 0 & 0 \\ 0 & 1 & 0 \\ 2 & 1 & 0 \end{bmatrix}$$

Thus $\begin{bmatrix} 1 \\ 5 \\ 7 \end{bmatrix}$ and $\begin{bmatrix} 0 \\ 1 \\ 1 \end{bmatrix}$ as well as $\begin{bmatrix} 1 \\ 0 \\ 2 \end{bmatrix}$ and $\begin{bmatrix} 0 \\ 1 \\ 1 \end{bmatrix}$ form a basis for the column space and therefore also for the range of T.

(b) To investigate the solution space of A, we look for conditions on x, y, and z such that

$$\begin{bmatrix} 1 & -1 & 3 \\ 5 & 6 & -4 \\ 7 & 4 & 2 \end{bmatrix} \begin{bmatrix} x \\ y \\ z \end{bmatrix} = \begin{bmatrix} 0 \\ 0 \\ 0 \end{bmatrix}$$

The solution of the resulting system of equations is $x = -14t$, $y = 19t$, and $z = 11t$

where t is arbitrary. Thus one basis for $\ker(T)$ is the vector $\begin{bmatrix} -14 \\ 19 \\ 11 \end{bmatrix}$.

(c) By Parts (a) and (b), we have $\text{rank}(T) = 2$ and $\text{nullity}(T) = 1$.

(d) Since $\text{rank}(A) = 2$ and A has 3 columns, nullity $(A) = 3 - 2 = 1$. This also follows from Theorem 8.2.2 and Part (c) above.

12. **(a)** Since the range of T is the column space of A, we reduce A using column operations. This yields

$$\begin{bmatrix} 1 & 0 & 0 & 0 \\ 0 & 1 & 0 & 0 \end{bmatrix}$$

Thus $\begin{bmatrix} 1 \\ 0 \end{bmatrix}$ and $\begin{bmatrix} 0 \\ 1 \end{bmatrix}$ form a basis for the range of T.

(b) If

$$\begin{bmatrix} 4 & 1 & 5 & 2 \\ 1 & 2 & 3 & 0 \end{bmatrix} \begin{bmatrix} x \\ y \\ z \\ w \end{bmatrix} = \begin{bmatrix} 0 \\ 0 \end{bmatrix}$$

then $x = -s - 4r$, $y = -s + 2r$, $z = s$, and $w = 7r$ where s and r are arbitrary. Thus,

$$\begin{bmatrix} x \\ y \\ z \\ w \end{bmatrix} = \begin{bmatrix} -1 \\ -1 \\ 1 \\ 0 \end{bmatrix} s + \begin{bmatrix} -4 \\ 2 \\ 0 \\ 7 \end{bmatrix} r$$

That is, the set $\{[-1, -1, 1, 0]^T, [-4, 2, 0, 7]^T\}$ is one basis for $\ker(T)$.

12. **(c)** By Parts (a) and (b), rank(T) = nullity$(T) = 2$.

   (d) Since A has 4 columns and rank$(A) = 2$, we have nullity$(A) = 4 - 2 = 2$. This also follows from Theorem 8.2.2 and Part (c) above.

14. **(a)** The orthogonal projection on the xz-plane maps R^3 to the entire xz-plane. Hence its range is the xz-plane. It maps only the y-axis to $(0, 0, 0)$, so that its nullspace is the y-axis.

16. **(a)** The nullity of T is $5 - 3 = 2$.

   (c) The nullity of T is $6 - 3 = 3$.

18. **(a)** The dimension of the solution space of $A\mathbf{x} = \mathbf{0}$ is, by virtue of the Dimension Theorem, $7 - 4 = 3$.

   (b) Since the range space of A has dimension 4, the range cannot be R^5, which has dimension 5.

19. By Theorem 8.2.1, the kernel of T is a subspace of R^3. Since the only subspaces of R^3 are the origin, a line through the origin, a plane through the origin, or R^3 itself, the result follows. It is clear that all of these possibilities can actually occur.

21. **(a)** If

$$\begin{bmatrix} 1 & 3 & 4 \\ 3 & 4 & 7 \\ -2 & 2 & 0 \end{bmatrix} \begin{bmatrix} x \\ y \\ z \end{bmatrix} = \begin{bmatrix} 0 \\ 0 \\ 0 \end{bmatrix}$$

then $x = -t$, $y = -t$, $z = t$. These are parametric equations for a line through the origin.

   (b) Using elementary column operations, we reduce the given matrix to

$$\begin{bmatrix} 1 & 0 & 0 \\ 3 & -5 & 0 \\ -2 & 8 & 0 \end{bmatrix}$$

Thus, $(1, 3, -2)^T$ and $(0, -5, 8)^T$ form a basis for the range. That range, which we can interpret as a subspace of R^3, is a plane through the origin. To find a normal to that plane, we compute

$$(1, 3, -2) \times (0, -5, 8) = (14, -8, -5)$$

Therefore, an equation for the plane is

$$14x - 8y - 5z = 0$$

Alternatively, but more painfully, we can use elementary row operations to reduce the matrix

$$\begin{bmatrix} 1 & 3 & 4 & x \\ 3 & 4 & 7 & y \\ -2 & 2 & 0 & z \end{bmatrix}$$

to the matrix

$$\begin{bmatrix} 1 & 0 & 1 & (-4x + 3y)/5 \\ 0 & 1 & 1 & (3x - y)/5 \\ 0 & 0 & 0 & 14x - 8y - 5z \end{bmatrix}$$

Thus the vector $(x, y, z)^T$ is in the range of T if and only if $14x - 8y - 5z = 0$.

22. Suppose that $\mathbf{v}$ is any vector in V and write

(∗) $$\mathbf{v} = a_1\mathbf{v}_1 + a_2\mathbf{v}_2 + \cdots + a_n\mathbf{v}_n$$

Now define a transformation $T : V \rightarrow W$ by

$$T(\mathbf{v}) = a_1\mathbf{w}_1 + a_2\mathbf{w}_2 + \cdots + a_n\mathbf{w}_n$$

Note that T is well-defined because, by Theorem 5.4.1, the constants $a_1, a_2, \ldots, a_n$ in (∗) are unique.

In order to complete the problem, you must show (i) that T is linear and (ii) that $T(\mathbf{v}_i) = \mathbf{w}_i$ for $i = 1, 2, \ldots, n$.

23. **(a)** Suppose that $\{v_1, v_2, \ldots, v_n\}$ is a basis for V and let

(∗) $v = a_1 v_1 + a_2 v_2 + \cdots + a_n v_n$

be an arbitrary vector in V. Since T is linear, we have

(∗∗) $T(v) = a_1 T(v_1) + \cdots + a_n T(v_n)$

The hypothesis that $\dim(\ker(T)) = 0$ implies that $T(v) = 0$ if and only if $v = 0$. Since $\{v_1, \ldots, v_n\}$ is a linearly independent set, by (∗), $v = 0$ if and only if

$$a_1 = a_2 = \cdots = a_n = 0$$

Thus, by (∗∗), $T(v) = 0$ if and only if

$$a_1 = a_2 = \cdots = a_n = 0$$

That is, the vectors $T(v_1), T(v_2), \ldots, T(v_n)$ form a linearly independent set in $R(T)$. Since they also span $R(T)$, then $\dim(R(T)) = n$.

25. If $J(p) = \displaystyle\int_{-1}^{1} (ax + b)\,dx = (ax^2/2 + bx)\Big]_{-1}^{1} = 2b$, then the kernel of J is the set of all $p(x) = ax + b$ for which $b = 0$ except the y-axis.

26. If $f(x)$ is in the kernel of $D \circ D$, then $f''(x) = 0$ or $f(x) = ax + b$. Since these are the only eligible functions $f(x)$ for which $f''(x) = 0$ (Why?), the kernel of $D \circ D$ is the set of all functions $f(x) = ax + b$, or all straight lines in the plane. Similarly, the kernel of $D \circ D \circ D$ is the set of all functions $f(x) = ax^2 + bx + c$, or all straight lines except the y-axis and certain parabolas in the plane.

27. **(a)** kernel, range

   (b) $(1, 1, 1)$

   (c) the dimension of V

   (d) 3

28. **(a)** Since the range of T has dimension 3 minus the nullity of T, then the range of T has dimension 2. Therefore it is a plane through the origin.

(b) As in Part (a), if the range of T has dimension 2, then the kernel must have dimension 1. Hence, it is a line through the origin.

29. **(a)** If $T(f(x)) = f''''(x)$, then $T(ax^3 + bx^2 + cx + d) - 0$ for all polynomials in P_3. Moreover, these are the only functions which are mapped to the zero vector, that is, the function which is identically zero.

  (b) If T transforms $f(x)$ to its $(n+1)$st derivative, then kernel $(T) = P_n$.

EXERCISE SET 8.3

1. **(a)** Clearly $\ker(T) = \{(0,0)\}$, so T is one-to-one.

   (c) Since $T(x,y) = (0,0)$ if and only if $x = y$ and $x = -y$, the kernel is $(0,0)$ and T is one-to-one.

   (e) Here $T(x,y) = (0,0,0)$ if and only if x and y satisfy the equations $x-y = 0$, $-x+y = 0$, and $2x - 2y = 0$. That is, (x,y) is in $\ker(T)$ if and only if $x = y$, so the kernel of T is this line and T is not one-to-one.

2. **(a)** Since $\det(A) = 1 \neq 0$, then T has an inverse. By direct calculation

$$A^{-1} = \begin{bmatrix} 1 & -2 \\ -2 & 5 \end{bmatrix}$$

   and so

$$T^{-1}\begin{bmatrix} x_1 \\ x_2 \end{bmatrix} = A^{-1}\begin{bmatrix} x_1 \\ x_2 \end{bmatrix} = \begin{bmatrix} x_1 - 2x_2 \\ -2x_1 + 5x_2 \end{bmatrix}$$

3. **(a)** Since $\det(A) = 0$, or equivalently, $\operatorname{rank}(A) < 3$, T has no inverse.

3. **(c)** Since A is invertible, we have

$$
T^{-1} \begin{bmatrix} x_1 \\ x_2 \\ x_3 \end{bmatrix} = A^{-1} \begin{bmatrix} x_1 \\ x_2 \\ x_3 \end{bmatrix} = \begin{bmatrix} \frac{1}{2} & -\frac{1}{2} & \frac{1}{2} \\ -\frac{1}{2} & \frac{1}{2} & \frac{1}{2} \\ \frac{1}{2} & \frac{1}{2} & -\frac{1}{2} \end{bmatrix} \begin{bmatrix} x_1 \\ x_2 \\ x_3 \end{bmatrix}
$$

$$
= \begin{bmatrix} \frac{1}{2}\left(x_1 - x_2 + x_3 \right) \\ \frac{1}{2}\left(-x_1 + x_2 + x_3 \right) \\ \frac{1}{2}\left(x_1 + x_2 - x_3 \right) \end{bmatrix}
$$

4. **(a)** Since the kernel of this transformation is the line $x = 2y$, the transformation is not one-to-one.

(c) Since the kernel of the transformation is $(0, 0)$, the transformation is one-to-one.

5. **(a)** The kernel of T is the line $y = -x$ since all points on this line (and only those points) map to the origin.

(b) Since the kernel is not $(0, 0)$, the transformation is not one-to-one.

7. **(b)** Since nullity$(T) = n-$ rank$(T) = 1$, T is not one-to-one.

(c) Here T cannot be one-to-one since rank$(T) \le n < m$, so nullity$(T) \ge 1$.

8. **(b)** If $p(x) = a_0 + a_1 x + a_2 x^2$, then

$$
T(p(x)) = a_0 + a_1(x + 1) + a_2(x + 1)^2
$$
$$
= (a_0 + a_1 + a_2) + (a_1 + 2a_2)x + a_2 x^2
$$

Thus $T(p(x)) = 0$ if and only if $a_2 = a_1 + 2a_2 = a_0 + a_1 + a_2 = 0$, or $a_2 = a_1 = a_0 = 0$. That is, the nullity of T is zero so T is one-to-one.

10. **(a)** Since infinitely many vectors $(0, 0, \ldots, 0, x_n)$ map to the zero vector, T is not one-to-one.

(c) Here T is one-to-one and $T^{-1}(x_1, x_2, \ldots, x_n) = (x_n, x_1, \ldots, x_{n-1})$.

11. **(a)** We know that T will have an inverse if and only if its kernel is the zero vector, which means if and only if none of the numbers $a_i = 0$.

12. Note that T_1 and T_2 can be represented by the matrices

$$A_{T_1} = \begin{bmatrix} 1 & 1 \\ 1 & -1 \end{bmatrix} \quad \text{and} \quad A_{T_2} = \begin{bmatrix} 2 & 1 \\ 1 & -2 \end{bmatrix}$$

that is,

$$T_1 \begin{bmatrix} x \\ y \end{bmatrix} = A_{T_1} \begin{bmatrix} x \\ y \end{bmatrix} \quad \text{and} \quad T_2 \begin{bmatrix} x \\ y \end{bmatrix} = A_{T_2} \begin{bmatrix} x \\ y \end{bmatrix}$$

(a) Since $\det(A_{T_1}) = -2 \neq 0$ and $\det(A_{T_2}) = -5 \neq 0$, it follows that both T_1 and T_2 are one-to-one.

(b) The transformations T_1^{-1} and T_2^{-1} are represented by $A_{T_1}^{-1}$ and $A_{T_2}^{-1}$, respectively, where

$$A_{T_1}^{-1} = \frac{1}{2} \begin{bmatrix} 1 & 1 \\ 1 & -1 \end{bmatrix} \quad \text{and} \quad A_{T_2}^{-1} = \frac{1}{5} \begin{bmatrix} 2 & 1 \\ 1 & -2 \end{bmatrix}$$

Then $T_2 \circ T_1$ can be represented by

$$A_{T_2} A_{T_1} = \begin{bmatrix} 3 & 1 \\ -1 & 3 \end{bmatrix}$$

and $(T_2 \circ T_1)^{-1}$ can be represented by

$$(A_{T_2} A_{T_1})^{-1} = \frac{1}{10} \begin{bmatrix} 3 & -1 \\ 1 & 3 \end{bmatrix}$$

that is

$$(T_2 \circ T_1)^{-1} \begin{bmatrix} x \\ y \end{bmatrix} = \frac{1}{10} \begin{bmatrix} 3 & -1 \\ 1 & 3 \end{bmatrix} \begin{bmatrix} x \\ y \end{bmatrix} = \frac{1}{10} \begin{bmatrix} 3x - y \\ x + 3y \end{bmatrix}$$

12. **(c)** It is easy to verify that

$$A_{T_1}^{-1} A_{T_2}^{-1} = (A_{T_2} A_{T_1})^{-1}$$

<u>Alternative Solution:</u> If you want to do this the hard way, notice that if $T_1(x, y) = (x + y, \ x - y)$, then $T_1^{-1}(x, y) = \left(\dfrac{x + y}{2}, \ \dfrac{x - y}{2} \right)$ and if $T_2(x, y) = (2x + y, \ x - 2y)$ then $T_2^{-1}(x, y) = \left(\dfrac{2x + y}{5}, \ \dfrac{x - 2y}{5} \right)$. Also $T_2 \circ T_1(x, y) = T_2(x + y, \ x - y) = (3x + y, \ -x + 3y)$ and $(T_1 \circ T_2)^{-1}(x, y) = \left(\dfrac{3x - y}{10}, \ \dfrac{x + 3y}{10} \right)$.

13. **(a)** By inspection, $T_1^{-1}(p(x)) = p(x)/x$, where $p(x)$ must, of course, be in the range of T_1 and hence have constant term zero. Similarly $T_2^{-1}(p(x)) = p(x-1)$, where, again, $p(x)$ must be in the range of T_2. Therefore $(T_2 \circ T_1)^{-1}(p(x)) = p(x - 1)/x$ for appropriate polynomials $p(x)$.

16. If there were such a transformation T, it would have nullity zero (because it would be one-to-one) and rank equal to the dimension of W (because its range would be W). But the Dimension Theorem guarantees that $\text{rank}(T) + \text{nullity}(T) = \dim(V)$, which is, by hypothesis, greater than $\dim(W)$. Thus there can be no such transformation.

17. **(a)** Since T sends the nonzero matrix $\begin{bmatrix} 0 & 1 \\ 0 & 0 \end{bmatrix}$ to the zero matrix, it is not one-to-one.

(c) Since T sends only the zero matrix to the zero matrix, it is one-to-one. By inspection, $T^{-1}(A) = T(A)$.

<u>Alternative Solution:</u> T can be represented by the matrix

$$T_B = \begin{bmatrix} 0 & 0 & 0 & 1 \\ 0 & -1 & 0 & 0 \\ 0 & 0 & -1 & 0 \\ 1 & 0 & 0 & 0 \end{bmatrix}$$

By direct calculation, $T_B = (T_B)^{-1}$, so $T = T^{-1}$.

18. If $T(x, y) = (x + ky, -y) = (0, 0)$, then $x = y = 0$. Hence $\ker(T) = \{(0,0)\}$ and therefore T is one-to-one. Since $T(T(x, y)) = T(x + ky, -y) = (x + ky + k(-y), -(-y)) = (x, y)$, we have $T^{-1} = T$.

19. Suppose that $\mathbf{w}_1$ and $\mathbf{w}_2$ are in $R(T)$. We must show that

$$T^{-1}(\mathbf{w}_1 + \mathbf{w}_2) = T^{-1}(\mathbf{w}_1) + T^{-1}(\mathbf{w}_2)$$

and

$$T^{-1}(k\mathbf{w}_1) = kT^{-1}(\mathbf{w}_1)$$

Because T is one-to-one, the above equalities will hold if and only if the results of applying T to both sides are indeed valid equalities. This follows immediately from the linearity of the transformation T.

21. It is easy to show that T is linear. However, T is not one-to-one, since, for instance, it sends the function $f(x) = x - 5$ to the zero vector.

22. (a) False. Since T is not one-to-one, it has no inverse. For T^{-1} to exist, T must map each point in its domain to a *unique* point in its range.

   (b) True. If $T_1(\mathbf{v}_1) = T_1(\mathbf{v}_2)$ where $\mathbf{v}_1 \neq \mathbf{v}_2$, then $T_2 \circ T_1(\mathbf{v}_1) = T_2 \circ T_1(\mathbf{v}_2)$ where $\mathbf{v}_1 \neq \mathbf{v}_2$.

   (c) True, since such rotations and reflections are one-to-one linear transformations.

23. Yes. The transformation is linear and only $(0, 0, 0)$ maps to the zero polynomial. Clearly distinct triples in R^3 map to distinct polynomials in P_2.

24.   It does. T is a linear operator on M_{22}. If $EA = O$ for any matrix A, then $A = E^{-1}O = O$. Thus T is one-to-one. Alternatively, if $EA_1 = EA_2$, then $E(A, -A_2) = O$, so $A_1 = A_2 = E^{-1}O = O$. That is, $A_1 = A_2$.

25.   No. T is a linear operator by Theorem 3.4.2. However, it is not one-to-one since $T(\mathbf{a}) = \mathbf{a} \times \mathbf{a} = \mathbf{0} = T(\mathbf{0})$. That is, T maps $\mathbf{a}$ to the zero vector, so if T is one-to-one, $\mathbf{a}$ must be the zero vector. But then T would be the zero transformation, which is certainly not one-to-one.

EXERCISE SET 8.4

2. **(a)** Let $B = \{1, x, x^2\} = \{\mathbf{u}_1, \mathbf{u}_2, \mathbf{u}_3\}$ and $B' = \{1, x\} = \{\mathbf{v}_1, \mathbf{v}_2\}$. Observe that

$$
\begin{aligned}
T(\mathbf{u}_1) = T(1) &= 1 &&= 1\mathbf{v}_1 \\
T(\mathbf{u}_2) = T(x) &= 1 - 2x &&= 1\mathbf{v}_1 - 2\mathbf{v}_2 \\
T(\mathbf{u}_3) = T(x^2) &= \quad\;\; -3x &&= \quad\quad -3\mathbf{v}_2
\end{aligned}
$$

Hence the matrix of T with respect to B and B' is

$$
\begin{bmatrix}
1 & 1 & 0 \\
0 & -2 & -3
\end{bmatrix}
$$

(b) We have

$$
[T]_{B',B}[\mathbf{x}]_B =
\begin{bmatrix}
1 & 1 & 0 \\
0 & -2 & -3
\end{bmatrix}
\begin{bmatrix}
c_0 \\
c_1 \\
c_2
\end{bmatrix}
=
\begin{bmatrix}
c_0 + c_1 \\
-2c_1 - 3c_2
\end{bmatrix}
$$

and

$$
\left[T\left(c_0 + c_1 x + c_2 x^2\right)\right]_{B'} = \left[(c_0 + c_1) - (2c_1 + 3c_2)x\right]_{B'}
$$

$$
=
\begin{bmatrix}
c_0 + c_1 \\
-2c_1 - 3c_2
\end{bmatrix}
$$

4. (a) Since

$$T(\mathbf{u}_1) = \begin{bmatrix} 0 \\ 2 \end{bmatrix} = 2\mathbf{u}_1 + 2\mathbf{u}_2$$

and

$$T(\mathbf{u}_2) = \begin{bmatrix} -1 \\ -1 \end{bmatrix} = -\mathbf{u}_1$$

we have

$$[T]_B = \begin{bmatrix} 2 & -1 \\ 2 & 0 \end{bmatrix}$$

(b) Since

$$[T]_B[\mathbf{x}]_B = \begin{bmatrix} 2 & -1 \\ 2 & 0 \end{bmatrix} \begin{bmatrix} x_1 \\ x_2 \end{bmatrix}_B$$

$$= \begin{bmatrix} 2 & -1 \\ 2 & 0 \end{bmatrix} [x_1(-\mathbf{u}_2) + x_2(\mathbf{u}_1 + \mathbf{u}_2)]_B$$

$$= \begin{bmatrix} 2 & -1 \\ 2 & 0 \end{bmatrix} \begin{bmatrix} x_2 \\ x_2 - x_1 \end{bmatrix} = \begin{bmatrix} x_1 + x_2 \\ 2x_2 \end{bmatrix}$$

and

$$[T(\mathbf{x})]_B = \begin{bmatrix} x_1 - x_2 \\ x_1 + x_2 \end{bmatrix}_B = (x_1 + x_2)\mathbf{u}_1 + 2x_2\mathbf{u}_2$$

Formula (5a) does, indeed, hold.

6. (a) Observe that

$$T(\mathbf{v}_1) = (1, -1, 0) = \mathbf{v}_1 - \mathbf{v}_2$$

$$T(\mathbf{v}_2) = (-1, 1, -1) = -\frac{3}{2}\mathbf{v}_1 + \frac{1}{2}\mathbf{v}_2 + \frac{1}{2}\mathbf{v}_3$$

$$T(\mathbf{v}_3) = (0, 0, 1) = \frac{1}{2}\mathbf{v}_1 + \frac{1}{2}\mathbf{v}_2 - \frac{1}{2}\mathbf{v}_3$$

Thus the matrix of T with respect to B is

$$\begin{bmatrix} 1 & -\dfrac{3}{2} & \dfrac{1}{2} \\[2ex] -1 & \dfrac{1}{2} & \dfrac{1}{2} \\[2ex] 0 & \dfrac{1}{2} & -\dfrac{1}{2} \end{bmatrix}$$

(b) If we solve for $(x_1, x_2, x_3)_B = a_1\mathbf{v}_1 + a_2\mathbf{v}_2 + a_3\mathbf{v}_3$, we find that

$$[\mathbf{x}]_B = \begin{bmatrix} \dfrac{x_1 - x_2 + x_3}{2} & \dfrac{-x_1 + x_2 + x_3}{2} & \dfrac{x_1 + x_2 - x_3}{2} \end{bmatrix}^T$$

Thus we have

$$[T]_B[\mathbf{x}]_B = \begin{bmatrix} 1 & -\dfrac{3}{2} & \dfrac{1}{2} \\[2ex] -1 & \dfrac{1}{2} & \dfrac{1}{2} \\[2ex] 0 & \dfrac{1}{2} & -\dfrac{1}{2} \end{bmatrix} \begin{bmatrix} \dfrac{x_1 - x_2 + x_3}{2} \\[2ex] \dfrac{-x_1 + x_2 + x_3}{2} \\[2ex] \dfrac{x_1 + x_2 - x_3}{2} \end{bmatrix}$$

$$= \begin{bmatrix} \dfrac{3x_1 - 2x_2 - x_3}{2} \\[2ex] \dfrac{-x_1 + 2x_2 - x_3}{2} \\[2ex] \dfrac{-x_1 + x_3}{2} \end{bmatrix}$$

and

$$[T(\mathbf{x})]_B = [(x_1 - x_2, \; x_2 - x_1, \; x_1 - x_3)]_B$$

$$= \begin{bmatrix} \dfrac{(x_1 - x_2) - (x_2 - x_1) + (x_1 - x_3)}{2} \\[2ex] \dfrac{-(x_1 - x_2) + (x_2 - x_1) + (x_1 - x_3)}{2} \\[2ex] \dfrac{(x_1 - x_2) + (x_2 - x_1) - (x_1 - x_3)}{2} \end{bmatrix} = \begin{bmatrix} \dfrac{3x_1 - 2x_2 - x_3}{2} \\[2ex] \dfrac{-x_1 + 2x_2 - x_3}{2} \\[2ex] \dfrac{-x_1 + x_3}{2} \end{bmatrix}$$

8. **(a)** Since

$$T(1) = x$$

$$T(x) = x(x - 3) = -3x + x^2$$

$$T(x^2) = x(x - 3)^2 = 9x - 6x^2 + x^3$$

we have

$$[T]_{B',B} = \begin{bmatrix} 0 & 0 & 0 \\ 1 & -3 & 9 \\ 0 & 1 & -6 \\ 0 & 0 & 1 \end{bmatrix}$$

(b) Since we are working with standard bases, we have

$$[T]_{B',B}[\mathbf{x}]_B = \begin{bmatrix} 0 & 0 & 0 \\ 1 & -3 & 9 \\ 0 & 1 & -6 \\ 0 & 0 & 1 \end{bmatrix} \begin{bmatrix} 1 \\ 1 \\ -1 \end{bmatrix} = \begin{bmatrix} 0 \\ -11 \\ 7 \\ -1 \end{bmatrix}$$

so that $T(1 + x - x^2) = -11x + 7x^2 - x^3$.

(c) We have $T(1 + x - x^2) = x(1 + (x - 3) - (x - 3)^2) = -11x + 7x^2 - x^3$.

9. **(a)** Since A is the matrix of T with respect to B, then we know that the first and second columns of A must be $[T(\mathbf{v}_1)]_B$ and $[T(\mathbf{v}_2)]_B$, respectively. That is

$$[T(\mathbf{v}_1)]_B = \begin{bmatrix} 1 \\ -2 \end{bmatrix}$$

$$[T(\mathbf{v}_2)]_B = \begin{bmatrix} 3 \\ 5 \end{bmatrix}$$

Alternatively, since $\mathbf{v}_1 = 1\mathbf{v}_1 + 0\mathbf{v}_2$ and $\mathbf{v}_2 = 0\mathbf{v}_1 + 1\mathbf{v}_2$, we have

$$[T(\mathbf{v}_1)]_B = A \begin{bmatrix} 1 \\ 0 \end{bmatrix} = \begin{bmatrix} 1 \\ -2 \end{bmatrix}$$

and

$$[T(\mathbf{v}_2)]_B = A \begin{bmatrix} 0 \\ 1 \end{bmatrix} = \begin{bmatrix} 3 \\ 5 \end{bmatrix}$$

(b) From Part (a),

$$T(\mathbf{v}_1) = \mathbf{v}_1 - 2\mathbf{v}_2 = \begin{bmatrix} 3 \\ -5 \end{bmatrix}$$

and

$$T(\mathbf{v}_2) = 3\mathbf{v}_1 + 5\mathbf{v}_2 = \begin{bmatrix} -2 \\ 29 \end{bmatrix}$$

(c) Since we already know $T(\mathbf{v}_1)$ and $T(\mathbf{v}_2)$, all we have to do is express $[x_1 \quad x_2]^T$ in terms of $\mathbf{v}_1$ and $\mathbf{v}_2$. If

$$\begin{bmatrix} x_1 \\ x_2 \end{bmatrix} = a\mathbf{v}_1 + b\mathbf{v}_2 = a \begin{bmatrix} 1 \\ 3 \end{bmatrix} + b \begin{bmatrix} -1 \\ 4 \end{bmatrix}$$

then

$$x_1 = a - b$$
$$x_2 = 3a + 4b$$

or

$$a = (4x_1 + x_2)/7$$
$$b = (-3x_1 + x_2)/7$$

Thus

$$T\left(\begin{bmatrix} x_1 \\ x_2 \end{bmatrix}\right) = \frac{4x_1 + x_2}{7} \begin{bmatrix} 3 \\ -5 \end{bmatrix} + \frac{-3x_1 + x_2}{7} \begin{bmatrix} -2 \\ 29 \end{bmatrix}$$

$$= \begin{bmatrix} \dfrac{18x_1 + x_2}{7} \\ \dfrac{-107x_1 + 24x_2}{7} \end{bmatrix}$$

9. **(d)** By the above formula,

$$T\left(\begin{bmatrix} 1 \\ 1 \end{bmatrix}\right) = \begin{bmatrix} 19/7 \\ -83/7 \end{bmatrix}$$

11. **(a)** The columns of A, by definition, are $[T(\mathbf{v}_1)]_B$, $[T(\mathbf{v}_2)]_B$, and $[T(\mathbf{v}_3)]_B$, respectively.

(b) From Part (a),

$$
\begin{aligned}
T(\mathbf{v}_1) &= \quad \mathbf{v}_1 + 2\mathbf{v}_2 + 6\mathbf{v}_3 = 16 + 51x + 19x^2 \\
T(\mathbf{v}_2) &= 3\mathbf{v}_1 \qquad\quad - 2\mathbf{v}_3 = -6 - \quad 5x + \quad 5x^2 \\
T(\mathbf{v}_3) &= -\mathbf{v}_1 + 5\mathbf{v}_2 + 4\mathbf{v}_3 = \quad 7 + 40x + 15x^2
\end{aligned}
$$

(c) Let $a_0 + a_1x + a_2x^2 = b_0\mathbf{v}_1 + b_1\mathbf{v}_2 + b_2\mathbf{v}_3$. Then

$$
\begin{aligned}
a_0 &= \quad - b_1 + 3b_2 \\
a_1 &= 3b_0 + 3b_1 + 7b_2 \\
a_2 &= 3b_0 + 2b_1 + 2b_2
\end{aligned}
$$

This system of equations has the solution

$$
\begin{aligned}
b_0 &= (a_0 - a_1 + 2a_2)/3 \\
b_1 &= (-5a_0 + 3a_1 - 3a_2)/8 \\
b_2 &= (a_0 + a_1 - a_2)/8
\end{aligned}
$$

Thus

$$T(a_0 + a_1x + a_2x^2) = b_0 T(\mathbf{v}_1) + b_1 T(\mathbf{v}_2) + b_2 T(\mathbf{v}_3)$$

$$= \frac{239a_0 - 161a_1 + 289a_2}{24}$$

$$+ \frac{201a_0 - 111a_1 + 247a_2}{8}x$$

$$+ \frac{61a_0 - 31a_1 + 107a_2}{12}x^2$$

(d) By the above formula,

$$T(1 + x^2) = 22 + 56x + 14x^2$$

13. **(a)** Since

$$T_1(1) = 2 \quad \text{and} \quad T_1(x) = -3x^2$$

$$T_2(1) = 3x \quad T_2(x) = 3x^2 \quad \text{and} \quad T_2(x^2) = 3x^3$$

$$T_2 \circ T_1(1) = 6x \quad \text{and} \quad T_2 \circ T_1(x) = -9x^3$$

we have

$$[T_1]_{B'',B} = \begin{bmatrix} 2 & 0 \\ 0 & 0 \\ 0 & -3 \end{bmatrix} \qquad [T_2]_{B',B''} = \begin{bmatrix} 0 & 0 & 0 \\ 3 & 0 & 0 \\ 0 & 3 & 0 \\ 0 & 0 & 3 \end{bmatrix}$$

and

$$[T_2 \circ T_1]_{B',B} = \begin{bmatrix} 0 & 0 \\ 6 & 0 \\ 0 & 0 \\ 0 & -9 \end{bmatrix}$$

(b) We observe that here

$$[T_2 \circ T_1]_{B',B} = [T_2]_{B',B''} \, [T_1]_{B'',B}$$

15. If T is a contraction or a dilation of V, then T maps any basis $B = \{v_1, \ldots, v_n\}$ of V to $\{kv_1, \ldots, kv_n\}$ where k is a nonzero constant. Therefore the matrix of T with respect to B is

$$
\begin{bmatrix}
k & 0 & 0 & \cdots & 0 \\
0 & k & 0 & \cdots & 0 \\
0 & 0 & k & \cdots & 0 \\
\vdots & \vdots & \vdots & & \vdots \\
0 & 0 & 0 & \cdots & k
\end{bmatrix}
$$

17. The standard matrix for T is just the $m \times n$ matrix whose columns are the transforms of the standard basis vectors. But since B is indeed the standard basis for R^n, the matrices are the same. Moreover, since B' is the standard basis for R^m, the resulting transformation will yield vector components relative to the standard basis, rather than to some other basis.

18. **(a)** Since $D(1) = 0$, $D(x) = 1$, and $D(x^2) = 2x$, then

$$
\begin{bmatrix}
0 & 1 & 0 \\
0 & 0 & 2 \\
0 & 0 & 0
\end{bmatrix}
$$

is the matrix of D with respect to B.

(b) Since $D(2) = 0$, $D(2 - 3x) = -3 = -\dfrac{3}{2}p_1$ and $D(2 - 3x + 8x^2) = -3 + 16x = \dfrac{23}{6}p_1 - \dfrac{16}{3}p_2$, the matrix of D with respect to B is

$$
\begin{bmatrix}
0 & -\dfrac{3}{2} & \dfrac{23}{6} \\
0 & 0 & -\dfrac{16}{3} \\
0 & 0 & 0
\end{bmatrix}
$$

(c) Using the matrix of Part (a), we obtain

$$D(6 - 6x + 24x^2) = \begin{bmatrix} 0 & 1 & 0 \\ 0 & 0 & 2 \\ 0 & 0 & 0 \end{bmatrix} \begin{bmatrix} 6 \\ -6 \\ 24 \end{bmatrix} - \begin{bmatrix} -6 \\ 48 \\ 0 \end{bmatrix} - -6 + 48x$$

(d) Since $6 - 6x + 24x^2 = \mathbf{p}_1 - \mathbf{p}_2 + 3\mathbf{p}_3$, we have

$$[D(6 - 6x + 24x^2)]_B = \begin{bmatrix} 0 & -\dfrac{3}{2} & \dfrac{23}{6} \\ 0 & 0 & -\dfrac{16}{3} \\ 0 & 0 & 0 \end{bmatrix} \begin{bmatrix} 1 \\ -1 \\ 3 \end{bmatrix} = \begin{bmatrix} 13 \\ -16 \\ 0 \end{bmatrix}_B$$

or

$$D(6 - 6x + 24x^2) = 13(2) - 16(2 - 3x) = -6 + 48x$$

19. (c) Since $D(\mathbf{f}_1) = 2\mathbf{f}_1$, $D(\mathbf{f}_2) = \mathbf{f}_1 + 2\mathbf{f}_2$, and $D(\mathbf{f}_3) = 2\mathbf{f}_2 + 2\mathbf{f}_3$, we have the matrix

$$\begin{bmatrix} 2 & 1 & 0 \\ 0 & 2 & 2 \\ 0 & 0 & 2 \end{bmatrix}$$

20. The upper left-hand corner is all vectors in the space V. The upper right-hand corner is all vectors in the range of T. Thus it is a subspace of W. Since V is a real vector space, the lower left-hand corner is all of R^4. The lower right-hand corner is the range of the transformation obtained by multiplying every element of R^4 by the matrix $[T]_{B',B}$. It is thus a subspace of R^7.

EXERCISE SET 8.5

1. First, we find the matrix of T with respect to B. Since

$$T(\mathbf{u}_1) = \begin{bmatrix} 1 \\ 0 \end{bmatrix}$$

and

$$T(\mathbf{u}_2) = \begin{bmatrix} -2 \\ -1 \end{bmatrix}$$

then

$$A = [T]_B = \begin{bmatrix} 1 & -2 \\ 0 & -1 \end{bmatrix}$$

In order to find P, we note that $\mathbf{v}_1 = 2\mathbf{u}_1 + \mathbf{u}_2$ and $\mathbf{v}_2 = -3\mathbf{u}_1 + 4\mathbf{u}_2$. Hence the transition matrix from B' to B is

$$P = \begin{bmatrix} 2 & -3 \\ 1 & 4 \end{bmatrix}$$

Thus

$$P^{-1} = \begin{bmatrix} \dfrac{4}{11} & \dfrac{3}{11} \\ -\dfrac{1}{11} & \dfrac{2}{11} \end{bmatrix}$$

and therefore

$$A' = [T]_{B'} = P^{-1}[T]_B P = \frac{1}{11} \begin{bmatrix} 4 & 3 \\ -1 & 2 \end{bmatrix} \begin{bmatrix} 1 & -2 \\ 0 & -1 \end{bmatrix} \begin{bmatrix} 2 & -3 \\ 1 & 4 \end{bmatrix}$$

$$= \begin{bmatrix} -\dfrac{3}{11} & -\dfrac{56}{11} \\ -\dfrac{2}{11} & \dfrac{3}{11} \end{bmatrix}$$

2. In order to compute $A = [T]_B$, we note that

$$T(\mathbf{u}_1) = \begin{bmatrix} 16 \\ -2 \end{bmatrix} = (0.8)\mathbf{u}_1 + (3.6)\mathbf{u}_2$$

and

$$T(\mathbf{u}_2) = \begin{bmatrix} -3 \\ 16 \end{bmatrix} = (6.1)\mathbf{u}_1 + (-3.8)\mathbf{u}_2$$

Hence

$$A = [T]_B = \begin{bmatrix} 0.8 & 6.1 \\ 3.6 & -3.8 \end{bmatrix}$$

In order to find P, we note that

$$\mathbf{v}_1 = (1.3)\mathbf{u}_1 + (-0.4)\mathbf{u}_2$$
$$\mathbf{v}_2 = (-0.5)\mathbf{u}_1$$

Hence

$$P = \begin{bmatrix} 1.3 & -0.5 \\ -0.4 & 0 \end{bmatrix}$$

and

$$P^{-1} = \begin{bmatrix} 0 & -2.5 \\ -2 & -6.5 \end{bmatrix}$$

It then follows that

$$A' = [T]_{B'} = P^{-1}AP = \begin{bmatrix} -15.5 & 4.5 \\ -37.5 & 12.5 \end{bmatrix}$$

3. Since $T(\mathbf{u_1}) = (1/\sqrt{2}, \ 1/\sqrt{2})$ and $T(\mathbf{u_2}) = (-1/\sqrt{2}, \ 1/\sqrt{2})$, then the matrix of T with respect to B is

$$A = [T]_B = \begin{bmatrix} 1/\sqrt{2} & -1/\sqrt{2} \\ 1/\sqrt{2} & 1/\sqrt{2} \end{bmatrix}$$

From Exercise 1, we know that

$$P = \begin{bmatrix} 2 & -3 \\ 1 & 4 \end{bmatrix} \quad \text{and} \quad P^{-1} = \frac{1}{11}\begin{bmatrix} 4 & 3 \\ -1 & 2 \end{bmatrix}$$

Thus

$$A' = [T]_{B'} = P^{-1}AP = \frac{1}{11\sqrt{2}}\begin{bmatrix} 13 & -25 \\ 5 & 9 \end{bmatrix}$$

5. Since $T(\mathbf{e_1}) = (1,0,0)$, $T(\mathbf{e_2}) = (0,1,0)$, and $T(\mathbf{e_3}) = (0,0,0)$, we have

$$A = [T]_B = \begin{bmatrix} 1 & 0 & 0 \\ 0 & 1 & 0 \\ 0 & 0 & 0 \end{bmatrix}$$

In order to compute P, we note that $\mathbf{v_1} = \mathbf{e_1}$, $\mathbf{v_2} = \mathbf{e_1} + \mathbf{e_2}$, and $\mathbf{v_3} = \mathbf{e_1} + \mathbf{e_2} + \mathbf{e_3}$. Hence,

$$P = \begin{bmatrix} 1 & 1 & 1 \\ 0 & 1 & 1 \\ 0 & 0 & 1 \end{bmatrix}$$

and

$$P^{-1} = \begin{bmatrix} 1 & -1 & 0 \\ 0 & 1 & -1 \\ 0 & 0 & 1 \end{bmatrix}$$

Thus

$$[T]_{B'} = \begin{bmatrix} 1 & -1 & 0 \\ 0 & 1 & -1 \\ 0 & 0 & 1 \end{bmatrix} \begin{bmatrix} 1 & 0 & 0 \\ 0 & 1 & 0 \\ 0 & 0 & 0 \end{bmatrix} \begin{bmatrix} 1 & 1 & 1 \\ 0 & 1 & 1 \\ 0 & 0 & 1 \end{bmatrix} = \begin{bmatrix} 1 & 0 & 0 \\ 0 & 1 & 1 \\ 0 & 0 & 0 \end{bmatrix}$$

7. Since

$$T(\mathbf{p}_1) = 9 + 3x = \frac{2}{3}\mathbf{p}_1 + \frac{1}{2}\mathbf{p}_2$$

and

$$T(\mathbf{p}_2) = 12 + 2x = -\frac{2}{9}\mathbf{p}_1 + \frac{4}{3}\mathbf{p}_2$$

we have

$$[T]_B = \begin{bmatrix} \dfrac{2}{3} & -\dfrac{2}{9} \\[2ex] \dfrac{1}{2} & \dfrac{4}{3} \end{bmatrix}$$

We note that $\mathbf{q}_1 = -\dfrac{2}{9}\mathbf{p}_1 + \dfrac{1}{3}\mathbf{p}_2$ and $\mathbf{q}_2 = \dfrac{7}{9}\mathbf{p}_1 - \dfrac{1}{6}\mathbf{p}_2$. Hence

$$P = \begin{bmatrix} -\dfrac{2}{9} & \dfrac{7}{9} \\ \dfrac{1}{3} & -\dfrac{1}{6} \end{bmatrix}$$

and

$$P^{-1} = \begin{bmatrix} \dfrac{3}{4} & \dfrac{7}{2} \\ \dfrac{3}{2} & 1 \end{bmatrix}$$

Therefore

$$[T]_{B'} = \begin{bmatrix} \dfrac{3}{4} & \dfrac{7}{2} \\ \dfrac{3}{2} & 1 \end{bmatrix} \begin{bmatrix} \dfrac{2}{3} & -\dfrac{2}{9} \\ \dfrac{1}{2} & \dfrac{4}{3} \end{bmatrix} \begin{bmatrix} -\dfrac{2}{9} & \dfrac{7}{9} \\ \dfrac{1}{3} & -\dfrac{1}{6} \end{bmatrix} = \begin{bmatrix} 1 & 1 \\ 0 & 1 \end{bmatrix}$$

8. **(a)** Since $T(1,0) = (3,-1)$ and $T(0,1) = (-4,7)$, we have

$$\det(T) = \begin{vmatrix} 3 & -4 \\ -1 & 7 \end{vmatrix} = 17$$

(c) Since $T(1) = 1$, $T(x) = x - 1$, and $T(x^2) = (x-1)^2$, we have

$$\det(T) = \begin{vmatrix} 1 & -1 & 1 \\ 0 & 1 & -2 \\ 0 & 0 & 1 \end{vmatrix} = 1$$

9. **(a)** If A and C are similar $n \times n$ matrices, then there exists an invertible $n \times n$ matrix P such that $A = P^{-1}CP$. We can interpret P as being the transition matrix from a basis B' for R^n to a basis B. Moreover, C induces a linear transformation $T : R^n \to R^n$ where $C = [T]_B$. Hence $A = [T]_{B'}$. Thus A and C are matrices for the same transformation with respect to different bases. But from Theorem 8.2.2, we know that the rank of T is equal to the rank of C and hence to the rank of A.

Alternate Solution: We observe that if P is an invertible $n \times n$ matrix, then P represents a linear transformation of R^n *onto* R^n. Thus the rank of the transformation represented by the matrix CP is the same as that of C. Since P^{-1} is also invertible, its null space contains only the zero vector, and hence the rank of the transformation represented by the matrix $P^{-1}CP$ is also the same as that of C. Thus the ranks of A and C are equal. Again we use the result of Theorem 8.2.2 to equate the rank of a linear transformation with the rank of a matrix which represents it.

Second Alternative: Since the assertion that similar matrices have the same rank deals only with matrices and not with transformations, we outline a proof which involves only matrices. If $A = P^{-1}CP$, then P^{-1} and P can be expressed as products of elementary matrices. But multiplication of the matrix C by an elementary matrix is equivalent to performing an elementary row or column operation on C. From Section 5.5, we know that such operations do not change the rank of C. Thus A and C must have the same rank.

10. **(a)** We use the standard basis for P_4. Since $T(1) = 1$, $T(x) = 2x + 1$, $T(x^2) = (2x + 1)^2$, $T(x^3) = (2x + 1)^3$, and $T(x^4) = (2x + 1)^4$, we have

$$\det(T) = \begin{vmatrix} 1 & 1 & 1 & 1 & 1 \\ 0 & 2 & 4 & 6 & 8 \\ 0 & 0 & 4 & 12 & 24 \\ 0 & 0 & 0 & 8 & 32 \\ 0 & 0 & 0 & 0 & 16 \end{vmatrix} = 1024$$

(b) Thus $\text{rank}(T) = 5$, or $[T]$ is invertible, so, by Theorem 8.4.3, T is one-to-one.

11. **(a)** The matrix for T relative to the standard basis B is

$$[T]_B = \begin{bmatrix} 1 & -1 \\ 2 & 4 \end{bmatrix}$$

The eigenvalues of $[T]_B$ are $\lambda = 2$ and $\lambda = 3$, while corresponding eigenvectors are $(1, -1)$ and $(1, -2)$, respectively. If we let

$$P = \begin{bmatrix} 1 & 1 \\ -1 & -2 \end{bmatrix} \quad \text{then} \quad P^{-1} = \begin{bmatrix} 2 & 1 \\ -1 & -1 \end{bmatrix}$$

and

$$P^{-1}[T]_B P = \begin{bmatrix} 2 & 0 \\ 0 & 3 \end{bmatrix}$$

is diagonal. Since P represents the transition matrix from the basis B' to the standard basis B, we have

$$B' = \left\{ \begin{bmatrix} 1 \\ -1 \end{bmatrix}, \begin{bmatrix} 1 \\ -2 \end{bmatrix} \right\}$$

as a basis which produces a diagonal matrix for $[T]_{B'}$.

12. **(b)** The matrix for T relative to the standard basis B is

$$[T]_B = \begin{bmatrix} 0 & -1 & 1 \\ -1 & 0 & 1 \\ 1 & 1 & 0 \end{bmatrix}$$

The eigenvalues of $[T]_B$ are $\lambda = 1$ and $\lambda = -2$. The eigenspace corresponding to $\lambda = 1$ is spanned by the vectors $\begin{bmatrix} -1 \\ 1 \\ 0 \end{bmatrix}$ and $\begin{bmatrix} 1 \\ 0 \\ 1 \end{bmatrix}$ and the eigenspace corresponding to $\lambda = -2$ is spanned by the vector $\begin{bmatrix} 1 \\ 1 \\ -1 \end{bmatrix}$. If we let

$$P = \begin{bmatrix} -1 & 1 & 1 \\ 1 & 0 & 1 \\ 0 & 1 & -1 \end{bmatrix}$$

then $P^{-1}[T]_B P$ will be diagonal. Hence the basis

$$B' = \left\{ \begin{bmatrix} -1 \\ 1 \\ 0 \end{bmatrix}, \begin{bmatrix} 1 \\ 0 \\ 1 \end{bmatrix}, \begin{bmatrix} 1 \\ 1 \\ -1 \end{bmatrix} \right\}$$

will produce a diagonal matrix for $[T]_{B'}$.

13. **(a)** The matrix of T with respect to the standard basis for P_2 is

$$A = \begin{bmatrix} 5 & 6 & 2 \\ 0 & -1 & -8 \\ 1 & 0 & -2 \end{bmatrix}$$

The characteristic equation of A is

$$\lambda^3 - 2\lambda^2 - 15\lambda + 36 = (\lambda - 3)^2(\lambda + 4) = 0$$

and the eigenvalues are therefore $\lambda = -4$ and $\lambda = 3$.

(b) If we set $\lambda = -4$, then $(\lambda I - A)\mathbf{x} = \mathbf{0}$ becomes

$$\begin{bmatrix} -9 & -6 & -2 \\ 0 & -3 & 8 \\ -1 & 0 & -2 \end{bmatrix} \begin{bmatrix} x_1 \\ x_2 \\ x_3 \end{bmatrix} = \begin{bmatrix} 0 \\ 0 \\ 0 \end{bmatrix}$$

The augmented matrix reduces to

$$\begin{bmatrix} 1 & 0 & 2 & 0 \\ 0 & 1 & -8/3 & 0 \\ 0 & 0 & 0 & 0 \end{bmatrix}$$

and hence $x_1 = -2s$, $x_2 = \dfrac{8}{3}s$, and $x_3 = s$. Therefore the vector

$$\begin{bmatrix} -2 \\ 8/3 \\ 1 \end{bmatrix}$$

is a basis for the eigenspace associated with $\lambda = -4$. In P^2, this vector represents the polynomial $-2 + \dfrac{8}{3}x + x^2$.

  If we set $\lambda = 3$ and carry out the above procedure, we find that $x_1 = 5s$, $x_2 = -2s$, and $x_3 = s$. Thus the polynomial $5 - 2x + x^2$ is a basis for the eigenspace associated with $\lambda = 3$.

14. **(a)** We look for values of λ such that

$$T\left(\begin{bmatrix} a & b \\ c & d \end{bmatrix}\right) = \lambda \begin{bmatrix} a & b \\ c & d \end{bmatrix}$$

or

$$\begin{bmatrix} 2c & a+c \\ b-2c & d \end{bmatrix} = \begin{bmatrix} \lambda a & \lambda b \\ \lambda c & \lambda d \end{bmatrix}$$

If we equate corresponding entries, we find that

$$(*)\qquad \begin{aligned} \lambda a \qquad\qquad -2c \qquad\qquad &= 0 \\ -a + \lambda b \qquad - c \qquad\qquad &= 0 \\ -\ b + (\lambda + 2)c \qquad\qquad &= 0 \\ (\lambda - 1)d &= 0 \end{aligned}$$

This system of equations has a nontrivial solution for a, b, c, and d only if

$$\det \begin{bmatrix} \lambda & 0 & -2 & 0 \\ -1 & \lambda & -1 & 0 \\ 0 & -1 & \lambda+2 & 0 \\ 0 & 0 & 0 & \lambda-1 \end{bmatrix} = 0$$

or

$$(\lambda - 1)(\lambda + 2)(\lambda^2 - 1) = 0$$

Therefore the eigenvalues are $\lambda = 1$, $\lambda = -2$, and $\lambda = -1$.

(b) We find a basis only for the eigenspace of T associated with $\lambda = 1$. The bases associated with the other eigenvalues are found in a similar way. If $\lambda = 1$, the equation $T(\mathbf{x}) = \lambda\mathbf{x}$ becomes $T(\mathbf{x}) = \mathbf{x}$ and the augmented matrix for the system of equations $(*)$ above is

$$\begin{bmatrix} 1 & 0 & -2 & 0 & 0 \\ -1 & 1 & -1 & 0 & 0 \\ 0 & -1 & 3 & 0 & 0 \\ 0 & 0 & 0 & 0 & 0 \end{bmatrix}$$

This reduces to

$$\begin{bmatrix} 1 & 0 & -2 & 0 & 0 \\ 0 & 1 & -3 & 0 & 0 \\ 0 & 0 & 0 & 0 & 0 \\ 0 & 0 & 0 & 0 & 0 \end{bmatrix}$$

and hence $a = 2t$, $b = 3t$, $c = t$, and $d = s$. Therefore the matrices

$$\begin{bmatrix} 2 & 3 \\ 1 & 0 \end{bmatrix} \quad \text{and} \quad \begin{bmatrix} 0 & 0 \\ 0 & 1 \end{bmatrix}$$

form a basis for the eigenspace associated with $\lambda = 1$.

15. If $\mathbf{v}$ is an eigenvector of T corresponding to λ, then $\mathbf{v}$ is a nonzero vector which satisfies the equation $T(\mathbf{v}) = \lambda\mathbf{v}$ or $(\lambda I - T)\mathbf{v} = \mathbf{0}$. Thus $\lambda I - T$ maps $\mathbf{v}$ to $\mathbf{0}$, or $\mathbf{v}$ is in the kernel of $\lambda I - T$.

17. Since $C[\mathbf{x}]_B = D[\mathbf{x}]_B$ for all $\mathbf{x}$ in V, we can, in particular, let $\mathbf{x} = \mathbf{v}_i$ for each of the basis vectors $\mathbf{v}_1, \ldots, \mathbf{v}_n$ of V. Since $[\mathbf{v}_i]_B = \mathbf{e}_i$ for each i where $\{\mathbf{e}_1, \ldots, \mathbf{e}_n\}$ is the standard basis for R^n, this yields $C\mathbf{e}_i = D\mathbf{e}_i$ for $i = 1, \ldots, n$. But $C\mathbf{e}_i$ and $D\mathbf{e}_i$ are just the ith columns of C and D, respectively. Since corresponding columns of C and D are all equal, we have $C = D$.

18. Let B be the standard basis for R^2, and let

$$B' = \{(\cos\theta, \sin\theta), (-\sin\theta, \cos\theta)\} = \{\mathbf{v}_1, \mathbf{v}_2\}$$

be the basis consisting of the unit vector $\mathbf{v}_1$ lying along ℓ and the unit vector $\mathbf{v}_2$ perpendicular to ℓ. Note that to change basis from B to B', we rotate through an angle $-\theta$. Now relative to B', T is just the orthogonal projection on the $\mathbf{v}_1$-axis, which is accomplished by setting the $\mathbf{v}_2$-coordinate equal to zero. Then to return to the basis B, we rotate through an angle θ. Thus

$$T\left(\begin{bmatrix} x \\ y \end{bmatrix}\right) = [T]_{B,B'}[T]_{B'}[I]_{B',B}\begin{bmatrix} x \\ y \end{bmatrix}$$

$$= \begin{bmatrix} \cos\theta & -\sin\theta \\ \sin\theta & \cos\theta \end{bmatrix}\begin{bmatrix} 1 & 0 \\ 0 & 0 \end{bmatrix}\begin{bmatrix} \cos\theta & \sin\theta \\ -\sin\theta & \cos\theta \end{bmatrix}\begin{bmatrix} x \\ y \end{bmatrix}$$

$$= \begin{bmatrix} \cos^2\theta & \sin\theta\,\cos\theta \\ \sin\theta\,\cos\theta & \sin^2\theta \end{bmatrix}\begin{bmatrix} x \\ y \end{bmatrix}$$

19.   (a)   False. Every matrix is similar to itself, since $A = I^{-1}AI$.

(b)   True. Suppose that $A = P^{-1}BP$ and $B = Q^{-1}CQ$. Then

$$A = P^{-1}(Q^{-1}CQ)P = (P^{-1}Q^{-1})C(QP) = (QP)^{-1}C(QP)$$

Therefore A and C are similar.

(c)   True. By Table 1, A is invertible if and only if B is invertible, which guarantees that A is singular if and only if B is singular.
      Alternatively, if $A = P^{-1}BP$, then $B = PAP^{-1}$. Thus, if B is singular, then so is A. Otherwise, B would be the product of 3 invertible matrices.

(d)   True. If $A = P^{-1}BP$, then $A^{-1}(P^{-1}BP)^{-1} = P^{-1}B^{-1}(P^{-1})^{-1} = P^{-1}B^{-1}P$, so A^{-1} and B^{-1} are similar.

20.   If A is invertible and B is singular, then it follows from Exercise 19(c) that A and B cannot be similar. Hence

$$A = \begin{bmatrix} 1 & 0 \\ 0 & 1 \end{bmatrix} \quad \text{and} \quad B = \begin{bmatrix} 1 & 0 \\ 0 & 0 \end{bmatrix}$$

are not similar.

   Moreover, if A and B are both invertible but have different determinants, then it follows from Table 1 that they cannot be similar. Hence.

$$A = \begin{bmatrix} 1 & 0 \\ 0 & 1 \end{bmatrix} \quad \text{and} \quad B = \begin{bmatrix} 2 & 0 \\ 0 & 1 \end{bmatrix}$$

are not similar.

22. Let $B = P^{-1}AP$ so that $A = PBP^{-1}$. Let λ be an eigenvalue common to A and B and let $\mathbf{x}$ be an eigenvector of A corresponding to λ. Then $A\mathbf{x} = \lambda\mathbf{x}$, or

$$Ax = (PBP^{-1})x = \lambda x$$

so that

$$BP^{-1}x = P^{-1}\lambda x$$

Thus

$$B(P^{-1}x) = \lambda(P^{-1}x)$$

That is, $P^{-1}\mathbf{x}$ is an eigenvector of B corresponding to λ.

SUPPLEMENTARY EXERCISES 8

3. By the properties of an inner product, we have

$$T(\mathbf{v} + \mathbf{w}) = \langle \mathbf{v} + \mathbf{w}, \mathbf{v}_0 \rangle \mathbf{v}_0$$

$$= (\langle \mathbf{v}, \mathbf{v}_0 \rangle + \langle \mathbf{w}, \mathbf{v}_0 \rangle) \mathbf{v}_0$$

$$= \langle \mathbf{v}, \mathbf{v}_0 \rangle \mathbf{v}_0 + \langle \mathbf{w}, \mathbf{v}_0 \rangle \mathbf{v}_0$$

$$= T(\mathbf{v}) + T(\mathbf{w})$$

and

$$T(k\mathbf{v}) = \langle k\mathbf{v}, \mathbf{v}_0 \rangle \mathbf{v}_0 = k\langle \mathbf{v}, \mathbf{v}_0 \rangle \mathbf{v}_0 = kT(\mathbf{v})$$

Thus T is a linear operator on V.

4. **(a)** By direct computation, we have

$$T(\mathbf{x} + \mathbf{y}) = ((\mathbf{x} + \mathbf{y}) \cdot \mathbf{v}_1, \ldots, (\mathbf{x} + \mathbf{y}) \cdot \mathbf{v}_m)$$

$$= (\mathbf{x} \cdot \mathbf{v}_1 + \mathbf{y} \cdot \mathbf{v}_1, \ldots, \mathbf{x} \cdot \mathbf{v}_m + \mathbf{y} \cdot \mathbf{v}_m)$$

$$= (\mathbf{x} \cdot \mathbf{v}_1, \ldots, \mathbf{x} \cdot \mathbf{v}_m) + (\mathbf{y} \cdot \mathbf{v}_1, \ldots, \mathbf{y} \cdot \mathbf{v}_m)$$

$$= T(\mathbf{x}) + T(\mathbf{y})$$

and

$$T(k\mathbf{x}) = ((k\mathbf{x}) \cdot \mathbf{v}_1, \ldots, (k\mathbf{x}) \cdot \mathbf{v}_m)$$

$$= (k(\mathbf{x} \cdot \mathbf{v}_1), \ldots, k(\mathbf{x} \cdot \mathbf{v}_m))$$

$$= kT(\mathbf{x})$$

4. **(b)** If $\{e_1, \ldots, e_n\}$ is the standard basis for R^n, and if $v_i = (a_{1i}, a_{2i}, \ldots, a_{ni})$ for $i = 1, \ldots, m$, then

$$T(e_j) = (a_{j1}, a_{j2}, \ldots, a_{jm})$$

But $T(e_j)$, interpreted as a column vector, is just the jth column of the standard matrix for T. Thus the ith row of this matrix is $(a_{1i}, a_{2i}, \ldots, a_{ni})$, which is just v_i.

5. **(a)** The matrix for T with respect to the standard basis is

$$A = \begin{bmatrix} 1 & 0 & 1 & 1 \\ 2 & 1 & 3 & 1 \\ 1 & 0 & 0 & 1 \end{bmatrix}$$

We first look for a basis for the range of T; that is, for the space of vectors $\mathbf{b}$ such that $A\mathbf{x} = \mathbf{b}$. If we solve the system of equations

$$x \quad\quad + \ z + w = b_1$$
$$2x + y + 3z + w = b_2$$
$$x \quad\quad\quad + w = b_3$$

we find that $z = b_1 - b_3$ and that any one of x, y, or w will determine the other two. Thus, $T(e_3)$ and any two of the remaining three columns of A is a basis for $R(T)$.

Alternate Solution: We can use the method of Section 5.5 to find a basis for the column space of A by reducing A^T to row-echelon form. This yields

$$\begin{bmatrix} 1 & 2 & 1 \\ 0 & 1 & 0 \\ 0 & 0 & 1 \\ 0 & 0 & 0 \end{bmatrix}$$

so that the three vectors

$$\begin{bmatrix} 1 \\ 2 \\ 1 \end{bmatrix} \quad \begin{bmatrix} 0 \\ 1 \\ 0 \end{bmatrix} \quad \begin{bmatrix} 0 \\ 0 \\ 1 \end{bmatrix}$$

form a basis for the column space of T and hence for its range.

Second Alternative: Note that since rank$(A) = 3$, then $R(T)$ is a 3-dimensional subspace of R^3 and hence is all of R^3. Thus the standard basis for R^3 is also a basis for $R(T)$.

To find a basis for the kernel of T, we consider the solution space of $A\mathbf{x} = \mathbf{0}$. If we set $b_1 = b_2 = b_3 = 0$ in the above system of equations, we find that $z = 0$, $x = -w$, and $y = w$. Thus the vector $(-1, 1, 0, 1)$ forms a basis for the kernel.

7. **(a)** We know that T can be thought of as multiplication by the matrix

$$[T]_B = \begin{bmatrix} 1 & 1 & 2 & -2 \\ 1 & -1 & -4 & 6 \\ 1 & 2 & 5 & -6 \\ 3 & 2 & 3 & -2 \end{bmatrix}$$

where reduction to row-echelon form easily shows that rank$([T]_B) = 2$. Therefore the rank of T is 2 and the nullity of T is $4 - 2 = 2$.

(b) Since $[T]_B$ is not invertible, T is not one-to-one.

9. **(a)** If $A = P^{-1}BP$, then

$$A^T = (P^{-1}BP)^T$$
$$= P^T B^T (P^{-1})^T$$
$$= \left((P^T)^{-1}\right)^{-1} B^T (P^{-1})^T$$
$$= \left((P^{-1})^T\right)^{-1} B^T (P^{-1})^T$$

Therefore A^T and B^T are similar. You should verify that if P is invertible, then so is P^T and that $(P^T)^{-1} = (P^{-1})^T$.

10. If statement (i) holds, then the range of T is V and hence rank$(T) = n$. Therefore, by the Dimension Theorem, the nullity of T is zero, so that statement (ii) cannot hold. Thus (i) and (ii) cannot hold simultaneously.

Now if statement (i) does *not* hold, then the range of T is a proper subspace of V and rank$(T) < n$. The Dimension Theorem then implies that the nullity of T is greater than zero. Thus statement (ii) *must* hold. Hence exactly one of the two statements must always hold.

11. If we let $X = \begin{bmatrix} a & b \\ c & d \end{bmatrix}$, then we have

$$T\left(\begin{bmatrix} a & b \\ c & d \end{bmatrix}\right) = \begin{bmatrix} a+c & b+d \\ 0 & 0 \end{bmatrix} + \begin{bmatrix} b & b \\ d & d \end{bmatrix}$$

$$= \begin{bmatrix} a+b+c & 2b+d \\ d & d \end{bmatrix}$$

The matrix X is in the kernel of T if and only if $T(X) = \mathbf{0}$, i.e., if and only if

$$a + b + c \quad = 0$$
$$2b \quad + d = 0$$
$$d = 0$$

Hence

$$X = \begin{bmatrix} a & 0 \\ -a & 0 \end{bmatrix}$$

The space of all such matrices X is spanned by the matrix $\begin{bmatrix} 1 & 0 \\ -1 & 0 \end{bmatrix}$, and therefore has dimension 1. Thus the nullity is 1. Since the dimension of M_{22} is 4, the rank of T must therefore be 3.

Alternate Solution. Using the computations done above, we have that the matrix for this transformation with respect to the standard basis in M_{22} is

$$\begin{bmatrix} 1 & 1 & 1 & 0 \\ 0 & 2 & 0 & 1 \\ 0 & 0 & 0 & 1 \\ 0 & 0 & 0 & 1 \end{bmatrix}$$

Since this matrix has rank 3, the rank of T is 3, and therefore the nullity must be 1.

12. We are given that there exist invertible matrices P and Q such that $A = P^{-1}BP$ and $B = Q^{-1}CQ$. Therefore

$$A = P^{-1}(Q^{-1}CQ)P = (QP)^{-1}C(QP)$$

That is, A and C are similar.

13. The standard basis for M_{22} is the set of matrices

$$\begin{bmatrix} 1 & 0 \\ 0 & 0 \end{bmatrix}, \begin{bmatrix} 0 & 1 \\ 0 & 0 \end{bmatrix}, \begin{bmatrix} 0 & 0 \\ 1 & 0 \end{bmatrix}, \begin{bmatrix} 0 & 0 \\ 0 & 1 \end{bmatrix}$$

If we think of the above matrices as the vectors

$$[1 \ 0 \ 0 \ 0]^T, \quad [0 \ 1 \ 0 \ 0]^T, \quad [0 \ 0 \ 1 \ 0]^T, \quad [0 \ 0 \ 0 \ 1]^T$$

then L takes these vectors to

$$[1 \ 0 \ 0 \ 0]^T, \quad [0 \ 0 \ 1 \ 0]^T, \quad [0 \ 1 \ 0 \ 0]^T, \quad [0 \ 0 \ 0 \ 1]^T$$

Therefore the desired matrix for L is

$$\begin{bmatrix} 1 & 0 & 0 & 0 \\ 0 & 0 & 1 & 0 \\ 0 & 1 & 0 & 0 \\ 0 & 0 & 0 & 1 \end{bmatrix}$$

14. **(a)** Reading directly from P, we have

$$\mathbf{v}_1 = 2\mathbf{u}_1 + \mathbf{u}_2$$
$$\mathbf{v}_2 = -\mathbf{u}_1 + \mathbf{u}_2 + \mathbf{u}_3$$
$$\mathbf{v}_3 = 3\mathbf{u}_1 + 4\mathbf{u}_2 + 2\mathbf{u}_3$$

14. **(b)** Since

$$P^{-1} = \begin{bmatrix} -2 & 5 & -7 \\ -2 & 4 & -5 \\ 1 & -2 & 3 \end{bmatrix}$$

by direct calculation we have

$$\mathbf{u}_1 = -2\mathbf{v}_1 - 2\mathbf{v}_2 + \mathbf{v}_3$$

$$\mathbf{u}_2 = 5\mathbf{v}_1 + 4\mathbf{v}_2 - 2\mathbf{v}_3$$

$$\mathbf{u}_3 = -7\mathbf{v}_1 - 5\mathbf{v}_2 + 3\mathbf{v}_3$$

15. The transition matrix P from B' to B is

$$P = \begin{bmatrix} 1 & 1 & 1 \\ 0 & 1 & 1 \\ 0 & 0 & 1 \end{bmatrix}$$

Therefore, by Theorem 8.5.2, we have

$$[T]_{B'} = P^{-1}[T]_B P = \begin{bmatrix} -4 & 0 & 9 \\ 1 & 0 & -2 \\ 0 & 1 & 1 \end{bmatrix}$$

<u>Alternate Solution</u>: We compute the above result more directly. It is easy to show that $\mathbf{u}_1 = \mathbf{v}_1$, $\mathbf{u}_2 = -\mathbf{v}_1 + \mathbf{v}_2$, and $\mathbf{u}_3 = -\mathbf{v}_2 + \mathbf{v}_3$. So

$$T(\mathbf{v}_1) = T(\mathbf{u}_1) = -3\mathbf{u}_1 + \mathbf{u}_2 = -4\mathbf{v}_1 + \mathbf{v}_2$$

$$T(\mathbf{v}_2) = T(\mathbf{u}_1 + \mathbf{u}_2) = T(\mathbf{u}_1) + T(\mathbf{u}_2)$$

$$= \mathbf{u}_1 + \mathbf{u}_2 + \mathbf{u}_3 = \mathbf{v}_3$$

$$T(\mathbf{v}_3) = T(\mathbf{u}_1 + \mathbf{u}_2 + \mathbf{u}_3) = T(\mathbf{u}_1) + T(\mathbf{u}_2) + T(\mathbf{u}_3)$$

$$= 8\mathbf{u}_1 - \mathbf{u}_2 + \mathbf{u}_3$$

$$= 9\mathbf{v}_1 - 2\mathbf{v}_2 + \mathbf{v}_3$$

16. Let $P = \begin{bmatrix} a & b \\ c & d \end{bmatrix}$ and solve for a, b, c, and d so that

$$\begin{bmatrix} 1 & 1 \\ -1 & 4 \end{bmatrix} = P^{-1} \begin{bmatrix} 2 & 1 \\ 1 & 3 \end{bmatrix} P \quad \text{or} \quad P \begin{bmatrix} 1 & 1 \\ -1 & 4 \end{bmatrix} = \begin{bmatrix} 2 & 1 \\ 1 & 3 \end{bmatrix} P$$

This yields the system of equations

$$a - b = 2a + c$$
$$a + 4b = 2b + d$$
$$c - d = a + 3c$$
$$c + 4d = b + 3c$$

which has solution $b = -a$, $c = 0$, and $d = -a$. Thus, for instance,

$$P = \begin{bmatrix} 1 & -1 \\ 0 & -1 \end{bmatrix} \quad \text{and} \quad P^{-1} = \begin{bmatrix} 1 & -1 \\ 0 & -1 \end{bmatrix}$$

will work.

However, if we try the same procedure on the other pair of matrices, we find that $a = 3d$, $b = d$, and $c = 3d$, so that $\det(P) = 3d^2 - 3d^2 = 0$. Thus P is not invertible.

Alternatively, if there exists an invertible matrix P such that

$$\begin{bmatrix} 3 & 1 \\ -6 & -2 \end{bmatrix} = P^{-1} \begin{bmatrix} -1 & 2 \\ 1 & 0 \end{bmatrix} P$$

then we have a matrix with zero determinant on the left, and one with nonzero determinant on the right. Why? Thus the two matrices cannot be similar.

17. Since

$$T \left(\begin{bmatrix} 1 \\ 0 \\ 0 \end{bmatrix} \right) = \begin{bmatrix} 1 \\ 0 \\ 1 \end{bmatrix}, \quad T \left(\begin{bmatrix} 0 \\ 1 \\ 0 \end{bmatrix} \right) = \begin{bmatrix} -1 \\ 1 \\ 0 \end{bmatrix}, \quad \text{and} \quad T \left(\begin{bmatrix} 0 \\ 0 \\ 1 \end{bmatrix} \right) = \begin{bmatrix} 1 \\ 0 \\ -1 \end{bmatrix}$$

we have

$$[T]_B = \begin{bmatrix} 1 & -1 & 1 \\ 0 & 1 & 0 \\ 1 & 0 & -1 \end{bmatrix}$$

In fact, this result can be read directly from $[T(X)]_B$.

18. We know that $\det(T) \neq 0$ if and only if the matrix of T relative to *any* basis B has nonzero determinant; that is, if and only if *every* such matrix is invertible. Choose a particular basis B for V and let V have dimension n. Then the matrix $[T]_B$ represents T as a matrix transformation from R^n to R^n. By Theorem 4.3.1, such a transformation is one-to-one if and only if $[T]_B$ is invertible. Thus T is one-to-one if and only if $\det(T) \neq 0$.

19. **(a)** Recall that $D(\mathbf{f+g}) = (f(x)+g(x))'' = f''(x)+g''(x)$ and $D(c\mathbf{f}) = (cf(x))'' = cf''(x)$.

   (b) Recall that $D(\mathbf{f}) = \mathbf{0}$ if and only if $f'(x) = a$ for some constant a if and only if $f(x) = ax + b$ for constants a and b. Since the functions $f(x) = x$ and $g(x) = 1$ are linearly independent, they form a basis for the kernel of D.

   (c) Since $D(\mathbf{f}) = f(x)$ if and only if $f''(x) = f(x)$ if and only if $f(x) = ae^x + be^{-x}$ for a and b arbitrary constants, the functions $f(x) = e^x$ and $g(x) = e^{-x}$ span the set of all such functions. This is clearly a subspace of $C^2(-\infty, \infty)$ (Why?), and to show that it has dimension 2, we need only check that e^x and e^{-x} are linearly independent functions. To this end, suppose that there exist constants c_1 and c_2 such that $c_1e^x + c_2e^{-x} = 0$. If we let $x = 0$ and $x = 1$, we obtain the equations $c_1 + c_2 = 0$ and $c_1e + c_2e^{-1} = 0$. These imply that $c_1 = c_2 = 0$, so e^x and e^{-x} are linearly independent.

21. **(a)** We have

$$T(p(x) + q(x)) = \begin{bmatrix} p(x_1) + q(x_1) \\ p(x_2) + q(x_2) \\ p(x_3) + q(x_3) \end{bmatrix} = \begin{bmatrix} p(x_1) \\ p(x_2) \\ p(x_3) \end{bmatrix} + \begin{bmatrix} q(x_1) \\ q(x_2) \\ q(x_3) \end{bmatrix} = T(p(x)) + T(q(x))$$

and

$$T(kp(x)) = \begin{bmatrix} kp(x_1) \\ kp(x_2) \\ kp(x_3) \end{bmatrix} = k \begin{bmatrix} p(x_1) \\ p(x_2) \\ p(x_3) \end{bmatrix} = kT(p(x))$$

(b) Since T is defined for quadratic polynomials only, and the numbers x_1, x_2, and x_3 are distinct, we can have $p(x_1) = p(x_2) = p(x_3) = 0$ if and only if p is the zero polynomial. (Why?) Thus $\ker(T) = \{\mathbf{0}\}$, so T is one-to-one.

(c) We have

$$T(a_1 P_1(x) + a_2 P_2(x) + a_3 P_3(x)) = a_1 T(P_1(x)) + a_2 T(P_2(x)) + a_3 T(P_3(x))$$

$$= a_1 \begin{bmatrix} 1 \\ 0 \\ 0 \end{bmatrix} + a_2 \begin{bmatrix} 0 \\ 1 \\ 0 \end{bmatrix} + a_3 \begin{bmatrix} 0 \\ 0 \\ 1 \end{bmatrix}$$

$$= \begin{bmatrix} a_1 \\ a_2 \\ a_3 \end{bmatrix}$$

(d) From the above calculations, we see that the points must lie on the curve.

23. Since

$$D(x^k) = \begin{cases} 0 & \text{if } k = 0 \\ kx^{k-1} & \text{if } k = 1, 2, \ldots, n \end{cases}$$

then

$$[D(x^k)]_B = \begin{cases} (0, \ldots, 0) & \text{if } k = 0 \\ (0, \ldots, k, \ldots, 0) & \text{if } k = 1, 2, \ldots, n \end{cases}$$

$$\uparrow$$
$$k^{th} \text{ component}$$

where the above vectors all have $n + 1$ components. Thus the matrix of D with respect to B is

$$
\begin{bmatrix}
0 & 1 & 0 & 0 & \cdots & 0 \\
0 & 0 & 2 & 0 & \cdots & 0 \\
0 & 0 & 0 & 3 & \cdots & 0 \\
\vdots & \vdots & \vdots & \vdots & & \vdots \\
0 & 0 & 0 & 0 & \cdots & n \\
0 & 0 & 0 & 0 & \cdots & 0
\end{bmatrix}
$$

24. Call the basis B and the vectors $\mathbf{v}_0, \mathbf{v}_1, \ldots, \mathbf{v}_n$. Notice that $D(\mathbf{v}_i) = \mathbf{v}_{i-1}$ for $i = 1, \ldots, n$, while $D(\mathbf{v}_0) = \mathbf{0}$. That is,

$$
[D(\mathbf{v}_k)]_B =
\begin{cases}
(0, \ldots, 0) & \text{if } k = 0 \\
(0, \ldots, 1, \ldots, 0) & \text{if } k = 1, 2, \ldots, n \\
\qquad\quad \uparrow & \\
\quad k^{th} \text{ component} &
\end{cases}
$$

where the above vectors all have $n + 1$ components. Thus the matrix of D with respect to B is

$$
\begin{bmatrix}
0 & 1 & 0 & 0 & \cdots & 0 \\
0 & 0 & 1 & 0 & \cdots & 0 \\
0 & 0 & 0 & 1 & \cdots & 0 \\
\vdots & \vdots & \vdots & \vdots & & \vdots \\
0 & 0 & 0 & 0 & \cdots & 1 \\
0 & 0 & 0 & 0 & \cdots & 0
\end{bmatrix}
$$

In fact, the differentiation operator maps P_n to P_{n-1}.

25. Let B_n and B_{n+1} denote the bases for P_n and P_{n+1}, respectively. Since

$$
J(x^k) = \frac{x^{k+1}}{k+1} \quad \text{for } k = 0, \ldots, n
$$

we have

$$[J(x^k)]_{B_{n+1}} = \left(0,\ldots,\frac{1}{k+1},\ldots,0\right) \qquad (n+2 \text{ components})$$

$$\uparrow$$
$$k + 2^{nd} \text{ component}$$

where $[x^k]_{B_n} = [0,\ldots,1,\ldots,0]^T$ with the entry 1 as the $(k+1)^{st}$ component out of a total of $n+1$ components. Thus the matrix of J with respect to B_{n+1} is

$$\begin{bmatrix} 0 & 0 & 0 & \cdots & 0 \\ 1 & 0 & 0 & \cdots & 0 \\ 0 & 1/2 & 0 & \cdots & 0 \\ 0 & 0 & 1/3 & \cdots & 0 \\ \vdots & \vdots & \vdots & & \vdots \\ 0 & 0 & 0 & \cdots & 1/(n+1) \end{bmatrix}$$

with $n+2$ rows and $n+1$ columns.

EXERCISE SET 9.1

1. (a) The system is of the form $\mathbf{y}' = A\mathbf{y}$ where

$$A = \begin{bmatrix} 1 & 4 \\ 2 & 3 \end{bmatrix}$$

The eigenvalues of A are $\lambda = 5$ and $\lambda = -1$ and the corresponding eigenspaces are spanned by the vectors

$$\begin{bmatrix} 1 \\ 1 \end{bmatrix} \qquad \text{and} \qquad \begin{bmatrix} -2 \\ 1 \end{bmatrix}$$

respectively. Thus if we let

$$P = \begin{bmatrix} 1 & -2 \\ 1 & 1 \end{bmatrix}$$

we have

$$D = P^{-1}AP = \begin{bmatrix} 5 & 0 \\ 0 & -1 \end{bmatrix}$$

Let $\mathbf{y} = P\mathbf{u}$ and hence $\mathbf{y}' = P\mathbf{u}'$. Then

$$\mathbf{u}' = \begin{bmatrix} 5 & 0 \\ 0 & -1 \end{bmatrix}\mathbf{u}$$

or

$$u_1' = 5u_1$$
$$u_2' = -u_2$$

405

Therefore

$$u_1 = c_1 e^{5x}$$

$$u_2 = c_2 e^{-x}$$

Thus the equation $\mathbf{y} = P\mathbf{u}$ is

$$\begin{bmatrix} y_1 \\ y_2 \end{bmatrix} = \begin{bmatrix} 1 & -2 \\ 1 & 1 \end{bmatrix} \begin{bmatrix} c_1 e^{5x} \\ c_2 e^{-x} \end{bmatrix} = \begin{bmatrix} c_1 e^{5x} - 2c_2 e^{-x} \\ c_1 e^{5x} + c_2 e^{-x} \end{bmatrix}$$

or

$$y_1 = c_1 e^{5x} - 2c_2 e^{-x}$$

$$y_2 = c_1 e^{5x} + c_2 e^{-x}$$

1.   (b)   If $y_1(0) = y_2(0) = 0$, then

$$c_1 - 2c_2 = 0$$

$$c_1 + c_2 = 0$$

so that $c_1 = c_2 = 0$. Thus $y_1 = 0$ and $y_2 = 0$.

3.   (a)   The system is of the form $\mathbf{y}' = A\mathbf{y}$ where

$$A = \begin{bmatrix} 4 & 0 & 1 \\ -2 & 1 & 0 \\ -2 & 0 & 1 \end{bmatrix}$$

The eigenvalues of A are $\lambda = 1$, $\lambda = 2$, and $\lambda = 3$ and the corresponding eigenspaces are spanned by the vectors

$$\begin{bmatrix} 0 \\ 1 \\ 0 \end{bmatrix} \quad \begin{bmatrix} -1/2 \\ 1 \\ 1 \end{bmatrix} \quad \begin{bmatrix} -1 \\ 1 \\ 1 \end{bmatrix}$$

respectively. Thus, if we let

$$P = \begin{bmatrix} 0 & -1/2 & -1 \\ 1 & 1 & 1 \\ 0 & 1 & 1 \end{bmatrix}$$

then

$$D = P^{-1}AP = \begin{bmatrix} 1 & 0 & 0 \\ 0 & 2 & 0 \\ 0 & 0 & 3 \end{bmatrix}$$

Let $\mathbf{y} = P\mathbf{u}$ and hence $\mathbf{y}' = P\mathbf{u}'$. Then

$$\mathbf{u}' = \begin{bmatrix} 1 & 0 & 0 \\ 0 & 2 & 0 \\ 0 & 0 & 3 \end{bmatrix} \mathbf{u}$$

so that

$$u_1' = u_1$$
$$u_2' = 2u_2$$
$$u_3' = 3u_3$$

Therefore

$$u_1 = c_1 e^x$$
$$u_2 = c_2 e^{2x}$$
$$u_3 = c_3 e^{3x}$$

Thus the equation $\mathbf{y} = P\mathbf{u}$ is

$$\begin{bmatrix} y_1 \\ y_2 \\ y_3 \end{bmatrix} = \begin{bmatrix} 0 & -1/2 & -1 \\ 1 & 1 & 1 \\ 0 & 1 & 1 \end{bmatrix} \begin{bmatrix} c_1 e^x \\ c_2 e^{2x} \\ c_3 e^{3x} \end{bmatrix}$$

or

$$y_1 = -\frac{1}{2}c_2 e^{2x} - c_3 e^{3x}$$
$$y_2 = c_1 e^x + c_2 e^{2x} + c_3 e^{3x}$$
$$y_3 = c_2 e^{2x} + c_3 e^{3x}$$

Note: If we use

$$\begin{bmatrix} 0 \\ 1 \\ 0 \end{bmatrix} \quad \begin{bmatrix} -1 \\ 2 \\ 2 \end{bmatrix} \quad \begin{bmatrix} 1 \\ -1 \\ -1 \end{bmatrix}$$

as basis vectors for the eigenspaces, then

$$P = \begin{bmatrix} 0 & -1 & 1 \\ 1 & 2 & -1 \\ 0 & 2 & -1 \end{bmatrix}$$

and

$$y_1 = -c_2 e^{2x} + c_3 e^{3x}$$

$$y_2 = c_1 e^x + 2c_2 e^{2x} - c_3 e^{3x}$$

$$y_3 = 2c_2 e^{2x} - c_3 e^{3x}$$

There are, of course, infinitely many other ways of writing the answer, depending upon what bases you choose for the eigenspaces. Since the numbers c_1, c_2, and c_3 are arbitrary, the "different" answers do, in fact, represent the same functions.

3. **(b)** If we set $x = 0$, then the initial conditions imply that

$$-\frac{1}{2}c_2 - c_3 = -1$$

$$c_1 + c_2 + c_3 = 1$$

$$c_2 + c_3 = 0$$

or, equivalently, that $c_1 = 1$, $c_2 = -2$, and $c_3 = 2$. If we had used the "different" solution we found in Part (a), then we would have found that $c_1 = 1$, $c_2 = -1$, and $c_3 = -2$. In either case, when we substitute these values into the appropriate equations, we find that

$$y_1 = e^{2x} - 2e^{3x}$$

$$y_2 = e^x - 2e^{2x} + 2e^{3x}$$

$$y_3 = -2e^{2x} + 2e^{3x}$$

5. Following the hint, let $y = f(x)$ be a solution to $y' = ay$, so that $f'(x) = af(x)$. Now consider the function $g(x) = f(x)e^{-ax}$. Observe that

$$g'(x) = f'(x)e^{-ax} - af(x)e^{-ax}$$
$$= af(x)e^{-ax} - af(x)e^{-ax}$$
$$= 0$$

Thus $g(x)$ must be a constant; say $g(x) = c$. Therefore,

$$f(x)e^{-ax} = c$$

or

$$f(x) = ce^{ax}$$

That is, every solution of $y' = ay$ has the form $y = ce^{ax}$.

6. Suppose that $\mathbf{y}' = A\mathbf{y}$ where A is diagonalizable. Then there is an invertible matrix P such that $P^{-1}AP = D$ where D is a diagonal matrix whose diagonal entries are the eigenvalues of A. If we let $\mathbf{y} = P\mathbf{u}$ so that $\mathbf{y}' = P\mathbf{u}'$, then $\mathbf{y}' = A\mathbf{y}$ becomes $P\mathbf{u}' = AP\mathbf{u}$ or

$$\mathbf{u}' = P^{-1}AP\mathbf{u}$$
$$= D\mathbf{u}$$

That is,

$$u_1' = \lambda_1 u_1$$
$$\vdots \qquad \vdots$$
$$u_n' = \lambda_n u_n$$

and hence

$$u_1 = c_1 e^{\lambda_1 x}$$
$$\vdots \qquad \vdots$$
$$u_n = c_n e^{\lambda_n x}$$

Therefore $\mathbf{y} = P\mathbf{u}$ can be written

$$y_1 = p_{11}c_1e^{\lambda_1 x} + p_{12}c_2e^{\lambda_2 x} + \cdots + p_{1n}c_ne^{\lambda_n x}$$

$$\vdots \qquad \vdots \qquad \vdots \qquad \qquad \vdots$$

$$y_n = p_{n1}c_1e^{\lambda_1 x} + p_{n2}c_2e^{\lambda_2 x} + \cdots + p_{nn}c_ne^{\lambda_n x}$$

where p_{ij} is the ij^{th} entry in the matrix P. That is, each of the functions y_i is a linear combination of $e^{\lambda_1 x}, \ldots, e^{\lambda_n x}$.

7. If $y_1 = y$ and $y_2 = y'$, then $y_1' = y_2$ and $y_2' = y'' = y' + 6y = y_2 + 6y_1$. That is,

$$y_1' = \qquad y_2$$

$$y_2' = 6y_1 + y_2$$

or $\mathbf{y}' = A\mathbf{y}$ where

$$A = \begin{bmatrix} 0 & 1 \\ 6 & 1 \end{bmatrix}$$

The eigenvalues of A are $\lambda = -2$ and $\lambda = 3$ and the corresponding eigenspaces are spanned by the vectors

$$\begin{bmatrix} -1 \\ 2 \end{bmatrix} \quad \text{and} \quad \begin{bmatrix} 1 \\ 3 \end{bmatrix}$$

respectively. Thus, if we let

$$P = \begin{bmatrix} -1 & 1 \\ 2 & 3 \end{bmatrix}$$

then

$$P^{-1}AP = \begin{bmatrix} -2 & 0 \\ 0 & 3 \end{bmatrix}$$

Let $\mathbf{y} = P\mathbf{u}$ and hence $\mathbf{y}' = P\mathbf{u}'$. Then

$$\mathbf{u'} = \begin{bmatrix} -2 & 0 \\ 0 & 3 \end{bmatrix} \mathbf{u}$$

or

$$y_1 = -c_1 e^{-2x} + c_2 e^{3x}$$
$$y_2 = 2c_1 e^{-2x} + 3c_2 e^{3x}$$

Therefore

$$u_1 = c_1 e^{-2x}$$
$$u_2 = c_2 e^{3x}$$

Thus the equation $\mathbf{y} = P\mathbf{u}$ is

$$\begin{bmatrix} y_1 \\ y_2 \end{bmatrix} = \begin{bmatrix} -1 & 1 \\ 2 & 3 \end{bmatrix} \begin{bmatrix} c_1 e^{-2x} \\ c_2 e^{3x} \end{bmatrix}$$

or

$$y_1 = -c_1 e^{-2x} + c_2 e^{3x}$$
$$y_2 = 2c_1 e^{-2x} + 3c_2 e^{3x}$$

Note that $y_1' = y_2$, as required, and, since $y_1 - y$, then

$$y = -c_1 e^{-2x} + c_2 e^{3x}$$

Since c_1 and c_2 are arbitrary, any answer of the form $y = ae^{-2x} + be^{3x}$ is correct.

8.   If we let $y_1 = y$, $y_2 = y'$, and $y_3 = y''$, then we obtain the system

$$y_1' = y_2$$
$$y_2' = y_3$$
$$y_3' = 6y_1 - 11y_2 + 6y_3$$

The associated matrix is therefore

$$A = \begin{bmatrix} 0 & 1 & 0 \\ 0 & 0 & 1 \\ 6 & -11 & 6 \end{bmatrix}$$

The eigenvalues of A are $\lambda = 1$, $\lambda = 2$, and $\lambda = 3$ and the corresponding eigenvectors are

$$\begin{bmatrix} 1 \\ 1 \\ 1 \end{bmatrix}, \begin{bmatrix} 1 \\ 2 \\ 4 \end{bmatrix} \quad \text{and} \quad \begin{bmatrix} 1 \\ 3 \\ 9 \end{bmatrix}$$

The solution is, after some computation,

$$y = c_1 e^x + c_2 e^{2x} + c_3 e^{3x}$$

9. **(b)** Suppose that the coefficient matrix A of the system $\mathbf{y}' = A\mathbf{y}$ is diagonalizable and let $D = P^{-1}AP$ where D and P are as given in the problem. Let $\mathbf{y} = P\mathbf{u}$ and $\mathbf{y}' = P\mathbf{u}'$ so that $\mathbf{u}' = D\mathbf{u}$. Thus $u_i' = \lambda_i u_i$ from which it follows that $u_i = c_i e^{\lambda_i x}$ for $i = 1, \ldots, n$. Since $\mathbf{y} = P\mathbf{u}$, we have

$$\mathbf{y} = [\mathbf{x}_1 \mid \mathbf{x}_2 \mid \cdots \mid \mathbf{x}_n] \begin{bmatrix} c_1 e^{\lambda_1 x} \\ c_2 e^{\lambda_2 x} \\ \vdots \\ c_n e^{\lambda_n x} \end{bmatrix}$$

which is the desired result.

EXERCISE SET 9.2

1. **(a)** Since $T(x, y) = (-y, -x)$, the standard matrix is

$$\begin{bmatrix} 0 & -1 \\ -1 & 0 \end{bmatrix}$$

(c) Since $T(x, y) = (x, 0)$, the standard matrix is

$$\begin{bmatrix} 1 & 0 \\ 0 & 0 \end{bmatrix}$$

3. **(b)** Since $T(x, y, z) = (x, -y, z)$, the standard matrix is

$$\begin{bmatrix} 1 & 0 & 0 \\ 0 & -1 & 0 \\ 0 & 0 & 1 \end{bmatrix}$$

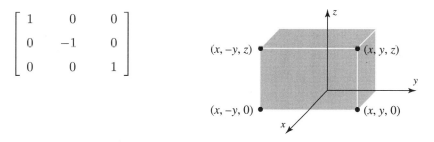

5. **(a)** This transformation leaves the z-coordinate of every point fixed. However it sends $(1, 0, 0)$ to $(0, 1, 0)$ and $(0, 1, 0)$ to $(-1, 0, 0)$. The standard matrix is therefore

$$\begin{bmatrix} 0 & -1 & 0 \\ 1 & 0 & 0 \\ 0 & 0 & 1 \end{bmatrix}$$

5. **(c)** This transformation leaves the y-coordinate of every point fixed. However it sends $(1,0,0)$ to $(0,0,-1)$ and $(0,0,1)$ to $(1,0,0)$. The standard matrix is therefore

$$\begin{bmatrix} 0 & 0 & 1 \\ 0 & 1 & 0 \\ -1 & 0 & 0 \end{bmatrix}$$

12. **(b)** To reduce this matrix to the identity matrix, we add -2 times Row 1 to Row 2 and then add -4 times Row 2 to Row 1. These operations when performed on I yield

$$\begin{bmatrix} 1 & 0 \\ -2 & 1 \end{bmatrix} \quad \text{and} \quad \begin{bmatrix} 1 & -4 \\ 0 & 1 \end{bmatrix}$$

respectively. The inverses of the above two matrices are

$$\begin{bmatrix} 1 & 0 \\ 2 & 1 \end{bmatrix} \quad \text{and} \quad \begin{bmatrix} 1 & 4 \\ 0 & 1 \end{bmatrix}$$

respectively. Thus,

$$\begin{bmatrix} 1 & 4 \\ 2 & 9 \end{bmatrix} = \begin{bmatrix} 1 & 0 \\ 2 & 1 \end{bmatrix}\begin{bmatrix} 1 & 4 \\ 0 & 1 \end{bmatrix}$$

and therefore the transformation represents a shear by a factor of 4 in the x-direction followed by a shear by a factor of 2 in the y-direction.

(c) To reduce this matrix to the identity matrix, we multiply Row 1 by $-1/2$, multiply Row 2 by $1/4$, and interchange Rows 1 and 2. These operations when performed on I yield

$$\begin{bmatrix} -1/2 & 0 \\ 0 & 1 \end{bmatrix} \quad \begin{bmatrix} 1 & 0 \\ 0 & 1/4 \end{bmatrix} \quad \begin{bmatrix} 0 & 1 \\ 1 & 0 \end{bmatrix}$$

respectively. The inverses of the above matrices are

$$\begin{bmatrix} -2 & 0 \\ 0 & 1 \end{bmatrix} \quad \begin{bmatrix} 1 & 0 \\ 0 & 4 \end{bmatrix} \quad \begin{bmatrix} 0 & 1 \\ 1 & 0 \end{bmatrix}$$

respectively. Thus,

$$\begin{bmatrix} 0 & -2 \\ 4 & 0 \end{bmatrix} = \begin{bmatrix} -2 & 0 \\ 0 & 1 \end{bmatrix} \begin{bmatrix} 1 & 0 \\ 0 & 4 \end{bmatrix} \begin{bmatrix} 0 & 1 \\ 1 & 0 \end{bmatrix}$$

Therefore, the transformation represents a reflection about the line $y = x$ followed by expansions by factors of 4 and -2 in the y- and x-directions, respectively. Note that the order in which the two expansion matrices occur is immaterial. However the position of the reflection matrix cannot be changed.

We return to the original matrix and note that we could, of course, interchange Rows 1 and 2 first and then multiply Row 1 by 1/4 and Row 2 by $-1/2$. This yields the factorization

$$\begin{bmatrix} 0 & -2 \\ 4 & 0 \end{bmatrix} = \begin{bmatrix} 0 & 1 \\ 1 & 0 \end{bmatrix} \begin{bmatrix} 4 & 0 \\ 0 & 1 \end{bmatrix} \begin{bmatrix} 1 & 0 \\ 0 & -2 \end{bmatrix}$$

Here the order of the two expansion matrices can also be interchanged. Note that there are two other factorizations, for a grand total of six.

13. **(a)** $\begin{bmatrix} 1 & 0 \\ 0 & 5 \end{bmatrix} \begin{bmatrix} 1/2 & 0 \\ 0 & 1 \end{bmatrix} = \begin{bmatrix} 1/2 & 0 \\ 0 & 5 \end{bmatrix}$

(c) $\begin{bmatrix} -1 & 0 \\ 0 & -1 \end{bmatrix} \begin{bmatrix} 0 & 1 \\ 1 & 0 \end{bmatrix} = \begin{bmatrix} 0 & -1 \\ -1 & 0 \end{bmatrix}$

14. **(a)** $\begin{bmatrix} 0 & 1 \\ 1 & 0 \end{bmatrix} \begin{bmatrix} 5 & 0 \\ 0 & 1 \end{bmatrix} \begin{bmatrix} -1 & 0 \\ 0 & 1 \end{bmatrix} = \begin{bmatrix} 0 & 1 \\ -5 & 0 \end{bmatrix}$

15. **(b)** The matrices which represent compressions along the x- and y-axes are $\begin{bmatrix} k & 0 \\ 0 & 1 \end{bmatrix}$ and $\begin{bmatrix} 1 & 0 \\ 0 & k \end{bmatrix}$, respectively, where $0 < k < 1$. But

$$\begin{bmatrix} k & 0 \\ 0 & 1 \end{bmatrix}^{-1} = \begin{bmatrix} 1/k & 0 \\ 0 & 1 \end{bmatrix}$$

and

$$\begin{bmatrix} 1 & 0 \\ 0 & k \end{bmatrix}^{-1} = \begin{bmatrix} 1 & 0 \\ 0 & 1/k \end{bmatrix}$$

Since $0 < k < 1$ implies that $1/k > 1$, the result follows.

15. (c) The matrices which represent reflections about the x- and y-axes are $\begin{bmatrix} 1 & 0 \\ 0 & -1 \end{bmatrix}$ and $\begin{bmatrix} -1 & 0 \\ 0 & 1 \end{bmatrix}$, respectively. Since these matrices are their own inverses, the result follows.

16. Since $A^{-1} = \begin{bmatrix} -2 & 3 \\ -3 & 4 \end{bmatrix}$, it follows that points (x', y') on the image line must satisfy the equations

$$x = -2x' + 3y'$$

$$y = -3x' + 4y'$$

where $y = -4x + 3$. Hence $-3x' + 4y' = 8x' - 12y' + 3$, or $11x' - 16y' + 3 = 0$. That is, the image of $y = -4x + 3$ has the equation $11x - 16y + 3 = 0$.

17. (a) The matrix which represents this shear is $\begin{bmatrix} 1 & 3 \\ 0 & 1 \end{bmatrix}$; its inverse is $\begin{bmatrix} 1 & -3 \\ 0 & 1 \end{bmatrix}$. Thus, points (x', y') on the image line must satisfy the equations

$$x = x' - 3y'$$

$$y = \qquad y'$$

where $y = 2x$. Hence $y' = 2x' - 6y'$, or $2x' - 7y' = 0$. That is, the equation of the image line is $2x - 7y = 0$.

Alternatively, we could note that the transformation leaves $(0,0)$ fixed and sends $(1, 2)$ to $(7, 2)$. Thus $(0, 0)$ and $(7, 2)$ determine the image line which has the equation $2x - 7y = 0$.

(c) The reflection and its inverse are both represented by the matrix $\begin{bmatrix} 0 & 1 \\ 1 & 0 \end{bmatrix}$. Thus the point (x', y') on the image line must satisfy the equations

$$x = y'$$

$$y = x'$$

where $y = 2x$. Hence $x' = 2y'$, so the image line has the equation $x - 2y = 0$.

(e) The rotation can be represented by the matrix $\begin{bmatrix} 1/2 & -\sqrt{3}/2 \\ \sqrt{3}/2 & 1/2 \end{bmatrix}$. This sends the origin to itself and the point $(1, 2)$ to the point $((1-2\sqrt{3})/2, (2+\sqrt{3})/2)$. Since both $(0, 0)$ and $(1, 2)$ lie on the line $y = 2x$, their images determine the image of the line under the required rotation. Thus, the image line is represented by the equation $(2+\sqrt{3})x + (2\sqrt{3} - 1)y = 0$.

Alternatively, we could find the inverse of the matrix, $\begin{bmatrix} 1/2 & \sqrt{3}/2 \\ -\sqrt{3}/2 & 1/2 \end{bmatrix}$, and proceed as we did in Parts (a) and (c).

18. Since the shear is represented by the matrix $\begin{bmatrix} 1 & k \\ 0 & 1 \end{bmatrix}$, it will send $(2, 1)$ to $(2 + k, k)$ and $(3, 0)$ to $(3, 0)$. The origin, of course, remains fixed. Thus we must find a value of k for which the vectors $(2 + k, k)$ and $(3, 0)$ are orthogonal; that is,

$$(2 + k, k) \cdot (3, 0) = 3(2 + k) = 0$$

Clearly $k = -2$.

20. The equation of a line in the plane is $Ax + By + C = 0$ where not both A and B are zero. Moreover

$$\begin{bmatrix} a & b \\ c & d \end{bmatrix}^{-1} = \frac{1}{ad - bc} \begin{bmatrix} d & -b \\ -c & a \end{bmatrix}$$

where $ad - bc \neq 0$. Thus (x, y) is transformed into (x', y') where

$$x = \frac{dx' - by'}{ad - bc}$$

$$y = \frac{-cx' + ay'}{ad - bc}$$

Therefore, the line $Ax + By + C = 0$ is transformed into

$$A\frac{dx - by}{ad - bc} + B\frac{-cx + ay}{ad - bc} + C = 0$$

or

$$\frac{dA - cB}{ad - bc}x + \frac{-bA + aB}{ad - bc}y + C = 0$$

This is the equation of a line provided $dA - cB$ and $-bA + aB$ are not both zero. But if both numbers are zero, then we have

$$bdA - bcB = 0$$

$$-bdA + adB = 0$$

or

$$(ad - bc)B = 0$$

This implies that $B = 0$ since $ad - bc \neq 0$. However, $B = 0$ implies that $dA = bA = 0$. Thus either $A = 0$ or $d = b = 0$. But since $ad - bc \neq 0$, b and d cannot both be zero. Hence A must equal zero. Finally, since not both A and B can equal zero, then not both $dA - cB$ and $-bA + aB$ can equal zero.

21. We use the notation and the calculations of Exercise 20. If the line $Ax + By + C = 0$ passes through the origin, then $C = 0$, and the equation of the image line reduces to $(dA - cB)x + (-bA + aB)y = 0$. Thus it also must pass through the origin.

   The two lines $A_1x + B_1y + C_1 = 0$ and $A_2x + B_2y + C_2 = 0$ are parallel if and only if $A_1B_2 = A_2B_1$. Their image lines are parallel if and only if

$$(dA_1 - cB_1)(-bA_2 + aB_2) = (dA_2 - cB_2)(-bA_1 + aB_1)$$

or

$$bcA_2B_1 + adA_1B_2 = bcA_1B_2 + adA_2B_1$$

or

$$(ad - bc)(A_1B_2 - A_2B_1) = 0$$

or

$$A_1B_2 - A_2B_1 = 0$$

Thus the image lines are parallel if and only if the given lines are parallel.

22. (a) Since x is fixed, but y and z are interchanged, we have

$$\begin{bmatrix} 1 & 0 & 0 \\ 0 & 0 & 1 \\ 0 & 1 & 0 \end{bmatrix}$$

23. **(a)** The matrix which transforms (x, y, z) to $(x + kz,\ y + kz,\ z)$ is

$$\begin{bmatrix} 1 & 0 & k \\ 0 & 1 & k \\ 0 & 0 & 1 \end{bmatrix}$$

24. **(a)** A reflection about the x-axis will not change the length of any vector. Thus the only possible eigenvalues are 1 and -1. The only two unit vectors which are transformed to vectors with the same or opposite direction are the standard basis vectors, $\mathbf{e}_1$ and $\mathbf{e}_2$. The vector $\mathbf{e}_1$ goes to itself and $\mathbf{e}_2$ goes to $-\mathbf{e}_2$. Therefore $\mathbf{e}_1$ and $\mathbf{e}_2$ are linearly independent eigenvectors with corresponding eigenvalues 1 and -1.

This is easily checked using the matrix

$$A = \begin{bmatrix} 1 & 0 \\ 0 & -1 \end{bmatrix}$$

with characteristic equation $(\lambda - 1)(\lambda + 1) = 0$. The eigenspaces will be the x and y axes.

(c) This is similar to Part (a) above only the two lines through the origin which remain fixed will be $y = x$ and $y = -x$. Thus 2 linearly independent eigenvectors will be $(1,\ 1)$ and $(-1, 1)$. The corresponding eigenvalues will be 1 and -1.

Again, this is easily checked using the matrix

$$A = \begin{bmatrix} 0 & 1 \\ 1 & 0 \end{bmatrix}$$

with characteristic equation $\lambda^2 - 1 = 0$.

(e) The only line through the origin which remains fixed when $k \neq 0$ is the y-axis. Since distance along the y-axis is not changed, the only possible eigenvalue is $\lambda = 1$. A corresponding eigenvector is $\mathbf{e}_2 = (0, 1)$.

(f) The only angles through which we can rotate R^2 and keep a line through the origin fixed are integer multiples of 0, π, and 2π. Since rotations by integer multiples of 0 and 2π leave the plane unchanged, we consider only rotations by odd multiples of π. These leave all lines through the origin fixed but reverse their directions. Thus such rotations have -1 as an eigenvalue, and $(1, 0)$ and $(0, 1)$ form a basis for the eigenspace.

EXERCISE SET 9.3

1. We have

$$\begin{bmatrix} a \\ b \end{bmatrix} = \left(\begin{bmatrix} 1 & 0 \\ 1 & 1 \\ 1 & 2 \end{bmatrix}^T \begin{bmatrix} 1 & 0 \\ 1 & 1 \\ 1 & 2 \end{bmatrix} \right)^{-1} \begin{bmatrix} 1 & 0 \\ 1 & 1 \\ 1 & 2 \end{bmatrix}^T \begin{bmatrix} 0 \\ 2 \\ 7 \end{bmatrix}$$

$$= \begin{bmatrix} 3 & 3 \\ 3 & 5 \end{bmatrix}^{-1} \begin{bmatrix} 9 \\ 16 \end{bmatrix} = \begin{bmatrix} \dfrac{5}{6} & -\dfrac{1}{2} \\ -\dfrac{1}{2} & \dfrac{1}{2} \end{bmatrix} \begin{bmatrix} 9 \\ 16 \end{bmatrix}$$

$$= \begin{bmatrix} -1/2 \\ 7/2 \end{bmatrix}$$

Thus the desired line is $y = -1/2 + (7/2)x$.

3. Here

$$M = \begin{bmatrix} 1 & 2 & 4 \\ 1 & 3 & 9 \\ 1 & 5 & 25 \\ 1 & 6 & 36 \end{bmatrix}$$

and

$$
\begin{bmatrix} a_1 \\ a_2 \\ a_3 \end{bmatrix} = (M^T M)^{-1} M^T \begin{bmatrix} 0 \\ -10 \\ -48 \\ -76 \end{bmatrix}
$$

$$
= \begin{bmatrix} 4 & 16 & 74 \\ 16 & 74 & 376 \\ 74 & 376 & 2018 \end{bmatrix}^{-1} \begin{bmatrix} -134 \\ -726 \\ -4026 \end{bmatrix}
$$

$$
= \begin{bmatrix} \dfrac{221}{10} & -\dfrac{62}{5} & \dfrac{3}{2} \\[2mm] -\dfrac{62}{5} & \dfrac{649}{90} & -\dfrac{8}{9} \\[2mm] \dfrac{3}{2} & -\dfrac{8}{9} & \dfrac{1}{9} \end{bmatrix} \begin{bmatrix} -134 \\ -726 \\ -4026 \end{bmatrix}
$$

$$
= \begin{bmatrix} 2 \\ 5 \\ -3 \end{bmatrix}
$$

Thus the desired quadratic is $y = 2 + 5x - 3x^2$.

5. The two column vectors of M are linearly independent if and only if neither is a nonzero multiple of the other. Since all of the entries in the first column are equal, the columns are linearly independent if and only if the second column has at least two different entries, or if and only if at least two of the numbers x_i are distinct.

6. Since we have

$$
M = \begin{bmatrix} 1 & x_1 & x_1^2 & \cdots & x_1^m \\ \vdots & \vdots & \vdots & & \vdots \\ 1 & x_n & x_n^2 & \cdots & x_n^m \end{bmatrix}
$$

the columns of M will be linearly independent if and only if $a_0 = a_1 = \cdots = a_m = 0$ is the only solution to the equation

$$a_0 \begin{bmatrix} 1 & \cdots & 1 \end{bmatrix}^T \mid a_1 \begin{bmatrix} x_1 & \cdots & x_n \end{bmatrix}^T + a_2 \begin{bmatrix} x_1^2 & \cdots & x_n^2 \end{bmatrix}^T + \cdots + a_m \begin{bmatrix} x_1^m & \cdots & x_n^m \end{bmatrix}^T = \begin{bmatrix} 0 & \cdots & 0 \end{bmatrix}^T$$

That is, if and only if the system of equations

$$a_0 + a_1 x_1 + a_2 x_1^2 + \cdots + a_m x_1^m = 0$$

$$\vdots \qquad \vdots \qquad \vdots \qquad \qquad \vdots \qquad \vdots$$

$$a_0 + a_1 x_n + a_2 x_n^2 + \cdots + a_m x_n^m = 0$$

has only the trivial solution for the numbers $a_0, \ldots, a_m$. But if any of the numbers $a_i \neq 0$, then there are at least $m + 1$ distinct numbers x_i which must satisfy the equation

$$a_0 + a_1 x + a_2 x^2 + \cdots + a_m x^m = 0$$

This is a polynomial of degree at most m and can therefore have at most m distinct real roots. Thus, indeed, $a_0 = \cdots = a_m = 0$.

8. Let x_i denote the month, starting with January $= 1$, and y_i denote the sales for that month. Then we have

$$\begin{bmatrix} a_0 \\ a_1 \\ a_2 \end{bmatrix} = \left(M^T M \right)^{-1} M^T \begin{bmatrix} 4.0 \\ 4.4 \\ 5.2 \\ 6.4 \\ 8.0 \end{bmatrix}$$

where

$$M = \begin{bmatrix} 1 & 1 & 1^2 \\ 1 & 2 & 2^2 \\ 1 & 3 & 3^2 \\ 1 & 4 & 4^2 \\ 1 & 5 & 5^2 \end{bmatrix} \qquad M^T M = \begin{bmatrix} 5 & 15 & 55 \\ 15 & 55 & 225 \\ 55 & 225 & 979 \end{bmatrix}$$

$$(M^T M)^{-1} = \begin{bmatrix} \dfrac{23}{5} & -\dfrac{33}{10} & \dfrac{1}{2} \\[2mm] -\dfrac{33}{10} & \dfrac{187}{70} & -\dfrac{3}{7} \\[2mm] \dfrac{1}{2} & -\dfrac{3}{7} & \dfrac{1}{14} \end{bmatrix} \quad \text{and} \quad M^T y = \begin{bmatrix} 28.0 \\ 94.0 \\ 370.8 \end{bmatrix}$$

so that the desired quadratic is $y = 4 - (.2)x + (.2)x^2$. When $x = 12$, $y = 30.4$, so the projected sales for December are $30,400.

EXERCISE SET 9.4

1. **(a)** Since $f(x) = 1 + x$, we have

$$a_0 = \frac{1}{\pi} \int_0^{2\pi} (1+x)dx = 2 + 2\pi$$

Using Example 1 and some simple integration, we obtain

$$a_k = \frac{1}{\pi} \int_0^{2\pi} (1+x)\cos(kx)dx = 0$$

$$k = 1, 2, \ldots$$

$$b_k = \frac{1}{\pi} \int_0^{2\pi} (1+x)\sin(kx)dx = -\frac{2}{k}$$

Thus, the least squares approximation to $1+x$ on $[0, 2\pi]$ by a trigonometric polynomial of order ≤ 2 is

$$1 + x \simeq (1+\pi) - 2\sin x - \sin 2x$$

2. **(a)** Since $f(x) = x^2$,

$$a_0 = \frac{1}{\pi} \int_0^{2\pi} x^2 dx = \frac{8}{3}\pi^2$$

Using Example 1 and integration by parts or a table of integrals, we find that

$$a_k = \frac{1}{\pi} \int_0^{2\pi} x^2 \cos(kx)dx = -\frac{2}{k\pi} \int_0^{2\pi} x\sin(kx)dx = \frac{4}{k^2}$$

$$k = 1, 2, \ldots$$

$$b_k = \frac{1}{\pi} \int_0^{2\pi} x^2 \sin(kx)dx = -\frac{(2\pi)^2}{k\pi} + \frac{2}{k\pi} \int_0^{2\pi} x\cos(kx)dx = -\frac{4\pi}{k}$$

Thus, the least squares approximation to x^2 on $[0, 2\pi]$ by a trigonometric polynomial of order at most 3 is

$$x^2 \simeq \frac{4}{3}\pi^2 + 4\cos x + \cos 2x + \frac{4}{9}\cos 3x - 4\pi\sin x - 2\pi\sin 2x - \frac{4\pi}{3}\sin 3x$$

3. (a) The space W of continuous functions of the form $a + be^x$ over $[0, 1]$ is spanned by the functions $\mathbf{u}_1 = 1$ and $\mathbf{u}_2 = e^x$. First we use the Gram-Schmidt process to find an orthonormal basis $\{\mathbf{v}_1, \mathbf{v}_2\}$ for W.

Since $\langle \mathbf{f}, \mathbf{g} \rangle = \displaystyle\int_0^1 f(x)g(x)dx$, then $\|\mathbf{u}_1\| = 1$ and hence

$$\mathbf{v}_1 = 1$$

Thus

$$\mathbf{v}_2 = \frac{e^x - \langle e^x, 1 \rangle 1}{\|e^x - \langle e^x, 1 \rangle 1\|} = \frac{e^x - e + 1}{\alpha}$$

where α is the constant

$$\alpha = \|e^x - e + 1\| = \left[\int_0^1 (e^x - e + 1)^2 dx \right]^{1/2}$$

$$= \left[\frac{(3 - e)(e - 1)}{2} \right]^{1/2}$$

Therefore the orthogonal projection of x on W is

$$\operatorname{proj}_W x = \langle x, 1 \rangle 1 + \left\langle x, \frac{e^x - e + 1}{\alpha} \right\rangle \frac{e^x - e + 1}{\alpha}$$

$$= \int_0^1 x dx + \frac{e^x - e + 1}{\alpha} \int_0^1 \frac{x(e^x - e + 1)}{\alpha} dx$$

$$= \frac{1}{2} + \frac{e^x - e + 1}{\alpha} \left[\frac{3 - e}{2\alpha} \right]$$

$$= \frac{1}{2} + \left[\frac{1}{e - 1} \right] (e^x - e + 1)$$

$$= -\frac{1}{2} + \left[\frac{1}{e - 1} \right] e^x$$

(b) The mean square error is

$$\int_0^1 \left[x - \left(-\frac{1}{2} + \left[\frac{1}{e-1} \right] e^x \right) \right]^2 dx = \frac{13}{12} + \frac{1+e}{2(1-e)} = \frac{1}{12} + \frac{3-e}{2(1-e)}$$

The answer above is deceptively short since a great many calculations are involved.

To shortcut some of the work, we derive a different expression for the mean square error (m.s.e.). By definition,

$$\text{m.s.e.} = \int_a^b [f(x) - g(x)]^2 dx$$

$$= \|\mathbf{f} - \mathbf{g}\|^2$$

$$= \langle \mathbf{f} - \mathbf{g}, \ \mathbf{f} - \mathbf{g} \rangle$$

$$= \langle \mathbf{f}, \ \mathbf{f} - \mathbf{g} \rangle - \langle \mathbf{g}, \ \mathbf{f} - \mathbf{g} \rangle$$

Recall that $\mathbf{g} = \text{proj}_W \mathbf{f}$, so that $\mathbf{g}$ and $\mathbf{f} - \mathbf{g}$ are orthogonal (see Figure 3). Therefore,

$$\text{m.s.e.} = \langle \mathbf{f}, \ \mathbf{f} - \mathbf{g} \rangle$$

$$= \langle \mathbf{f}, \mathbf{f} \rangle - \langle \mathbf{fg} \rangle$$

But $\mathbf{g} = \langle \mathbf{f}, \mathbf{v}_1 \rangle \mathbf{v}_1 + \langle \mathbf{f}, \mathbf{v}_2 \rangle \mathbf{v}_2$, so that

$(*)$
$$\text{m.s.e.} = \langle \mathbf{f}, \mathbf{f} \rangle - \langle \mathbf{f}, \mathbf{v}_1 \rangle^2 - \langle \mathbf{f}, \mathbf{v}_2 \rangle^2$$

Now back to the problem at hand. We know $\langle \mathbf{f}, \mathbf{v}_1 \rangle$ and $\langle \mathbf{f}, \mathbf{v}_2 \rangle$ from Part (a). Thus, in this case,

$$\text{m.s.e.} = \int_0^1 x^2 dx - \left(\frac{1}{2} \right)^2 - \left(\frac{3-e}{2\alpha} \right)^2$$

$$= \frac{1}{12} - \frac{3-e}{2(e-1)} \approx .0014$$

Clearly the formula $(*)$ above can be generalized. If W is an n-dimensional space with orthonormal basis $\{\mathbf{v}_1, \mathbf{v}_2, \ldots, \mathbf{v}_n\}$, then

$(**)$
$$\text{m.s.e.} = \| \mathbf{f} \|^2 - \langle \mathbf{f}, \mathbf{v}_1 \rangle^2 - \cdots - \langle \mathbf{f}, \mathbf{v}_n \rangle^2$$

4.   **(a)** The space W of polynomials of the form $a_0 + a_1 x$ over $[0, 1]$ has the basis $\{1, x\}$. The Gram-Schmidt process yields the orthonormal basis $\{1, \sqrt{3}(2x - 1)\}$. Therefore, the orthogonal projection of e^x on W is

$$\text{proj}_W e^x = \langle e^x, 1 \rangle 1 + \langle e^x, \sqrt{3}(2x - 1) \rangle \sqrt{3}(2x - 1)$$

$$= e - 1 + \sqrt{3}(3 - e)[\sqrt{3}(2x - 1)]$$

$$= (4e - 10) + (18 - 6e)x$$

(b) From Part (a) and ($*$) in the solution to 3(b), we have

$$\text{m.s.e.} = \int_0^1 e^{2x}\, dx - (e - 1)^2 - [\sqrt{3}(3 - e)]^2$$

$$= \frac{(3 - e)(7e - 19)}{2} \approx .004$$

5.   **(a)** The space W of polynomials of the form $a_0 + a_1 x + a_2 x^2$ over $[-1, 1]$ has the basis $\{1, x, x^2\}$. Using the inner product $\langle \mathbf{u}, \mathbf{v} \rangle = \displaystyle\int_{-1}^{1} u(x)v(x)\, dx$ and the Gram-Schmidt process, we obtain the orthonormal basis

$$\left(\frac{1}{\sqrt{2}},\ \sqrt{\frac{3}{2}}x,\ \frac{1}{2}\sqrt{\frac{5}{2}}(3x^2 - 1) \right)$$

(See Exercise 29, Section 6.3.) Thus

$$\langle \sin \pi x,\ \mathbf{v}_1 \rangle = \frac{1}{\sqrt{2}} \int_{-1}^{1} \sin(\pi x)\, dx = 0$$

$$\langle \sin \pi x,\ \mathbf{v}_2 \rangle = \sqrt{\frac{3}{2}} \int_{-1}^{1} x \sin(\pi x)\, dx = \frac{2}{\pi}\sqrt{\frac{3}{2}}$$

$$\langle \sin \pi x,\ \mathbf{v}_3 \rangle = \frac{1}{2}\sqrt{\frac{5}{2}} \int_{-1}^{1} (3x^2 - 1) \sin(\pi x)\, dx = 0$$

Therefore,

$$\sin \pi x \simeq \frac{3}{\pi}x$$

(b) From Part (a) and (**) in the solution to 3(b), we have

$$\text{m.s.e.} = \int_{-1}^{1} \sin^2(\pi x)dx - \frac{6}{\pi^2} = 1 - \frac{6}{\pi^2} \approx .39$$

8. The Fourier series of $\pi - x$ is

$$\frac{a_0}{2} + \sum_{k=1}^{\infty}(a_k \cos kx + b_k \sin kx)$$

where

$$a_0 = \frac{1}{\pi}\int_0^{2\pi}(\pi - x)dx = 0$$

and

$$a_k = \frac{1}{\pi}\int_0^{2\pi}(\pi - x)\cos(kx)dx = 0$$

$$k = 1, 2, \ldots$$

$$b_k = \frac{1}{\pi}\int_0^{2\pi}(\pi - x)\sin(kx)dx = \frac{2}{k}$$

Thus the required Fourier series is $\sum_{k=1}^{\infty}\frac{2}{k}\sin kx$. Notice that this is consistent with the result of Example 1, Part (b).

EXERCISE SET 9.5

1. **(d)** Since one of the terms in this expression is the product of 3 rather than 2 variables, the expression is not a quadratic form.

5. **(b)** The quadratic form can be written as

$$[x_1 \quad x_2] \begin{bmatrix} 7 & 1/2 \\ 1/2 & 4 \end{bmatrix} \begin{bmatrix} x_1 \\ x_2 \end{bmatrix} = \mathbf{x}^T A \mathbf{x}$$

The characteristic equation of A is $(\lambda - 7)(\lambda - 4) - 1/4 = 0$ or

$$4\lambda^2 - 44\lambda + 111 = 0$$

which gives us

$$\lambda = \frac{11 \pm \sqrt{10}}{2}$$

If we solve for the eigenvectors, we find that for $\lambda = \dfrac{11 + \sqrt{10}}{2}$

$$x_1 = (3 + \sqrt{10})x_2$$

and for $\lambda = \dfrac{11 - \sqrt{10}}{2}$

$$x_1 = (3 - \sqrt{10})x_2$$

Therefore the normalized eigenvectors are

$$\begin{bmatrix} \dfrac{3 + \sqrt{10}}{\sqrt{20 + 6\sqrt{10}}} \\ \dfrac{1}{\sqrt{20 + 6\sqrt{10}}} \end{bmatrix} \quad \text{and} \quad \begin{bmatrix} \dfrac{3 - \sqrt{10}}{\sqrt{20 - 6\sqrt{10}}} \\ \dfrac{1}{\sqrt{20 - 6\sqrt{10}}} \end{bmatrix}$$

or, if we simplify,

$$\begin{bmatrix} \dfrac{1}{\sqrt{20 - 6\sqrt{10}}} \\[3mm] \dfrac{1}{\sqrt{20 + 6\sqrt{10}}} \end{bmatrix} \quad \text{and} \quad \begin{bmatrix} \dfrac{-1}{\sqrt{20 + 6\sqrt{10}}} \\[3mm] \dfrac{1}{\sqrt{20 - 6\sqrt{10}}} \end{bmatrix}$$

Thus the maximum value of the given form with its constraint is

$$\frac{11 + \sqrt{10}}{2} \quad \text{at} \quad x_1 = \frac{1}{\sqrt{20 - 6\sqrt{10}}} \quad \text{and} \quad x_2 = \frac{1}{\sqrt{20 + 6\sqrt{10}}}$$

The minimum value is

$$\frac{11 - \sqrt{10}}{2} \quad \text{at} \quad x_1 = \frac{-1}{\sqrt{20 + 6\sqrt{10}}} \quad \text{and} \quad x_2 = \frac{1}{\sqrt{20 - 6\sqrt{10}}}$$

6. (b) We write the quadratic form as

$$\begin{bmatrix} x_1 & x_2 & x_3 \end{bmatrix} \begin{bmatrix} 2 & 1 & 1 \\ 1 & 1 & 0 \\ 1 & 0 & 1 \end{bmatrix} \begin{bmatrix} x_1 \\ x_2 \\ x_3 \end{bmatrix} = \mathbf{x}^T A \mathbf{x}$$

The characteristic equation of A is

$$\lambda^3 - 4\lambda^2 + 3\lambda = \lambda(\lambda - 3)(\lambda - 1) = 0$$

with roots $\lambda = 0, 1, 3$. The corresponding normalized eigenvectors are

$$\begin{bmatrix} 1/\sqrt{3} \\ -1/\sqrt{3} \\ -1/\sqrt{3} \end{bmatrix} \quad \begin{bmatrix} 0 \\ 1/\sqrt{2} \\ -1/\sqrt{2} \end{bmatrix} \quad \begin{bmatrix} 2/\sqrt{6} \\ 1/\sqrt{6} \\ 1/\sqrt{6} \end{bmatrix}$$

respectively. Thus the maximum value of the given form with its constraint is

$$3 \text{ at } x_1 = 2/\sqrt{6}, \ x_2 = 1/\sqrt{6}, \ x_3 = 1/\sqrt{6}$$

The minimum value is

$$0 \text{ at } x_1 = 1/\sqrt{3}, \ x_2 = -1/\sqrt{3}, \ x_3 = -1/\sqrt{3}$$

7. **(b)** The eigenvalues of this matrix are the roots of the equation $\lambda^2 - 10\lambda + 24 = 0$. They are $\lambda = 6$ and $\lambda = 4$ which are both positive. Therefore the matrix is positive definite.

8. **(b)** The principal submatrices are

$$[5] \quad \text{and} \quad \begin{bmatrix} 5 & -1 \\ -1 & 5 \end{bmatrix}$$

Their determinants are 5 and 24, so the matrix is positive definite.

9. **(b)** The characteristic equation of this matrix is $\lambda^3 - 3\lambda + 2 = (\lambda + 1)^2(\lambda - 2)$. Since two of the eigenvalues are negative, the matrix is not positive definite.

11. **(a)** Since $x_1^2 + x_2^2 > 0$ unless $x_1 = x_2 = 0$, the form is positive definite.

(c) Since $(x_1 - x_2)^2 \geq 0$, the form is positive semidefinite. It is not positive definite because it can equal zero whenever $x_1 = x_2$ even when $x_1 = x_2 \neq 0$.

(e) If $|x_1| > |x_2|$, then the form has a positive value, but if $|x_1| < |x_2|$, then the form has a negative value. Thus it is indefinite.

12. **(a)** The related quadratic form is $3x_1^2 - 2x_2^2 + x_3^2$ which is indefinite since its values can be both positive and negative. Hence the matrix is indefinite.

(c) The related quadratic form is $6x_1^2 + 9x_2^2 + x_3^2 + 14x_1x_2 + 2x_1x_3 + 4x_2x_3$. The possible signs of the values which this form can assume are not obvious and its characteristic equation is rather messy. However, the determinant of the matrix is zero while the determinants of the remaining principal submatrices are positive. Hence the matrix is positive semidefinite.

13. **(a)** By definition,

$$T(\mathbf{x}+\mathbf{y}) = (\mathbf{x}+\mathbf{y})^T A(\mathbf{x}+\mathbf{y})$$

$$= (\mathbf{x}^T + \mathbf{y}^T)A(\mathbf{x}+\mathbf{y})$$

$$= \mathbf{x}^T A\mathbf{x} + \mathbf{x}^T A\mathbf{y} + \mathbf{y}^T A\mathbf{x} + \mathbf{y}^T A\mathbf{y}$$

$$= T(\mathbf{x}) + \mathbf{x}^T A\mathbf{y} + \left(\mathbf{x}^T A^T \mathbf{y}\right)^T + T(\mathbf{y})$$

$$= T(\mathbf{x}) + \mathbf{x}^T A\mathbf{y} + \mathbf{x}^T A^T \mathbf{y} + T(\mathbf{y})$$

(The transpose of a 1×1 matrix is itself.)

$$= T(\mathbf{x}) + 2\mathbf{x}^T A\mathbf{y} + T(\mathbf{y})$$

(Assuming that A is symmetric, $A^T = A$.)

(b) We have

$$T(k\mathbf{x}) = (k\mathbf{x})^T A(k\mathbf{x})$$

$$= k^2 \mathbf{x}^T A\mathbf{x} \qquad \text{(Every term has a factor of } k^2.)$$

$$= k^2 T(\mathbf{x})$$

(c) The transformation is not linear because $T(k\mathbf{x}) \neq kT(\mathbf{x})$ unless $k = 0$ or 1 by Part (b).

14. **(b)** The symmetric matrix associated with the quadratic form is

$$\begin{bmatrix} 5 & 2 & -1 \\ 2 & 1 & -1 \\ -1 & -1 & k \end{bmatrix}$$

The determinants of the principal submatrices are 5, 1, and $k - 2$. These are positive provided $k > 2$. Thus the quadratic form is positive definite if and only if $k > 2$.

15. If we expand the quadratic form, it becomes

$$c_1^2 x_1^2 + c_2^2 x_2^2 + \cdots + c_n^2 x_n^2 + 2c_1 c_2 x_1 x_2 + 2c_1 c_3 x_1 x_3 +$$

$$\cdots + 2c_1 c_n x_1 x_n + 2c_2 c_3 x_2 x_3 + \cdots + 2c_{n-1} c_n x_{n-1} x_n$$

Thus we have

$$
A = \begin{bmatrix}
c_1^2 & c_1 c_2 & c_1 c_3 & \cdots & c_1 c_n \\
c_1 c_2 & c_2^2 & c_2 c_3 & \cdots & c_2 c_n \\
c_1 c_3 & c_2 c_3 & c_3^2 & \cdots & c_3 c_n \\
\vdots & \vdots & \vdots & & \vdots \\
c_1 c_n & c_2 c_n & c_3 c_n & \cdots & c_n^2
\end{bmatrix}
$$

and the quadratic form is given by $\mathbf{x}^T A \mathbf{x}$ where $\mathbf{x} = [x_1 \ x_2 \ \cdots \ x_n]^T$.

16. **(a)** We have

$$
s_{\mathbf{x}}^2 = \frac{1}{n-1}\left[x_1^2 + x_2^2 + \cdots + x_n^2 - 2\bar{x}\left(x_1 + \cdots + x_n\right) + n\bar{x}^2 \right]
$$

$$
= \frac{1}{n-1}\left[x_1^2 + \cdots + x_n^2 - \frac{2}{n}\left(x_1 + \cdots + x_n\right)^2 + \frac{1}{n}\left(x_1 + \cdots + x_n\right)^2 \right]
$$

$$
= \frac{1}{n-1}\left[x_1^2 + \cdots + x_n^2 - \frac{1}{n}\left(x_1^2 + \cdots + x_n^2 + 2\left(x_1 x_2 + \cdots + x_{n-1} x_n\right)\right) \right]
$$

$$
= \frac{1}{n-1}\left[\frac{n-1}{n}\left(x_1^2 + \cdots + x_n^2\right) - \frac{2}{n}\left(x_1 x_2 + \cdots + x_{n-1} x_n\right) \right]
$$

Thus $s_{\mathbf{x}}^2 = \mathbf{x}^T A \mathbf{x}$ where

$$
A = \begin{bmatrix}
\dfrac{1}{n} & \dfrac{-1}{n(n-1)} & \dfrac{-1}{n(n-1)} & \cdots & \dfrac{-1}{n(n-1)} \\[2ex]
\dfrac{-1}{n(n-1)} & \dfrac{1}{n} & \dfrac{-1}{n(n-1)} & \cdots & \dfrac{-1}{n(n-1)} \\[2ex]
\vdots & \vdots & \vdots & & \vdots \\[2ex]
\dfrac{-1}{n(n-1)} & \dfrac{-1}{n(n-1)} & \dfrac{-1}{n(n-1)} & \cdots & \dfrac{1}{n}
\end{bmatrix}
$$

(b) Since $s_{\mathbf{x}}^2$ is $1/(n-1)$ times a sum of squares, it cannot be negative. In fact, it can only be zero if $(x_i - \bar{x}) = 0$ for $i = 1, \ldots, n$, which can only be true in case $x_1 = x_2 = \cdots = x_n$. If we let all the values of x_i be equal but not zero, we have that $s_{\mathbf{x}}^2 = 0$ while $\mathbf{x} \neq \mathbf{0}$. Thus the form is positive semidefinite.

17. To show that $\lambda_n \leq \mathbf{x}^T A \mathbf{x}$ if $\|\mathbf{x}\| = 1$, we use the equation from the proof dealing with λ_1 and the fact that λ_n is the smallest eigenvalue. This gives

$$\mathbf{x}^T A \mathbf{x} = \langle \mathbf{x}, A\mathbf{x} \rangle = \lambda_1 \langle \mathbf{x}, \mathbf{v}_1 \rangle^2 + \lambda_2 \langle \mathbf{x}, \mathbf{v}_2 \rangle^2 + \cdots + \lambda_n \langle \mathbf{x}, \mathbf{v}_n \rangle^2$$

$$\geq \lambda_n \langle \mathbf{x}, \mathbf{v}_1 \rangle^2 + \lambda_n \langle \mathbf{x}, \mathbf{v}_2 \rangle^2 + \cdots + \lambda_n \langle \mathbf{x}, \mathbf{v}_n \rangle^2$$

$$= \lambda_n \left(\langle \mathbf{x}, \mathbf{v}_1 \rangle^2 + \cdots + \langle \mathbf{x}, \mathbf{v}_n \rangle^2 \right)$$

$$= \lambda_n$$

Now suppose that $\mathbf{x}$ is an eigenvector of A corresponding to λ_n. As in the proof dealing with λ_1, we have

$$\mathbf{x}^T A \mathbf{x} = \langle \mathbf{x}, A\mathbf{x} \rangle = \langle \mathbf{x}, \lambda_n \mathbf{x} \rangle = \lambda_n \langle \mathbf{x}, \mathbf{x} \rangle = \lambda_n \|\mathbf{x}\|^2 = \lambda_n$$

EXERCISE SET 9.6

1. **(a)** The quadratic form $\mathbf{x}^T A \mathbf{x}$ can be written as

$$[x_1 \quad x_2] \begin{bmatrix} 2 & -1 \\ -1 & 2 \end{bmatrix} \begin{bmatrix} x_1 \\ x_2 \end{bmatrix}$$

The characteristic equation of A is $\lambda^2 - 4\lambda + 3 = 0$. The eigenvalues are $\lambda = 3$ and $\lambda = 1$. The corresponding eigenspaces are spanned by the vectors

$$\begin{bmatrix} 1 \\ -1 \end{bmatrix} \quad \text{and} \quad \begin{bmatrix} 1 \\ 1 \end{bmatrix}$$

respectively. These vectors are orthogonal. If we normalize them, we can use the result to obtain a matrix P such that the substitution $\mathbf{x} = P\mathbf{y}$ or

$$\begin{bmatrix} x_1 \\ x_2 \end{bmatrix} = \begin{bmatrix} 1/\sqrt{2} & 1/\sqrt{2} \\ -1/\sqrt{2} & 1/\sqrt{2} \end{bmatrix} \begin{bmatrix} y_1 \\ y_2 \end{bmatrix}$$

will eliminate the cross-product term. This yields the new quadratic form

$$[y_1 \quad y_2] \begin{bmatrix} 3 & 0 \\ 0 & 1 \end{bmatrix} \begin{bmatrix} y_1 \\ y_2 \end{bmatrix}$$

or $3y_1^2 + y_2^2$.

2. **(a)** The quadratic form can be written as $\mathbf{x}^T A \mathbf{x}$ where

$$\mathbf{x}^T = [x_1 \quad x_2 \quad x_3]$$

and

$$A = \begin{bmatrix} 3 & 2 & 0 \\ 2 & 4 & -2 \\ 0 & -2 & 5 \end{bmatrix}$$

The characteristic equation of A is

$$\lambda^3 - 12\lambda^2 + 39\lambda - 28 = (\lambda - 7)(\lambda - 4)(\lambda - 1) = 0$$

The eigenspaces corresponding to $\lambda = 7$, $\lambda = 4$, and $\lambda = 1$ are spanned by the vectors

$$\begin{bmatrix} 1 \\ 2 \\ -2 \end{bmatrix} \qquad \begin{bmatrix} 2 \\ 1 \\ 2 \end{bmatrix} \qquad \begin{bmatrix} 2 \\ -2 \\ -1 \end{bmatrix}$$

respectively. If we normalize these vectors, we obtain the matrix

$$P = \begin{bmatrix} 1/3 & 2/3 & 2/3 \\ 2/3 & 1/3 & -2/3 \\ -2/3 & 2/3 & -1/3 \end{bmatrix}$$

By direct calculation, we have

$$P^T A P = \begin{bmatrix} 7 & 0 & 0 \\ 0 & 4 & 0 \\ 0 & 0 & 1 \end{bmatrix} = B$$

so the substitution $\mathbf{x} = P\mathbf{y}$ will yield the quadratic form

$$\mathbf{y}^T B \mathbf{y} \qquad \text{or} \qquad 7y_1^2 + 4y_2^2 + y_3^2$$

7. (a) If we complete the squares, then the equation $9x^2 + 4y^2 - 36x - 24y + 36 = 0$ becomes

$$9(x^2 - 4x + 4) + 4(y^2 - 6y + 9) = -36 + 9(4) + 4(9)$$

or

$$9(x - 2)^2 + 4(y - 3)^2 = 36$$

or

$$\frac{(x')^2}{4} + \frac{(y')^2}{9} = 1$$

This is an ellipse.

(c) If we complete the square, then $y^2 - 8x - 14y + 49 = 0$ becomes

$$y^2 - 14y + 49 = 8x$$

or

$$(y - 7)^2 = 8x$$

This is the parabola $(y')^2 = 8x'$.

(e) If we complete the squares, then $2x^2 - 3y^2 + 6x + 20y = -41$ becomes

$$2\left(x^2 + 3x + \frac{9}{4}\right) - 3\left(y^2 - \frac{20}{3}y + \frac{100}{9}\right) = -41 + \frac{9}{2} - \frac{100}{3}$$

or

$$2\left(x + \frac{3}{2}\right)^2 - 3\left(y - \frac{10}{3}\right)^2 = -\frac{419}{6}$$

or

$$12(x')^2 - 18(y')^2 = -419$$

This is the hyperbola

$$\frac{(y')^2}{\frac{419}{18}} - \frac{(x')^2}{\frac{419}{12}} = 1$$

8. **(a)** The matrix form for the conic $2x^2 - 4xy - y^2 + 8 = 0$ is

$(*)$ $\qquad\qquad\qquad \mathbf{x}^T A \mathbf{x} + 8 = 0$

where

$$A = \begin{bmatrix} 2 & -2 \\ -2 & -1 \end{bmatrix}$$

The eigenvalues of A are $\lambda_1 = 3$ and $\lambda_2 = -2$ and the eigenspaces are spanned by the vectors

$$\begin{bmatrix} -2 \\ 1 \end{bmatrix} \quad \text{and} \quad \begin{bmatrix} 1 \\ 2 \end{bmatrix}$$

respectively. Normalizing these vectors yields

$$\begin{bmatrix} \dfrac{-2}{\sqrt{5}} \\ \dfrac{1}{\sqrt{5}} \end{bmatrix} \quad \text{and} \quad \begin{bmatrix} \dfrac{1}{\sqrt{5}} \\ \dfrac{2}{\sqrt{5}} \end{bmatrix}$$

We choose

$$P = \begin{bmatrix} \dfrac{1}{\sqrt{5}} & \dfrac{-2}{\sqrt{5}} \\ \dfrac{2}{\sqrt{5}} & \dfrac{1}{\sqrt{5}} \end{bmatrix}$$

Note that $\det(P) = 1$. (Interchanging columns yields a matrix with determinant -1.) If we substitute $\mathbf{x} = P\mathbf{x}'$ into $(*)$, we have

$$(P\mathbf{x}')^T A (P\mathbf{x}') + 8 = 0$$

or

$$(\mathbf{x}')^T (P^T A P)\mathbf{x}' + 8 = 0$$

Since

$$P^T A P = \begin{bmatrix} -2 & 0 \\ 0 & 3 \end{bmatrix}$$

this yields the hyperbola

$$-2(x')^2 + 3(y')^2 + 8 = 0$$

Note that if we had let

$$P = \begin{bmatrix} \dfrac{2}{\sqrt{5}} & \dfrac{1}{\sqrt{5}} \\[3mm] \dfrac{-1}{\sqrt{5}} & \dfrac{2}{\sqrt{5}} \end{bmatrix}$$

then $\det(P) = 1$ and the rotation would have produced the diagonal matrix

$$P^T A P = \begin{bmatrix} 3 & 0 \\ 0 & -2 \end{bmatrix}$$

and hence the hyperbola

$$3(x')^2 - 2(y')^2 + 8 = 0$$

That is, there are two different rotations which will put the hyperbola into the two different standard positions.

8.   **(b)**   The matrix form for the conic $5x^2 + 4xy + 5y^2 = 9$ is

$$\mathbf{x}^T A \mathbf{x} = 9$$

where

$$A = \begin{bmatrix} 5 & 2 \\ 2 & 5 \end{bmatrix}$$

The eigenvalues of A are $\lambda_1 = 3$ and $\lambda_2 = 7$ and the eigenspaces are spanned by the vectors

$$\begin{bmatrix} -1 \\ 1 \end{bmatrix} \quad \text{and} \quad \begin{bmatrix} 1 \\ 1 \end{bmatrix}$$

If we normalize these vectors, we obtain

$$\begin{bmatrix} \dfrac{-1}{\sqrt{2}} \\[3mm] \dfrac{1}{\sqrt{2}} \end{bmatrix} \quad \text{and} \quad \begin{bmatrix} \dfrac{1}{\sqrt{2}} \\[3mm] \dfrac{1}{\sqrt{2}} \end{bmatrix}$$

Hence we can let

$$P = \begin{bmatrix} \dfrac{1}{\sqrt{2}} & \dfrac{-1}{\sqrt{2}} \\[2mm] \dfrac{1}{\sqrt{2}} & \dfrac{1}{\sqrt{2}} \end{bmatrix}$$

Note that $\det(P) = 1$. Thus if we let $\mathbf{x} = P\mathbf{x}'$, then we have

$$(\mathbf{x}')^T (P^T A P)\mathbf{x}' = 9$$

where

$$P^T A P = \begin{bmatrix} 7 & 0 \\ 0 & 3 \end{bmatrix}$$

This yields the ellipse,

$$7(x')^2 + 3(y')^2 = 9$$

Had we let

$$P = \begin{bmatrix} \dfrac{1}{\sqrt{2}} & \dfrac{1}{\sqrt{2}} \\[2mm] \dfrac{-1}{\sqrt{2}} & \dfrac{1}{\sqrt{2}} \end{bmatrix}$$

then we would have obtained

$$3(x')^2 + 7(y')^2 = 9$$

which is the same ellipse rotated $90°$.

9.   The matrix form for the conic $9x^2 - 4xy + 6y^2 - 10x - 20y = 5$ is

$$\mathbf{x}^T A \mathbf{x} + K\mathbf{x} = 5$$

where

$$A = \begin{bmatrix} 9 & -2 \\ -2 & 6 \end{bmatrix} \qquad \text{and} \qquad K = [\,-10 \quad -20\,]$$

The eigenvalues of A are $\lambda_1 = 5$ and $\lambda_2 = 10$ and the eigenspaces are spanned by the vectors

$$\begin{bmatrix} 1 \\ 2 \end{bmatrix} \quad \text{and} \quad \begin{bmatrix} -2 \\ 1 \end{bmatrix}$$

Thus we can let

$$P = \begin{bmatrix} \dfrac{1}{\sqrt{5}} & \dfrac{-2}{\sqrt{5}} \\[2mm] \dfrac{2}{\sqrt{5}} & \dfrac{1}{\sqrt{5}} \end{bmatrix}$$

Note that $\det(P) = 1$. If we let $\mathbf{x} = P\mathbf{x}'$, then

$$(\mathbf{x}')^T (P^T AP)\mathbf{x}' + KP\mathbf{x}' = 5$$

where

$$P^T AP = \begin{bmatrix} 5 & 0 \\ 0 & 10 \end{bmatrix} \quad \text{and} \quad KP = \begin{bmatrix} -10\sqrt{5} & 0 \end{bmatrix}$$

Thus we have the equation

$$5(x')^2 + 10(y')^2 - 10\sqrt{5}x' = 5$$

If we complete the square, we obtain the equation

$$5\left((x')^2 - 2\sqrt{5}x' + 5\right) + 10(y')^2 = 5 + 25$$

or

$$(x'')^2 + 2(y'')^2 = 6$$

where $x'' = x' - \sqrt{5}$ and $y'' = y'$. This is the ellipse

$$\frac{(x'')^2}{6} + \frac{(y'')^2}{3} = 1$$

Of course we could also rotate to obtain the same ellipse in the form $2(x'')^2 + (y'')^2 = 6$, which is just the other standard position.

11. The matrix form for the conic $2x^2 - 4xy - y^2 - 4x - 8y = -14$ is

$$\mathbf{x}^T A \mathbf{x} + K \mathbf{x} = -14$$

where

$$A = \begin{bmatrix} 2 & -2 \\ -2 & -1 \end{bmatrix} \quad \text{and} \quad K = \begin{bmatrix} -4 & -8 \end{bmatrix}$$

The eigenvalues of A are $\lambda_1 = 3$, $\lambda_2 = -2$ and the eigenspaces are spanned by the vectors

$$\begin{bmatrix} -2 \\ 1 \end{bmatrix} \quad \text{and} \quad \begin{bmatrix} 1 \\ 2 \end{bmatrix}$$

Thus we can let

$$P = \begin{bmatrix} \dfrac{2}{\sqrt{5}} & \dfrac{1}{\sqrt{5}} \\ \dfrac{-1}{\sqrt{5}} & \dfrac{2}{\sqrt{5}} \end{bmatrix}$$

Note that $\det(P) = 1$. If we let $\mathbf{x} = P\mathbf{x}'$, then

$$(\mathbf{x}')^T (P^T A P)\mathbf{x}' + KP\mathbf{x}' = -14$$

where

$$P^T A P = \begin{bmatrix} 3 & 0 \\ 0 & -2 \end{bmatrix} \quad \text{and} \quad KP = \begin{bmatrix} 0 & -4\sqrt{5} \end{bmatrix}$$

Thus we have the equation

$$3(x')^2 - 2(y')^2 - 4\sqrt{5}y' = -14$$

If we complete the square, then we obtain

$$3(x')^2 - 2\left((y')^2 + 2\sqrt{5}y' + 5\right) = -14 - 10$$

or

$$3(x'')^2 - 2(y'')^2 = -24$$

where $x'' = x'$ and $y'' = y' + \sqrt{5}$. This is the hyperbola

$$\frac{(y'')^2}{12} - \frac{(x'')^2}{8} = 1$$

We could also rotate to obtain the same hyperbola in the form $2(x'')^2 - 3(y'')^2 = 24$.

14. The matrix form for the conic $4x^2 - 20xy + 25y^2 - 15x - 6y = 0$ is

$$\mathbf{x}^T A\mathbf{x} + K\mathbf{x} = 0$$

where

$$A = \begin{bmatrix} 4 & -10 \\ -10 & 25 \end{bmatrix} \quad \text{and} \quad K = \begin{bmatrix} -15 & -6 \end{bmatrix}$$

The eigenvalues of A are $\lambda_1 = 29$ and $\lambda_2 = 0$ and the eigenspaces are spanned by the vectors

$$\begin{bmatrix} -2 \\ 5 \end{bmatrix} \quad \text{and} \quad \begin{bmatrix} 5 \\ 2 \end{bmatrix}$$

Thus we can let

$$P = \begin{bmatrix} \dfrac{2}{\sqrt{29}} & \dfrac{5}{\sqrt{29}} \\ \dfrac{-5}{\sqrt{29}} & \dfrac{2}{\sqrt{29}} \end{bmatrix}$$

Note that $\det(P) = 1$. If we let $\mathbf{x} = P\mathbf{x}'$, then

$$(\mathbf{x}')^T (P^T AP)\mathbf{x}' + KP\mathbf{x}' = 0$$

where

$$P^T AP = \begin{bmatrix} 29 & 0 \\ 0 & 0 \end{bmatrix} \quad \text{and} \quad KP = \begin{bmatrix} 0 & -87/\sqrt{29} \end{bmatrix}$$

Thus we have the equation

$$29(x')^2 - \frac{87}{\sqrt{29}}y' = 0$$

which is the parabola $(x')^2 = \frac{3}{\sqrt{29}}y'$. The other standard positions for this parabola are represented by the equations

$$(x')^2 = -\frac{3}{\sqrt{29}}y' \quad \text{and} \quad (y')^2 = \pm\frac{3}{\sqrt{29}}x'$$

15. **(a)** The equation $x^2 - y^2 = 0$ can be written as $(x-y)(x+y) = 0$. Thus it represents the two intersecting lines $x \pm y = 0$.

(b) The equation $x^2 + 3y^2 + 7 = 0$ can be written as $x^2 + 3y^2 = -7$. Since the left side of this equation cannot be negative, then there are no points (x, y) which satisfy the equation.

(c) If $8x^2 + 7y^2 = 0$, then $x = y = 0$. Thus the graph consists of the single point $(0, 0)$.

(d) This equation can be rewritten as $(x - y)^2 = 0$. Thus it represents the single line $y = x$.

(e) The equation $9x^2 + 12xy + 4y^2 - 52 = 0$ can be written as $(3x + 2y)^2 = 52$ or $3x + 2y = \pm\sqrt{52}$. Thus its graph is the two parallel lines $3x + 2y \pm 2\sqrt{13} = 0$.

(f) The equation $x^2 + y^2 - 2x - 4y = -5$ can be written as $x^2 - 2x + 1 + y^2 - 4y + 4 = 0$ or $(x - 1)^2 + (y - 2)^2 = 0$. Thus it represents the point $(1,2)$.

EXERCISE SET 9.7

5. (a) If we complete the squares, the quadric becomes

$$9(x^2 - 2x + 1) + 36(y^2 - 4y + 4) + 4(z^2 - 6z + 9)$$

$$= -153 + 9 + 144 + 36$$

or

$$9(x - 1)^2 + 36(y - 2)^2 + 4(z - 3)^2 = 36$$

or

$$\frac{(x')^2}{4} + \frac{(y')^2}{1} + \frac{(z')^2}{9} = 1$$

This is an ellipsoid.

(c) If we complete the square, the quadric becomes

$$3(x^2 + 14x + 49) - 3y^2 - z^2 = -144 + 147$$

or

$$3(x + 7)^2 - 3y^2 - z^2 = 3$$

or

$$\frac{(x')^2}{1} - \frac{(y')^2}{1} - \frac{(z')^2}{3} = 1$$

This is a hyperboloid of two sheets.

5. **(e)** If we complete the squares, the quadric becomes

$$(x^2 + 2x + 1) + 16(y^2 - 2y + 1) - 16z = 15 + 1 + 16$$

or

$$(x + 1)^2 + 16(y - 1)^2 - 16(z + 2) = 0$$

or

$$\frac{(x')^2}{16} + \frac{(y')^2}{1} - z' = 0$$

This is an elliptic paraboloid.

(g) If we complete the squares, the quadric becomes

$$(x^2 - 2x + 1) + (y^2 + 4y + 4) + (z^2 - 6z + 9) = 11 + 1 + 4 + 9$$

or

$$(x - 1)^2 + (y + 2)^2 + (z - 3)^2 = 25$$

or

$$\frac{(x')^2}{25} + \frac{(y')^2}{25} + \frac{(z')^2}{25} = 1$$

This is a sphere.

6. **(a)** The matrix form for the quadric is $\mathbf{x}^T A \mathbf{x} + 150 = 0$ where

$$A = \begin{bmatrix} 2 & 0 & 36 \\ 0 & 3 & 0 \\ 36 & 0 & 23 \end{bmatrix}$$

The eigenvalues of A are $\lambda_1 = -25$, $\lambda_2 = 3$, and $\lambda_3 = 50$. The corresponding eigenspaces are spanned by the orthogonal vectors

$$\begin{bmatrix} 4 \\ 0 \\ -3 \end{bmatrix} \quad \begin{bmatrix} 0 \\ 1 \\ 0 \end{bmatrix} \quad \begin{bmatrix} 3 \\ 0 \\ 4 \end{bmatrix}$$

Thus we can let $\mathbf{x} = P\mathbf{x}'$ where

$$P = \begin{bmatrix} 4/5 & 0 & 3/5 \\ 0 & 1 & 0 \\ -3/5 & 0 & 4/5 \end{bmatrix} \quad \text{and} \quad P^T A P = \begin{bmatrix} -25 & 0 & 0 \\ 0 & 3 & 0 \\ 0 & 0 & 50 \end{bmatrix}$$

Note that $\det(P) = 1$. (This is the only reason for actually calculating P in this problem.) Using the above transformation, we obtain the quadric

$$-25(x')^2 + 3(y')^2 + 50(z')^2 + 150 = 0$$

or

$$25(x')^2 - 3(y')^2 - 50(z')^2 - 150 = 0$$

This is a hyperboloid of two sheets.

(c) The matrix form for the quadric is $\mathbf{x}^T A\mathbf{x} + K\mathbf{x} = 0$ where

$$A = \begin{bmatrix} 144 & 0 & -108 \\ 0 & 100 & 0 \\ -108 & 0 & 81 \end{bmatrix} \quad \text{and} \quad K = \begin{bmatrix} -540 & 0 & -720 \end{bmatrix}$$

The eigenvalues of A are $\lambda_1 = 225$, $\lambda_2 = 100$, and $\lambda_3 = 0$. The eigenspaces are spanned by the vectors

$$\begin{bmatrix} 4 \\ 0 \\ -3 \end{bmatrix} \quad \begin{bmatrix} 0 \\ 1 \\ 0 \end{bmatrix} \quad \begin{bmatrix} 3 \\ 0 \\ 4 \end{bmatrix}$$

Thus we can let $\mathbf{x} = P\mathbf{x}'$ where

$$P = \begin{bmatrix} 4/5 & 0 & 3/5 \\ 0 & 1 & 0 \\ -3/5 & 0 & 4/5 \end{bmatrix} \quad \text{and} \quad P^T A P = \begin{bmatrix} 225 & 0 & 0 \\ 0 & 100 & 0 \\ 0 & 0 & 0 \end{bmatrix}$$

Note that $\det(P) = 1$. Hence the quadric becomes

$$\mathbf{x}^T (P^T A P)\mathbf{x} + K P\mathbf{x}' = 0$$

or

$$225(x')^2 + 100(y')^2 - 900z' = 0$$

This is the elliptic paraboloid

$$9(x')^2 + 4(y')^2 - 36z' = 0$$

7.   The matrix form for the quadric is $\mathbf{x}^T A\mathbf{x} + K\mathbf{x} = -9$ where

$$A = \begin{bmatrix} 0 & 1 & 1 \\ 1 & 0 & 1 \\ 1 & 1 & 0 \end{bmatrix} \quad \text{and} \quad K = \begin{bmatrix} -6 & -6 & -4 \end{bmatrix}$$

The eigenvalues of A are $\lambda_1 = \lambda_2 = -1$ and $\lambda_3 = 2$, and the vectors

$$\mathbf{u}_1 = \begin{bmatrix} -1 \\ 0 \\ 1 \end{bmatrix} \qquad \mathbf{u}_2 = \begin{bmatrix} -1 \\ 1 \\ 0 \end{bmatrix} \quad \text{and} \quad \mathbf{u}_3 = \begin{bmatrix} 1 \\ 1 \\ 1 \end{bmatrix}$$

span the corresponding eigenspaces. Note that $\mathbf{u}_1 \cdot \mathbf{u}_3 = \mathbf{u}_2 \cdot \mathbf{u}_3 = 0$ but that $\mathbf{u}_1 \cdot \mathbf{u}_2 \neq 0$. Hence, we must apply the Gram-Schmidt process to $\{\mathbf{u}_1, \mathbf{u}_2\}$. We must also normalize $\mathbf{u}_3$. This gives the orthonormal set

$$\begin{bmatrix} \dfrac{-1}{\sqrt{2}} \\ 0 \\ \dfrac{1}{\sqrt{2}} \end{bmatrix} \qquad \begin{bmatrix} \dfrac{-1}{\sqrt{6}} \\ \dfrac{2}{\sqrt{6}} \\ \dfrac{-1}{\sqrt{6}} \end{bmatrix} \qquad \begin{bmatrix} \dfrac{1}{\sqrt{3}} \\ \dfrac{1}{\sqrt{3}} \\ \dfrac{-1}{\sqrt{3}} \end{bmatrix}$$

Thus we can let

$$P = \begin{bmatrix} \dfrac{1}{\sqrt{2}} & \dfrac{-1}{\sqrt{6}} & \dfrac{1}{\sqrt{3}} \\ 0 & \dfrac{2}{\sqrt{6}} & \dfrac{1}{\sqrt{3}} \\ \dfrac{-1}{\sqrt{2}} & \dfrac{-1}{\sqrt{6}} & \dfrac{1}{\sqrt{3}} \end{bmatrix}$$

Note that $\det(P) = 1$,

$$P^T AP = \begin{bmatrix} -1 & 0 & 0 \\ 0 & -1 & 0 \\ 0 & 0 & 2 \end{bmatrix} \quad \text{and} \quad KP = \begin{bmatrix} -\sqrt{2} & \dfrac{-2}{\sqrt{6}} & \dfrac{-16}{\sqrt{3}} \end{bmatrix}$$

Therefore the transformation $\mathbf{x} = P\mathbf{x}'$ reduces the quadric to

$$-(x')^2 - (y')^2 + 2(z')^2 - \sqrt{2}\,x' - \frac{2}{\sqrt{6}}y' - \frac{16}{\sqrt{3}}z' = -9$$

If we complete the squares, this becomes

$$\left[(x')^2 + \sqrt{2}\,x' + \frac{1}{2}\right] + \left[(y')^2 + \frac{2}{\sqrt{6}}y' + \frac{1}{6}\right]$$

$$- 2\left[(z')^2 - \frac{8}{\sqrt{3}}z' + \frac{16}{3}\right] = 9 + \frac{1}{2} + \frac{1}{6} - \frac{32}{3}$$

Letting $x'' = x' + \dfrac{1}{\sqrt{2}}$, $y'' = y' + \dfrac{1}{\sqrt{6}}$, and $z'' = z' - \dfrac{4}{\sqrt{3}}$ yields

$$(x'')^2 + (y'')^2 - 2(z'')^2 = -1$$

This is the hyperboloid of two sheets

$$\frac{(z'')^2}{1/2} - \frac{(x'')^2}{1} - \frac{(y'')^2}{1} = 1$$

9. The matrix form for the quadric is $\mathbf{x}^T A\mathbf{x} + K\mathbf{x} - 31 = 0$ where

$$A = \begin{bmatrix} 0 & 1 & 0 \\ 1 & 0 & 0 \\ 0 & 0 & 0 \end{bmatrix} \quad \text{and} \quad K = \begin{bmatrix} -6 & 10 & 1 \end{bmatrix}$$

The eigenvalues of A are $\lambda_1 = 1$, $\lambda_2 = -1$, and $\lambda_3 = 0$, and the corresponding eigenspaces are spanned by the orthogonal vectors

$$\begin{bmatrix} 1 \\ 1 \\ 0 \end{bmatrix} \quad \begin{bmatrix} -1 \\ 1 \\ 0 \end{bmatrix} \quad \begin{bmatrix} 0 \\ 0 \\ 1 \end{bmatrix}$$

Thus, we let $\mathbf{x} = P\mathbf{x}'$ where

$$P = \begin{bmatrix} \dfrac{1}{\sqrt{2}} & \dfrac{-1}{\sqrt{2}} & 0 \\ \dfrac{1}{\sqrt{2}} & \dfrac{1}{\sqrt{2}} & 0 \\ 0 & 0 & 1 \end{bmatrix}$$

Note that $\det(P) = 1$,

$$P^T A P = \begin{bmatrix} 1 & 0 & 0 \\ 0 & -1 & 0 \\ 0 & 0 & 0 \end{bmatrix} \quad \text{and} \quad KP = \begin{bmatrix} 2\sqrt{2} & 8\sqrt{2} & 1 \end{bmatrix}$$

Therefore, the equation of the quadric is reduced to

$$(x')^2 - (y')^2 + 2\sqrt{2}x' + 8\sqrt{2}y' + z' - 31 = 0$$

If we complete the squares, this becomes

$$\left[(x')^2 + 2\sqrt{2}x' + 2\right] - \left[(y')^2 - 8\sqrt{2}y' + 32\right] + z' = 31 + 2 - 32$$

Letting $x'' = x' + \sqrt{2}$, $y'' = y' - 4\sqrt{2}$, and $z'' = z' - 1$ yields

$$(x'')^2 - (y'')^2 + z'' = 0$$

This is a hyperbolic paraboloid.

11. We know that the equation of a general quadric Q can be put into the standard matrix form $\mathbf{x}^T A\mathbf{x} + K\mathbf{x} + j = 0$ where

$$A = \begin{bmatrix} a & d & e \\ d & b & f \\ e & f & c \end{bmatrix} \quad \text{and} \quad K = [\, g \quad h \quad i \,]$$

Since A is a symmetric matrix, then A is orthogonally diagonalizable by Theorem 7.3.1. Thus, by Theorem 7.2.1, A has 3 linearly independent eigenvectors. Now let T be the matrix whose column vectors are the 3 linearly independent eigenvectors of A. It follows from the proof of Theorem 7.2.1 and the discussion immediately following that theorem, that $T^{-1}AT$ will be a diagonal matrix whose diagonal entries are the eigenvalues λ_1, λ_2, and λ_3 of A. Theorem 7.3.2 guarantees that these eigenvalues are real.

As noted immediately after Theorem 7.3.2, we can, if necessary, transform the matrix T to a matrix S whose column vectors form an orthonormal set. To do this, orthonormalize the basis of each eigenspace before using its elements as column vectors of S.

Furthermore, by Theorem 6.5.1, we know that S is orthogonal. It follows from Theorem 6.5.2 that $\det(S) = \pm 1$.

In case $\det(S) = -1$, we interchange two columns in S to obtain a matrix P such that $\det(P) = 1$. If $\det(S) = 1$, we let $P = S$. Thus, P represents a rotation. Note that P is orthogonal, so that $P^{-1} = P^T$, and also, that P orthogonally diagonalizes A. In fact,

$$P^T A P = \begin{bmatrix} \lambda_1 & 0 & 0 \\ 0 & \lambda_2 & 0 \\ 0 & 0 & \lambda_3 \end{bmatrix}$$

Hence, if we let $\mathbf{x} = P\mathbf{x}'$, then the equation of the quadric Q becomes

$$(\mathbf{x}')^T \left(P^T A P\right) \mathbf{x}' + KP\mathbf{x}' + j = 0$$

or

$$\lambda_1 (x')^2 + \lambda_2 (y')^2 + \lambda_3 (z')^2 + g'x' + h'y' + i'z' + j = 0$$

where

$$[\, g' \quad h' \quad i' \,] = KP$$

Thus we have proved Theorem 9.7.1.

EXERCISE SET 9.8

1. If $AB = C$ where A is $m \times n$, B is $n \times p$, and C is $m \times p$, then C has mp entries, each of the form

$$c_{ij} = a_{i1}b_{1j} + a_{i2}b_{2j} + \cdots + a_{in}b_{nj}$$

Thus we need n multiplications and $n - 1$ additions to compute each of the numbers c_{ij}. Therefore we need mnp multiplications and $m(n - 1)p$ additions to compute C.

2. If A is an $n \times n$ matrix, then for any other $n \times n$ matrix, B, it will take (by Exercise 1) n^3 multiplications and $n^2(n - 1)$ additions to compute AB. To compute A^k, we must carry out $k - 1$ of these matrix multiplications, onc to compute A^2, a second to compute $A^3 = A(A^2)$, and in all, $k - 1$ to compute $A^k = A(A^{k-1})$. Therefore we must carry out a total of $n^3(k - 1)$ multiplications and $n^2(n - 1)(k - 1)$ additions to compute the product A^k.

5. Following the hint, we have

$$S_n = 1 + \quad 2 \quad + \quad 3 \quad + \cdots + n$$
$$S_n = n + (n - 1) + (n - 2) + \cdots + 1$$

or

$$2S_n = (n + 1) + (n + 1) + (n + 1) + \cdots + (n + 1)$$

Thus

$$S_n = \frac{n(n + 1)}{2}$$

7. **(a)** By direct computation,

$$(k+1)^3 - k^3 = k^3 + 3k^2 + 3k + 1 - k^3 = 3k^2 + 3k + 1$$

(b) The sum "telescopes". That is,

$$[2^3 - 1^3] + [3^3 - 2^3] + [4^3 - 3^3] + \cdots + [(n+1)^3 - n^3]$$

$$= 2^3 - 1^3 + 3^3 - 2^3 + 4^3 - 3^3 + \cdots + (n+1)^3 - n^3$$

$$= (n+1)^3 - 1$$

(c) By Parts (a) and (b), we have

$$3(1)^2 + 3(1) + 1 + 3(2)^2 + 3(2) + 1 + 3(3)^2 + 3(3) + 1 + \cdots + 3n^2 + 3n + 1$$

$$= 3(1^2 + 2^2 + 3^2 + \cdots + n^2) + 3(1 + 2 + 3 + \cdots + n) + n$$

$$= (n+1)^3 - 1$$

(d) Thus, by Part (c) and Exercise 6, we have

$$1^2 + 2^2 + 3^2 + \cdots + n^2 = \frac{1}{3}\left[(n+1)^3 - 1 - 3\frac{n(n+1)}{2} - n\right]$$

$$= \frac{(n+1)^3}{3} - \frac{1}{3} - \frac{n(n+1)}{2} - \frac{n}{3}$$

$$= \frac{2(n+1)^3 - 2 - 3n(n+1) - 2n}{6}$$

$$= \frac{(n+1)[2(n+1)^2 - 3n - 2]}{6}$$

$$= \frac{(n+1)(2n^2 + 4n + 2 - 3n - 2)}{6}$$

$$= \frac{(n+1)(2n^2 + n)}{6}$$

$$= \frac{n(n+1)(2n+1)}{6}$$

9. Since R is a row-echelon form of an invertible $n \times n$ matrix, it has ones down the main diagonal and nothing but zeros below. If, as usual, we let $\mathbf{x} = [x_1 \quad x_2 \quad \cdots \quad x_n]^T$ and $\mathbf{b} = [b_1 \quad b_2 \quad \cdots \quad b_n]^T$, then we have $x_n = b_n$ with no computations. However, since $x_{n-1} = b_{n-1} - cx_n$ for some number c, it will require one multiplication and one addition to find $x_{n\,1}$. In general,

$$x_i = b_i - \text{some linear combination of } x_{i+1}, x_{i+2}, \cdots, x_n$$

Therefore it will require two multiplications and two additions to find x_{n-2}, three of each to find x_{n-3}, and finally, $n-1$ of each to find x_1. That is, it will require

$$1 + 2 + 3 + \cdots + (n-1) = \frac{(n-1)n}{2}$$

multiplications and the same number of additions to solve the system by back substitution.

10. To reduce an invertible $n \times n$ matrix to I_n, we first divide Row 1 by a_{11} (which we are assuming is not zero since no row interchanges are required). This requires $n-1$ multiplications because we can ignore the first row, first column position which must be one. We then subtract a_{i1} times Row 1 from Row i for $i = 2, 3, \cdots, n$ to reduce the first column to that of I_n. This requires $(n-1)^2$ multiplications and the same number of additions because we can ignore those entries which we know must be zero. Thus we have used $n - 1 + (n-1)^2 = n(n-1)$ multiplications and $(n-1)^2$ additions.

We then repeat the process on the $(n-1) \times (n-1)$ submatrix obtained from the original by deleting the first row and the first column. This will require $(n-1)(n-2)$ multiplicaitons and $(n-2)^2$ additions.

We continue in this fashion until we have reduced every diagonal entry to one and every entry below the diagonal to zero. The number of multiplications required so far is

$$n(n-1) + (n-1)(n-2) + (n-2)(n-3) + \cdots + 3 \cdot 2 + 2 \cdot 1$$

$$= n^2 - n + (n-1)^2 - (n-1) + (n-2)^2 - (n-2) + \cdots + 3^2 - 3 + 2^2 - 2$$

$$= [n^2 + (n-1)^2 + \cdots + 3^2 + 2^2] - [n + (n-1) + \cdots + 3 + 2]$$

$$= \frac{n(n+1)(2n+1)}{6} - 1 - \left[\frac{n(n+1)}{2} - 1 \right]$$

$$= \frac{n^3 - n}{3}$$

and the number of additions is

$$(n-1)^2 + (n-2)^2 + \cdots + 2^2 + 1^2 = \frac{(n-1)(n)2n-1)}{6}$$

$$= \frac{n^3}{3} - \frac{n^2}{2} + \frac{n}{6}$$

Finally, we must reduce the entries above the main diagonal to zero. We start with the nth column. We multiply Row n by a_{in} for $i = 1, 2, \ldots, n-1$ and subtract the result from Row i. This would appear to involve $n - 1$ multiplications and the same number of additions since we can ignore everything but the entries in the last column. However, the result is a foregone conclusion. In fact, once we have reduced the matrix to row-echelon form, the final reduction is automatic. Thus the above formulas give all the multiplications and additions required.

11. To solve a linear system whose coefficient matrix is an invertible $n \times n$ matrix, A, we form the $n \times (n+1)$ matrix $[A \mid b]$ and reduce A to I_n. Thus we first divide Row 1 by a_{11}, using n multiplications (ignoring the multiplication whose result must be one and assuming that $a_{11} \neq 0$ since no row interchanges are required). We then subtract a_{i1} times Row 1 from Row i for $i = 2, \ldots, n$ to reduce the first column to that of I_n. This requires $n(n-1)$ multiplications and the same number of additions (again ignoring the operations whose results we already know). The total number of multiplications so far is n^2 and the total number of additions is $n(n-1)$.

To reduce the second column to that of I_n, we repeat the procedure, starting with Row 2 and ignoring Column 1. Thus $n - 1$ multiplications assure us that there is a one on the main diagonal, and $(n-1)^2$ multiplications and additions will make all $n - 1$ of the remaining column entries zero. This requires $n(n-1)$ new multiplications and $(n-1)^2$ new additions.

In general, to reduce Column i to the ith column of I_n, we require $n + 1 - i$ multiplications followed by $(n+1-i)(n-1)$ multiplications and additions, for a total of $n(n+1-i)$ multiplications and $(n+1-i)(n-1)$ additions.

If we add up all these numbers, we find that we need

$$n^2 + n(n-1) + n(n-2) + \cdots + n(2) + n(1) = n\left(n + (n-1) + \cdots + 2 + 1\right)$$

$$= \frac{n^2(n+1)}{2}$$

$$= \frac{n^3}{2} + \frac{n^2}{2}$$

multiplications and

$$n(n-1) + (n-1)^2 + (n-2)(n-1) + \cdots + 2(n-1) + (n-1)$$

$$= (n-1)(n + (n-1) + \cdots + 2 + 1)$$

$$= \frac{(n-1)(n)(n+1)}{2}$$

$$= \frac{n^3}{2} - \frac{n}{2}$$

additions to compute the reduction.

12. Since $\lim\limits_{x \to +\infty} (1/x) = 0$, we have

$$\lim_{x \to +\infty} \frac{P(x)}{a_k x^k} = \lim_{x \to +\infty} \frac{a_0 + a_1 x + \cdots + a_k x^k}{a_k x^k}$$

$$= \lim_{x \to +\infty} \left(\frac{a_0}{a_k x^k} + \frac{a_1}{a_k x^{k-1}} + \cdots + \frac{a_{k-1}}{a_k x} + 1 \right)$$

$$= 0 + 0 + \cdots + 0 + 1$$

$$= 1$$

EXERCISE SET 9.9

1. The system in matrix form is

$$
\begin{bmatrix} 3 & -6 \\ -2 & 5 \end{bmatrix} \begin{bmatrix} x_1 \\ x_2 \end{bmatrix} = \begin{bmatrix} 3 & 0 \\ -2 & 1 \end{bmatrix} \begin{bmatrix} 1 & -2 \\ 0 & 1 \end{bmatrix} \begin{bmatrix} x_1 \\ x_2 \end{bmatrix} = \begin{bmatrix} 0 \\ 1 \end{bmatrix}
$$

This reduces to two matrix equations

$$
\begin{bmatrix} 1 & -2 \\ 0 & 1 \end{bmatrix} \begin{bmatrix} x_1 \\ x_2 \end{bmatrix} = \begin{bmatrix} y_1 \\ y_2 \end{bmatrix}
$$

and

$$
\begin{bmatrix} 3 & 0 \\ -2 & 1 \end{bmatrix} \begin{bmatrix} y_1 \\ y_2 \end{bmatrix} = \begin{bmatrix} 0 \\ 1 \end{bmatrix}
$$

The second matrix equation yields the system

$$
3y_1 \qquad = 0
$$
$$
-2y_1 + y_2 = 1
$$

which has $y_1 = 0$, $y_2 = 1$ as its solution. If we substitute these values into the first matrix equation, we obtain the system

$$
x_1 - 2x_2 = 0
$$
$$
x_2 = 1
$$

This yields the final solution $x_1 = 2$, $x_2 = 1$.

3. To reduce the matrix of coefficients to a suitable upper triangular matrix, we carry out the following operations:

$$\begin{bmatrix} 2 & 8 \\ -1 & -1 \end{bmatrix} \rightarrow \begin{bmatrix} 1 & 4 \\ -1 & -1 \end{bmatrix} \rightarrow \begin{bmatrix} 1 & 4 \\ 0 & 3 \end{bmatrix} \rightarrow \begin{bmatrix} 1 & 4 \\ 0 & 1 \end{bmatrix} = U$$

These operations involve multipliers $1/2$, 1, and $1/3$. Thus the corresponding lower triangular matrix is

$$L = \begin{bmatrix} 2 & 0 \\ -1 & 3 \end{bmatrix}$$

We therefore have the two matrix equations

$$\begin{bmatrix} 1 & 4 \\ 0 & 1 \end{bmatrix} \begin{bmatrix} x_1 \\ x_2 \end{bmatrix} = \begin{bmatrix} y_1 \\ y_2 \end{bmatrix}$$

and

$$\begin{bmatrix} 2 & 0 \\ -1 & 3 \end{bmatrix} \begin{bmatrix} y_1 \\ y_2 \end{bmatrix} = \begin{bmatrix} -2 \\ -2 \end{bmatrix}$$

The second matrix equation yields the system

$$2y_1 \qquad = -2$$
$$-y_1 + 3y_2 = -2$$

which has $y_1 = -1$, $y_2 = -1$ as its solution. If we substitute these values into the first matrix equation, we obtain the system

$$x_1 + 4x_2 = -1$$
$$x_2 = -1$$

This yields the final solution $x_1 = 3$, $x_2 = -1$.

5. To reduce the matrix of coefficients to a suitable upper triangular matrix, we carry out the following operations:

$$
\begin{bmatrix}
2 & -2 & -2 \\
0 & -2 & 2 \\
-1 & 5 & 2
\end{bmatrix}
\rightarrow
\begin{bmatrix}
1 & -1 & -1 \\
0 & -2 & 2 \\
-1 & 5 & 2
\end{bmatrix}
\rightarrow
\begin{bmatrix}
1 & 1 & 1 \\
0 & -2 & 2 \\
0 & 4 & 1
\end{bmatrix}
\rightarrow
$$

$$
\begin{bmatrix}
1 & -1 & -1 \\
0 & 1 & -1 \\
0 & 4 & 1
\end{bmatrix}
\rightarrow
\begin{bmatrix}
1 & -1 & -1 \\
0 & 1 & -1 \\
0 & 0 & 5
\end{bmatrix}
\rightarrow
\begin{bmatrix}
1 & -1 & -1 \\
0 & 1 & -1 \\
0 & 0 & 1
\end{bmatrix}
= U
$$

These operations involve multipliers of $1/2$, 0 (for the 2nd row), 1, $-1/2$, -4, and $1/5$. Thus the corresponding lower triangular matrix is

$$
L =
\begin{bmatrix}
2 & 0 & 0 \\
0 & -2 & 0 \\
-1 & 4 & 5
\end{bmatrix}
$$

We therefore have the matrix equations

$$
\begin{bmatrix}
1 & -1 & -1 \\
0 & 1 & -1 \\
0 & 0 & 1
\end{bmatrix}
\begin{bmatrix}
x_1 \\
x_2 \\
x_3
\end{bmatrix}
=
\begin{bmatrix}
y_1 \\
y_2 \\
y_3
\end{bmatrix}
$$

and

$$
\begin{bmatrix}
2 & 0 & 0 \\
0 & -2 & 0 \\
-1 & 4 & 5
\end{bmatrix}
\begin{bmatrix}
y_1 \\
y_2 \\
y_3
\end{bmatrix}
=
\begin{bmatrix}
-4 \\
-2 \\
6
\end{bmatrix}
$$

The second matrix equation yields the system

$$
\begin{aligned}
2y_1 \qquad\qquad\quad &= -4 \\
-2y_2 \qquad\quad &= -2 \\
-y_1 + 4y_2 + 5y_3 &= 6
\end{aligned}
$$

which has $y_1 = -2$, $y_2 = 1$ and $y_3 = 0$ as its solution. If we substitute these values into the first matrix equation, we obtain the system

$$
\begin{aligned}
x_1 - x_2 - x_3 &= -2 \\
x_2 - x_3 &= 1 \\
x_3 &= 0
\end{aligned}
$$

This yields the final solution $x_1 = -1$, $x_2 = 1$, $x_3 = 0$.

11. **(a)** To reduce A to row-echelon form, we carry out the following operations:

$$
\begin{bmatrix} 2 & 1 & -1 \\ -2 & -1 & 2 \\ 2 & 1 & 0 \end{bmatrix} \rightarrow
\begin{bmatrix} 1 & 1/2 & -1/2 \\ -2 & -1 & 2 \\ 2 & 1 & 0 \end{bmatrix} \rightarrow
\begin{bmatrix} 1 & 1/2 & -1/2 \\ 0 & 0 & 1 \\ 2 & 1 & 0 \end{bmatrix}
$$

$$
\rightarrow
\begin{bmatrix} 1 & 1/2 & -1/2 \\ 0 & 0 & 1 \\ 0 & 0 & 1 \end{bmatrix} \rightarrow
\begin{bmatrix} 1 & 1/2 & -1/2 \\ 0 & 0 & 1 \\ 0 & 0 & 0 \end{bmatrix} = U
$$

This involves multipliers $1/2$, 2, -2, 1 (for the 2nd diagonal entry), and -1. Where no multiplier is needed in the second entry of the last row, we use the multiplier 1, thus obtaining the lower triangular matrix

$$
L = \begin{bmatrix} 2 & 0 & 0 \\ -2 & 1 & 0 \\ 2 & 1 & 1 \end{bmatrix}
$$

In fact, if we compute LU, we see that it will equal A no matter what entry we choose for the lower right-hand corner of L.

If we stop just before we reach row-echelon form, we obtain the matrices

$$
U = \begin{bmatrix} 1 & 1/2 & -1/2 \\ 0 & 0 & 1 \\ 0 & 0 & 1 \end{bmatrix} \qquad
L = \begin{bmatrix} 2 & 0 & 0 \\ -2 & 1 & 0 \\ 2 & 0 & 1 \end{bmatrix}
$$

which will also serve.

(b) We have that $A = LU$ where

$$L = \begin{bmatrix} 2 & 0 & 0 \\ -2 & 1 & 0 \\ 2 & 1 & 1 \end{bmatrix} \qquad U = \begin{bmatrix} 1 & 1/2 & -1/2 \\ 0 & 0 & 1 \\ 0 & 0 & 0 \end{bmatrix}$$

If we let

$$L_1 = \begin{bmatrix} 1 & 0 & 0 \\ -1 & 1 & 0 \\ 1 & 1 & 1 \end{bmatrix} \qquad \text{and} \quad D = \begin{bmatrix} 2 & 0 & 0 \\ 0 & 1 & 0 \\ 0 & 0 & 1 \end{bmatrix}$$

then $A = L_1 DU$ as desired. (See the matrices at the very end of Section 9.9.)

(c) Let $U_2 = DU$ and note that this matrix is upper triangular. Then $A = L_1 U_2$ is of the required form.

12. Suppose that

$$\begin{bmatrix} 0 & 1 \\ 1 & 0 \end{bmatrix} = \begin{bmatrix} a & 0 \\ b & c \end{bmatrix}\begin{bmatrix} d & e \\ 0 & f \end{bmatrix} = \begin{bmatrix} ad & ae \\ bd & be + cf \end{bmatrix}$$

This implies that $ad = 0$ but that $ae = bd = 1$. That is, if the matrix has an LU-decomposition, then we must have 4 numbers which satisfy the above equations. Now if $ad = 0$, then either a or d must be 0. However, this cannot be true if $ae = bd = 1$. Thus the matrix has no such decomposition.

13. **(a)** If A has such an LU-decomposition, we can write it as

$$\begin{bmatrix} a & b \\ c & d \end{bmatrix} = \begin{bmatrix} 1 & 0 \\ w & 1 \end{bmatrix}\begin{bmatrix} x & y \\ 0 & z \end{bmatrix} = \begin{bmatrix} x & y \\ wx & yw + z \end{bmatrix}$$

This yields the system of equations

$$x = a \qquad y = b \qquad wx = c \qquad yw + z = d$$

Since $a \neq 0$, this has the unique solution

$$x = a \qquad y = b \qquad w = c/a \qquad z = (ad - bc)/a$$

The uniqueness of the solution guarantees the uniqueness of the LU-decomposition.

13. **(b)** By the above,

$$\begin{bmatrix} a & b \\ c & d \end{bmatrix} = \begin{bmatrix} 1 & 0 \\ c/a & 1 \end{bmatrix} \begin{bmatrix} a & b \\ 0 & (ad - bc)/a \end{bmatrix}$$

15. We have that $L = E_1^{-1} E_2^{-1} \cdots E_k^{-1}$ where each of the matrices E_i is an elementary matrix which does not involve interchanging rows. By Exercise 27 of Section 2.4, we know that if E is an invertible lower triangular matrix, then E^{-1} is also lower triangular. Now the matrices E_i are all lower triangular and invertible by their construction. Therefore for $i = 1, \ldots, k$ we have that E_i^{-1} is lower triangular. Hence L, as the product of lower triangular matrices, must also be lower triangular.

17. Let A be any $n \times n$ matrix. We know that A can be reduced to row-echelon form and that this may require row interchanges. If we perform these interchanges (if any) first, we reduce A to the matrix

$$E_k \cdots E_1 A = B$$

where E_i is the elementary matrix corresponding to the ith such interchange. Now we know that B has an LU-decomposition, call it LU where U is a row-echelon form of A. That is,

$$E_k \cdots E_1 A = LU$$

where each of the matrices E_i is elementary and hence invertible. (In fact, $E_i^{-1} = E_i$ for all E_i. Why?) If we let

$$P = (E_k \cdots E_1)^{-1} = E_1^{-1} \cdots E_k^{-1} \qquad \text{if } k > 0$$

and $P = I$ if no row interchanges are required, then we have $A = PLU$ as desired.

18.   First we interchange Row 2 and Row 3 of A to obtain the equation

$$
B = \begin{bmatrix} 3 & -1 & 0 \\ 0 & 2 & 1 \\ 3 & -1 & 1 \end{bmatrix} = \begin{bmatrix} 1 & 0 & 0 \\ 0 & 0 & 1 \\ 0 & 1 & 0 \end{bmatrix} \begin{bmatrix} 3 & -1 & 0 \\ 3 & -1 & 1 \\ 0 & 2 & 1 \end{bmatrix} = EA
$$

Next we find an LU-decomposition of B as follows

$$
B \rightarrow \begin{bmatrix} 1 & -1/3 & 0 \\ 0 & 2 & 1 \\ 3 & -1 & 1 \end{bmatrix} \rightarrow \begin{bmatrix} 1 & -1/3 & 0 \\ 0 & 2 & 1 \\ 0 & 0 & 1 \end{bmatrix}
$$

$$
\rightarrow \begin{bmatrix} 1 & -1/3 & 0 \\ 0 & 1 & 1/2 \\ 0 & 0 & 1 \end{bmatrix} = U
$$

This gives us

$$
L = \begin{bmatrix} 3 & 0 & 0 \\ 0 & 2 & 0 \\ 3 & 0 & 1 \end{bmatrix}
$$

so that $B = EA = LU$ or $A = E^{-1}LU$. Since

$$
E^{-1} = \begin{bmatrix} 1 & 0 & 0 \\ 0 & 0 & 1 \\ 0 & 1 & 0 \end{bmatrix} = E
$$

we have $A = ELU$, or

$$
A = \begin{bmatrix} 1 & 0 & 0 \\ 0 & 0 & 1 \\ 0 & 1 & 0 \end{bmatrix} \begin{bmatrix} 3 & 0 & 0 \\ 0 & 2 & 0 \\ 3 & 0 & 1 \end{bmatrix} \begin{bmatrix} 1 & -1/3 & 0 \\ 0 & 1 & 1/2 \\ 0 & 0 & 1 \end{bmatrix}
$$

EXERCISE SET 10.1

3. **(b)** Since two complex numbers are equal if and only if both their real and imaginary parts are equal, we have

$$x + y = 3$$

and

$$x - y = 1$$

Thus $x = 2$ and $y = 1$.

4. **(a)** $$z_1 + z_2 = 1 - 2i + 4 + 5i = 5 + 3i$$

(c) $$4z_1 = 4(1 - 2i) = 4 - 8i$$

(e) $$3z_1 + 4z_2 = 3(1 - 2i) + 4(4 + 5i)$$
$$= 3 - 6i + 16 + 20i$$
$$= 19 + 14i$$

5. **(a)** Since complex numbers obey all the usual rules of algebra, we have

$$z = 3 + 2i - (1 - i) = 2 + 3i$$

(c) Since $(i - z) + (2z - 3i) = -2 + 7i$, we have

$$i + (-z + 2z) - 3i = -2 + 7i$$

or

$$z = -2 + 7i - i + 3i = -2 + 9i$$

6. (a) $z_1 + z_2 = 4 + 5i$ and $z_1 - z_2 = 2 - 3i$

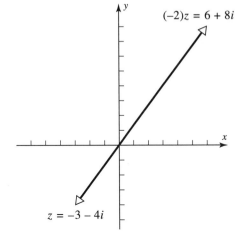

7. (b) $-2z = 6 + 8i$

8. (a) We have $k_1 i + k_2(1 + i) = k_2 + (k_1 + k_2)\,i = 3 - 2i$. Hence, $k_2 = 3$ and $k_1 + k_2 = -2$ or $k_1 = -5$.

9. (c)

$$z_1 z_2 = \frac{1}{6}(2+4i)(1-5i) = \frac{2}{6}(1+2i)(1-5i) = \frac{1}{3}(1-3i+10) = \frac{11}{3} - i$$

$$z_1^2 = \frac{4}{9}(1+2i)^2 = \frac{4}{9}(-3+4i)$$

$$z_2^2 = \frac{1}{4}(1-5i)^2 = \frac{1}{4}(-24-10i) = -6 - \frac{5}{2}i$$

10. (a) First, $z_1 z_2 = (2-5i)(-1-i) = -7+3i$. Thus,

$$z_1 - z_1 z_2 = (2-5i) - (-7+3i) = 9 - 8i$$

(c) Since $1 + z_2 = -i$, then $z_1 + (1+z_2) = 2 - 6i$. Thus,

$$[z_1 + (1+z_2)]^2 = (2-6i)^2 = 2^2(1-3i)^2 = 4(-8-6i) = -32 - 24i$$

11. Since $(4-6i)^2 = 2^2(2-3i)^2 = 4(-5-12i) = -4(5+12i)$, then

$$(1+2i)(4-6i)^2 = -4(1+2i)(5+12i) = -4(-19+22i) = 76 - 88i$$

13. Since $(1-3i)^2 = -8 - 6i = -2(4+3i)$, then

$$(1-3i)^3 = -2(1-3i)(4+3i) = -2(13-9i)$$

15. Since $(2+i)\left(\frac{1}{2} + \frac{3}{4}i\right) = \frac{1}{4} + 2i$, then

$$\left[(2+i)\left(\frac{1}{2} + \frac{3}{4}i\right)\right]^2 = \left(\frac{1}{4} + 2i\right)^2 = -\frac{63}{16} + i$$

17. Since $i^2 = -1$ and $i^3 = -i$, then $1 + i + i^2 + i^3 = 0$. Thus $\left(1 + i + i^2 + i^3\right)^{100} = 0$.

19. **(a)**

$$A + 3iB = \begin{bmatrix} 1 & i \\ -i & 3 \end{bmatrix} + \begin{bmatrix} 6i & -3 + 6i \\ 3 + 9i & 12i \end{bmatrix}$$

$$= \begin{bmatrix} 1 + 6i & -3 + 7i \\ 3 + 8i & 3 + 12i \end{bmatrix}$$

(d) $A^2 = \begin{bmatrix} 2 & 4i \\ -4i & 10 \end{bmatrix}$ and $B^2 = \begin{bmatrix} 11 + i & 12 + 6i \\ 18 - 6i & 23 + i \end{bmatrix}$

Hence

$$B^2 - A^2 = \begin{bmatrix} 9 + i & 12 + 2i \\ 18 - 2i & 13 + i \end{bmatrix}$$

20. **(a)**

$$A(BC) = \begin{bmatrix} 3 + 2i & 0 \\ -i & 2 \\ 1 + i & 1 - i \end{bmatrix} \begin{bmatrix} 5 + i & 4i & -11 \\ 3i & -2 & -5i \end{bmatrix}$$

$$= \begin{bmatrix} 13 + 13i & -8 + 12i & -33 - 22i \\ 1 + i & 0 & i \\ 7 + 9i & -6 + 6i & -16 - 16i \end{bmatrix}$$

(c) We have

$$CA = \begin{bmatrix} -6i & -1 - i \\ 6 + i & -5 + 9i \end{bmatrix}$$ and $B^2 = \begin{bmatrix} -1 & 0 \\ 0 & -1 \end{bmatrix}$

Hence

$$(CA)B^2 = \begin{bmatrix} 6i & 1 + i \\ -6 - i & 5 - 9i \end{bmatrix}$$

21. **(a)** Let $z = x + iy$. Then

$$\text{Im}(iz) = \text{Im}[i(x + iy)] = \text{Im}(-y + ix) = x$$
$$= \text{Re}(x + iy) = \text{Re}(z)$$

22. **(a)** We have

$$z^2 + 2z + 2 = 0 \quad \Longleftrightarrow \quad z = \frac{-2 \pm \sqrt{4 - 8}}{2} = -1 \pm i$$

Let $z_1 = -1 + i$ and $z_2 = -1 - i$. Then

$$z_1^2 + 2z_1 + 2 = (-1 + i)^2 + 2(-1 + i) + 2$$
$$= -2i - 2 + 2i + 2$$
$$= 0$$

Thus, z_1 is a root of $z^2 + 2z + 2$. The verification that z_2 is also a root is similar.

23. **(a)** We know that $i^1 = i$, $i^2 = -1$, $i^3 = -i$, and $i^4 = 1$. We also know that $i^{m+n} = i^m i^n$ and $i^{mn} = (i^m)^n$ where m and n are positive integers. The proof can be broken into four cases:

$$\begin{aligned}
&\textbf{1.} \quad n = 1, 5, \ 9, \cdots \quad \text{or} \quad n = 4k + 1 \\
&\textbf{2.} \quad n = 2, 6, 10, \cdots \quad \text{or} \quad n = 4k + 2 \\
&\textbf{3.} \quad n = 3, 7, 11, \cdots \quad \text{or} \quad n = 4k + 3 \\
&\textbf{4.} \quad n = 4, 8, 12, \cdots \quad \text{or} \quad n = 4k + 4
\end{aligned}$$

where $k = 0, 1, 2, \ldots$. In each case, $i^n = i^{4k+\ell}$ for some integer ℓ between 1 and 4. Thus

$$i^n = i^{4k} i^\ell = \left(i^4\right)^k i^\ell = 1^k i^\ell = i^\ell$$

This completes the proof.

(b) Since $2509 = 4 \cdot 627 + 1$, Case 1 of Part (a) applies, and hence $i^{2509} = i$.

24. Let $z_1 = x_1 + iy_1$ and $z_2 = x_2 + iy_2$ and observe that

$$z_1 z_2 = 0 \quad \Longleftrightarrow \quad x_1 x_2 = y_1 y_2 \quad \text{and} \quad x_1 y_2 = -x_2 y_1$$

If we suppose that $x_1 \neq 0$, then we have

$$z_1 z_2 = 0 \quad \Longleftrightarrow \quad x_2 = \frac{y_1 y_2}{x_1} \quad \text{and} \quad y_2 = \frac{-x_2 y_1}{x_1}$$

This implies that

$$y_2 = -\left(\frac{y_1}{x_1}\right)^2 y_2$$

But since $-(y_1/x_1)^2 \leq 0$, the above equation can hold if and only if $y_2 = 0$. Recall that $x_2 = (y_1 y_2)/x_1$; then $y_2 = 0$ implies that $x_2 = 0$ and hence that $z_2 = 0$. Thus if $x_1 \neq 0$, then $z_2 = 0$. A similar argument shows that if $y_1 \neq 0$, then $z_2 = 0$. Therefore, if $z_1 \neq 0$, then $z_2 = 0$. Similarly, if $z_2 \neq 0$, then $z_1 = 0$.

25. Observe that $zz_1 = zz_2 \Longleftrightarrow zz_1 - zz_2 = 0 \Longleftrightarrow z(z_1 - z_2) = 0$. Since $z \neq 0$ by hypothesis, it follows from Exercise 24 that $z_1 - z_2 = 0$, *i.e.*, that $z_1 = z_2$.

26. **(a)** Let $z_1 = x_1 + iy_1$ and $z_2 = x_2 + iy_2$. Then

$$\begin{aligned} z_1 + z_2 &= (x_1 + iy_1) + (x_2 + iy_2) \\ &= (x_1 + x_2) + i(y_1 + y_2) \\ &= (x_2 + x_1) + i(y_2 + y_1) \\ &= (x_2 + iy_2) + (x_1 + iy_2) \\ &= z_2 + z_1 \end{aligned}$$

27. **(a)** Let $z_1 = x_1 + iy_1$ and $z_2 = x_2 + iy_2$. Then

$$
\begin{aligned}
z_1 z_2 &= (x_1 + iy_1)(x_2 + iy_2) \\
&= (x_1 x_2 - y_1 y_2) + i(x_1 y_2 + x_2 y_1) \\
&= (x_2 x_1 - y_2 y_1) + i(y_2 x_1 + y_1 x_2) \\
&= (x_2 + iy_2)(x_1 + iy_1) \\
&= z_2 z_1
\end{aligned}
$$

EXERCISE SET 10.2

2. **(a)** Since $i = 0 + 1i$, then

$$|i| = \sqrt{0^2 + 1^2} = 1$$

(c) $|-3 - 4i| = \sqrt{(-3)^2 + (-4)^2} = \sqrt{25} = 5$

(e) $|-8| = |-8 + 0i| = \sqrt{(-8)^2 + 0^2} = 8$

3. **(a)** We have

$$z\bar{z} = (2 - 4i)(2 + 4i) = 20$$

On the other hand,

$$|z|^2 = 2^2 + (-4)^2 = 20$$

(b) We have

$$z\bar{z} = (-3 + 5i)(-3 - 5i) = 34$$

On the other hand,

$$|z|^2 = (-3)^2 + 5^2 = 34$$

4. **(a)** From Equation (5), we have

$$\frac{z_1}{z_2} = \frac{1 - 5i}{3 + 4i} = \frac{(1 - 5i)(3 - 4i)}{3^2 + 4^2} = \frac{-17 - 19i}{25}$$

4. **(c)** Again using Equation (5), we have

$$\frac{z_1}{z_2} = \frac{1 - 5i}{3 - 4i} = \frac{(1 - 5i)(3 + 4i)}{3^2 + (-4)^2} = \frac{23 - 11i}{25}$$

(e) Since $|z_2| = 5$, we have

$$\frac{z_1}{|z_2|} = \frac{1 - 5i}{5} = \frac{1}{5} - i$$

5. **(a)** Equation (5) with $z_1 = 1$ and $z_2 = i$ yields

$$\frac{1}{i} = \frac{1(-i)}{1} = -i$$

(c)
$$\frac{1}{z} = \frac{7}{-i} = \frac{7(i)}{1} = 7i$$

6. **(a)** Since

$$\frac{z_1}{z_2} = \frac{1 + i}{1 - 2i} = \frac{(1 + i)(1 + 2i)}{5} = -\frac{1}{5} + \frac{3}{5}i$$

then

$$z_1 - \frac{z_1}{z_2} = (1 + i) - \left(-\frac{1}{5} + \frac{3}{5}i\right) = \frac{6}{5} + \frac{2}{5}i$$

(c) Since

$$\frac{iz_1}{z_2} = \frac{i(1 + i)}{1 - 2i} = \frac{-1 + i}{1 - 2i} = \frac{(-1 + i)(1 + 2i)}{(1 - 2i)(1 + 2i)} = -\frac{3}{5} - \frac{1}{5}i$$

and

$$\bar{z}_1^2 = 2i$$

then

$$\bar{z}_1^2 - \frac{iz_1}{z_2} = 2i - \left(-\frac{3}{5} - \frac{1}{5}i\right) = \frac{3}{5} + \frac{11}{5}i$$

7. Equation (5) with $z_1 = i$ and $z_2 = 1 + i$ gives

$$\frac{i}{1+i} = \frac{i(1-i)}{2} = \frac{1}{2} + \frac{1}{2}i$$

9. Since $(3 + 4i)^2 = -7 + 24i$, we have

$$\frac{1}{(3+4i)^2} = \frac{-7-24i}{(-7)^2 + (-24)^2} = \frac{-7-24i}{625}$$

11. Since

$$\frac{\sqrt{3}+i}{\sqrt{3}-i} = \frac{(\sqrt{3}+i)^2}{4} = \frac{2+2\sqrt{3}i}{4} = \frac{1}{2} + \frac{\sqrt{3}}{2}i$$

then

$$\frac{\sqrt{3}+i}{(1-i)(\sqrt{3}-i)} = \frac{\frac{1}{2}+\frac{\sqrt{3}}{2}i}{1-i} = \frac{\left(\frac{1}{2}+\frac{\sqrt{3}}{2}i\right)(1+i)}{2} = \frac{1-\sqrt{3}}{4} + \left(\frac{1+\sqrt{3}}{4}\right)i$$

13. We have

$$\frac{i}{1-i} = \frac{i(1+i)}{2} = -\frac{1}{2} + \frac{1}{2}i$$

and

$$(1-2i)(1+2i) = 5$$

Thus

$$\frac{i}{(1-i)(1-2i)(1+2i)} = \frac{-\frac{1}{2}+\frac{1}{2}i}{5} = -\frac{1}{10} + \frac{1}{10}i$$

15. **(a)** If $iz = 2 - i$, then

$$z = \frac{2 - i}{i} = \frac{(2 - i)(-i)}{1} = -1 - 2i$$

16. **(a)** $\overline{z + 5i} = \bar{z} + \overline{5i} = z - 5i$

(c) The result follows easily from the fact that $\overline{i + \bar{z}} = \bar{i} + \bar{\bar{z}} = z - i$. This result holds whenever $z \neq i$.

17. **(a)** The set of points satisfying the equation $|z| = 2$ is the set of all points representing vectors of length 2. Thus, it is a circle of radius 2 and center at the origin.

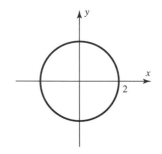

Analytically, if $z = x + iy$, then

$$|z| = 2 \quad \Longleftrightarrow \quad \sqrt{x^2 + y^2} = 2$$
$$\Longleftrightarrow \quad x^2 + y^2 = 4$$

which is the equation of the above circle.

(c) The values of z which satisfy the equation $|z - i| = |z + i|$ are just those z whose distance from the point i is equal to their distance from the point $-i$. Geometrically, then, z can be any point on the real axis.

We now show this result analytically. Let $z = x + iy$. Then

$$|z - i| = |z + i| \quad \Longleftrightarrow \quad |z - i|^2 = |z + i|^2$$
$$\Longleftrightarrow \quad |x + i(y - 1)|^2 = |x + i(y + 1)|^2$$
$$\Longleftrightarrow \quad x^2 + (y - 1)^2 = x^2 + (y + 1)^2$$
$$\Longleftrightarrow \quad -2y = 2y$$
$$\Longleftrightarrow \quad y = 0$$

18. **(a)** This inequality represents the set of all points whose distance from the point $-i$ is at most 1.

Analytically let

$z = x + iy$. Then

$$|z + i| \leq 1 \iff |z + i|^2 \leq 1$$

$$\iff x^2 + (y + 1)^2 \leq 1$$

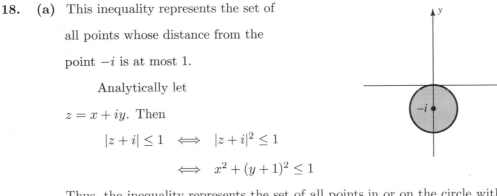

Thus, the inequality represents the set of all points in or on the circle with center at $-i$ and radius 1.

(c) Since $|2z - 4i| < 1 \iff |z - 2i| < \dfrac{1}{2}$, this inequality represents the set of all points whose distance from the point $2i$ is less than $1/2$.

Analytically, if $z = x + iy$, then

$$|2z - 4i| < 1 \iff |2z - 4i|^2 < 1$$

$$\iff (2x)^2 + (2y - 4)^2 < 1$$

$$\iff x^2 + (y - 2)^2 < \frac{1}{4}$$

19. **(a)** $\operatorname{Re}(\overline{iz}) = \operatorname{Re}(\bar{i}\,\bar{z}) = \operatorname{Re}[(-i)(x - iy)] = \operatorname{Re}(-y - ix) = -y$

(c) $\operatorname{Re}(i\bar{z}) = \operatorname{Re}[i(x - iy)] = \operatorname{Re}(y + ix) = y$

20. **(a)** Since $1/i = -i$ and $(-i)^n = (-1)^n(i)^n$, this problem is a variation on Exercise 23 of Section 10.1.

(b) Since $2509 = 4\,(627) + 1$ and $(i)^4 = 1$, then $(i)^{2509} = i$ and therefore $(1/i)^{2509} = (-1)^{2509} \cdot i = -i$.

21. **(a)** Let $z = x + iy$. Then

$$\frac{1}{2}(z + \bar{z}) = \frac{1}{2}[(x + iy) + (x - iy)] = \frac{1}{2}(2x) = x = \text{Re}(z)$$

22. Let $z = x + iy$. Then

$$z = \bar{z} \quad \Longleftrightarrow \quad x + iy = x - iy \quad \Longleftrightarrow \quad y = 0 \quad \Longleftrightarrow \quad z \text{ is real.}$$

23. **(a)** Equation (5) gives

$$\frac{z_1}{z_2} = \frac{1}{|z_2|^2} z_1 \bar{z}_2$$

$$= \frac{1}{x_2^2 + y_2^2} (x_1 + iy_1)(x_2 - iy_2)$$

$$= \frac{1}{x_2^2 + y_2^2} [(x_1 x_2 + y_1 y_2) + i(x_2 y_1 - x_1 y_2)]$$

Thus

$$\text{Re}\left(\frac{z_1}{z_2}\right) = \frac{x_1 x_2 + y_1 y_2}{x_2^2 + y_2^2}$$

25. $|z| = \sqrt{x^2 + y^2} = \sqrt{x^2 + (-y)^2} = |\bar{z}|$

26. Let $z_1 = x_1 + iy_1$ and $z_2 = x_2 + iy_2$.

(a)

$$\overline{z_1 - z_2} = \overline{(x_1 + iy_1) - (x_2 + iy_2)}$$

$$= \overline{(x_1 - x_2) + i(y_1 - y_2)}$$

$$= (x_1 - x_2) - i(y_1 - y_2)$$

$$= (x_1 - iy_1) - (x_2 - iy_2)$$

$$= \bar{z}_1 - \bar{z}_2$$

(c)

$$\overline{(z_1/z_2)} = \frac{1}{|z_2|^2} z_1 \bar{z}_2$$

$$= \frac{1}{|z_2|^2} \bar{z}_1 \cdot \bar{\bar{z}}_2 \qquad \text{(Part (b))}$$

$$= \bar{z}_1/\bar{z}_2 \qquad \text{(Part (d))}$$

27. (a) $\overline{z^2} = \overline{zz} = \bar{z}\bar{z} = (\bar{z})^2$

(b) We use mathematical induction. In Part (a), we verified that the result holds when $n = 2$. Now, assume that $(\bar{z})^n = \overline{z^n}$. Then

$$(\bar{z})^{n+1} = (\bar{z})^n \bar{z}$$

$$= \overline{z^n}\,\bar{z}$$

$$= \overline{z^{n+1}}$$

and the result is proved.

28. Let A denote the matrix of the system, i.e.,

$$A = \begin{bmatrix} i & -i \\ 2 & 1 \end{bmatrix}$$

Thus $\det(A) = i - (-2i) = 3i$. Hence

$$x_1 = \frac{1}{3i} \begin{vmatrix} -2 & -i \\ i & 1 \end{vmatrix} = \frac{1}{3i}\left(-2 + i^2\right) = \frac{-1}{i} = i$$

and

$$x_2 = \frac{1}{3i} \begin{vmatrix} i & -2 \\ 2 & i \end{vmatrix} = \frac{1}{3i}\left(i^2 + 4\right) = \frac{1}{i} = -i$$

30.   Let A denote the matrix of the system, i.e.,

$$A = \begin{bmatrix} 1 & 1 & 1 \\ 1 & 1 & -1 \\ 1 & -1 & 1 \end{bmatrix}$$

Then $\det(A) = -4$ and hence

$$x_1 = -\frac{1}{4} \begin{vmatrix} 3 & 1 & 1 \\ 2+2i & 1 & -1 \\ -1 & -1 & 1 \end{vmatrix} = \frac{1}{2} + i$$

$$x_2 = -\frac{1}{4} \begin{vmatrix} 1 & 3 & 1 \\ 1 & 2+2i & -1 \\ 1 & -1 & 1 \end{vmatrix} = 2$$

$$x_3 = -\frac{1}{4} \begin{vmatrix} 1 & 1 & 3 \\ 1 & 1 & 2+2i \\ 1 & -1 & -1 \end{vmatrix} = \frac{1}{2} - i$$

32. $\begin{bmatrix} -1 & -1-i \\ -1+i & -2 \end{bmatrix} \rightarrow \begin{bmatrix} 1 & 1+i \\ -1+i & -2 \end{bmatrix} \rightarrow \begin{bmatrix} 1 & 1+i \\ 0 & 0 \end{bmatrix}$

Therefore $x_1 = -(1+i)t$ and $x_2 = t$ where t is arbitrary.

34.

$$\begin{bmatrix} 1 & i & -i \\ -1 & 1-i & 2i \\ 2 & -1+2i & -3i \end{bmatrix} \rightarrow \begin{bmatrix} 1 & i & -i \\ 0 & 1 & i \\ 0 & -1 & -i \end{bmatrix} \rightarrow$$

$$\begin{bmatrix} 1 & i & -i \\ 0 & 1 & i \\ 0 & 0 & 0 \end{bmatrix} \rightarrow \begin{bmatrix} 1 & 0 & 1-i \\ 0 & 1 & i \\ 0 & 0 & 0 \end{bmatrix}$$

If we let $x_3 = t$, where t is arbitrary, then $x_2 = -it$ and $x_1 = -(1-i)t$.

35. (a)

$$A^{-1} = \frac{1}{i^2 + 2} \begin{bmatrix} i & 2 \\ -1 & i \end{bmatrix} = \begin{bmatrix} i & 2 \\ -1 & i \end{bmatrix}$$

It is easy to verify that $AA^{-1} = A^{-1}A = I$.

36. From Exercise 27(b), we have that $(z)^n = \overline{z^n}$. Since a_k is real, it follows that $a_k(\overline{z^k}) = \overline{a_k z^k}$. Thus,

$$p(\bar{z}) = a_0 + a_1\bar{z} + a_2(\bar{z})^2 + \cdots + a_n(\bar{z})^n$$

$$= \bar{a}_0 + \overline{a_1 z} + \overline{a_2 z^2} + \cdots + \overline{a_n z^n}$$

$$= \overline{a_0 + a_1 z + a_2 z^2 + \cdots + a_n z^n}$$

$$= \overline{p(z)}$$

Hence, $p(z) = 0 \implies p(\bar{z}) = \bar{0} = 0$.

39. (a)

$$\left[\begin{array}{ccc|ccc} 1 & 1+i & 0 & 1 & 0 & 0 \\ 0 & 1 & i & 0 & 1 & 0 \\ -i & 1-2i & 2 & 0 & 0 & 1 \end{array}\right]$$

$$\left[\begin{array}{ccc|ccc} 1 & 1+i & 0 & 1 & 0 & 0 \\ 0 & 1 & i & 0 & 1 & 0 \\ 0 & -i & 2 & i & 0 & 1 \end{array}\right] \qquad \boxed{R_3 \to R_3 + iR_1}$$

$$\left[\begin{array}{ccc|ccc} 1 & 1+i & 0 & 1 & 0 & 0 \\ 0 & 1 & i & 0 & 1 & 0 \\ 0 & 0 & 1 & i & i & 1 \end{array}\right] \qquad \boxed{R_3 \to R_3 + iR_2}$$

$$\left[\begin{array}{ccc|ccc} 1 & 1+i & 0 & 1 & 0 & 0 \\ 0 & 1 & 0 & 1 & 2 & -i \\ 0 & 0 & 1 & i & i & 1 \end{array}\right] \qquad \boxed{R_2 \to R_2 - iR_3}$$

$$\left[\begin{array}{ccc|ccc} 1 & 0 & 0 & -i & -2-2i & -1+i \\ 0 & 1 & 0 & 1 & 2 & -i \\ 0 & 0 & 1 & i & i & 1 \end{array}\right] \qquad \boxed{R_2 \to R_1 - (1+i)R_2}$$

Thus

$$A^{-1} = \left[\begin{array}{ccc} -i & -2-2i & -1+i \\ 1 & 2 & -i \\ i & i & 1 \end{array}\right]$$

40. We have

$$|z - 1|^2 = (z - 1)(\overline{z - 1}) = (z - 1)(\bar{z} - 1)$$

and

$$|\bar{z} - 1|^2 = (\bar{z} - 1)(\overline{\bar{z} - 1}) = (\bar{z} - 1)(\bar{\bar{z}} - 1) = (\bar{z} - 1)(z - 1)$$

Thus, $|z - 1| = |\bar{z} - 1|$.

Since z and $\bar{z}$ are symmetric about the real axis, they are, in fact, equidistant from any point on that axis.

41. **(a)** We have $|z_1 - z_2| = \sqrt{(a_1 - a_2)^2 + (b_1 - b_2)^2}$, which is just the distance between the two numbers z_1 and z_2 when they are considered as points in the complex plane.

(b) Let $z_1 = 12$, $z_2 = 6 + 2i$, and $z_3 = 8 + 8i$. Then

$$|z_1 - z_2|^2 = 6^2 + (-2)^2 = 40$$
$$|z_1 - z_3|^2 = 4^2 + (-8)^2 = 80$$
$$|z_2 - z_3|^2 = (-2)^2 + (-6)^2 = 40$$

Since the sum of the squares of the lengths of two sides is equal to the square of the third side, the three points determine a right triangle.

EXERCISE SET 10.3

1. **(a)** If $z = 1$, then $\arg z = 2k\pi$ where $k = 0, \pm 1, \pm 2, \cdots$. Thus, $\text{Arg}(1) = 0$.

   (c) If $z = -i$, then $\arg z = \dfrac{3\pi}{2} + 2k\pi$ where $k = 0, \pm 1, \pm 2, \cdots$. Thus, $\text{Arg}(-i) = -\pi/2$.

   (e) If $z = -1 + \sqrt{3}i$, then $\arg z = \dfrac{2\pi}{3} + 2k\pi$ where $k = 0, \pm 1, \pm 2, \cdots$. Thus, $\text{Arg}(-1 + \sqrt{3}i) = \dfrac{2\pi}{3}$.

2. We have

$$\arg(1 - \sqrt{3}i) = \frac{5\pi}{3} + 2k\pi \qquad k = 0, \pm 1, \pm 2, \cdots$$

   (a) Put $k = 0$ in the above equation.

   (b) Put $k = -1$ in the above equation.

3. **(a)** Since $|2i| = 2$ and $\text{Arg}(2i) = \pi/2$, we have

$$2i = 2\left[\cos\left(\frac{\pi}{2}\right) + i\sin\left(\frac{\pi}{2}\right)\right]$$

   (c) Since $|5 + 5i| = \sqrt{50} = 5\sqrt{2}$ and $\text{Arg}(5 + 5i) = \pi/4$, we have

$$5 + 5i = 5\sqrt{2}\left[\cos\left(\frac{\pi}{4}\right) + i\sin\left(\frac{\pi}{4}\right)\right]$$

3. **(e)** since $|-3-3i| = \sqrt{18} = 3\sqrt{2}$ and $\text{Arg}(-3-3i) = -\dfrac{3\pi}{4}$, we have

$$-3-3i = 3\sqrt{2}\left[\cos\left(-\frac{3\pi}{4}\right) + i\sin\left(-\frac{3\pi}{4}\right)\right]$$

4. We have $|z_1| = 2$, $\text{Arg}(z_1) = \dfrac{\pi}{4}$, $|z_2| = 3$, and $\text{Arg}(z_2) = \dfrac{\pi}{6}$.

(a) Here $|z_1 z_2| = |z_1||z_2| = 6$ and $\text{Arg}(z) = \text{Arg}(z_1) + \text{Arg}(z_2) = \dfrac{5\pi}{12}$. Hence

$$z_1 z_2 = 6\left[\cos\left(\frac{5\pi}{12}\right) + i\sin\left(\frac{5\pi}{12}\right)\right]$$

(c) Here $|z_2/z_1| = |z_2|/|z_1| = 3/2$ and $\text{Arg}(z_2/z_1) = \text{Arg}(z_2) - \text{Arg}(z_1) = -\dfrac{\pi}{12}$. Hence

$$\frac{z_2}{z_1} = \frac{3}{2}\left[\cos\left(-\frac{\pi}{12}\right) + i\sin\left(-\frac{\pi}{12}\right)\right]$$

5. We have $|z_1| = 1$, $\text{Arg}(z_1) = \dfrac{\pi}{2}$, $|z_2| = 2$, $\text{Arg}(z_2) = -\dfrac{\pi}{3}$, $|z_3| = 2$, and $\text{Arg}(z_3) = \dfrac{\pi}{6}$. So

$$\left|\frac{z_1 z_2}{z_3}\right| = \frac{|z_1||z_2|}{|z_3|} = 1$$

and

$$\text{Arg}\left(\frac{z_1 z_2}{z_3}\right) = \text{Arg}(z_1) + \text{Arg}(z_2) - \text{Arg}(z_3) = 0$$

Therefore

$$\frac{z_1 z_2}{z_3} = \cos(0) + i\sin(0) = 1$$

6. **(a)** We have $r = \sqrt{2}$, $\theta = \dfrac{\pi}{4}$, and $n = 12$. Thus

$$(1+i)^{12} = 2^6[\cos(3\pi) + i\sin(3\pi)] = -64$$

(c) We have $r = 2$, $\theta = \dfrac{\pi}{6}$, and $n = 7$. Thus

$$(\sqrt{3} + i)^7 = 2^7 \left[\cos\left(\frac{7\pi}{6}\right) + i\sin\left(\frac{7\pi}{6}\right) \right]$$

$$= 2^7 \left[-\frac{\sqrt{3}}{2} - i\frac{1}{2} \right] = -64\sqrt{3} - 64i$$

7. We use Formula (10).

(a) We have $r = 1$, $\theta = -\dfrac{\pi}{2}$, and $n = 2$. Thus

$$(-i)^{1/2} = \cos\left(-\frac{\pi}{4} + k\pi\right) + i\sin\left(-\frac{\pi}{4} + k\pi\right) \qquad k = 0, 1$$

Thus, the two square roots of $-i$ are:

$$\cos\left(-\frac{\pi}{4}\right) + i\sin\left(-\frac{\pi}{4}\right) = \frac{1}{\sqrt{2}} - \frac{1}{\sqrt{2}}i$$

$$\cos\left(\frac{3\pi}{4}\right) + i\sin\left(\frac{3\pi}{4}\right) = -\frac{1}{\sqrt{2}} + \frac{1}{\sqrt{2}}i$$

(c) We have $r = 27$, $\theta = \pi$, and $n = 3$. Thus

$$(-27)^{1/3} = 3 \left[\cos\left(\frac{\pi}{3} + \frac{2k\pi}{3}\right) + i\sin\left(\frac{\pi}{3} + \frac{2k\pi}{3}\right) \right] \qquad k = 0, 1, 2$$

Therefore, the three cube roots of -27 are:

$$3\left[\cos\left(\frac{\pi}{3}\right) + i\sin\left(\frac{\pi}{3}\right)\right] = \frac{3}{2} + \frac{3\sqrt{3}}{2}i$$

$$3[\cos(\pi) + i\sin(\pi)] = -3$$

$$3\left[\cos\left(\frac{5\pi}{3}\right) + i\sin\left(\frac{5\pi}{3}\right)\right] = \frac{3}{2} - \frac{3\sqrt{3}}{2}i$$

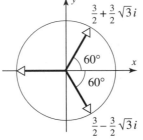

7. (e) Here $r = 1$, $\theta = \pi$, and $n = 4$. Thus

$$(-1)^{1/4} = \cos\left(\frac{\pi}{4} + \frac{k\pi}{2}\right) + i\sin\left(\frac{\pi}{4} + \frac{k\pi}{2}\right) \qquad k = 0, 1, 2, 3$$

Therefore the four fourth roots of -1 are:

$$\cos\frac{\pi}{4} + i\sin\frac{\pi}{4} = \frac{1}{\sqrt{2}} + \frac{1}{\sqrt{2}}i$$

$$\cos\frac{3\pi}{4} + i\sin\frac{3\pi}{4} = -\frac{1}{\sqrt{2}} + \frac{1}{\sqrt{2}}i$$

$$\cos\frac{5\pi}{4} + i\sin\frac{5\pi}{4} = -\frac{1}{\sqrt{2}} - \frac{1}{\sqrt{2}}i$$

$$\cos\frac{7\pi}{4} + i\sin\frac{7\pi}{4} = \frac{1}{\sqrt{2}} - \frac{1}{\sqrt{2}}i$$

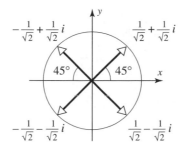

9. We observe that $w = 1$ is one sixth root of 1. Since the remaining 5 must be equally spaced around the unit circle, any two roots must be separated from one another by an angle of $\dfrac{2\pi}{6} = \dfrac{\pi}{3} = 60°$. We show all six sixth roots in the diagram.

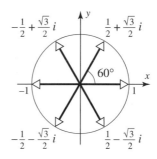

11. We have $z^4 = 16 \iff z = 16^{1/4}$. The fourth roots of 16 are 2, $2i$, -2, and $-2i$.

12. The fourth roots of -8 may be found by using Formula (10) with $r = 8$, $\theta = \pi$, and $n = 4$. This yields

$$z_1 = 8^{1/4}\left[\cos\left(\frac{\pi}{4}\right) + i\sin\left(\frac{\pi}{4}\right)\right] = 2^{1/4} + 2^{1/4}i$$

$$z_2 = 8^{1/4}\left[\cos\left(\frac{3\pi}{4}\right) + i\sin\left(\frac{3\pi}{4}\right)\right] = -2^{1/4} + 2^{1/4}i$$

$$z_3 = 8^{1/4}\left[\cos\left(\frac{5\pi}{4}\right) + i\sin\left(\frac{5\pi}{4}\right)\right] = -2^{1/4} - 2^{1/4}i$$

$$z_4 = 8^{1/4}\left[\cos\left(\frac{7\pi}{4}\right) + i\sin\left(\frac{7\pi}{4}\right)\right] = 2^{1/4} - 2^{1/4}i$$

Since $z^4 + 8$ has exactly four zeros, it follows that

$$z^4 + 8 = (z - z_1)(z - z_2)(z - z_3)(z - z_4)$$

But

$$(z - z_1)(z - z_4) = z^2 + 2^{5/4}z + 2^{3/2}$$

and

$$(z - z_2)(z - z_3) = z^2 + 2^{5/4}z + 2^3/2$$

Thus,

$$z^4 + 8 = \left(z^2 - 2^{5/4}z + 2^{3/2}\right)\left(z^2 + 2^{5/4}z + 2^{3/2}\right)$$

14. (a) We have $r = \sqrt{2}$, $\theta = \dfrac{\pi}{4}$, and $n = 8$. Thus

$$(1 + i)^8 = 2^4[\cos(2\pi) + i\sin(2\pi)] = 16$$

15. (a) Since

$$z = 3e^{i\pi} = 3[\cos(\pi) + i\sin(\pi)] = -3$$

then $\text{Re}(z) = -3$ and $\text{Im}(z) = 0$.

(c) Since

$$\bar{z} = \sqrt{2}e^{i\pi/2} = \sqrt{2}\left[\cos\left(\frac{\pi}{2}\right) + i\sin\left(\frac{\pi}{2}\right)\right] = \sqrt{2}i$$

then $z = -\sqrt{2}i$ and hence $\text{Re}(z) = 0$ and $\text{Im}(z) = -\sqrt{2}$.

16. The values of $z^{1/n}$ given in Formula (10) are all of the form

$$z_k = r^{1/n}[\cos\theta_k + i\sin\theta_k]$$

where

$$\theta_k = \frac{\theta}{n} + \frac{2k\pi}{n}$$

Thus, two roots, z_ℓ and z_m, can be equal if and only if $\cos \theta_\ell + i \sin \theta_\ell = \cos \theta_m + i \sin \theta_m$. This can occur if and only if $\cos \theta_\ell = \cos \theta_m$ and $\sin \theta_\ell = \sin \theta_m$, which, in turn, can occur if and only if θ_ℓ and θ_m differ by an integral multiple of 2π. But

$$\theta_\ell - \theta_m = \frac{2\ell\pi}{n} - \frac{2m\pi}{n} = \frac{(\ell - m)}{n}(2\pi)$$

Thus we have:

(a) If $\ell \neq m$ and ℓ and m are integers between 0 and $n - 1$, then $0 < \left| \dfrac{\ell - m}{n} \right| < 1$ so that

$\dfrac{(\ell - m)}{n}(2\pi)$ cannot be an integral multiple of 2π and hence the two resulting values

of $z^{1/n}$ are different.

(b) If, say, ℓ is not in the range 0 to $n - 1$, then we can write ℓ in the form

$$\ell = nk + m$$

for some integer m between 0 and $n - 1$ and for some integer $k \neq 0$. Thus

$$\theta_\ell - \theta_m = \frac{(\ell - m)}{n}(2\pi) = k(2\pi)$$

which guarantees that $z_\ell = z_m$.

17. <u>Case 1.</u> Suppose that $n = 0$. Then

$$(\cos \theta + i \sin \theta)^n = 1 = \cos(0) + i \sin(0)$$

So Formula (7) is valid if $n = 0$.

<u>Case 2.</u> In order to verify that Formula (7) holds if n is a negative integer, we first let $n = -1$. Then

$$(\cos \theta + i \sin \theta)^{-1} = \frac{1}{\cos \theta + i \sin \theta}$$

$$= \cos \theta - i \sin \theta$$

$$= \cos(-\theta) + i \sin(-\theta)$$

Thus, Formula (7) is valid if $n = -1$.

Now suppose that n is a positive integer (and hence that $-n$ is a negative integer). Then

$$(\cos \theta + i \sin \theta)^{-n} = \left[(\cos \theta + i \sin, \ \theta)^{-1} \right]^n$$

$$= [\cos(-\theta) + i \sin(-\theta)]^n$$

$$= \cos(-n\theta) + i \sin(-n\theta) \quad [\text{By Formula (7)}]$$

This completes the proof.

19.   We have $z_1 = r_1 e^{i\theta_1}$ and $z_2 = r_2 e^{i\theta_2}$. But (see Exercise 17)

$$\frac{1}{z_2} = \frac{1}{r_2 e^{i\theta_2}} = \frac{1}{r_2} e^{-i\theta_2}$$

If we replace z_2 by $1/z_2$ in Formula (3), we obtain

$$\frac{z_1}{z_2} = z_1 \left(\frac{1}{z_2} \right)$$

$$= \frac{r_1}{r_2} \left[\cos \left(\theta_1 + (-\theta_2) \right) + i \sin \left(\theta_1 + (-\theta_2) \right) \right]$$

$$= \frac{r_1}{r_2} \left[\cos \left(\theta_1 - \theta_2 \right) + i \sin \left(\theta_1 - \theta_2 \right) \right]$$

which is Formula (5).

20.   The first equation, when multiplied out, becomes

$$\cos^2 \theta + 2i \cos \theta \sin \theta - \sin^2 \theta = \cos 2\theta + i \sin 2\theta$$

If we equate real and imaginary parts, we obtain the identities

$$\cos^2 \theta - \sin^2 \theta = \cos 2\theta$$

and

$$2 \cos \theta \sin \theta = \sin 2\theta$$

Similarly, the second equation yields

$$\cos^3 \theta - 3 \cos \theta \sin^2 \theta = \cos 3\theta$$

and

$$3 \cos^2 \theta \sin \theta - \sin^3 \theta = \sin 3\theta$$

21. If

$$e^{i\theta} = \cos \theta + i \sin \theta$$

then replacing θ with $-\theta$ yields

$$e^{-i\theta} = \cos(-\theta) + i \sin(-\theta) = \cos \theta - i \sin \theta$$

If we then compute $e^{i\theta} + e^{-i\theta}$ and $e^{i\theta} - e^{-i\theta}$, the results will follow.

22. If $(a + bi)^3 = 8$, then $|a + bi| = 2$. That is, $a^2 + b^2 = 4$.

23. Let $z = r(\cos \theta + i \sin \theta)$. Formula (5) guarantees that $1/z = r^{-1}(\cos(-\theta) + i \sin(-\theta))$ since $z \neq 0$. Applying Formula (6) for n a positive integer to the above equation yields

$$z^{-n} = \left(\frac{1}{z}\right)^n = r^{-n}(\cos(-n\theta) + i \sin(-n\theta))$$

which is just Formula (6) for $-n$ a negative integer.

EXERCISE SET 10.4

1. **(a)** $\mathbf{u} - \mathbf{v} = (2i - (-i),\ 0 - i,\ -1 - (1 + i),\ 3 - (-1))$

   $\qquad = (3i,\ -i,\ -2 - i,\ 4)$

   (c) $-\mathbf{w} + \mathbf{v} = (-(1 + i) - i,\ i + i,\ -(-1 + 2i) + (1 + i),\ 0 + (-1))$

   $\qquad = (-1 - 2i,\ 2i,\ 2 - i,\ -1)$

   (e) $-i\mathbf{v} = (-1,\ 1,\ 1 - i,\ i)$ and $2i\mathbf{w} = (-2 + 2i,\ 2,\ -4 - 2i,\ 0)$. Thus

   $$-i\mathbf{v} + 2i\mathbf{w} = (-3 + 2i,\ 3,\ -3 - 3i,\ i)$$

2. Observe that $i\mathbf{x} = \mathbf{u} - \mathbf{v} - \mathbf{w}$. Thus

   $$\mathbf{x} = -i(\mathbf{u} - \mathbf{v} - \mathbf{w})$$

3. Consider the equation $c_1\mathbf{u}_1 + c_2\mathbf{u}_2 + c_3\mathbf{u}_3 = (-3 + i,\ 3 + 2i,\ 3 - 4i)$. The augmented matrix for this system of equations is

$$
\begin{bmatrix}
1 - i & 2i & 0 & -3 + i \\
i & 1 + i & 2i & 3 + 2i \\
0 & 1 & 2 - i & 3 - 4i
\end{bmatrix}
$$

The row-echelon form for the above matrix is

$$\begin{bmatrix} 1 & -1+i & 0 & -2-i \\ 0 & 1 & \dfrac{1}{2}+\dfrac{1}{2}i & \dfrac{3}{2}+\dfrac{1}{2}i \\ 0 & 0 & 1 & 1-i \end{bmatrix}$$

Hence, $c_3 = 2-i$, $c_2 = \dfrac{3}{2}+\dfrac{1}{2}i - \left(\dfrac{1}{2}+\dfrac{1}{2}i\right)c_3 = 0$, and $c_1 = -2-i$.

5.   (a) $\|\mathbf{v}\| = \sqrt{|1|^2 + |i|^2} = \sqrt{2}$

      (c) $\|\mathbf{v}\| = \sqrt{|2i|^2 + |0|^2 + |2i+1|^2 + |(-1)|^2} = \sqrt{4+0+5+1} = \sqrt{10}$

6.   (a)   Since $\mathbf{u}+\mathbf{v} = (3i,\ 3+4i,\ -3i)$, then

$$\|\mathbf{u}+\mathbf{v}\| = \sqrt{|3i|^2 + |3+4i|^2 + |-3i|^2} = \sqrt{9+25+9} = \sqrt{43}$$

      (c)   Here $\|\mathbf{u}\| = \sqrt{|3i|^2 + |-i|^2} = \sqrt{9+1} = \sqrt{10}$, $-i\mathbf{u} = (3,0,-1)$, and $\|-i\mathbf{u}\| = \sqrt{10}$. It follows that

$$\|-i\mathbf{u}\| + i\|\mathbf{u}\| = \sqrt{10} + i\sqrt{10}$$

      (e)   Since $\|\mathbf{w}\| = \sqrt{|1+i|^2 + |2i|^2} = \sqrt{2+4} = \sqrt{6}$, then

$$\frac{1}{\|\mathbf{w}\|}\mathbf{w} = \left(\frac{1+i}{\sqrt{6}},\ \frac{2i}{\sqrt{6}},\ 0\right)$$

8.   Since $k\mathbf{v} = (3ki,\ 4ki)$, then

$$\|k\mathbf{v}\| = \sqrt{|3ki|^2 + |4ki|^2} = 5|k|.$$

Thus $\|k\mathbf{v}\| = 1 \Longleftrightarrow |k| = \pm 1/5$.

9.   **(a)** $\mathbf{u} \cdot \mathbf{v} = (-i)(-3i) + (3i)(-2i) = -3 + 6 = 3.$

(c) $\mathbf{u} \cdot \mathbf{v} = (1 - i)(4 - 6i) + (1 + i)(5i) + (2i)(-1 - i) + (3)(-i)$

$$= (-2 - 10i) + (-5 + 5i) + (2 - 2i) + (-3i)$$

$$= -5 - 10i$$

11.   Let V denote the set and let

$$\mathbf{u} = \begin{bmatrix} u & 0 \\ 0 & \bar{u} \end{bmatrix} \quad \text{and} \quad \mathbf{v} = \begin{bmatrix} v & 0 \\ 0 & \bar{v} \end{bmatrix}$$

We check the axioms listed in the definition of a vector space (see Section 5.1).

(1) $$\mathbf{u} + \mathbf{v} = \begin{bmatrix} u + v & 0 \\ 0 & \bar{u} + \bar{v} \end{bmatrix} = \begin{bmatrix} u + v & 0 \\ 0 & \overline{u + v} \end{bmatrix}$$

So $\mathbf{u} + \mathbf{v}$ belongs to V.

(2)   Since $u + v = v + u$ and $\bar{u} + \bar{v} = \bar{v} + \bar{u}$, it follows that $\mathbf{u} + \mathbf{v} = \mathbf{v} + \mathbf{u}$.

(3)   Axiom (3) follows by a routine, if tedious, check.

(4)   The matrix $\begin{bmatrix} 0 & 0 \\ 0 & 0 \end{bmatrix}$ serves as the zero vector.

(5)   Let $-\mathbf{u} = \begin{bmatrix} -u & 0 \\ 0 & -\bar{u} \end{bmatrix} = \begin{bmatrix} -u & 0 \\ 0 & \overline{-u} \end{bmatrix}$. Then $\mathbf{u} + (-\mathbf{u}) = \mathbf{0}$.

(6)   Since $k\mathbf{u} = \begin{bmatrix} ku & 0 \\ 0 & k\bar{u} \end{bmatrix}$, $k\mathbf{u}$ will be in V if and only if $k\bar{u} = \overline{ku}$, which is true if and only if k is real or $u = 0$. Thus Axiom (6) fails.

(7)–(9)   These axioms all hold by virtue of the properties of matrix addition and scalar multiplication. However, as seen above, the closure property of scalar multiplication may fail, so the vectors need not be in V.

(10)   Clearly $1\mathbf{u} = \mathbf{u}$.

Thus, this set is not a vector space because Axiom (6) fails.

12. **(a)** Since

$$(z_1, 0, 0) + (z_2, 0, 0) = (z_1 + z_2, \ 0, \ 0)$$

and

$$k(z, 0, 0) = (kz, 0, 0)$$

this set is closed under both addition and scalar multiplication. Hence it is a subspace.

(b) Since

$$(z_1, \ i, \ i) + (z_2, \ i, \ i) = (z_1 + z_2, \ 2i, \ 2i)$$

this is not a subspace. Part (b) of Theorem 5.2.1 also is not satisfied.

13. Suppose that $T(\mathbf{x}) = A\mathbf{x} = \mathbf{0}$. It is easy to show that the reduced row echelon form for A is

$$\begin{bmatrix} 1 & 0 & (1+3i)/2 \\ 0 & 1 & (1+i)/2 \\ 0 & 0 & 0 \end{bmatrix}$$

Hence, $x_1 = (-(1+3i)/2)x_3$ and $x_2 = (-(1+i)/2)x_3$ where x_3 is an arbitrary complex number. That is,

$$\mathbf{x} = \begin{bmatrix} -(1+3i)/2 \\ -(1+i)/2 \\ 1 \end{bmatrix}$$

spans the kernel of T and hence T has nullity one.

Alternatively, the equation $A\mathbf{x} = \mathbf{0}$ yields the system

$$ix_1 - ix_2 - x_3 = 0$$
$$x_1 - ix_2 + (1+i)x_3 = 0$$
$$0 + (1-i)x_2 + x_3 = 0$$

The third equation implies that $x_3 = -(1-i)x_2$. If we substitute this expression for x_3 into the first equation, we obtain $x_1 = (2+i)x_2$. The second equation will then be valid for all such x_1 and x_3. That is, x_2 is arbitrary. Thus the kernel of T is also spanned by the vector

$$\begin{bmatrix} x_1 \\ x_2 \\ x_3 \end{bmatrix} = \begin{bmatrix} 2+i \\ 1 \\ -1+i \end{bmatrix}$$

If we multiply this vector by $-(1+i)/2$, then this answer agrees with the previous one.

14. **(a)** While this set is closed under vector addition, it is not closed under scalar multiplication since the real entries, when multiplied by a non-real scalar, need not remain real. Hence it does not form a subspace.

(b) This set *does* form a subspace since

$$\begin{bmatrix} u_1 & u_2 \\ u_3 & -u_1 \end{bmatrix} + \begin{bmatrix} v_1 & v_2 \\ v_3 & -v_1 \end{bmatrix} = \begin{bmatrix} u_1 + v_1 & u_2 + v_2 \\ u_3 + v_3 & -(u_1 + v_1) \end{bmatrix}$$

and

$$k \begin{bmatrix} z_1 & z_2 \\ z_3 & -z_1 \end{bmatrix} = \begin{bmatrix} kz_1 & kz_2 \\ kz_3 & -(kz_1) \end{bmatrix}$$

That is, the set is closed under both operations.

15. **(a)** Since

$$(f+g)(1) = f(1) + g(1) = 0 + 0 = 0$$

and

$$kf(1) = k(0) = 0$$

for all functions f and g in the set and for all scalars k, this set forms a subspace.

(c) Since

$$(f+g)(-x) = f(-x) + g(-x) = \overline{f(x)} + \overline{g(x)}$$
$$= \overline{f(x) + g(x)} = \overline{(f+g)(x)}$$

the set is closed under vector addition. It is closed under scalar multiplication by a real scalar, but not by a complex scalar. For instance, if $f(x) = xi$, then $f(x)$ is in the set but $if(x)$ is not.

16. **(a)** Consider the equation $k_1\mathbf{u} + k_2\mathbf{v} = (3i,\ 3i,\ 3i)$. Equating components yields

$$k_1 i + k_2 2i = 3i$$

$$-k_1 i + k_2 4i = 3i$$

$$k_1 3i \qquad\quad = 3i$$

The above system has the solution $k_1 = k_2 = 1$. Thus, $(3i,\ 3i,\ 3i)$ is a linear combination of $\mathbf{u}$ and $\mathbf{v}$.

(c) Consider the equation $k_1\mathbf{u} + k_2\mathbf{v} = (i, 5i, 6i)$. Equating components yields

$$k_1 i + k_2 2i = i$$

$$-k_1 i + k_2 4i = 5i$$

$$k_1 3i \qquad\quad = 6i$$

The above system is inconsistent. Hence, $(i, 5i, 6i)$ is not a linear combination of $\mathbf{u}$ and $\mathbf{v}$.

17. **(a)** Consider the equation $k_1\mathbf{u} + k_2\mathbf{v} + k_3\mathbf{w} = (1,\ 1,\ 1)$. Equating components yields

$$k_1 + (1+i)k_2 \qquad\quad = 1$$

$$k_2 + ik_3 = 1$$

$$-ik_1 + (1-2i)k_2 + 2k_3 = 1$$

Solving the system yields $k_1 = -3 - 2i$, $k_2 = 3 - i$, and $k_3 = 1 + 2i$.

(c) Let A be the matrix whose first, second and third columns are the components of $\mathbf{u}$, $\mathbf{v}$, and $\mathbf{w}$, respectively. By Part (a), we know that $\det(A) \neq 0$. Hence, $k_1 = k_2 = k_3 = 0$.

18. We let A denote the matrix whose first, second and third columns are the components of $\mathbf{v}_1$, $\mathbf{v}_2$, and $\mathbf{v}_3$, respectively.

(a) Since $\det(A) = 6i \neq 0$, it follows that $\mathbf{v}_1$, $\mathbf{v}_2$, and $\mathbf{v}_3$ span C^3.

(d) Since $\det(A) = 0$, it follows that $\mathbf{v}_1$, $\mathbf{v}_2$, and $\mathbf{v}_3$ do not span C^3.

19.   **(a)** Recall that $e^{ix} = \cos x + i \sin x$ and that $e^{-ix} = \cos(-x) + i \sin(-x) = \cos x - i \sin x$. Therefore,

$$\cos x = \frac{e^{ix} + e^{-ix}}{2} = \frac{1}{2}\mathbf{f} + \frac{1}{2}\mathbf{g}$$

and so $\cos x$ lies in the space spanned by $\mathbf{f}$ and $\mathbf{g}$.

  (b) If $a\mathbf{f} + b\mathbf{g} = \sin x$, then (see Part (a))

$$(a + b)\cos x + (a - b)i \sin x = \sin x$$

Thus, since the sine and cosine functions are linearly independent, we have

$$a + b = 0$$

and

$$a - b = -i$$

This yields $a = -i/2$, $b = i/2$, so again the vector lies in the space spanned by $\mathbf{f}$ and $\mathbf{g}$.

  (c) If $a\mathbf{f} + b\mathbf{g} = \cos x + 3i \sin x$, then (see Part (a))

$$a + b = 1$$

and

$$a - b = 3$$

Hence $a = 2$ and $b = -1$ and thus the given vector does lie in the space spanned by $\mathbf{f}$ and $\mathbf{g}$.

20.   **(a)** Note that $i\mathbf{u}_1 = \mathbf{u}_2$.

21. Let A denote the matrix whose first, second, and third columns are the components of $\mathbf{u}_1$, $\mathbf{u}_2$, and $\mathbf{u}_3$, respectively.

  (a) Since the last row of A consists entirely of zeros, it follows that $\det(A) = 0$ and hence $\mathbf{u}_1$, $\mathbf{u}_2$, and $\mathbf{u}_3$, are linearly dependent.

  (c) Since $\det(A) = i \neq 0$, then $\mathbf{u}_1$, $\mathbf{u}_2$, and $\mathbf{u}_3$ are linearly independent.

22. Observe that $\mathbf{f} - 3\mathbf{g} - 3\mathbf{h} = \mathbf{0}$.

23. **(a)** Since the dimension of C^2 is two, any basis for C^2 must contain precisely two vectors.

24. **(a)** Since $\begin{vmatrix} 2i & 4i \\ -i & 0 \end{vmatrix} = -4 \neq 0$, the vectors are linearly independent and hence form a basis for C^2.

(d) Since $\begin{vmatrix} 2 - 3i & 3 + 2i \\ i & -1 \end{vmatrix} = 0$, the vectors are linearly dependent and hence are not a basis for C^2.

25. Since the number of vectors is the same as the dimension of C^3, the vectors will form a basis if and only if they are linearly independent.

(a) Since $\begin{vmatrix} i & i & i \\ 0 & i & i \\ 0 & 0 & i \end{vmatrix} = -i \neq 0$, the vectors are linearly independent. Hence, they form a basis.

(c) From Problem 21(c), we know that the vectors are linearly independent. Hence, they form a basis.

26. The row-echelon form of the matrix of the system is

$$\begin{bmatrix} 1 & 1 + i \\ 0 & 0 \end{bmatrix}$$

So x_2 is arbitrary and $x_1 = -(1 + i)x_2$. Hence, the dimension of the solution space is one and $\begin{bmatrix} -(1 + i) \\ 1 \end{bmatrix}$ is a basis for that space.

28. The reduced row-echelon form of the matrix of the system is

$$
\begin{bmatrix}
1 & 0 & -3 - 6i \\
0 & 1 & 3i \\
0 & 0 & 0
\end{bmatrix}
$$

So x_3 is arbitrary, $x_2 = (-3i)x_3$, and $x_1 = (3 + 6i)x_3$. Hence, the dimension of the solution space is one and $\begin{bmatrix} 3 + 6i \\ -3i \\ 1 \end{bmatrix}$ is a basis for that space.

30. Let $\mathbf{u} = (u_1, u_2, \ldots, u_n)$ and $\mathbf{v} = (v_1, v_2, \ldots, v_n)$. From the definition of the Euclidean inner product in C^n, we have

$$
\begin{aligned}
\mathbf{u} \cdot (k\mathbf{v}) &= u_1(\overline{kv_1}) + u_2(\overline{kv_2}) + \cdots + u_n(\overline{kv_n}) \\
&= u_1(\bar{k}\bar{v}_1) + u_2(\bar{k}\bar{v}_2) + \cdots + u_n(\bar{k}\bar{v}_n) \\
&= \bar{k}(u_1\bar{v}_1) + \bar{k}(u_2\bar{v}_2) + \cdots + \bar{k}(u_n\bar{v}_n) \\
&= \bar{k}[u_1\bar{v}_1 + u_2\bar{v}_2 + \cdots + u_n\bar{v}_n] \\
&= \bar{k}(\mathbf{u} \cdot \mathbf{v})
\end{aligned}
$$

31. (a) Let $\mathbf{u} = (u_1, u_2, \ldots, u_n)$, $\mathbf{v} = (v_1, v_2, \ldots, v_n)$ and $\mathbf{w} = (w_1, w_2, \ldots, w_n)$. Then, since we have $\mathbf{u} + \mathbf{v} = (u_1 + v_1, u_2 + v_2, \ldots, u_n + v_n)$,

$$
\begin{aligned}
(\mathbf{u} + \mathbf{v}) \cdot \mathbf{w} &= (u_1 + v_1)\bar{w}_1 + (u_2 + v_2)\bar{w}_2 + \cdots + (u_n + v_n)\bar{w}_n \\
&= [u_1\bar{w}_1 + u_2\bar{w}_2 + \cdots + u_n\bar{w}_n] + [v_1\bar{w}_1 + v_2\bar{w}_2 + \cdots + v_n\bar{w}_n] \\
&= \mathbf{u} \cdot \mathbf{w} + \mathbf{v} \cdot \mathbf{w}
\end{aligned}
$$

32. Hint: Show that

$$
\|\mathbf{u} + k\mathbf{v}\|^2 = \|\mathbf{u}\|^2 + k(\mathbf{v} \cdot \mathbf{u}) + \bar{k}(\mathbf{u} \cdot \mathbf{v}) + k\bar{k}\|\mathbf{v}\|^2
$$

and apply this result to each term on the right-hand side of the identity.

33. Let $f(x) = f_1(x) + if_2(x)$ and $g(x) = g_1(x) + ig_2(x)$. Then

$$(f + g)(x) = f(x) + g(x)$$
$$= [f_1(x) + g_1(x)] + i[f_2(x) + g_2(x)]$$

and

$$(kf)(x) = kf(x)$$
$$= kf_1(x) + ikf_2(x)$$

Thus, if f and g are continuous, then $f + g$ is continuous; similarly, if f is continuous, then kf is continuous. Hence, the result follows from Theorem 5.2.1.

EXERCISE SET 10.5

1. Let $\mathbf{u} = (u_1,\, u_2)$, $\mathbf{v} = (v_1,\, v_2)$, and $\mathbf{w} = (w_1,\, w_2)$. We proceed to check the four axioms

(1)
$$\overline{\langle \mathbf{v}, \mathbf{u} \rangle} = \overline{3v_1\bar{u}_1 + 2v_2\bar{u}_2}$$

$$= 3u_1\bar{v}_1 + 2u_2\bar{v}_2 = \langle \mathbf{u}, \mathbf{v} \rangle$$

(2)
$$\langle \mathbf{u} + \mathbf{v}, \mathbf{w} \rangle = 3(u_1 + v_1)\bar{w}_1 + 2(u_2 + v_2)\bar{w}_2$$

$$= [3u_1\bar{w}_1 + 2u_2\bar{w}_2] + [3v_1\bar{w}_1 + 2v_2\bar{w}_2]$$

$$= \langle \mathbf{u}, \mathbf{w} \rangle + \langle \mathbf{v}, \mathbf{w} \rangle$$

(3)
$$\langle k\mathbf{u}, \mathbf{v} \rangle = 3(ku_1)\bar{v}_1 + 2(ku_2)\bar{v}_2$$

$$= k[3u_1\bar{v}_1 + 2u_2\bar{v}_2] = k\langle \mathbf{u}, \mathbf{v} \rangle$$

(4)
$$\langle \mathbf{u}, \mathbf{u} \rangle = 3u_1\bar{u}_1 + 2u_2\bar{u}_2$$

$$= 3|u_1|^2 + 2|u_2|^2 \qquad \text{(Theorem 10.2.1)}$$

$$\geq 0$$

Indeed, $\langle \mathbf{u}, \mathbf{u} \rangle = 0 \Longleftrightarrow u_1 = u_2 = 0 \Longleftrightarrow \mathbf{u} = \mathbf{0}$.

Hence, this is an inner product on C^2.

2. (a) $\langle \mathbf{u}, \mathbf{v} \rangle = 3(2i)(\overline{-i}) + 2(-i)(\overline{3i}) = -6 - 6 = -12$

   (c) $\langle \mathbf{u}, \mathbf{v} \rangle = 3(1 + i)(\overline{1 - i}) + 2(1 - i)(\overline{1 + i})$

   $$= 3(2i) + 2(-2i) = 2i$$

3. Let $\mathbf{u} = (u_1, u_2)$ and $\mathbf{v} = (v_1, v_2)$. We check Axioms 1 and 4, leaving 2 and 3 to you.

(1)
$$\overline{\langle \mathbf{v}, \mathbf{u} \rangle} = \overline{v_1 \bar{u}_1} + \overline{(1+i)v_1 \bar{u}_2} + \overline{(1-i)v_2 \bar{u}_1} + \overline{3 v_2 \bar{u}_2}$$

$$= u_1 \bar{v}_1 + (1-i)u_2 \bar{v}_1 + (1+i)u_1 \bar{v}_2 + 3 u_2 \bar{v}_2$$

$$= \langle \mathbf{u}, \mathbf{v} \rangle$$

(4) Recall that $|\text{Re}(z)| \leq |z|$ by Problem 29 of Section 10.1. Now

$$\langle \mathbf{u}, \mathbf{u} \rangle = u_1 \bar{u}_1 + (1+i)u_1 \bar{u}_2 + (1-i)u_2 \bar{u}_1 + 3 u_2 \bar{u}_2$$

$$= |u_1|^2 + (1+i)u_1 \bar{u}_2 + \overline{(1+i)u_1 \bar{u}_2} + 3|u_2|^2$$

$$= |u_1|^2 + 2\text{Re}((1+i)u_1 \bar{u}_2) + 3|u_2|^2$$

$$\geq |u_1|^2 - 2|(1+i)u_1 \bar{u}_2| + 3|u_2|^2$$

$$= |u_1|^2 - 2\sqrt{2}\,|u_1||u_2| + 3|u_2|^2$$

$$= \left(|u_1| - \sqrt{2}\,|u_2| \right)^2 + |u_2|^2$$

$$\geq 0$$

Moreover, $\langle \mathbf{u}, \mathbf{u} \rangle = 0$ if and only if both $|u_2|$ and $|u_1| - \sqrt{2}\,|u_2| = 0$, or $\mathbf{u} = \mathbf{0}$.

4. **(a)** $\langle \mathbf{u}, \mathbf{v} \rangle = (2i)i + (1+i)(2i)(-3i) + (1-i)(-i)(i) + 3(-i)(-3i)$

$$= -4 + 5i$$

(c) $\langle \mathbf{u}, \mathbf{v} \rangle = (1+i)(1+i) + (1+i)(1+i)(1-i) + (1-i)(1-i)(1+i) + 3(1-i)(1-i)$

$$= 4 - 4i$$

5. **(a)** This is *not* an inner product on C^2. Axioms 1–3 are easily checked. Moreover,

$$\langle \mathbf{u}, \mathbf{u} \rangle = u_1 \bar{u}_1 = |u_1| \geq 0$$

However, $\langle \mathbf{u}, \mathbf{u} \rangle = 0 \iff u_1 = 0 \not\Longleftrightarrow \mathbf{u} = \mathbf{0}$. For example, $\langle i, i \rangle = 0$ although $i \neq \mathbf{0}$. Hence, Axiom 4 fails.

(c) This is *not* an inner product on C^2. Axioms 1 and 4 are easily checked. However, for $\mathbf{w} = (w_1, w_2)$, we have

$$\langle \mathbf{u} + \mathbf{v}, \mathbf{w} \rangle = |u_1 + v_1|^2 |w_1|^2 + |u_2 + v_2|^2 |w_2|^2$$

$$\neq \left(|u_1|^2 + |v_1|^2 \right) |w_1|^2 + \left(|u_2|^2 + |v_2|^2 \right) |w_2|^2$$

$$= \langle \mathbf{u}, \mathbf{w} \rangle + \langle \mathbf{v}, \mathbf{w} \rangle$$

For instance, $\langle 1+1, 1 \rangle = 4$, but $\langle 1, 1 \rangle + \langle 1, 1 \rangle = 2$. Moreover, $\langle k\mathbf{u}, \mathbf{v} \rangle = |k|^2 \langle \mathbf{u}, \mathbf{v} \rangle$, so that $\langle k\mathbf{u}, \mathbf{v} \rangle \neq k \langle \mathbf{u}, \mathbf{v} \rangle$ for most values of $k, \mathbf{u}$, and $\mathbf{v}$. Thus both Axioms 2 and 3 fail.

(e) Axiom 1 holds since

$$\overline{\langle \mathbf{v}, \mathbf{u} \rangle} = \overline{2v_1\bar{u}_1 + iv_1\bar{u}_2 - iv_2\bar{u}_1 + 2v_2\bar{u}_2}$$

$$= 2u_1\bar{v}_1 - iu_2\bar{v}_1 + iu_1\bar{v}_2 + 2u_2\bar{v}_2$$

$$= \langle \mathbf{u}, \mathbf{v} \rangle$$

A similar argument serves to verify Axiom 2 and Axiom 3 holds by inspection. Finally, using the result of Problem 29 of Section 10.1, we have

$$\langle \mathbf{u}, \mathbf{u} \rangle = 2u_1\bar{u}_1 + iu_1\bar{u}_2 - iu_2\bar{u}_1 + 2u_2\bar{u}_2$$

$$= 2|u_1|^2 + 2\mathrm{Re}(iu_1\bar{u}_2) + 2|u_2|^2$$

$$\geq 2|u_1|^2 - 2|iu_1\bar{u}_2| + 2|u_2|^2$$

$$= (|u_1| - |u_2|)^2 + |u_1|^2 + |u_2|^2$$

$$\geq 0$$

Moreover, $\langle \mathbf{u}, \mathbf{u} \rangle = 0 \iff u_1 = u_2 = 0$, or $\mathbf{u} = \mathbf{0}$. Thus all four axioms hold.

6. $\langle \mathbf{u}, \mathbf{v} \rangle = (-i)(3) + (1+i)(-2+3i) + (1-i)(-4i) + i(1)$

$$= -9 - 5i$$

8. First,

$$\overline{\langle \mathbf{g}, \mathbf{f} \rangle} = \overline{(g_1(0) + ig_2(0))(\overline{f_1(0) + if_2(0)})}$$

$$= (f_1(0) + if_2(0))\overline{(g_1(0) + ig_2(0))}$$

$$= \langle \mathbf{f}, \mathbf{g} \rangle$$

Second, if $\mathbf{h} = h_1(x) + ih_2(x)$, then

$$\langle \mathbf{f} + \mathbf{g}, \mathbf{h} \rangle = (f_1(0) + g_1(0) + i[f_2(0) + g_2(0)]) \cdot \overline{(h_1(0) + ih_2(0))}$$

$$= \langle \mathbf{f}, \mathbf{h} \rangle + \langle \mathbf{g}, \mathbf{h} \rangle$$

Third, it is easily checked that $\langle k\mathbf{f}, \mathbf{g} \rangle = k\langle \mathbf{f}, \mathbf{g} \rangle$. And finally,

$$\langle \mathbf{f}, \mathbf{f} \rangle = (f_1(0) + if_2(0))\overline{(f_1(0) + if_2(0))}$$

$$= |f_1(0)|^2 + |f_2(0)|^2$$

$$\geq 0$$

However, $\langle \mathbf{f}, \mathbf{f} \rangle = 0 \iff f_1(0) = f_2(0) = 0 \not\Longleftrightarrow \mathbf{f} = \mathbf{0}$. For instance, if $\mathbf{f} = ix$, then $\langle \mathbf{f}, \mathbf{f} \rangle = 0$, but $\mathbf{f} \neq \mathbf{0}$. Therefore, this is not an inner product.

9. **(a)** $\|\mathbf{w}\| = [3(-i)(i) + 2(3i)(-3i)]^{1/2} = \sqrt{21}$

   (c) $\|\mathbf{w}\| = [3(0)(0) + 2(2 - i)(2 + i)]^{1/2} = \sqrt{10}$

10. **(a)** $\|\mathbf{w}\| = [|-i|^2 + |3i|^2]^{1/2} = \sqrt{10}$

   (c) $\|\mathbf{w}\| = [|0|^2 + |2 - i|^2]^{1/2} = \sqrt{5}$

11. **(a)** $\|\mathbf{w}\| = [(1)(1) + (1 + i)(1)(i) + (1 - i)(-i)(1) + 3(-i)(i)]^{1/2} = \sqrt{2}$

   (c) $\|\mathbf{w}\| = [(3 - 4i)(3 + 4i)]^{1/2} = 5$

12. **(a)** $\|A\| = [(-i)(i) + (7i)(-7i) + (6i)(-6i) + (2i)(-2i)]^{1/2} = 3\sqrt{10}$

13. **(a)** Since $\mathbf{u} - \mathbf{v} = (1 - i, 1 + i)$, then

$$d(\mathbf{u}, \mathbf{v}) = [3(1-i)(1+i) + 2(1+i)(1-i)]^{1/2} = \sqrt{10}$$

14. **(a)** Since $\mathbf{u} - \mathbf{v} = (1 - i, 1 + i)$,

$$d(\mathbf{u}, \mathbf{v}) = [(1-i)(1+i) + (1+i)(1-i)]^{1/2} = 2$$

15. **(a)** Since $\mathbf{u} - \mathbf{v} = (1 - i, 1 + i)$,

$$d(\mathbf{u}, \mathbf{v}) = [(1-i)(1+i) + (1+i)(1-i)(1-i)$$
$$+ (1-i)(1+i)(1+i) + 3(1+i)(1-i)]^{1/2}$$
$$= 2\sqrt{3}$$

16. **(a)** Since $A - B = \begin{bmatrix} 6i & 5i \\ i & 6i \end{bmatrix}$,

$$d(A, B) = [(6i)(-6i) + (5i)(-5i) + (i)(-i) + (6i)(-6i)]^{1/2} = \sqrt{98} = 7\sqrt{2}$$

17. **(a)** Since $\mathbf{u} \cdot \mathbf{v} = (2i)(-i) + (i)(-6i) + (3i)(\bar{k})$, then $\mathbf{u} \cdot \mathbf{v} = 0 \iff 8 + 3i\bar{k} = 0$ or $k = -8i/3$.

18. **(a)** Let B denote the given matrix. Then

$$\langle A, B \rangle = (2i)(-3) + (i)(1+i) + (-i)(1+i) + (3i)(2) = 0$$

Thus, A and B are orthogonal.

(d) Let B denote the given matrix. Then

$$\langle A, B \rangle = i + (-i)(3+i) = 1 - 2i$$

Thus, A and B are not orthogonal.

19. Since $\mathbf{x} = \dfrac{1}{\sqrt{3}}e^{i\theta}(i, 1, 1)$, we have

$$\|\mathbf{x}\| = \frac{1}{\sqrt{3}}\|e^{i\theta}\|\,\|(i, 1, 1)\| = \frac{1}{\sqrt{3}}(1)(1+1+1)^{1/2} = 1$$

Also

$$\langle \mathbf{x}, (1, i, 0)\rangle = \frac{1}{\sqrt{3}}e^{i\theta}[(i, 1, 1)\cdot(1, i, 0)]$$

$$= \frac{1}{\sqrt{3}}e^{i\theta}(i - i + 0)$$

$$= 0$$

and

$$\langle \mathbf{x}, (0, i, -i)\rangle = \frac{1}{\sqrt{3}}e^{i\theta}[(i, 1, 1)\cdot(0, i, -i)]$$

$$= \frac{1}{\sqrt{3}}e^{i\theta}(0 - i + i)$$

$$= 0$$

20. **(a)** Since $\|(0, 1 - i)\| = \sqrt{2}$, the set of vectors is not orthonormal. (However, the set is orthogonal.)

(b) Since $\left(\dfrac{i}{\sqrt{2}}, -\dfrac{i}{\sqrt{2}}\right)\cdot\left(\dfrac{i}{\sqrt{2}}, \dfrac{i}{\sqrt{2}}\right) = -\dfrac{i^2}{2} + \dfrac{i^2}{2} = 0$ and the norm of each vector is 1, the set is orthonormal.

21. **(a)** Call the vectors $\mathbf{u}_1$, $\mathbf{u}_2$, and $\mathbf{u}_3$, respectively. Then $\|\mathbf{u}_1\| = \|\mathbf{u}_2\| = \|\mathbf{u}_3\| = 1$ and $\mathbf{u}_1\cdot\mathbf{u}_2 = \mathbf{u}_1\cdot\mathbf{u}_3 = 0$. However, $\mathbf{u}_2\cdot\mathbf{u}_3 = \dfrac{i^2}{\sqrt{6}} + \dfrac{(-i)^2}{\sqrt{6}} = -\dfrac{2}{\sqrt{6}} \neq 0$. Hence the set is not orthonormal.

22. If we use the Euclidean inner product, then $\|\mathbf{x}\| = \sqrt{2/5} \neq 1$. Hence, the vectors cannot be orthonormal with respect to the Euclidean inner product. In fact, neither vector is normal and the vectors are not orthogonal.

We now let $\langle \mathbf{u}, \mathbf{v} \rangle = 3u_1\bar{v}_1 + 2u_2\bar{v}_2$. Then

$$\|\mathbf{x}\|^2 = 3\left(\frac{i}{\sqrt{5}}\right)\left(-\frac{i}{\sqrt{5}}\right) + 2\left(-\frac{i}{\sqrt{5}}\right)\left(\frac{i}{\sqrt{5}}\right) = 1$$

$$\|\mathbf{y}\|^2 = 3\left(\frac{2i}{\sqrt{30}}\right)\left(-\frac{2i}{\sqrt{30}}\right) + 2\left(\frac{3i}{\sqrt{30}}\right)\left(-\frac{3i}{\sqrt{30}}\right) = 1$$

and

$$\langle \mathbf{x}, \mathbf{y} \rangle = 3\left(\frac{i}{\sqrt{5}}\right)\left(-\frac{2i}{\sqrt{30}}\right) + 2\left(-\frac{i}{\sqrt{5}}\right)\left(-\frac{3i}{\sqrt{30}}\right) = 0$$

Hence $\mathbf{x}$ and $\mathbf{y}$ are orthonormal with respect to this inner product.

24. **(a)** We have

$$\mathbf{v}_1 = \frac{\mathbf{u}_1}{\|\mathbf{u}_1\|} = \frac{(i, -3i)}{\sqrt{10}} = \left(\frac{i}{\sqrt{10}}, -\frac{3i}{\sqrt{10}}\right)$$

and, since $\mathbf{u}_2 \cdot \mathbf{v}_1 = -4/\sqrt{10}$, it follows that

$$\mathbf{u}_2 - (\mathbf{u}_2 \cdot \mathbf{v}_1)\mathbf{v}_1 = (2i,\ 2i) + \frac{4}{\sqrt{10}}\left(\frac{i}{\sqrt{10}}, \frac{-3i}{\sqrt{10}}\right)$$

$$= \left(\frac{12i}{5}, \frac{4i}{5}\right)$$

Because $\left\|\left(\dfrac{12i}{5}, \dfrac{4i}{5}\right)\right\| = \dfrac{4\sqrt{10}}{5}$, then

$$\mathbf{v}_2 = \left(\frac{3i}{\sqrt{10}}, \frac{i}{\sqrt{10}}\right)$$

25. **(a)** We have

$$\mathbf{v}_1 = \left(\frac{i}{\sqrt{3}}, \frac{i}{\sqrt{3}}, \frac{i}{\sqrt{3}}\right)$$

and since $\mathbf{u}_2 \cdot \mathbf{v}_1 = 0$, then $\mathbf{u}_2 - (\mathbf{u}_2 \cdot \mathbf{v}_1)\mathbf{v}_1 = \mathbf{u}_2$. Thus,

$$\mathbf{v}_2 = \left(-\frac{i}{\sqrt{2}}, \; \frac{i}{\sqrt{2}}, \; 0 \right)$$

Also, $\mathbf{u}_3 \cdot \mathbf{v}_1 = 4/\sqrt{3}$ and $\mathbf{u}_3 \cdot \mathbf{v}_2 = 1/\sqrt{2}$ and hence

$$\mathbf{u}_3 - (\mathbf{u}_3 \cdot \mathbf{v}_1)\mathbf{v}_1 - (\mathbf{u}_3 \cdot \mathbf{v}_2)\mathbf{v}_2 = \left(\frac{i}{6}, \; \frac{i}{6}, \; \frac{-i}{3} \right)$$

Since the norm of the above vector is $1/\sqrt{6}$, we have

$$\mathbf{v}_3 = \left(\frac{i}{\sqrt{6}}, \; \frac{i}{\sqrt{6}}, \; \frac{-2i}{\sqrt{6}} \right)$$

27. Let $\mathbf{u}_1 = (0, \; i, \; 1-i)$ and $\mathbf{u}_2 = (-i, \; 0, \; 1+i)$. We shall apply the Gram-Schmidt process to $\{\mathbf{u}_1, \mathbf{u}_2\}$. Since $\|\mathbf{u}_1\| = \sqrt{3}$, it follows that

$$\mathbf{v}_1 = \left(0, \; \frac{i}{\sqrt{3}}, \; \frac{1-i}{\sqrt{3}} \right)$$

Because $\mathbf{u}_2 \cdot \mathbf{v}_1 = 2i/\sqrt{3}$, then

$$\mathbf{u}_2 - (\mathbf{u}_2 \cdot \mathbf{v}_1)\mathbf{v}_1 = (-i, \; 0, \; 1+i) - \left(0, \; -\frac{2}{3}, \; \frac{2}{3} + \frac{2}{3}i \right)$$

$$= \left(-i, \; \frac{2}{3}, \; \frac{1}{3} + \frac{1}{3}i \right)$$

and because the norm of the above vector is $\sqrt{15}/3$, we have

$$\mathbf{v}_2 = \left(\frac{-3i}{\sqrt{15}}, \; \frac{2}{\sqrt{15}}, \; \frac{1+i}{\sqrt{15}} \right)$$

28. Following Theorem 6.3.6, we orthonormalize the set $\{\mathbf{u}_1, \mathbf{u}_2\}$. This gives us

$$\mathbf{v}_1 = \frac{\mathbf{u}_1}{\|\mathbf{u}_1\|} = \left(-\frac{i}{\sqrt{6}}, \; 0, \; \frac{i}{\sqrt{6}}, \; \frac{2i}{\sqrt{6}} \right)$$

and, since $\mathbf{u}_2 \cdot \mathbf{v}_1 = 2/\sqrt{6}$,

$$\mathbf{u}_2 - (\mathbf{u}_2 \cdot \mathbf{v}_1)\mathbf{v}_1 = (0,\ i,\ 0,\ i) - \frac{1}{3}(-i,\ 0,\ i,\ 2i)$$

$$= \left(\frac{i}{3},\ i,\ -\frac{i}{3},\ \frac{i}{3}\right)$$

so that

$$\mathbf{v}_2 = \left(\frac{i}{2\sqrt{3}},\ \frac{3i}{2\sqrt{3}},\ -\frac{i}{2\sqrt{3}},\ \frac{i}{2\sqrt{3}}\right)$$

Therefore,

$$\mathbf{w}_1 = (\mathbf{w} \cdot \mathbf{v}_1)\mathbf{v}_1 + (\mathbf{w} \cdot \mathbf{v}_2)\mathbf{v}_2$$

$$= \frac{7}{\sqrt{6}}\mathbf{v}_1 + \frac{-1}{2\sqrt{3}}\mathbf{v}_2$$

$$= \left(-\frac{5}{4}i,\ -\frac{1}{4}i,\ \frac{5}{4}i,\ \frac{9}{4}i\right)$$

and

$$\mathbf{w}_2 = \mathbf{w} - \mathbf{w}_1$$

$$= \left(\frac{1}{4}i,\ \frac{9}{4}i,\ \frac{19}{4}i,\ -\frac{9}{4}i\right)$$

29. (a) By Axioms (2) and (3) for inner products,

$$\langle \mathbf{u} - k\mathbf{v},\ \mathbf{u} - k\mathbf{v} \rangle = \langle \mathbf{u},\ \mathbf{u} - k\mathbf{v} \rangle + \langle -k\mathbf{v},\ \mathbf{u} - k\mathbf{v} \rangle$$

$$= \langle \mathbf{u},\ \mathbf{u} - k\mathbf{v} \rangle - k\langle \mathbf{v},\ \mathbf{u} - k\mathbf{v} \rangle$$

If we use Properties (ii) and (iii) of inner products, then we obtain

$$\langle \mathbf{u} - k\mathbf{v},\ \mathbf{u} - k\mathbf{v} \rangle = \langle \mathbf{u}, \mathbf{u} \rangle - \bar{k}\langle \mathbf{u}, \mathbf{v} \rangle - k\langle \mathbf{v}, \mathbf{u} \rangle + k\bar{k}\langle \mathbf{v}, \mathbf{v} \rangle$$

Finally, Axiom (1) yields

$$\langle \mathbf{u} - k\mathbf{v},\ \mathbf{u} - k\mathbf{v} \rangle = \langle \mathbf{u}, \mathbf{u} \rangle - \bar{k}\langle \mathbf{u}, \mathbf{v} \rangle - k\overline{\langle \mathbf{u}, \mathbf{v} \rangle} + k\bar{k}\langle \mathbf{v}, \mathbf{v} \rangle$$

and the result is proved.

(b) This follows from Part (a) and Axiom (4) for inner products.

30. Suppose that $\mathbf{v} \neq \mathbf{0}$ and hence that $\langle \mathbf{v}, \mathbf{v} \rangle \neq 0$. Using the hint, we obtain

$$0 \leq \langle \mathbf{u}, \mathbf{u} \rangle - \left(\frac{\overline{\langle \mathbf{u}, \mathbf{v} \rangle}}{\langle \mathbf{v}, \mathbf{v} \rangle} \right) \langle \mathbf{u}, \mathbf{v} \rangle - \left(\frac{\langle \mathbf{u}, \mathbf{v} \rangle}{\langle \mathbf{v}, \mathbf{v} \rangle} \right) \overline{\langle \mathbf{u}, \mathbf{v} \rangle} + \left| \frac{\langle \mathbf{u}, \mathbf{v} \rangle}{\langle \mathbf{v}, \mathbf{v} \rangle} \right|^2 \langle \mathbf{v}, \mathbf{v} \rangle$$

Since $\langle \mathbf{v}, \mathbf{v} \rangle$ is real, this yields the inequality

$$0 \leq \langle \mathbf{u}, \mathbf{u} \rangle - \frac{|\langle \mathbf{u}, \mathbf{v} \rangle|^2}{\langle \mathbf{v}, \mathbf{v} \rangle} - \frac{|\langle \mathbf{u}, \mathbf{v} \rangle|^2}{\langle \mathbf{v}, \mathbf{v} \rangle} + \frac{|\langle \mathbf{u}, \mathbf{v} \rangle|^2}{\langle \mathbf{v}, \mathbf{v} \rangle}$$

or

$$\frac{|\langle \mathbf{u}, \mathbf{v} \rangle|^2}{\langle \mathbf{v}, \mathbf{v} \rangle} \leq \langle \mathbf{u}, \mathbf{u} \rangle$$

or

$$|\langle \mathbf{u}, \mathbf{v} \rangle|^2 \leq \langle \mathbf{u}, \mathbf{u} \rangle \langle \mathbf{v}, \mathbf{v} \rangle$$

as desired.

In case $\mathbf{v} = \mathbf{0}$, Property (i) for inner products guarantees that both sides of the Cauchy-Schwarz inequality are zero, and thus equality holds.

32. If $\mathbf{u}$ and $\mathbf{v}$ are linearly dependent, then either $\mathbf{v} = \mathbf{0}$ or $\mathbf{u}$ is a scalar multiple of $\mathbf{v}$. In either case, a routine computation using properties of the inner product shows that equality holds.

Conversely, if equality holds, then we shall show that $\mathbf{u}$ and $\mathbf{v}$ must be linearly dependent. For suppose that they are not. Then $\mathbf{u} - k\mathbf{v} \neq \mathbf{0}$ for every scalar k, so it follows that $\langle \mathbf{u} - k\mathbf{v}, \mathbf{u} - k\mathbf{v} \rangle$ is greater than zero for all k. Thus, strict inequality must hold in Problem 29(b). Now if we choose k as in Problem 30 and rederive the Cauchy-Schwarz inequality, we find that strict inequality must hold there also. That is, if $\mathbf{u}$ and $\mathbf{v}$ are linearly independent, then the inequality must hold. Hence, if the equality holds, then we are forced to conclude that $\mathbf{u}$ and $\mathbf{v}$ are linearly dependent.

33. Hint: Let $\mathbf{v}$ be any nonzero vector, and consider the quantity $\langle \mathbf{v}, \mathbf{v} \rangle + \langle \mathbf{0}, \mathbf{v} \rangle$.

34. We have

$$\langle \mathbf{u}, \mathbf{v} + \mathbf{w} \rangle = \overline{\overline{\langle \mathbf{u}, \mathbf{v} + \mathbf{w} \rangle}}$$

$$= \overline{\langle \mathbf{v} + \mathbf{w}, \mathbf{u} \rangle} \qquad \text{(Axiom 1)}$$

$$= \overline{\langle \mathbf{v}, \mathbf{u} \rangle + \langle \mathbf{w}, \mathbf{u} \rangle} \qquad \text{(Axiom 2)}$$

$$= \overline{\langle \mathbf{v}, \mathbf{u} \rangle} + \overline{\langle \mathbf{w}, \mathbf{u} \rangle}$$

$$= \langle \mathbf{u}, \mathbf{v} \rangle + \langle \mathbf{u}, \mathbf{w} \rangle \qquad \text{(Axiom 1)}$$

35. **(d)** Observe that $\|\mathbf{u} + \mathbf{v}\|^2 = \langle \mathbf{u} + \mathbf{v},\ \mathbf{u} + \mathbf{v} \rangle$. As in Exercise 29,

$$\langle \mathbf{u} + \mathbf{v},\ \mathbf{u} + \mathbf{v} \rangle = \langle \mathbf{u}, \mathbf{u} \rangle + 2\operatorname{Re}(\langle \mathbf{u}, \mathbf{v} \rangle) + \langle \mathbf{v}, \mathbf{v} \rangle$$

Since (see Exercise 29 of Section 10.1)

$$|\operatorname{Re}(\langle \mathbf{u}, \mathbf{v} \rangle)| \leq |\langle \mathbf{u}, \mathbf{v} \rangle|$$

this yields

$$\langle \mathbf{u} + \mathbf{v},\ \mathbf{u} + \mathbf{v} \rangle \leq \langle \mathbf{u}, \mathbf{u} \rangle + 2|\langle \mathbf{u}, \mathbf{v} \rangle| + \langle \mathbf{v}, \mathbf{v} \rangle$$

By the Cauchy-Schwarz inequality and the definition of norm, this becomes

$$\|\mathbf{u} + \mathbf{v}\|^2 \leq \|\mathbf{u}\|^2 + 2\|\mathbf{u}\|\,\|\mathbf{v}\| + \|\mathbf{v}\|^2 = (\|\mathbf{u}\| + \|\mathbf{v}\|)^2$$

which yields the desired result.

(h) Replace $\mathbf{u}$ by $\mathbf{u} - \mathbf{w}$ and $\mathbf{v}$ by $\mathbf{w} - \mathbf{v}$ in Theorem 6.2.2, Part (d).

37. Observe that for any complex number k,

$$
\begin{aligned}
\|\mathbf{u} + k\mathbf{v}\|^2 &= \langle \mathbf{u} + k\mathbf{v},\ \mathbf{u} + k\mathbf{v} \rangle \\
&= \langle \mathbf{u}, \mathbf{u} \rangle + k\langle \mathbf{v}, \mathbf{u} \rangle + \bar{k}\langle \mathbf{u}, \mathbf{v} \rangle + k\bar{k}\langle \mathbf{v}, \mathbf{v} \rangle \\
&= \langle \mathbf{u}, \mathbf{u} \rangle + 2\operatorname{Re}(k\langle \mathbf{v}, \mathbf{u} \rangle) + |k|^2\langle \mathbf{v}, \mathbf{v} \rangle
\end{aligned}
$$

Therefore,

$$
\begin{aligned}
\|\mathbf{u} + \mathbf{v}\|^2 &- \|\mathbf{u} - \mathbf{v}\|^2 + i\|\mathbf{u} + i\mathbf{v}\|^2 - i\|\mathbf{u} - i\mathbf{v}\|^2 \\
&= (1 - 1 + i - i)\langle \mathbf{u}, \mathbf{u} \rangle + 2\operatorname{Re}(\langle \mathbf{v}, \mathbf{u} \rangle) - 2\operatorname{Re}(-\langle \mathbf{v}, \mathbf{u} \rangle) + 2i\operatorname{Re}(i\langle \mathbf{v}, \mathbf{u} \rangle) \\
&\quad -2i\operatorname{Re}(-i\langle \mathbf{v}, \mathbf{u} \rangle) + (1 - 1 + i - i)\langle \mathbf{v}, \mathbf{v} \rangle \\
&= 4\operatorname{Re}(\langle \mathbf{v}, \mathbf{u} \rangle) - 4i\operatorname{Im}(\langle \mathbf{v}, \mathbf{u} \rangle) \\
&= 4\overline{\langle \mathbf{v}, \mathbf{u} \rangle} \\
&= 4\langle \mathbf{u}, \mathbf{v} \rangle
\end{aligned}
$$

38. By the hint, we have $\mathbf{u} = \displaystyle\sum_{j=1}^{n} \langle \mathbf{u}, \mathbf{v}_j \rangle \mathbf{v}_j$ and $\mathbf{w} = \displaystyle\sum_{k=1}^{n} \langle \mathbf{w}, \mathbf{v}_k \rangle \mathbf{v}_k$. Therefore,

$$\langle \mathbf{u}, \mathbf{w} \rangle = \sum_{j,k=1}^{n} \langle \mathbf{u}, \mathbf{v}_j \rangle \overline{\langle \mathbf{w}, \mathbf{v}_k \rangle} \langle \mathbf{v}_j, \mathbf{v}_k \rangle$$

$$= \sum_{j=1}^{n} \langle \mathbf{u}, \mathbf{v}_j \rangle \overline{\langle \mathbf{w}, \mathbf{v}_j \rangle}$$

The last step uses the fact that the basis vectors $\mathbf{v}_j$ form an orthonormal set.

39. We check Axioms 2 and 4. For Axiom 2, we have

$$\langle \mathbf{f} + \mathbf{g}, \mathbf{h} \rangle = \int_a^b (\mathbf{f} + \mathbf{g}) \bar{\mathbf{h}} \, dx$$

$$= \int_a^b \mathbf{f} \, \bar{\mathbf{h}} \, dx + \int_a^b \mathbf{g} \, \bar{\mathbf{h}} \, dx$$

$$= \langle \mathbf{f}, \mathbf{h} \rangle + \langle \mathbf{g}, \mathbf{h} \rangle$$

For Axiom 4, we have

$$\langle \mathbf{f}, \mathbf{f} \rangle = \int_a^b \mathbf{f} \, \bar{\mathbf{f}} \, dx = \int_a^b |f|^2 dx$$

$$= \int_a^b \left[(f_1(x))^2 + (f_2(x))^2 \right] \, dx$$

Since $|\mathbf{f}|^2 \geq 0$ and $a < b$, then $\langle \mathbf{f}, \mathbf{f} \rangle \geq 0$. Also, since $\mathbf{f}$ is continuous, $\displaystyle\int_a^b |\mathbf{f}|^2 dx > 0$ unless $\mathbf{f} = \mathbf{0}$ on $[a, b]$. [That is, the integral of a nonnegative, real-valued, continuous function (which represents the area under that curve and above the x-axis from a to b) is positive unless the function is identically zero.]

40. **(a)** Since we are working in $C[0, 1]$,

$$\|\mathbf{g}\|^2 = \langle \mathbf{g}, \mathbf{g} \rangle = \int_0^1 \mathbf{g}\bar{\mathbf{g}}dx = \int_0^1 (1 + x^2)dx = \frac{4}{3}$$

Hence $\|\mathbf{g}\| = 2/\sqrt{3}$.

(b) Since we are again working in $C[0, 1]$,

$$\langle \mathbf{f}, \mathbf{g} \rangle = \int_0^1 x(\overline{1 + ix})dx$$

$$= \int_0^1 x(1 - ix)dx$$

$$= \frac{1}{2} - \frac{1}{3}i$$

41. Let $\mathbf{v}_m = e^{2\pi imx} = \cos(2\pi mx) + i\sin(2\pi mx)$. Then if $m \neq n$, we have

$$\langle \mathbf{v}_m, \mathbf{v}_n \rangle = \int_0^1 [\cos(2\pi mx) + i\sin(2\pi mx)] [\cos(2\pi nx) - i\sin(2\pi nx)] \, dx$$

$$= \int_0^1 [\cos(2\pi mx)\cos(2\pi nx) + \sin(2\pi mx)\sin(2\pi nx)] \, dx$$

$$+ i\int_0^1 [\sin(2\pi mx)\cos(2\pi nx) - \cos(2\pi mx)\sin(2\pi nx)] \, dx$$

$$= \int_0^1 \cos[2\pi(m - n)x] \, dx + i\int_0^1 \sin[2\pi(m - n)x] \, dx$$

$$= \frac{1}{2\pi(m - n)} \sin[2\pi(m - n)x]\Big]_0^1 - \frac{i}{2\pi(m - n)} \cos[2\pi(m - n)x]\Big]_0^1$$

$$= -\frac{i}{2\pi(m - n)} + \frac{i}{2\pi(m - n)}$$

$$= 0$$

Thus the vectors are orthogonal.

EXERCISE SET 10.6

4. **(b)** The row vectors of the matrix are

$$\mathbf{r}_1 = \left(\frac{i}{\sqrt{2}}, \frac{1}{\sqrt{2}} \right) \quad \text{and} \quad \mathbf{r}_2 = \left(-\frac{i}{\sqrt{2}}, \frac{1}{\sqrt{2}} \right)$$

Since $\|\mathbf{r}_1\| = \|\mathbf{r}_2\| = 1$ and $\mathbf{r}_1 \cdot \mathbf{r}_2 = 0$, it follows that the matrix is unitary.

(c) The row vectors of the matrix are

$$\mathbf{r}_1 = (1+i,\ 1+i) \quad \text{and} \quad \mathbf{r}_2 = (1-i,\ -1+i)$$

Although $\mathbf{r}_1 \cdot \mathbf{r}_2 = (1+i)(1+i) + (1+i)(-1-i) = 0$ (from which it follows that $\mathbf{r}_1$ and $\mathbf{r}_2$ are orthogonal), they are not orthonormal because $\|\mathbf{r}_1\| = \|\mathbf{r}_2\| = 2$. Hence, the matrix is not unitary.

5. **(b)** The row vectors of the matrix are

$$\mathbf{r}_1 = \left(\frac{1}{\sqrt{2}}, \frac{1}{\sqrt{2}} \right) \quad \text{and} \quad \mathbf{r}_2 = \left(-\frac{1+i}{2}, \frac{1+i}{2} \right)$$

Since $\|\mathbf{r}_1\| = \|\mathbf{r}_2\| = 1$ and

$$\mathbf{r}_1 \cdot \mathbf{r}_2 = \frac{1}{\sqrt{2}} \left(\frac{-1+i}{2} \right) + \frac{1}{\sqrt{2}} \left(\frac{1-i}{2} \right) = 0$$

the matrix is unitary by Theorem 10.6.2. Hence,

$$A^{-1} = A^* = \bar{A}^T = \begin{bmatrix} \dfrac{1}{\sqrt{2}} & \dfrac{-1+i}{2} \\ \dfrac{1}{\sqrt{2}} & \dfrac{1-i}{2} \end{bmatrix}$$

5. **(d)** The row vectors of the matrix are

$$\mathbf{r}_1 = \left(\frac{1+i}{2}, \ -\frac{1}{2}, \ \frac{1}{2} \right)$$

$$\mathbf{r}_2 = \left(\frac{i}{\sqrt{3}}, \ \frac{1}{\sqrt{3}}, \ \frac{-i}{\sqrt{3}} \right)$$

and

$$\mathbf{r}_3 = \left(\frac{3+i}{2\sqrt{15}}, \ \frac{4+3i}{2\sqrt{15}}, \ \frac{5i}{2\sqrt{15}} \right)$$

We have $\|\mathbf{r}_1\| = \|\mathbf{r}_2\| = \|\mathbf{r}_3\| = 1$,

$$\mathbf{r}_1 \cdot \mathbf{r}_2 = \left(\frac{1+i}{2} \right) \left(\frac{-i}{\sqrt{3}} \right) - \frac{1}{2} \cdot \frac{1}{\sqrt{3}} + \frac{1}{2} \cdot \frac{i}{\sqrt{3}} = 0$$

$$\mathbf{r}_2 \cdot \mathbf{r}_3 = \left(\frac{i}{\sqrt{3}} \right) \left(\frac{3-i}{2\sqrt{15}} \right) + \frac{1}{\sqrt{3}} \cdot \frac{4-3i}{2\sqrt{15}} - \frac{i}{\sqrt{3}} \cdot \frac{-5i}{2\sqrt{15}} = 0$$

and

$$\mathbf{r}_1 \cdot \mathbf{r}_3 = \left(\frac{1+i}{2} \right) \left(\frac{3-i}{2\sqrt{15}} \right) - \frac{1}{2} \cdot \frac{4-3i}{2\sqrt{15}} + \frac{1}{2} \cdot \frac{-5i}{2\sqrt{15}} = 0$$

Hence, by Theorem 10.6.2, the matrix is unitary and thus

$$A^{-1} = A^* = \bar{A}^T = \begin{bmatrix} \dfrac{1-i}{2} & \dfrac{-i}{\sqrt{3}} & \dfrac{3-i}{2\sqrt{15}} \\[2mm] -\dfrac{1}{2} & \dfrac{1}{\sqrt{3}} & \dfrac{4-3i}{2\sqrt{15}} \\[2mm] \dfrac{1}{2} & \dfrac{i}{\sqrt{3}} & \dfrac{-5i}{2\sqrt{15}} \end{bmatrix}$$

6. Using Formula (11) of Section 10.3, we see that $\overline{e^{i\theta}} = e^{-i\theta}$ and hence that $\overline{e^{-i\theta}} = e^{i\theta}$. Let

$$\mathbf{r}_1 = \left(\frac{e^{i\theta}}{\sqrt{2}}, \ \frac{e^{-i\theta}}{\sqrt{2}} \right) \text{ and } \mathbf{r}_2 = \left(\frac{ie^{i\theta}}{\sqrt{2}}, \ \frac{-ie^{-i\theta}}{\sqrt{2}} \right). \text{ Then}$$

$$\mathbf{r}_1 \cdot \mathbf{r}_1 = \mathbf{r}_2 \cdot \mathbf{r}_2 = \frac{1}{2} + \frac{1}{2} = 1$$

and

$$\mathbf{r}_1 \cdot \mathbf{r}_2 = \frac{1}{2} \left[e^{i\theta} \left(-ie^{-i\theta} \right) + e^{-i\theta} \left(ie^{i\theta} \right) \right] = \frac{1}{2}[-i+i] = 0$$

Hence, by Theorem 10.6.2, the row vectors form an orthonormal set and so the matrix is unitary.

7. The characteristic polynomial of A is

$$\det \begin{bmatrix} \lambda - 4 & -1+i \\ -1-i & \lambda - 5 \end{bmatrix} = (\lambda - 4)(\lambda - 5) - 2 = (\lambda - 6)(\lambda - 3)$$

Therefore, the eigenvalues are $\lambda = 3$ and $\lambda = 6$. To find the eigenvectors of A corresponding to $\lambda = 3$, we let

$$\begin{bmatrix} -1 & \\ -1-i & \end{bmatrix} \begin{bmatrix} x_1 \\ x_2 \end{bmatrix} = \begin{bmatrix} 0 \\ 0 \end{bmatrix}$$

This yields $x_1 = -(1-i)s$ and $x_2 = s$ where s is arbitrary. If we put $s = 1$, we see that

$\begin{bmatrix} -1+i \\ 1 \end{bmatrix}$ is a basis vector for the eigenspace corresponding to $\lambda = 3$. We normalize this

vector to obtain

$$\mathbf{p}_1 = \begin{bmatrix} \dfrac{-1+i}{\sqrt{3}} \\ \dfrac{1}{\sqrt{3}} \end{bmatrix}$$

To find the eigenvectors corresponding to $\lambda = 6$, we let

$$\begin{bmatrix} 2 & -1+i \\ -1-i & 1 \end{bmatrix} \begin{bmatrix} x_1 \\ x_2 \end{bmatrix} = \begin{bmatrix} 0 \\ 0 \end{bmatrix}$$

This yields $x_1 = \dfrac{1-i}{2}s$ and $x_2 = s$ where s is arbitrary. If we put $s = 1$, we have that

$\begin{bmatrix} (1-i)/2 \\ 1 \end{bmatrix}$ is a basis vector for the eigenspace corresponding to $\lambda = 6$. We normalize

this vector to obtain

$$\mathbf{p}_2 = \begin{bmatrix} \dfrac{1-i}{\sqrt{6}} \\ \dfrac{2}{\sqrt{6}} \end{bmatrix}$$

Thus

$$P = \begin{bmatrix} \dfrac{-1+i}{\sqrt{3}} & \dfrac{1-i}{\sqrt{6}} \\ \dfrac{1}{\sqrt{3}} & \dfrac{2}{\sqrt{6}} \end{bmatrix}$$

diagonalizes A and

$$P^{-1}AP = \begin{bmatrix} \dfrac{-1-i}{\sqrt{3}} & \dfrac{1}{\sqrt{3}} \\ \dfrac{1+i}{\sqrt{6}} & \dfrac{2}{\sqrt{6}} \end{bmatrix} \begin{bmatrix} 4 & 1-i \\ 1+i & 5 \end{bmatrix} \begin{bmatrix} \dfrac{-1+i}{\sqrt{3}} & \dfrac{1-i}{\sqrt{6}} \\ \dfrac{1}{\sqrt{3}} & \dfrac{2}{\sqrt{6}} \end{bmatrix}$$

$$= \begin{bmatrix} 3 & 0 \\ 0 & 6 \end{bmatrix}$$

9. The characteristic polynomial of A is

$$\det \begin{bmatrix} \lambda - 6 & -2 - 2i \\ -2 + 2i & \lambda - 4 \end{bmatrix} = (\lambda - 6)(\lambda - 4) - 8 = (\lambda - 8)(\lambda - 2)$$

Therefore the eigenvalues are $\lambda = 2$ and $\lambda = 8$. To find the eigenvectors of A corresponding to $\lambda = 2$, we let

$$\begin{bmatrix} -4 & -2 - 2i \\ -2 + 2i & -2 \end{bmatrix} \begin{bmatrix} x_1 \\ x_2 \end{bmatrix} = \begin{bmatrix} 0 \\ 0 \end{bmatrix}$$

This yields $x_1 = -\dfrac{1+i}{2}s$ and $x_2 = s$ where s is arbitrary. If we put $s = 1$, we have that $\begin{bmatrix} -(1+i)/2 \\ 1 \end{bmatrix}$ is a basis vector for the eigenspace corresponding to $\lambda = 2$. We normalize this vector to obtain

$$\mathbf{p}_1 = \begin{bmatrix} -\dfrac{1+i}{\sqrt{6}} \\ \dfrac{2}{\sqrt{6}} \end{bmatrix}$$

To find the eigenvectors corresponding to $\lambda = 8$, we let

$$\begin{bmatrix} 2 & -2-2i \\ -2+2i & 4 \end{bmatrix} \begin{bmatrix} x_1 \\ x_2 \end{bmatrix} = \begin{bmatrix} 0 \\ 0 \end{bmatrix}$$

This yields $x_1 = (1+i)s$ and $x_2 = s$ where s is arbitrary. If we set $s = 1$, we have that

$$\begin{bmatrix} 1+i \\ 1 \end{bmatrix}$$ is a basis vector for the eigenspace corresponding to $\lambda = 8$. We normalize this

vector to obtain

$$\mathbf{p}_2 = \begin{bmatrix} \dfrac{1+i}{\sqrt{3}} \\ \dfrac{1}{\sqrt{3}} \end{bmatrix}$$

Thus

$$P = \begin{bmatrix} -\dfrac{1+i}{\sqrt{6}} & \dfrac{1+i}{\sqrt{6}} \\ \dfrac{2}{\sqrt{6}} & \dfrac{1}{\sqrt{3}} \end{bmatrix}$$

diagonalizes A and

$$P^{-1}AP = \begin{bmatrix} \dfrac{-1+i}{\sqrt{6}} & \dfrac{2}{\sqrt{6}} \\ \dfrac{1-i}{\sqrt{3}} & \dfrac{1}{\sqrt{3}} \end{bmatrix} \begin{bmatrix} 6 & 2+2i \\ 2-2i & 4 \end{bmatrix} \begin{bmatrix} -\dfrac{1+i}{\sqrt{6}} & \dfrac{1+i}{\sqrt{3}} \\ \dfrac{2}{\sqrt{6}} & \dfrac{1}{\sqrt{3}} \end{bmatrix}$$

$$= \begin{bmatrix} 2 & 0 \\ 0 & 8 \end{bmatrix}$$

11. The characteristic polynomial of A is

$$\det \begin{bmatrix} \lambda - 5 & 0 & 0 \\ 0 & \lambda + 1 & 1 - i \\ 0 & 1 + i & \lambda \end{bmatrix} = (\lambda - 1)(\lambda - 5)(\lambda + 2)$$

Therefore, the eigenvalues are $\lambda = 1$, $\lambda = 5$, and $\lambda = -2$. To find the eigenvectors of A corresponding to $\lambda = 1$, we let

$$\begin{bmatrix} -4 & 0 & 0 \\ 0 & 2 & 1 - i \\ 0 & 1 + i & 1 \end{bmatrix} \begin{bmatrix} x_1 \\ x_2 \\ x_3 \end{bmatrix} = \begin{bmatrix} 0 \\ 0 \\ 0 \end{bmatrix}$$

This yields $x_1 = 0$, $x_2 = -\dfrac{1-i}{2}s$, and $x_3 = s$ where s is arbitrary. If we set $s = 1$, we have that $\begin{bmatrix} 0 \\ -(1-i)/2 \\ 1 \end{bmatrix}$ is a basis vector for the eigenspace corresponding to $\lambda = 1$. We normalize this vector to obtain

$$\mathbf{p}_1 = \begin{bmatrix} 0 \\ -\dfrac{1-i}{\sqrt{6}} \\ \dfrac{2}{\sqrt{6}} \end{bmatrix}$$

To find the eigenvectors corresponding to $\lambda = 5$, we let

$$\begin{bmatrix} 0 & 0 & 0 \\ 0 & 6 & 1 - i \\ 0 & 1 + i & 5 \end{bmatrix} \begin{bmatrix} x_1 \\ x_2 \\ x_3 \end{bmatrix} = \begin{bmatrix} 0 \\ 0 \\ 0 \end{bmatrix}$$

This yields $x_1 = s$ and $x_2 = x_3 = 0$ where s is arbitrary. If we let $s = 1$, we have that $\begin{bmatrix} 1 \\ 0 \\ 0 \end{bmatrix}$ is a basis vector for the eigenspace corresponding to $\lambda = 5$. Since this vector is already normal, we let

$$\mathbf{p}_2 = \begin{bmatrix} 1 \\ 0 \\ 0 \end{bmatrix}$$

To find the eigenvectors corresponding to $\lambda = -2$, we let

$$\begin{bmatrix} -7 & 0 & 0 \\ 0 & -1 & 1-i \\ 0 & 1+i & -2 \end{bmatrix} \begin{bmatrix} x_1 \\ x_2 \\ x_3 \end{bmatrix} = \begin{bmatrix} 0 \\ 0 \\ 0 \end{bmatrix}$$

This yields $x_1 = 0$, $x_2 = (1-i)s$, and $x_3 = s$ where s is arbitrary. If we let $s = 1$, we have

that $\begin{bmatrix} 0 \\ 1-i \\ 1 \end{bmatrix}$ is a basis vector for the eigenspace corresponding to $\lambda = -2$. We normalize

this vector to obtain

$$\mathbf{p}_3 = \begin{bmatrix} 0 \\ \dfrac{1-i}{\sqrt{3}} \\ \dfrac{1}{\sqrt{3}} \end{bmatrix}$$

Thus

$$P = \begin{bmatrix} 0 & 1 & 0 \\ -\dfrac{1-i}{\sqrt{6}} & 0 & \dfrac{1-i}{\sqrt{3}} \\ \dfrac{2}{\sqrt{6}} & 0 & \dfrac{1}{\sqrt{3}} \end{bmatrix}$$

diagonalizes A and

$$P^{-1}AP = \begin{bmatrix} 0 & -\dfrac{1+i}{\sqrt{6}} & \dfrac{2}{\sqrt{6}} \\ 1 & 0 & 0 \\ 0 & \dfrac{1+i}{\sqrt{3}} & \dfrac{1}{\sqrt{3}} \end{bmatrix} \begin{bmatrix} 5 & 0 & 0 \\ 0 & -1 & -1+i \\ 0 & -1-i & 0 \end{bmatrix} \begin{bmatrix} 0 & 1 & 0 \\ -\dfrac{1-i}{\sqrt{6}} & 0 & \dfrac{1-i}{\sqrt{3}} \\ \dfrac{2}{\sqrt{6}} & 0 & \dfrac{1}{\sqrt{3}} \end{bmatrix}$$

$$= \begin{bmatrix} 1 & 0 & 0 \\ 0 & 5 & 0 \\ 0 & 0 & -2 \end{bmatrix}$$

13. The eigenvalues of A are the roots of the equation

$$\det \begin{bmatrix} \lambda - 1 & -4i \\ -4i & \lambda - 3 \end{bmatrix} = \lambda^2 - 4\lambda + 19 = 0$$

The roots of this equation, which are $\lambda = \dfrac{4 \pm \sqrt{16 - 4(19)}}{2}$, are not real. This shows that the eigenvalues of a symmetric matrix with nonreal entries need not be real. Theorem 10.6.6 applies only to matrices with real entries.

15. We know that $\det(A)$ is the sum of all the signed elementary products $\pm a_{1j_1} a_{2j_2} \cdots a_{nj_n}$, where a_{ij} is the entry from the ith row and j^{th} column of A. Since the ij^{th} element of $\bar{A}$ is $\bar{a}_{ij}$, then $\det(\bar{A})$ is the sum of the signed elementary products $\pm \bar{a}_{1j_1} \bar{a}_{2j_2} \cdots \bar{a}_{nj_n}$ or $\pm \overline{a_{1j_1} a_{2j_2} \cdots a_{nj_n}}$. That is, $\det(\bar{A})$ is the sum of the conjugates of the terms in $\det(A)$. But since the sum of the conjugates is the conjugate of the sum, we have $\det(\bar{A}) = \overline{\det(A)}$.

16. (a) We have

$$\det(A^*) = \det(\bar{A}^T)$$

$$= \det(\bar{A}) \qquad\qquad (\text{since } \det(B) = \det(B^T))$$

$$= \overline{\det(A)} \qquad\qquad (\text{by Exercise 15})$$

(b) If A is Hermitian, then $A = A^*$. Therefore,

$$\det(A) = \det(A^*)$$
$$= \overline{\det(A)} \qquad \text{(by Part (a))}$$

But $z = \bar{z}$ if and only if z is real.

(c) If A is unitary, then $A^{-1} = A^*$. Therefore,

$$\det(A) = 1/\det(A^{-1})$$
$$= 1/\det(A^*)$$
$$= 1/\overline{\det(A)} \qquad \text{(by Part (a))}$$

Thus $|\det(A)|^2 = 1$. But since $|z| \geq 0$ for all z, it follows that $|\det(A)| = 1$.

18. **(b)** We have

$$(A + B)^* = \overline{(A + B)}^T$$
$$= (\bar{A} + \bar{B})^T$$
$$= (\bar{A}^T + \bar{B}^T)$$
$$= A^* + B^*$$

(d) We have

$$(AB)^* = \overline{(AB)}^T$$
$$= (\bar{A}\bar{B})^T$$
$$= \bar{B}^T \bar{A}^T$$
$$= B^* A^*$$

19. If A is invertible, then

$$A^*(A^{-1})^* = (A^{-1}A)^* \qquad \text{(by Exercise 18(d))}$$
$$= I^* = \bar{I}^T = I^T = I$$

Thus we have $(A^{-1})^* = (A^*)^{-1}$.

20. Hint: use Exercise 18(a) and Exercise 19.

21. Let $\mathbf{r}_i$ denote the ith row of A and let $\mathbf{c}_j$ denote the jth column of A^*. Then, since $A^* = \bar{A}^T = \overline{(A^T)}$, we have $\mathbf{c}_j = \bar{\mathbf{r}}_j$ for $j = 1, \ldots, n$. Finally, let

$$\delta_{ij} = \begin{cases} 0 & \text{if} \quad i \neq j \\ 1 & \text{if} \quad i = j \end{cases}$$

Then A is unitary $\Longleftrightarrow A^{-1} = A^*$. But

$$A^{-1} = A^* \quad \Longleftrightarrow \quad AA^* = I$$

$$\Longleftrightarrow \quad \mathbf{r}_i \cdot \mathbf{c}_j = \delta_{ij} \text{ for all } i, j$$

$$\Longleftrightarrow \quad \mathbf{r}_i \cdot \bar{\mathbf{r}}_j = \delta_{ij} \text{ for all } i, j$$

$$\Longleftrightarrow \quad \{\mathbf{r}_1, \ldots, \mathbf{r}_n\} \text{ is an orthonormal set}$$

22. We know that if A is unitary, then so is A^*. But since $(A^*)^* = A$, then if A^* is unitary, so is A. Moreover, the columns of A are the conjugates of the rows of A^*. Thus we have

$$A \text{ is unitary} \quad \Longleftrightarrow \quad A^* \text{ is unitary}$$

$$\Longleftrightarrow \quad \text{the rows of } A^* \text{ form an orthonormal set}$$

$$\Longleftrightarrow \quad \text{the conjugates of the columns of } A \text{ form an orthonormal set}$$

$$\Longleftrightarrow \quad \text{the columns of } A \text{ form an orthonormal set}$$

23. (a) We know that $A = A^*$, that $A\mathbf{x} = \lambda I \mathbf{x}$, and that $A\mathbf{y} = \mu I \mathbf{y}$. Therefore

$$\mathbf{x}^* A\mathbf{y} = \mathbf{x}^*(\mu I \mathbf{y}) = \mu(\mathbf{x}^* I \mathbf{y}) = \mu \mathbf{x}^* \mathbf{y}$$

and

$$\mathbf{x}^* A\mathbf{y} = [(\mathbf{x}^* A\mathbf{y})^*]^* = [\mathbf{y}^* A^* \mathbf{x}]^*$$

$$= [\mathbf{y}^* A\mathbf{x}]^* = [\mathbf{y}^*(\lambda I \mathbf{x})]^*$$

$$= [\lambda \mathbf{y}^* \mathbf{x}]^* = \lambda \mathbf{x}^* \mathbf{y}$$

The last step follows because λ, being the eigenvalue of an Hermitian matrix, is real.

(b) Subtracting the equations in Part (a) yields

$$(\lambda - \mu)(\mathbf{x}^*\mathbf{y}) = 0$$

Since $\lambda \neq \mu$, the above equation implies that $\mathbf{x}^*\mathbf{y}$ is the 1×1 zero matrix. Let $\mathbf{x} = (x_1, \ldots, x_n)$ and $\mathbf{y} = (y_1, \ldots, y_n)$. Then we have just shown that

$$\bar{x}_1 y_1 + \cdots + \bar{x}_n y_n = 0$$

so that

$$x_1 \bar{y}_1 + \cdots + x_n \bar{y}_n = \bar{0} = 0$$

and hence $\mathbf{x}$ and $\mathbf{y}$ are orthogonal.

SUPPLEMENTARY EXERCISES 10

2. If a and b are not both zero, then $|a|^2 + |b|^2 \neq 0$ and the inverse is

$$
\begin{bmatrix}
\dfrac{\bar{a}}{|a|^2 + |b|^2} & \dfrac{b}{|a|^2 + |b|^2} \\[3mm]
\dfrac{-\bar{b}}{|a|^2 + |b|^2} & \dfrac{a}{|a|^2 + |b|^2}
\end{bmatrix}
$$

This inverse is computed exactly as if the entries were real.

3. The system of equations has solution $x_1 = -is + t$, $x_2 = s$, $x_3 = t$. Thus

$$
\begin{bmatrix} x_1 \\ x_2 \\ x_3 \end{bmatrix} =
\begin{bmatrix} -i \\ 1 \\ 0 \end{bmatrix} s +
\begin{bmatrix} 1 \\ 0 \\ 1 \end{bmatrix} t
$$

where s and t are arbitrary. Hence

$$
\begin{bmatrix} -i \\ 1 \\ 0 \end{bmatrix} \qquad \text{and} \qquad \begin{bmatrix} 1 \\ 0 \\ 1 \end{bmatrix}
$$

form a basis for the solution space.

5. The eigenvalues are the solutions, λ, of the equation

$$\det \begin{bmatrix} \lambda & 0 & -1 \\ -1 & \lambda & -\omega - 1 - \dfrac{1}{\omega} \\ 0 & -1 & \lambda + \omega + 1 + \dfrac{1}{\omega} \end{bmatrix} = 0$$

or

$$\lambda^3 + \left(\omega + 1 + \frac{1}{\omega}\right)\lambda^2 - \left(\omega + 1 + \frac{1}{\omega}\right)\lambda - 1 = 0$$

But $\dfrac{1}{\omega} = \bar{\omega}$, so that $\omega + 1 + \dfrac{1}{\omega} = 2\,\mathrm{Re}(\omega) + 1 = 0$. Thus we have

$$\lambda^3 - 1 = 0$$

or

$$(\lambda - 1)(\lambda^2 + \lambda + 1) = 0$$

Hence $\lambda = 1$, ω, or $\bar{\omega}$. Note that $\bar{\omega} = \omega^2$.

6. (c) Following the hint, we let $z = \cos\theta + i\sin\theta = e^{i\theta}$ in Part (a). This yields

$$1 + e^{i\theta} + e^{2i\theta} + \cdots + e^{ni\theta} = \frac{1 - e^{(n+1)i\theta}}{1 - e^{i\theta}}$$

If we expand and equate real parts, we obtain

$$1 + \cos\theta + \cos 2\theta + \cdots + \cos n\theta = \mathrm{Re}\left(\frac{1 - e^{(n+1)i\theta}}{1 - e^{i\theta}}\right)$$

But

$$\text{Re}\left(\frac{1 - e^{(n+1)i\theta}}{1 - e^{i\theta}}\right) = \text{Re}\left(\frac{\left(1 - e^{(n+1)i\theta}\right)\left(1 - e^{-i\theta}\right)}{\left(1 - e^{i\theta}\right)\left(1 - e^{-i\theta}\right)}\right)$$

$$= \text{Re}\left(\frac{1 - e^{(n+1)i\theta} - e^{-i\theta} + e^{ni\theta}}{2 - 2\text{Re}(e^{i\theta})}\right)$$

$$= \frac{1}{2}\left(\frac{1 - \cos[(n+1)\theta] - \cos(-\theta) + \cos n\theta}{1 - \cos\theta}\right)$$

$$= \frac{1}{2}\left(\frac{1 - \cos\theta}{1 - \cos\theta} + \frac{\cos n\theta - \cos((n+1)\theta)}{1 - \cos\theta}\right)$$

$$= \frac{1}{2}\left(1 + \frac{\cos n\theta - [\cos n\theta \cos\theta - \sin n\theta \sin\theta]}{2\sin^2\frac{\theta}{2}}\right)$$

$$= \frac{1}{2}\left(1 + \frac{\cos n\theta(1 - \cos\theta) + 2\sin n\theta \sin\frac{\theta}{2}\cos\frac{\theta}{2}}{2\sin^2\frac{\theta}{2}}\right)$$

$$= \frac{1}{2}\left(1 + \frac{2\cos n\theta \sin^2\frac{\theta}{2} + 2\sin n\theta \sin\frac{\theta}{2}\cos\frac{\theta}{2}}{2\sin^2\frac{\theta}{2}}\right)$$

$$= \frac{1}{2}\left(1 + \frac{\cos n\theta \sin\frac{\theta}{2} + \sin n\theta \cos\frac{\theta}{2}}{\sin\frac{\theta}{2}}\right)$$

$$= \frac{1}{2}\left(1 + \frac{\sin\left[\left(n + \frac{1}{2}\right)\theta\right]}{\sin\frac{\theta}{2}}\right)$$

Observe that because $0 < \theta < 2\pi$, we have not divided by zero.

7. Observe that $\omega^3 = 1$. Hence by Exercise 6(b), we have $1 + \omega + \omega^2 = 0$. Therefore

$$\mathbf{v}_1 \cdot \mathbf{v}_2 = \frac{1}{3}(1 + \bar{\omega} + \bar{\omega}^2) = \frac{1}{3}\overline{(1 + \omega + \omega^2)} = 0$$

$$\mathbf{v}_1 \cdot \mathbf{v}_3 = \frac{1}{3}\left(1 + \bar{\omega}^2 + \bar{\omega}^4\right) = \frac{1}{3}\overline{(1 + \omega^2 + \omega)} = 0$$

$$\mathbf{v}_2 \cdot \mathbf{v}_3 = \frac{1}{3}\left(1 + \omega\bar{\omega}^2 + \omega^2\bar{\omega}^4\right)$$

$$= \frac{1}{3}\left(1 + \omega^5 + \omega^{10}\right) \qquad\qquad \text{(since } \bar{\omega} = \omega^2\text{)}$$

$$= \frac{1}{3}(1 + \omega^2 + \omega)$$

$$= 0$$

$$\mathbf{v}_1 \cdot \mathbf{v}_1 = \frac{1}{3}(1 + 1 + 1) = 1$$

$$\mathbf{v}_2 \cdot \mathbf{v}_2 = \frac{1}{3}\left(1 + \omega\bar{\omega} + \omega^2\bar{\omega}^2\right) = 1 \qquad\qquad \text{(since } \omega\bar{\omega} = 1\text{)}$$

$$\mathbf{v}_3 \cdot \mathbf{v}_3 = \frac{1}{3}\left(1 + \omega^2\bar{\omega}^2 + \omega^4\bar{\omega}^4\right) = 1 \qquad\qquad \text{(as above)}$$

Thus, the vectors form an orthonormal set in C^3.

8. Call the diagonal matrix D. Then

$$(UD)^* = (\overline{UD})^T = \bar{D}^T\bar{U}^T = \bar{D}U^* = \bar{D}U^{-1}$$

We need only check that $(UD)^* = (UD)^{-1}$. But

$$(UD)^*(UD) = (\bar{D}U^{-1})(UD) = \bar{D}D$$

and

$$\bar{D}D = \begin{bmatrix} |z_1|^2 & 0 & \cdots & 0 \\ 0 & |z_2|^2 & \cdots & 0 \\ \vdots & \vdots & & \vdots \\ 0 & 0 & \cdots & |z_n|^2 \end{bmatrix} - I$$

Hence $(UD)^* = (UD)^{-1}$ and so UD is unitary.

9. (a) Since $A^* = -A$, we have

$$(iA)^* = \bar{i}A^* = -i(-A) = iA$$

Hence iA is Hermitian.

(b) By Part (a), iA is Hermitian and hence unitarily diagonalizable with real eigenvalues. That is, there is a unitary matrix P and a diagonal matrix D with real entries such that

$$P^{-1}(iA)P = D$$

Thus

$$i(P^{-1}AP) = D$$

or

$$P^{-1}AP = -iD$$

That is, A is unitarily diagonalizable with eigenvalues which are $-i$ times the eigenvalues of iA, and hence pure imaginary.

EXERCISE SET 11.1

1. **(a)** Substituting the coordinates of the points into Eq. (4) yields

$$\begin{vmatrix} x & y & 1 \\ 1 & -1 & 1 \\ 2 & 2 & 1 \end{vmatrix} = 0$$

which, upon cofactor expansion along the first row, yields $-3x + y + 4 = 0$; that is, $y = 3x - 4$.

(b) As in (a),

$$\begin{vmatrix} x & y & 1 \\ 0 & 1 & 1 \\ 1 & -1 & 1 \end{vmatrix} = 0$$

yields $2x + y - 1 = 0$ or $y = -2x + 1$.

2. **(a)** Using Eq. (9) yields

$$\begin{vmatrix} x^2 + y^2 & x & y & 1 \\ 40 & 2 & 6 & 1 \\ 4 & 2 & 0 & 1 \\ 34 & 5 & 3 & 1 \end{vmatrix} = 0$$

which, upon first-row cofactor expansion, yields $18(x^2 + y^2) - 72x - 108y + 72 = 0$ or, dividing by 18, $x^2 + y^2 - 4x - 6y + 4 = 0$. Completing the squares in x and y yields the standard form $(x - 2)^2 + (y - 3)^2 = 9$.

2. **(b)** As in (a),

$$
\begin{vmatrix}
x^2 + y^2 & x & y & 1 \\
8 & 2 & -2 & 1 \\
34 & 3 & 5 & 1 \\
52 & -4 & 6 & 1
\end{vmatrix} = 0
$$

yields $50(x^2 + y^2) + 100x - 200y - 1000 = 0$; that is, $x^2 + y^2 + 2x - 4y - 20 = 0$. In standard form this is $(x+1)^2 + (y-2)^2 = 25$.

3. Using Eq. (10) we obtain

$$
\begin{vmatrix}
x^2 & xy & y^2 & x & y & 1 \\
0 & 0 & 0 & 0 & 0 & 1 \\
0 & 0 & 1 & 0 & -1 & 1 \\
4 & 0 & 0 & 2 & 0 & 1 \\
4 & -10 & 25 & 2 & -5 & 1 \\
16 & -4 & 1 & 4 & -1 & 1
\end{vmatrix} = 0
$$

which is the same as

$$
\begin{vmatrix}
x^2 & xy & y^2 & x & y \\
0 & 0 & 1 & 0 & -1 \\
4 & 0 & 0 & 2 & 0 \\
4 & -10 & 25 & 2 & -5 \\
16 & -4 & 1 & 4 & -1
\end{vmatrix} = 0
$$

by expansion along the second row (taking advantage of the zeros there). Add column five to column three and take advantage of another row of all but one zero to get

$$
\begin{vmatrix}
x^2 & xy & y^2 + y & x \\
4 & 0 & 0 & 2 \\
4 & -10 & 20 & 2 \\
16 & -4 & 0 & 4
\end{vmatrix} = 0.
$$

Now expand along the first row and get $160x^2 + 320xy + 160(y^2 + y) - 320x = 0$; that is, $x^2 + 2xy + y^2 - 2x + y = 0$, which is the equation of a parabola (since the discriminant $c_2^2 - 4c_1 c_3$ is zero).

4. **(a)** From Eq. (11), the equation of the plane is

$$\begin{vmatrix} x & y & z & 1 \\ 1 & 1 & -3 & 1 \\ 1 & -1 & 1 & 1 \\ 0 & -1 & 2 & 1 \end{vmatrix} = 0.$$

Expansion along the first row yields $-2x - 4y - 2z = 0$; that is, $x + 2y + z = 0$.

(b) As in (a),

$$\begin{vmatrix} x & y & z & 1 \\ 2 & 3 & 1 & 1 \\ 2 & -1 & -1 & 1 \\ 1 & 2 & 1 & 1 \end{vmatrix} = 0$$

yields $-2x + 2y - 4z + 2 = 0$; that is, $-x + y - 2z + 1 = 0$ for the equation of the plane.

5. **(a)** Using Eq. (12), the equation of the sphere is

$$\begin{vmatrix} x^2 + y^2 + z^2 & x & y & z & 1 \\ 14 & 1 & 2 & 3 & 1 \\ 6 & -1 & 2 & 1 & 1 \\ 2 & 1 & 0 & 1 & 1 \\ 6 & 1 & 2 & -1 & 1 \end{vmatrix} = 0.$$

Expanding by cofactors along the first row yields $16(x^2+y^2+z^2)-32x-64y-32z+32 = 0$; that is, $(x^2 + y^2 + z^2) - 2x - 4y - 2x + 2 = 0$. Completing the squares in each variable yields the standard form $(x - 1)^2 + (y - 2)^2 + (z - 1)^2 = 4$.

NOTE: When evaluating the cofactors, it is useful to take advantage of the column of ones and elementary row operations; for example, the cofactor of $x^2 + y^2 + z^2$ above can be evaluated as follows:

$$\begin{vmatrix} 1 & 2 & 3 & 1 \\ -1 & 2 & 1 & 1 \\ 1 & 0 & 1 & 1 \\ 1 & 2 & -1 & 1 \end{vmatrix} = \begin{vmatrix} 1 & 2 & 3 & 1 \\ -2 & 0 & -2 & 0 \\ 0 & -2 & -2 & 0 \\ 0 & 0 & -4 & 0 \end{vmatrix} = 16$$

by cofactor expansion of the latter determinant along the last column.

5. **(b)** As in (a),

$$\begin{vmatrix} x^2+y^2+z^2 & x & y & z & 1 \\ 5 & 0 & 1 & -2 & 1 \\ 11 & 1 & 3 & 1 & 1 \\ 5 & 2 & -1 & 0 & 1 \\ 11 & 3 & 1 & -1 & 1 \end{vmatrix} = 0$$

yields $-24(x^2+y^2+z^2)+48x+48y+72=0$; that is, $x^2+y^2+z^2-2x-2y-3=0$ or in standard form, $(x-1)^2+(y-1)^2+z^2=5$.

6. Substituting each of the points (x_1, y_1), (x_2, y_2), (x_3, y_3), (x_4, y_4), and (x_5, y_5) into the equation

$$c_1 x^2 + c_2 xy + c_3 y^2 + c_4 x + c_5 y + c_6 = 0$$

yields

$$c_1 x_1^2 + c_2 x_1 y_1 + c_3 y_1^2 + c_4 x_1 + c_5 y_1 + c_6 = 0.$$
$$\vdots \qquad\qquad \vdots \qquad\qquad \vdots$$
$$c_1 x_5^2 + c_2 x_5 y_5 + c_3 y_5^2 + c_4 x_5 + c_5 y_5 + c_6 = 0.$$

These together with the original equation form a homogeneous linear system with a non-trivial solution for c_1, c_2, ..., c_6. Thus the determinant of the coefficient matrix is zero, which is exactly Eq. (10).

7. As in the previous problem, substitute the coordinates (x_i, y_i, z_i) of each of the three points into the equation $c_1 x + c_2 y + c_3 z + c_4 = 0$ to obtain a homogeneous system with nontrivial solution for c_1, ..., c_4. Thus the determinant of the coefficient matrix is zero, which is exactly Eq. (11).

8. Substituting the coordinates (x_i, y_i, z_i) of the four points into the equation $c_1(x^2 + y^2 + z^2) + c_2 x + c_3 y + c_4 z + c_5 = 0$ of the sphere yields four equations, which together with the above sphere equation form a homogeneous linear system for c_1, ..., c_5 with a nontrivial solution. Thus the determinant of this system is zero, which is Eq. (12).

9. By substituting of the coordinates of the three points (x_1, y_1), (x_2, y_2) and (x_3, y_3) we obtain the equations:

$$c_1 y \ + c_2 x^2 + c_3 x \ + c_4 = 0$$
$$c_1 y_1 + c_2 x_1^2 + c_3 x_1 + c_4 = 0$$
$$c_1 y_2 + c_2 x_2^2 + c_3 x_2 + c_4 = 0$$
$$c_1 y_3 + c_2 x_3^2 + c_3 x_3 + c_4 = 0.$$

This is a homogeneous system with a nontrivial solution for c_1, c_2, c_3, c_4, so the determinant of the coefficient matrix is zero; that is,

$$\begin{vmatrix} y & x^2 & x & 1 \\ y_1 & x_1^2 & x_1 & 1 \\ y_2 & x_2^2 & x_2 & 1 \\ y_3 & x_3^2 & x_3 & 1 \end{vmatrix} = 0.$$

EXERCISE SET 11.2

1.

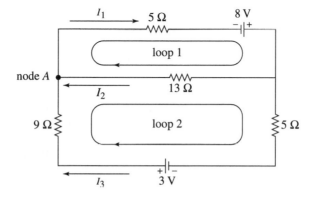

Applying Kirchhoff's current law to node A in the figure yields

$$I_1 = I_2 + I_3.$$

Applying Kirchhoff's voltage law and Ohm's law to Loops 1 and 2 yields

$$5I_1 + 13I_2 = 8$$

and

$$9I_3 - 13I_2 + 5I_3 = 3.$$

In matrix form these three equations are

$$
\begin{bmatrix}
1 & -1 & -1 \\
5 & 13 & 0 \\
0 & -13 & 14
\end{bmatrix}
\begin{bmatrix}
I_1 \\
I_2 \\
I_3
\end{bmatrix}
=
\begin{bmatrix}
0 \\
8 \\
3
\end{bmatrix}
$$

with solution $I_1 = \dfrac{255}{317}$, $I_2 = \dfrac{97}{317}$, $I_3 = \dfrac{158}{317}$.

2.

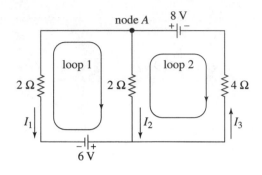

Node A gives $\quad I_1 + I_2 = I_3$.

Loop 1 gives $\quad -2I_1 + 2I_2 = -6$.

Loop 2 gives $\quad -2I_2 - 4I_3 = -8$.

In matrix form we have

$$
\begin{bmatrix}
1 & 1 & -1 \\
-2 & 2 & 0 \\
0 & -2 & -4
\end{bmatrix}
\begin{bmatrix}
I_1 \\
I_2 \\
I_3
\end{bmatrix}
\begin{bmatrix}
0 \\
-6 \\
-8
\end{bmatrix}
$$

with solution $I_1 = \dfrac{13}{5},\ I_2 = -\dfrac{2}{5},\ I_3 = \dfrac{11}{5}$.

3.

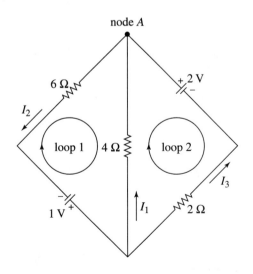

Node A gives $I_1 + I_3 = I_2$.

Loop 1 gives $-4I_1 - 6I_2 = -1$.

Loop 2 gives $4I_1 - 2I_3 = -2$.

In system form we have

$$
\begin{bmatrix}
1 & -1 & 1 \\
-4 & -6 & 0 \\
4 & 0 & -2
\end{bmatrix}
\begin{bmatrix}
I_1 \\
I_2 \\
I_3
\end{bmatrix}
=
\begin{bmatrix}
0 \\
-1 \\
-2
\end{bmatrix}
$$

with solution $I_1 = -\dfrac{5}{22}$, $I_2 = \dfrac{7}{22}$, $I_3 = \dfrac{6}{11}$.

4.

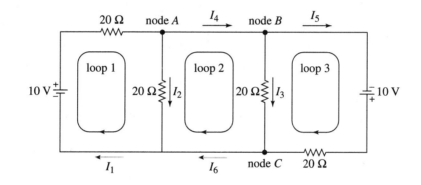

Node A gives $I_1 = I_2 + I_4$.

Node B gives $I_4 = I_3 + I_5$.

Node C gives $I_6 = I_3 + I_5$.

Loop 1 gives $20I_1 + 20I_2 = 10$.

Loop 2 gives $-20I_2 + 20I_3 = 0$.

Loop 3 gives $-20I_3 + 20I_5 = 10$.

In matrix form we have

$$
\begin{bmatrix}
1 & -1 & 0 & -1 & 0 & 0 \\
0 & 0 & 1 & -1 & 1 & 0 \\
0 & 0 & 1 & 0 & 1 & -1 \\
20 & 20 & 0 & 0 & 0 & 0 \\
0 & -20 & 20 & 0 & 0 & 0 \\
0 & 0 & -20 & 0 & 20 & 0
\end{bmatrix}
\begin{bmatrix}
I_1 \\ I_2 \\ I_3 \\ I_4 \\ I_5 \\ I_6
\end{bmatrix}
=
\begin{bmatrix}
0 \\ 0 \\ 0 \\ 10 \\ 0 \\ 10
\end{bmatrix}
$$

with solution $I_1 = I_4 = I_5 = I_6 = \dfrac{1}{2}$ and $I_2 = I_3 = 0$.

5.

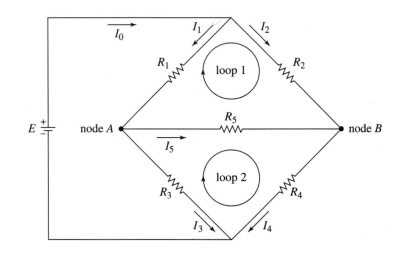

After setting $I_5 = 0$ we have that:

node A gives $I_1 = I_3$

node B gives $I_2 = I_4$

loop 1 gives $I_1 R_1 = I_2 R_2$

loop 2 gives $I_3 R_3 = I_4 R_4$.

From these four equations it easily follows that $R_4 = R_3 R_2 / R_1$.

6.

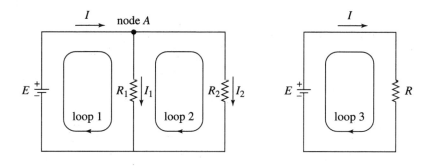

Node A in the first circuit gives $\quad I = I_1 + I_2$.

Loop 1 in the first circuit gives $I_1 R_1 = E$.

Loop 2 in the first circuit gives $I_1 R_1 = I_2 R_2$.

Loop 3 in the first circuit gives $\quad IR = E$.

From these four equations we obtain

$$R = \frac{E}{I} = \frac{E}{I_1 + I_2} = \frac{E}{\dfrac{E}{R_1} + \dfrac{E}{R_2}} = \frac{1}{\dfrac{1}{R_1} + \dfrac{1}{R_2}}.$$

EXERCISE SET 11.3

1.

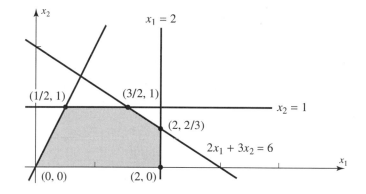

In the figure, the feasible region is shown and the extreme points are labelled. The values of the objective function are shown in the following table:

Extreme point (x_1, x_2)	Value of $z = 3x_1 + 2x_2$
$(0, 0)$	0
$(1/2, 1)$	3-1/2
$(3/2, 1)$	6-1/2
$(2, 2/3)$	7-1/3
$(2, 0)$	6

Thus the maximum, 7-1/3, is attained when $x_1 = 2$ and $x_2 = 2/3$.

2.

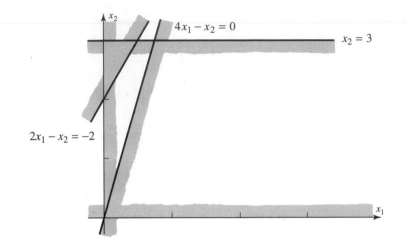

The intersection of the five half-planes defined by the constraints is empty. Thus this problem has no feasible solutions.

3.

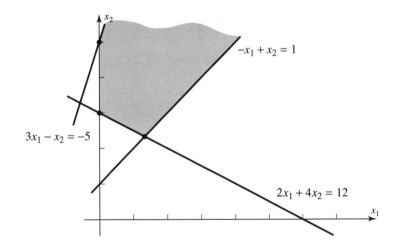

The feasible region for this problem, shown in the figure, is unbounded. The value of $z = -3x_1 + 2x_2$ cannot be minimized in this region since it becomes arbitrarily negative as we travel outward along the line $-x_1 + x_2 = 1$; i.e., the value of z is $-3x_1 + 2x_2 = -3x_1 + 2(x_1 + 1) = -x_1 + 2$ and x_1 can be arbitrarily large.

4.

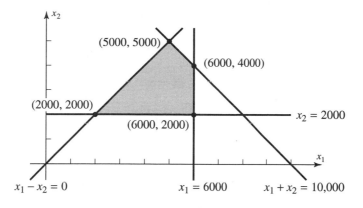

The feasible region and vertices for this problem are shown in the figure. The maximum value of the objective function $z = .1x_1 + .07x_2$ is attained when $x_1 = 6,000$ and $x_2 = 4,000$, then $z = 880$. In other words, she should invest $6,000$ in bond A and $4,000$ in bond B for an annual yield of 880.

5.

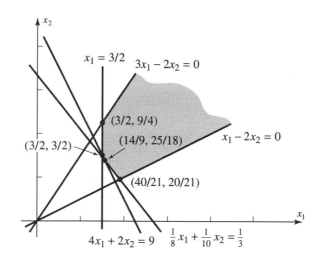

The feasible region and its extreme points are shown in the figure. Though the region is unbounded, x_1 and x_2 are always positive, so the objective function $z = 7.5x_1 + 5.0x_2$ is also. Thus, it has a minimum, which is attained at the point where $x_1 = 14/9$ and $x_2 = 25/18$. The value of z there is $335/18$. In the problem's terms, we use $7/9$ cups of milk and $25/18$ ounces of corn flakes, a minimum cost of $18.6\cancel{c}$ is realized.

6. Letting x_1 be the number of Company A's containers shipped and x_2 be the number of Company B's, the problem is

$$\text{Maximize} \quad z = 2.20x_1 + 3.00x_2$$

subject to

$$40x_1 + 50x_2 \leq 37,000$$

$$2x_1 + 3x_2 \leq 2,000$$

$$x_1 \geq 0$$

$$x_2 \geq 0.$$

The feasible region is shown in the figure. The vertex at which the maximum is attained is when $x_1 = 550$ and $x_2 = 300$, then $z = 2110$.

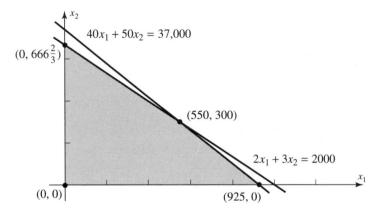

7. We must now maximize $z = 2.50x_1 + 3.00x_2$. The feasible region is as in 11.6, but now the maximum happens when $x_1 = 925$ and $x_2 = 0$, then $z = 2312.50$.

8. Let x_1 be the number of pounds of ingredient A used, and x_2 be the number of pounds of ingredient B. Then the problem is

$$\text{Minimize} \quad z = 8x_1 + 9x_2$$

subject to

$$2x_1 + 5x_2 \geq 10$$

$$2x_1 + 3x_2 \geq 8$$

$$6x_1 + 4x_2 \geq 12$$

$$x_1 \geq 0$$

$$x_2 \geq 0$$

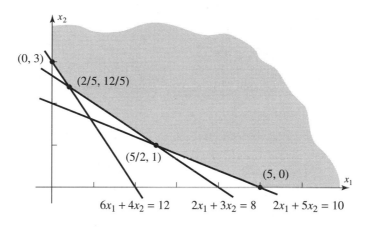

Though the feasible region shown in the figure is unbounded, the objective function is always positive there and hence must have a minimum. This minimum occurs at the vertex where $x_1 = 2/5$ and $x_2 = 12/5$. The minimum value of z is $124/5$ or $24.8\cancel{c}$.

9. It is sufficient to show that if a linear functions has the same value at two points, then it has that value along the line connecting the two points. Let $ax_1 + bx_2$ be the function, and $ax_1' + bx_2' = ax_1'' + bx_2'' = c$. Then $(tx_1' + (1-t)x_1'', tx_2' + (1-t)x_2'')$ is an arbitrary point on the line connecting (x_1', x_2') to (x_1'', x_2'') and $a(tx_1' + (1-t)x_1'') + b(tx_2' + (1-t)x_2'') = t(ax_1' + bx_2') + (1-t)(ax_1'' + bx_2'') = c$.

10.

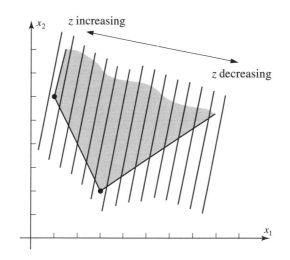

The level curves of the objective function $-5x_1 + x_2$ are shown in the figure, and it is readily seen that the objective function remains bounded in the region.

EXERCISE SET 11.4

1. **(a)** Step 1 yields

$$\begin{bmatrix} 13 & 0 & 6 \\ 10 & 0 & 3 \\ 11 & 0 & 4 \end{bmatrix}$$

and Step 2 yields

$$\begin{bmatrix} 3 & 0 & 3 \\ 0 & 0 & 0 \\ 1 & 0 & 1 \end{bmatrix}.$$

We see that two lines are needed to cover the zeros as shown. Thus the test for optimality fails. Applying Step 5 yields

$$\begin{bmatrix} 2 & 0 & 2 \\ 0 & 1 & 0 \\ 0 & 0 & 0 \end{bmatrix}.$$

Now we cannot cover the zeros with less than three lines, so the matrix contains an optimal assignment. Actually there are two such assignments:

$$\begin{bmatrix} 2 & \boxed{0} & 2 \\ \boxed{0} & 1 & 0 \\ 0 & 0 & \boxed{0} \end{bmatrix} \quad \text{and} \quad \begin{bmatrix} 2 & \boxed{0} & 2 \\ 0 & 1 & \boxed{0} \\ \boxed{0} & 0 & 0 \end{bmatrix};$$

from the original matrix, we see the cost for each is 30.

1. (b) Step 1 yields   and Step 2 yields

$$\begin{bmatrix} 5 & 0 & 2 & 3 \\ 8 & 6 & 0 & 7 \\ 0 & 5 & 3 & 1 \\ 7 & 0 & 2 & 3 \end{bmatrix} \qquad \begin{bmatrix} 5 & 0 & 2 & 2 \\ 8 & 6 & 0 & 6 \\ 0 & 5 & 3 & 0 \\ 7 & 0 & 2 & 2 \end{bmatrix}.$$

The zeros can be covered by three lines (as shown), so we proceed. Step 5 gives

$$\begin{bmatrix} 3 & 0 & 0 & 0 \\ 8 & 8 & 0 & 6 \\ 0 & 7 & 3 & 0 \\ 5 & 0 & 0 & 0 \end{bmatrix}.$$

Now we cannot cover the zeros with less than four lines, so there is an optimal assignment. There are two in fact:

$$\begin{bmatrix} 3 & \boxed{0} & 0 & 0 \\ 8 & 8 & \boxed{0} & 6 \\ \boxed{0} & 7 & 3 & 0 \\ 5 & 0 & 0 & \boxed{0} \end{bmatrix} \quad \text{and} \quad \begin{bmatrix} 3 & 0 & 0 & \boxed{0} \\ 8 & 8 & \boxed{0} & 6 \\ \boxed{0} & 7 & 3 & 0 \\ 5 & \boxed{0} & 0 & 0 \end{bmatrix},$$

each (from the original matrix) with a cost of -2.

(c) Step 1 yields

$$\begin{bmatrix} 5 & 2 & 0 & 0 & 3 \\ 7 & 3 & 0 & 5 & 6 \\ 4 & 1 & 0 & 7 & 7 \\ 0 & 3 & 7 & 1 & 4 \\ 0 & 5 & 4 & 1 & 5 \end{bmatrix}$$

and Step 2 yields

$$
\begin{bmatrix}
-5 & 1 & 0 & 0 & 0- \\
7 & 2 & 0 & 5 & 3 \\
-4 & 0 & 0 & 7 & 4- \\
0 & 2 & 7 & 1 & 1 \\
0 & 4 & 4 & 1 & 2
\end{bmatrix} .
$$

The zeros can be covered by four lines, so, Step 5 yields

$$
\begin{bmatrix}
6 & 1 & 1 & 0 & \boxed{0} \\
7 & 1 & \boxed{0} & 4 & 2 \\
5 & \boxed{0} & 1 & 7 & 4 \\
\boxed{0} & 1 & 7 & 0 & 0 \\
0 & 3 & 4 & \boxed{0} & 1
\end{bmatrix} .
$$

Now the zeros cannot be covered with fewer than five lines, so we have an optimal assignment. There are three such (one is shown), each having a cost of 39.

2. From the matrix (1), Step 1 yields

$$
\begin{bmatrix}
16 & 59 & 0 \\
6 & 46 & 0 \\
24 & 56 & 0
\end{bmatrix} .
$$

Step 2 yields

$$
\begin{bmatrix}
10 & 13 & 0 \\
-0 & 0 & 0- \\
18 & 10 & 0
\end{bmatrix}
$$

and this is not optimal. Step 5 yields

$$\begin{bmatrix} 0 & 3 & 0 \\ 0 & 0 & 10 \\ 8 & 0 & 0 \end{bmatrix}$$

clearly optimal, and we obtain the solutions given in the example.

3.   Another such covering is

$$\begin{bmatrix} -15 & 0 & 0 & 0- \\ 0 & 50 & 20 & 25 \\ 35 & 5 & 0 & 10 \\ 0 & 65 & 50 & 65 \end{bmatrix}.$$

Then Step 5 yields

$$\begin{bmatrix} -20 & 0 & 5 & 0- \\ 0 & 45 & 20 & 20 \\ -35 & 0 & 0 & 5- \\ 0 & 60 & 50 & 60 \end{bmatrix}$$

in which there is still a covering by three lines. Step 5 again yields

$$\begin{bmatrix} 40 & 0 & 5 & 0 \\ 0 & 25 & 0 & 0 \\ 55 & 0 & 0 & 5 \\ 0 & 40 & 30 & 40 \end{bmatrix}.$$

Since this matrix is the same as (8), we must get the same optimal assignments.

4. Another such assignment is

$$\begin{bmatrix} 0 & 3 & 0 & 4 & \boxed{0} \\ 4 & \boxed{0} & 6 & 1 & 5 \\ 3 & 1 & \boxed{0} & 4 & 0 \\ \boxed{0} & 1 & 6 & 2 & 1 \\ 0 & 0 & 0 & \boxed{0} & 3 \end{bmatrix}$$

giving the matches

Sue - Don	(rank = 10)
Ann - Tom	(rank = 9)
Bea - Joe	(rank = 6)
Fay - Bob	(rank = 6)
Hal - unmatched,	

a total score of 31.

5. **(a)** A dummy coin adds a column of zeros, then we negate to obtain a starting matrix:

$$\begin{bmatrix} -150 & -65 & -210 & -135 & 0 \\ -175 & -75 & -230 & -155 & 0 \\ -135 & -85 & -200 & -140 & 0 \\ -140 & -70 & -190 & -130 & 0 \\ -170 & -50 & -200 & -160 & 0 \end{bmatrix}.$$

Step 1 yields

$$\begin{bmatrix} 60 & 145 & 0 & 75 & 210 \\ 55 & 155 & 0 & 75 & 230 \\ 65 & 115 & 0 & 60 & 200 \\ 50 & 120 & 0 & 60 & 190 \\ 30 & 150 & 0 & 40 & 200 \end{bmatrix}$$

and Step 2 yields

$$\begin{bmatrix} 30 & 30 & 0 & 35 & 20 \\ 25 & 40 & 0 & 35 & 40 \\ 35 & 0 & 0 & 20 & 10 \\ 20 & 5 & 0 & 20 & 0 \\ 0 & 35 & 0 & 0 & 10 \end{bmatrix}$$

which is not optimal. Step 5 yields

$$\begin{bmatrix} 10 & 10 & 0 & 15 & 0 \\ 5 & 20 & 0 & 15 & 20 \\ 35 & 0 & 20 & 20 & 10 \\ 20 & 5 & 20 & 20 & 0 \\ 0 & 35 & 20 & 0 & 10 \end{bmatrix}$$

which is not optimal. Step 5 yields

$$\begin{bmatrix} 5 & 5 & \boxed{0} & 10 & 0 \\ \boxed{0} & 15 & 0 & 10 & 20 \\ 35 & \boxed{0} & 25 & 20 & 15 \\ 15 & 0 & 20 & 15 & \boxed{0} \\ 0 & 35 & 25 & \boxed{0} & 15 \end{bmatrix}.$$

This is optimal, containing only the optimal assignment shown. Thus, bidder 1 gets coin 3 ($210), bidder 2 gets coin 1 ($175), bidder 3 gets coin 2 ($85), bidder 5 gets coin 4 ($160), and bidder 4 gets nothing; a total of $630.

(b) We use bidder 2's row twice (so he has two chances at getting accepted), and add two dummy coin columns. Then negating yields the starting matrix

$$
\begin{bmatrix}
-150 & -65 & -210 & -135 & 0 & 0 \\
-175 & -75 & -230 & -155 & 0 & 0 \\
-175 & -75 & -230 & -155 & 0 & 0 \\
-135 & -85 & -200 & 140 & 0 & 0 \\
-140 & -70 & -190 & -130 & 0 & 0 \\
-170 & -50 & -200 & -160 & 0 & 0
\end{bmatrix}
$$

.

After Steps 1 and 2 we get

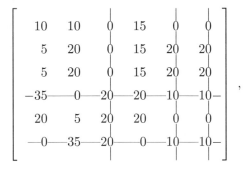

not optimal. Step 5 yields

$$
\begin{bmatrix}
10 & 10 & 0 & 15 & 0 & 0 \\
5 & 20 & 0 & 15 & 20 & 20 \\
5 & 20 & 0 & 15 & 20 & 20 \\
-35 & 0 & 20 & 20 & 10 & 10- \\
20 & 5 & 20 & 20 & 0 & 0 \\
0 & 35 & 20 & 0 & 10 & 10-
\end{bmatrix},
$$

still not optimal. Step 5 again

$$\begin{bmatrix} 5 & 5 & 0 & 10 & \boxed{0} & 0 \\ 0 & 15 & \boxed{0} & 10 & 20 & 20 \\ \boxed{0} & 15 & 0 & 10 & 20 & 20 \\ 35 & \boxed{0} & 25 & 20 & 15 & 15 \\ 15 & 0 & 20 & 15 & 0 & \boxed{0} \\ 0 & 35 & 25 & \boxed{0} & 15 & 15 \end{bmatrix},$$

which is optimal. The matrix contains several optimal assignments, but all result in: bidder 2 gets coins 1 and 3 ($405), bidder 3 gets coin 2 ($85), bidder 5 gets coin 4 ($160), and bidders 1 and 4 get nothing; this is a total of $650.

6. Negating the original matrix, and applying Steps 1 and 2 yield

$$\begin{bmatrix} -4 & 6 & 13 & 15 & 5 & 0 & 0 & 20 & 7 \\ -4 & 1 & 1 & 0 & 3 & 6 & 7 & 8 & 4 \\ -0 & 0 & 2 & 3 & 0 & 6 & 6 & 0 & 1 \\ 6 & 2 & 8 & 5 & 2 & 13 & 12 & 0 & 7 \\ 7 & 3 & 8 & 5 & 3 & 13 & 11 & 0 & 7 \\ 4 & 2 & 3 & 4 & 2 & 13 & 10 & 0 & 7 \\ 7 & 2 & 1 & 3 & 5 & 7 & 8 & 0 & 0 \\ 4 & 0 & 0 & 2 & 2 & 4 & 5 & 0 & 0 \\ 4 & 0 & 0 & 2 & 2 & 4 & 5 & 0 & 0 \end{bmatrix}.$$

This is not optimal, and Step 5 gives

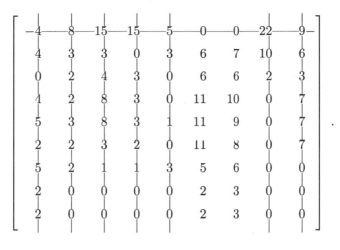

This is not optimal. Step 5 again yields

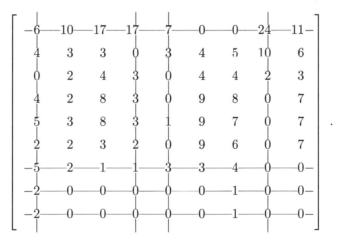

This is still not optimal. Step 5 again gives

$$
\begin{bmatrix}
8 & 10 & 17 & 19 & 9 & 0 & \boxed{0} & 26 & 11 \\
4 & 1 & 1 & \boxed{0} & 3 & 2 & 3 & 10 & 4 \\
\boxed{0} & 0 & 2 & 3 & 0 & 2 & 2 & 2 & 1 \\
4 & \boxed{0} & 6 & 3 & \boxed{0} & 7 & 6 & 0 & 5 \\
5 & 1 & 6 & 3 & 1 & 7 & 5 & \boxed{0} & 5 \\
2 & \boxed{0} & 1 & 2 & \boxed{0} & 7 & 4 & 0 & 5 \\
7 & 2 & 1 & 3 & 5 & 3 & 4 & 2 & \boxed{0} \\
4 & 0 & \boxed{0} & 2 & 2 & \boxed{0} & 1 & 2 & 0 \\
4 & 0 & \boxed{0} & 2 & 2 & \boxed{0} & 1 & 2 & 0 \\
\end{bmatrix}
$$

This is optimal, yielding the following lineup(s):

P - Pete	2B - Jill or Ann	LF - Herb
C - Liz	3B - Alex	CF - John or Lois
1B - Sam	SS - Ann or Jill	RF - Lois or John

All possibilities have a "score" of 125.

7. Proof of part (ii) of the test for optimality: Since each zero entry of an optimal assignment is in a different row, a horizontal line can cover at most one assigned zero; similarly, a vertical line can cover at most one assigned zero. Thus, as many lines as there are zeros is needed, that is, n lines.

Proof of part (i) of the test for optimality: Suppose we are given an $n \times n$ matrix that does not contain an optimal assignment of zeros. Then we cannot find a set of n zero entries, no two of which come from the same row or column. But we can always find a set of fewer than n zero entries, say k zero entries where $k < n$, no two of which come from the same row or column (possibly k is zero). A set of k such zero entries for which k is as large as possible is called a maximal assignment of zeros, and the zero entries in the set are called the assigned zeros of the maximal assignment. We shall prove the following theorem:

> **THEOREM**   *All of the zero entries in a matrix that contains a maximal assignment of zeros with k assigned zeros can be covered with k vertical and horizontal lines.*

PROOF. The following algorithm shows how the k lines in the statement of the theorem can be found:

Step 1. Search the matrix for unassigned zeros that do not contain any assigned zeros in their rows. Cover each of the columns in which these unassigned zeros lie with a vertical line. Label one such unassigned zero in each covered column as 0_1. Each column covered in this way must also contain an assigned zero, which we label as 0_1^*. (If not, then we could add the unassigned zero 0_1 in that column to the maximal assignment of zeros — a contradiction to the definition of a maximal assignment of zeros.)

Step 2. Search the matrix for uncovered, unassigned zeros that contain an assigned zero of the form 0_1^* in their rows. Cover each of the columns in which these uncovered, unassigned zeros lie with a vertical line. Label one such uncovered, unassigned zero in each newly-covered column as 0_2. Each column covered in this way must also contain an assigned zero, which we label as 0_2^*. (If not, then we could add the unassigned zero 0_2 in that column to the maximal assignment of zeros after we reassign the assigned zero of the form 0_1^* in its row to the corresponding unassigned zero 0_1.)

Step 3. Continue Step 2 iteratively, so that at the m-th step we search for uncovered, unassigned zeros that contain an assigned zero of the form 0_{m-1}^* in its row. Cover their columns with vertical lines and label one such uncovered, unassigned zero from each such covered column as 0_m. Each newly covered column must also contain an assigned zero, which we label as 0_m^* (otherwise we could increase the number of assigned zeros by one by replacing $m-1$ assigned zeros of the form 0_1^*, 0_2^*, ..., 0_{m-1}^* by m unassigned zeros of the form 0_1, 0_2, ..., 0_{m-1}, 0_m). We continue iteratively until we can find no more unassigned zeros which contain covered, assigned zeros in their rows.

Step 4. After Step 3, all uncovered, unassigned zeros must lie in rows that contain uncovered, assigned zeros. Passing horizontal lines through these remaining uncovered, assigned zeros completes the algorithm. All zeros in the matrix are covered and each of the k assigned zeros has either a vertical or a horizontal line through it. There are thus exactly k covering lines, which is what we wanted to prove. $\square$

As an example of this algorithm, the zero entries of the following 7×7 matrix containing a maximal assignment of zeros with six assigned zeros (labeled with asterisks):

$$\begin{bmatrix} 1 & 0^* & 1 & 0 & 1 & 1 & 1 \\ 1 & 1 & 0 & 0 & 1 & 1 & 0^* \\ 0^* & 1 & 1 & 0 & 1 & 0 & 0 \\ 1 & 0 & 1 & 0^* & 1 & 1 & 1 \\ 1 & 1 & 0^* & 1 & 0 & 0 & 1 \\ 1 & 1 & 1 & 0 & 0^* & 1 & 1 \\ 1 & 0 & 1 & 1 & 0 & 1 & 1 \end{bmatrix},$$

can be covered with six lines as follows:

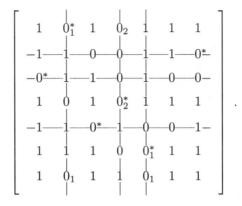

Remark: The above theorem was first proved in 1931 by two Hungarian mathematicians, D. König and E. Egerváry. It is after them that the Hungarian Method is named.

8.   **(a)**   Let q be the number of intersections of covering lines. Then there are $nm - q$ entries covered altogether (since each line covers n entries, but q are covered twice). There are thus $n^2 - (nm - q)$ uncovered entries. The total amount subtracted from C is thus $a(n^2 - nm + q)$, and the total amount added to C is aq. So, the sum of the entries of C' is equal to the sum of the entries of C plus $aq - a(n^2 - nm + q)$, i.e.,

$$\sum c_{ij} - \sum c'_{ij} = an(n - m).$$

(b)   We assume the original matrix contains integers only. Since the difference $\sum (c_{ij} - c'_{ij})$ is positive if $a > 0$ and $n > m$, the sum of the entries will decrease each time. If the process does not terminate, the sequence of sums (being a decreasing sequence of nonnegative integers) will reach zero. But then the matrix, non-negative, must be identically zero, so we must get an optimal assignment of zeros either way.

EXERCISE SET 11.5

1. We have $2b_1 = M_1$, $2b_2 = M_2$, ..., $2b_{n-1} = M_{n-1}$ from (6).
Inserting in (13) yields

$$6a_1h + M_1 = M_2$$
$$6a_2h + M_2 = M_3$$
$$\vdots$$
$$6a_{n-2}h + M_{n-2} = M_{n-1},$$

from which we obtain

$$a_1 = \frac{M_2 - M_1}{6h}$$

$$a_2 = \frac{M_3 - M_2}{6h}$$

$$\vdots$$

$$a_{n-2} = \frac{M_{n-1} - M_{n-2}}{6h}.$$

Now $S''\left(x_n\right) = M_n$, or from (6), $6a_{n-1}h + 2b_{n-1} = M_n$.

Also, $2b_{n-1} = M_{n-1}$ from (6) and so

$$6a_{n-1}h + M_{n-1} = M_n$$

or

$$a_{n-1} = \frac{M_n - M_{n-1}}{6h}.$$

Thus we have

$$a_i = \frac{M_{i+1} - M_i}{6h}$$

for $i = 1, 2, \ldots, n - 1$. From (9) and (11) we have

$$a_i h^3 + b_i h^2 + c_i h + d_i = y_{i+1}, \quad i = 1, 2, \ldots, n - 1.$$

Substituting the expressions for a_i, b_i, and d_i from (14) yields $\left(\dfrac{M_{i+1} - M_i}{6h}\right) h^3 + \dfrac{M_i}{2} h^2 +$ $c_i h + y_i = y_{i+1}, \quad i = 1, 2, \ldots, n - 1$. Solving for c_i gives

$$c_i = \frac{y_{i+1} - y_i}{h} - \left(\frac{M_{i+1} + 2M_i}{6}\right) h$$

for $i = 1, 2, \ldots, n - 1$.

2. **(a)** Set $h = .2$ and

$$\begin{aligned}
x_1 &= 0, & y_1 &= .00000 \\
x_2 &= .2, & y_2 &= .19867 \\
x_3 &= .4, & y_3 &= .38942 \\
x_4 &= .6, & y_4 &= .56464 \\
x_5 &= .8, & y_5 &= .71736 \\
x_6 &= 1.0, & y_6 &= .84147.
\end{aligned}$$

Then

$$6\left(y_1 - 2y_2 + y_3\right)/h^2 = -1.1880$$

$$6\left(y_2 - 2y_3 + y_4\right)/h^2 = -2.3295$$

$$6\left(y_3 - 2y_4 + y_5\right)/h^2 = -3.3750$$

$$6\left(y_4 - 2y_5 + y_6\right)/h^2 = -4.2915$$

and the linear system (21) for the parabolic runout spline becomes

$$
\begin{bmatrix}
5 & 1 & 0 & 0 \\
1 & 4 & 1 & 0 \\
0 & 1 & 4 & 1 \\
0 & 0 & 1 & 5
\end{bmatrix}
\begin{bmatrix}
M_2 \\
M_3 \\
M_4 \\
M_5
\end{bmatrix}
=
\begin{bmatrix}
-1.1880 \\
-2.3295 \\
-3.3750 \\
-4.2915
\end{bmatrix}.
$$

Solving this system yields

$$M_2 = -.15676$$
$$M_3 = -.40421$$
$$M_4 = -.55592$$
$$M_5 = -.74712.$$

From (19) and (20) we have

$$M_1 = M_2 = -0.15676$$
$$M_6 = M_5 = -0.74712.$$

The specified interval $.4 \leq x \leq .6$ is the third interval. Using (14) to solve for a_3, b_3, c_3 and d_3 gives

$$a_3 = \left(M_4 - M_3\right)/6h = -.12643$$

$$b_3 = M_3/2 = -.20211$$

$$c_3 = \left(y_4 - y_3\right)/h - \left(M_4 + 2M_3\right)h/6 = .92158$$

$$d_3 = y_3 = .38942 \quad .$$

The interpolating parabolic runout spline for $.4 \leq x \leq .6$ is thus

$$S(x) = -.12643(x - .4)^3 - .20211(x - .4)^2$$
$$+ .92158(x - .4) + .38942.$$

2.   (b)

$$S(.5) = -.12643(.1)^3 - .20211(.1)^2 + .92158(.1) + .38942$$

$$= .47943.$$

Since $\sin(.5) = S(.5) = .47943$ to five decimal places, the percentage error is zero.

3.   (a) Given that the points lie on a single cubic curve, the cubic runout spline will agree exactly with the single cubic curve.

(b) Set $h = 1$ and

$$\begin{array}{ll} x_1 = 0 \quad , & y_1 = 1 \\ x_2 = 1 \quad , & y_2 = 7 \\ x_3 = 2 \quad , & y_3 = 27 \\ x_4 = 3 \quad , & y_4 = 79 \\ x_5 = 4 \quad , & y_5 = 181. \end{array}$$

Then

$$6\left(y_1 - 2y_2 + y_3\right)/h^2 = 84$$

$$6\left(y_2 - 2y_3 + y_4\right)/h^2 = 192$$

$$6\left(y_3 - 2y_4 + y_5\right)/h^2 = 300$$

and the linear system (24) for the cubic runout spline becomes

$$\begin{bmatrix} 6 & 0 & 0 \\ 1 & 4 & 1 \\ 0 & 0 & 6 \end{bmatrix} \begin{bmatrix} M_2 \\ M_3 \\ M_4 \end{bmatrix} = \begin{bmatrix} 84 \\ 192 \\ 300 \end{bmatrix}.$$

Solving this system yields

$$M_2 = 14$$

$$M_3 = 32$$

$$M_4 = 50.$$

From (22) and (23) we have

$$M_1 = 2M_2 - M_3 = -4$$
$$M_5 = 2M_4 - M_3 = 68.$$

Using (14) to solve for the a_i's, b_i's, c_i's, and d_i's we have

$$a_1 = \left(M_2 - M_1\right)/6h = 3$$
$$a_2 = \left(M_3 - M_2\right)/6h = 3$$
$$a_3 = \left(M_4 - M_3\right)/6h = 3$$
$$a_4 = \left(M_5 - M_4\right)/6h = 3$$

$$b_1 = M_1/2 = -2$$
$$b_2 = M_2/2 = 7$$
$$b_3 = M_3/2 = 16$$
$$b_4 = M_4/2 = 25$$

$$c_1 = \left(y_2 - y_1\right)/h - \left(M_2 + 2M_1\right)h/6 = 5$$
$$c_2 = \left(y_3 - y_2\right)/h - \left(M_3 + 2M_2\right)h/6 = 10$$
$$c_3 = \left(y_4 - y_3\right)/h - \left(M_4 + 2M_3\right)h/6 = 33$$
$$c_4 = \left(y_5 - y_4\right)/h - \left(M_5 + 2M_4\right)h/6 = 74$$

$$d_1 = y_1 = 1$$
$$d_2 = y_2 = 7$$
$$d_3 = y_3 = 27$$
$$d_4 = y_4 = 79.$$

For $0 \leq x \leq 1$ we have

$$S(x) = S_1(x) = 3x^3 - 2x^2 + 5x + 1.$$

For $1 \leq x \leq 2$ we have

$$\begin{aligned}S(x) = S_2(x) &= 3(x-1)^3 + 7(x-1)^2 + 10(x-1) + 7 \\ &= 3x^3 - 2x^2 + 5x + 1.\end{aligned}$$

For $2 \leq x \leq 3$ we have

$$\begin{aligned}S(x) = S_3(x) &= 3(x-2)^3 + 16(x-2)^2 + 33(x-2) + 27 \\ &= 3x^3 - 2x^2 + 5x + 1.\end{aligned}$$

For $3 \leq x \leq 4$ we have

$$\begin{aligned}S(x) = S_4(x) &= 3(x-3)^3 + 25(x-3)^2 + 74(x-3) + 79 \\ &= 3x^3 - 2x^2 + 5x + 1.\end{aligned}$$

Thus $S_1(x) = S_2(x) = S_3(x) = S_4(x)$,

or $S(x) = 3x^3 - 2x^2 + 5x + 1 \quad$ for $0 \leq x \leq 4$.

4. The linear system (16) for the natural spline becomes

$$\begin{bmatrix} 4 & 1 & 0 \\ 1 & 4 & 1 \\ 0 & 1 & 4 \end{bmatrix} \begin{bmatrix} M_2 \\ M_3 \\ M_4 \end{bmatrix} = \begin{bmatrix} -.0001116 \\ -.0000816 \\ -.0000636 \end{bmatrix}.$$

Solving this system yields

$$M_2 = -.0000252$$
$$M_3 = -.0000108$$
$$M_4 = -.0000132.$$

From (17) and (18) we have

$$M_1 = 0$$
$$M_5 = 0.$$

Solving for the a_i's, b_i's, c_i's, and d_i's from Eqs. (14) we have

$$a_1 = \left(M_2 - M_1\right)/6h = -.00000042$$
$$a_2 = \left(M_3 - M_2\right)/6h = .00000024$$
$$a_3 = \left(M_4 - M_3\right)/6h = -.00000004$$
$$a_4 = \left(M_5 - M_4\right)/6h = -.00000022$$

$$b_1 = M_1/2 = 0$$
$$b_2 = M_2/2 = -.0000126$$
$$b_3 = M_3/2 = -.0000054$$
$$b_4 = M_4/2 = -.0000066$$
$$b_5 = M_5/2 = 0$$

$$c_1 = \left(y_2 - y_1\right)/h - \left(M_2 + 2M_1\right)h/6 = .000214$$
$$c_2 = \left(y_3 - y_2\right)/h - \left(M_3 + 2M_2\right)h/6 = .000088$$
$$c_3 = \left(y_4 - y_3\right)/h - \left(M_4 + 2M_3\right)h/6 = -.000092$$
$$c_4 = \left(y_5 - y_4\right)/h - \left(M_5 + 2M_4\right)h/6 = -.000212$$

$$d_1 = y_1 = .99815$$

$$d_2 = y_2 = .99987$$

$$d_3 = y_3 = .99973$$

$$d_4 = y_4 = .99823.$$

The resulting natural spline is

$$S(x) = \begin{cases} -.00000042(x+10)^3 & + .000214(x+10) + .99815, & -10 \le x \le 0 \\ .00000024(x)^3 & - .0000126(x)^2 & + .000088(x) & + .99987, & 0 \le x \le 10 \\ -.00000004(x-10)^3 - .0000054(x-10)^2 - .000092(x-10) + .99973, & 10 \le x \le 20 \\ -.00000022(x-20)^3 - .0000066(x-20)^2 - .000212(x-20) + .99823, & 20 \le x \le 30. \end{cases}$$

Assuming the maximum is attained in the interval $[0, 10]$ we set $S'(x)$ equal to zero in this interval:

$$S'(x) = .00000072x^2 - .0000252x + .000088 = 0.$$

To three significant digits the root of this quadratic equation in the interval $[0, 10]$ is $x = 3.93$, and

$$S(3.93) = 1.00004.$$

5. The linear system (24) for the cubic runout spline becomes

$$\begin{bmatrix} 6 & 0 & 0 \\ 1 & 4 & 1 \\ 0 & 0 & 6 \end{bmatrix} \begin{bmatrix} M_2 \\ M_3 \\ M_4 \end{bmatrix} = \begin{bmatrix} -.0001116 \\ -.0000816 \\ -.0000636 \end{bmatrix}.$$

Solving this system yields

$$M_2 = -.0000186$$

$$M_3 = -.0000131$$

$$M_4 = -.0000106.$$

From (22) and (23) we have

$$M_1 = 2M_2 - M_3 = -.0000241$$
$$M_5 = 2M_4 - M_3 = -.0000081.$$

Solving for the a_i's, b_i's, c_i's and d_i's from Eqs. (14) we have

$$a_1 = \left(M_2 - M_1\right)/6h = .00000009$$
$$a_2 = \left(M_3 - M_2\right)/6h = .00000009$$
$$a_3 = \left(M_4 - M_3\right)/6h = .00000004$$
$$a_4 = \left(M_5 - M_4\right)/6h = .00000004$$

$$b_1 = M_1/2 = -.0000121$$
$$b_2 = M_2/2 = -.0000093$$
$$b_3 = M_3/2 = -.0000066$$
$$b_4 = M_4/2 = -.0000053$$

$$c_1 = \left(y_2 - y_1\right)/h - \left(M_2 + 2M_1\right)h/6 = .000282$$
$$c_2 = \left(y_3 - y_2\right)/h - \left(M_3 + 2M_2\right)h/6 = .000070$$
$$c_3 = \left(y_4 - y_3\right)/h - \left(M_4 + 2M_3\right)h/6 = .000087$$
$$c_4 = \left(y_5 - y_4\right)/h - \left(M_5 + 2M_4\right)h/6 = .000207$$

$$d_1 = y_1 = .99815$$
$$d_2 = y_2 = .99987$$
$$d_3 = y_3 = .99973$$
$$d_4 = y_4 = .99823.$$

The resulting cubic runout spline is

$$
S(x) = \begin{cases}
.00000009(x+10)^3 - .0000121(x-10)^2 + .000282(x+10) + .99815, & -10 \le x \le 0 \\[2mm]
.00000009(x)^3 \qquad\;\; - .0000093(x)^2 \qquad + .000070(x) \qquad + .99987, & 0 \le x \le 10 \\[2mm]
.00000004(x-10)^3 - .0000066(x-10)^2 + .000087(x-10) + .99973, & 10 \le x \le 20 \\[2mm]
.00000004(x-20)^3 - .0000053(x-20)^2 + .000207(x-20) + .99823, & 20 \le x \le 30.
\end{cases}
$$

Assuming the maximum is attained in the interval $[0, 10]$, we set $S'(x)$ equal to zero in this interval:

$$ S'(x) = .00000027x^2 - .0000186x + .000070. $$

To three significant digits the root of this quadratic in the interval $[0, 10]$ is 4.00 and

$$ S(4.00) = 1.00001. $$

6. **(a)** Since $S\!\left(x_1\right) = y_1$ and $S\!\left(x_n\right) = y_n$, then from $S\!\left(x_1\right) = S\!\left(x_n\right)$ we have $y_1 = y_n$. By definition $S''\!\left(x_1\right) = M_1$ and $S''\!\left(x_n\right) = M_n$, and so from $S''\!\left(x_1\right) = S''\!\left(x_n\right)$ we have $M_1 = M_n$.

From (5) we have

$$
S'\!\left(x_1\right) = c_1
$$
$$
S'\!\left(x_n\right) = 3a_{n-1}h^2 + 2b_{n-1}h + c_{n-1}.
$$

Substituting for c_1, a_{n-1}, b_{n-1}, c_{n-1} from Eqs. (14) yields

$$
S'\!\left(x_1\right) = \left(y_2 - y_1\right)/h - \left(M_2 + 2M_1\right)h/6
$$
$$
S'\!\left(x_n\right) = \left(M_n - M_{n-1}\right)h/2 + M_{n-1}h
$$
$$
+ \left(y_n - y_{n-1}\right)/h - \left(M_n + 2M_{n-1}\right)h/6.
$$

Using $M_n = M_1$ and $y_n = y_1$, the last equation becomes

$$S'\left(x_n\right) = M_1 h/3 + M_{n-1} h/6 + \left(y_1 - y_{n-1}\right)/h.$$

From $S'\left(x_1\right) = S'\left(x_n\right)$ we obtain

$$\left(y_2 - y_1\right)/h - \left(y_1 - y_{n-1}\right)/h = M_1 h/3 + M_{n-1} h/6 + \left(M_2 + 2M_1\right)h/6$$

or

$$4M_1 + M_2 + M_{n-1} = 6\left(y_{n-1} - 2y_1 + y_2\right)/h^2.$$

(b) Eqs. (15) together with the three equations in part (a) of the exercise statement give

$$
\begin{aligned}
4M_1 \quad + \quad M_2 \quad + \quad M_{n-1} &= 6\left(y_{n-1} - 2y_1 + y_2\right)/h^2 \\
M_1 \quad + 4M_2 \quad + \quad M_3 \quad &= 6\left(y_1 \quad - 2y_2 + y_3\right)/h^2 \\
M_2 \quad + 4M_3 \quad + \quad M_4 \quad &= 6\left(y_2 \quad - 2y_3 + y_4\right)/h^2 \\
&\vdots \\
M_{n-3} + 4M_{n-2} + \quad M_{n-1} &= 6\left(y_{n-3} - 2y_{n-2} + y_{n-1}\right)/h^2 \\
M_1 \quad + \quad M_{n-2} + 4M_{n-1} &= 6\left(y_{n-2} - 2y_{n-1} + y_1 \quad\right)/h^2.
\end{aligned}
$$

This linear system for $M_1, M_2, \ldots, M_{n-1}$ in matrix form is

$$
\begin{bmatrix}
4 & 1 & 0 & 0 & \cdot & \cdot & \cdot & 0 & 0 & 0 & 1 \\
1 & 4 & 1 & 0 & \cdot & \cdot & \cdot & 0 & 0 & 0 & 0 \\
0 & 1 & 4 & 1 & \cdot & \cdot & \cdot & 0 & 0 & 0 & 0 \\
\vdots & \vdots & \vdots & \vdots & & & & \vdots & \vdots & \vdots & \vdots \\
0 & 0 & 0 & 0 & \cdot & \cdot & \cdot & 0 & 1 & 4 & 1 \\
1 & 0 & 0 & 0 & \cdot & \cdot & \cdot & 0 & 0 & 1 & 4
\end{bmatrix}
\begin{bmatrix}
M_1 \\
M_2 \\
M_3 \\
\vdots \\
M_{n-2} \\
M_{n-1}
\end{bmatrix}
= \frac{6}{h^2}
\begin{bmatrix}
y_{n-1} & - 2y_1 & + y_2 \\
y_1 & - 2y_2 & + y_3 \\
y_2 & - 2y_3 & + y_4 \\
& \vdots & \\
y_{n-3} & - 2y_{n-2} & + y_{n-1} \\
y_{n-2} & - 2y_{n-1} & + y_1
\end{bmatrix}.
$$

7. **(a)** Since $S'\left(x_1\right) = y_1'$ and $S'\left(x_n\right) = y_n'$, then from (5) we have

$$S'\left(x_1\right) = c_1$$
$$S'\left(x_n\right) = 3a_{n-1}h^2 + 2b_{n-1}h + c_{n-1}.$$

Substituting for c_1, a_{n-1}, b_{n-1}, c_{n-1} from Eqs. (14) yields

$$y_1' = \left(y_2 - y_1\right)/h - \left(M_2 + 2M_1\right)h/6$$
$$y_n' = \left(M_n - M_{n-1}\right)h/2 + M_{n-1}h$$
$$+ \left(y_n - y_{n-1}\right)/h - \left(M_n + 2M_{n-1}\right)h/6.$$

From the first equation we obtain

$$2M_1 + M_2 = 6\left(y_2 - y_1 - hy_1'\right)/h^2.$$

From the second equation we obtain

$$2M_n + M_{n-1} = 6\left(y_{n-1} - y_n + hy_n'\right)/h^2.$$

(b) Eqs. (15) together with the two equations in part (a) give

$$\begin{aligned}
2M_1 \;+\; M_2 \;\;&=\; 6\left(y_2 \;-\; y_1 \;-\; hy_1'\right)/h^2 \\
M_1 \;+\; 4M_2 \;+\; M_3 \;&=\; 6\left(y_1 \;-\; 2y_2 \;+\; y_3\right)/h^2 \\
M_2 \;+\; 4M_3 \;+\; M_4 \;&=\; 6\left(y_2 \;-\; 2y_3 \;+\; y_4\right)/h^2 \\
&\;\;\vdots \\
M_{n-2} + 4M_{n-1} + M_n \;&=\; 6\left(y_{n-2} - 2y_{n-1} + y_n\right)/h^2 \\
M_{n-1} \;+\; 2M_n \;&=\; 6\left(y_{n-1} - y_n + hy_n'\right)/h^2.
\end{aligned}$$

This linear system for $M_1, M_2, \ldots, M_n$ in matrix form is

$$\begin{bmatrix} 2 & 1 & 0 & 0 & \cdot & \cdot & \cdot & 0 & 0 & 0 & 0 \\ 1 & 4 & 1 & 0 & \cdot & \cdot & \cdot & 0 & 0 & 0 & 0 \\ 0 & 1 & 4 & 1 & \cdot & \cdot & \cdot & 0 & 0 & 0 & 0 \\ 0 & 0 & 1 & 4 & \cdot & \cdot & \cdot & 0 & 0 & 0 & 0 \\ \vdots & \vdots & \vdots & \vdots & & & & \vdots & \vdots & \vdots & \vdots \\ 0 & 0 & 0 & 0 & \cdot & \cdot & \cdot & 1 & 4 & 1 & 0 \\ 0 & 0 & 0 & 0 & \cdot & \cdot & \cdot & 0 & 1 & 4 & 1 \\ 0 & 0 & 0 & 0 & \cdot & \cdot & \cdot & 0 & 0 & 1 & 2 \end{bmatrix} \begin{bmatrix} M_1 \\ M_2 \\ M_3 \\ \vdots \\ M_{n-2} \\ M_{n-1} \\ M_n \end{bmatrix} = \frac{6}{h^2} \begin{bmatrix} -hy_1' & - & y_1 & + & y_2 \\ y_1 & - & 2y_2 & + & y_3 \\ y_2 & - & 2y_3 & + & y_4 \\ & & \vdots & & \\ y_{n-2} & - & 2y_{n-1} & + & y_n \\ y_{n-1} & - & y_n & + & hy_n' \end{bmatrix}.$$

EXERCISE SET 11.6

1. **(a)** $\mathbf{x}^{(1)} = P\mathbf{x}^{(0)} = \begin{bmatrix} .4 \\ .6 \end{bmatrix}$, $\quad \mathbf{x}^{(2)} = P\mathbf{x}^{(1)} = \begin{bmatrix} .46 \\ .54 \end{bmatrix}$.

   Continuing in this manner yields $\mathbf{x}^{(3)} = \begin{bmatrix} .454 \\ .546 \end{bmatrix}$,

   $\mathbf{x}^{(4)} = \begin{bmatrix} .4546 \\ .5454 \end{bmatrix}$ and $\mathbf{x}^{(5)} = \begin{bmatrix} .45454 \\ .54546 \end{bmatrix}$.

   (b) P is regular because all of the entries of P are positive. Its steady-state vector $\mathbf{q}$ solves $(P - I)\mathbf{q} = \mathbf{0}$; that is,

   $$\begin{bmatrix} -.6 & .5 \\ .6 & -.5 \end{bmatrix} \begin{bmatrix} q_1 \\ q_2 \end{bmatrix} = \begin{bmatrix} 0 \\ 0 \end{bmatrix}.$$

   This yields one independent equation, $.6q_1 - .5q_2 = 0$, or $q_1 = \dfrac{5}{6}q_2$. Solutions are thus

   of the form $\mathbf{q} = s \begin{bmatrix} 5/6 \\ 1 \end{bmatrix}$. Set $s = \dfrac{1}{\dfrac{5}{6} + 1} = \dfrac{6}{11}$ to obtain $\mathbf{q} = \begin{bmatrix} 5/11 \\ 6/11 \end{bmatrix}$.

2. **(a)** $\mathbf{x}^{(1)} = P\mathbf{x}^{(0)} = \begin{bmatrix} .7 \\ .2 \\ .1 \end{bmatrix}$; likewise $\mathbf{x}^{(2)} = \begin{bmatrix} .23 \\ .52 \\ .25 \end{bmatrix}$

   and $\mathbf{x}^{(3)} = \begin{bmatrix} .273 \\ .396 \\ .331 \end{bmatrix}$.

(b) P is regular because all of its entries are positive.

To solve $(P - I)\mathbf{q} = \mathbf{0}$, i.e.,

$$\begin{bmatrix} -.8 & .1 & .7 \\ .6 & -.6 & .2 \\ .2 & .5 & -.9 \end{bmatrix} \begin{bmatrix} q_1 \\ q_2 \\ q_3 \end{bmatrix} = \begin{bmatrix} 0 \\ 0 \\ 0 \end{bmatrix},$$

reduce the coefficient matrix to row-echelon form:

$$\begin{bmatrix} -8 & 1 & 7 \\ 6 & -6 & 2 \\ 2 & 5 & -9 \end{bmatrix} \Rightarrow \begin{bmatrix} 2 & 5 & -9 \\ 0 & -21 & 29 \\ 0 & 0 & 0 \end{bmatrix} \Rightarrow \begin{bmatrix} 1 & 0 & -\dfrac{22}{21} \\ 0 & 1 & -\dfrac{29}{21} \\ 0 & 0 & 0 \end{bmatrix}.$$

This yields solutions (setting $q_3 = s$) of the form $\begin{bmatrix} 22/21 \\ 29/21 \\ 1 \end{bmatrix} s.$

To obtain a probability vector, take $s = \dfrac{1}{\dfrac{22}{21} + \dfrac{29}{21} + 1} = \dfrac{21}{72},$

yielding $\mathbf{q} = \begin{bmatrix} \dfrac{22}{72} & \dfrac{29}{72} & \dfrac{21}{72} \end{bmatrix}^t.$

3. **(a)** Solve $(P - I)\mathbf{q} = \mathbf{0}$, i.e.,

$$\begin{bmatrix} -2/3 & 3/4 \\ 2/3 & -3/4 \end{bmatrix} \begin{bmatrix} q_1 \\ q_2 \end{bmatrix} = \begin{bmatrix} 0 \\ 0 \end{bmatrix}.$$

The only independent equation is $\dfrac{2}{3} q_1 = \dfrac{3}{4} q_2$, yielding $\mathbf{q} = \begin{bmatrix} 9/8 \\ 1 \end{bmatrix} s.$ Setting $s = \dfrac{8}{17}$

yields $\mathbf{q} = \begin{bmatrix} 9/17 \\ 8/17 \end{bmatrix}.$

(b) As in (a), solve

$$
\begin{bmatrix} -.19 & .26 \\ .19 & -.26 \end{bmatrix} \begin{bmatrix} q_1 \\ q_2 \end{bmatrix} = \begin{bmatrix} 0 \\ 0 \end{bmatrix}
$$

i.e., $.19q_1 = .26q_2$. Solutions have the form $\mathbf{q} = \begin{bmatrix} 26/19 \\ 1 \end{bmatrix} s$.

Set $s = \dfrac{19}{45}$ to get $\mathbf{q} = \begin{bmatrix} 26/45 \\ 19/45 \end{bmatrix}$.

(c) Again, solve

$$
\begin{bmatrix} -2/3 & 1/2 & 0 \\ 1/3 & -1 & 0 \\ 1/3 & 1/2 & -1/4 \end{bmatrix} \begin{bmatrix} q_1 \\ q_2 \\ q_3 \end{bmatrix} = \begin{bmatrix} 0 \\ 0 \\ 0 \end{bmatrix}
$$

by reducing the coefficient matrix to row-echelon form:

$$
\begin{bmatrix} 1 & 0 & -1/4 \\ 0 & 1 & -1/3 \\ 0 & 0 & 0 \end{bmatrix}
$$

yielding solutions of the form $\mathbf{q} = \begin{bmatrix} 1/4 \\ 1/3 \\ 1 \end{bmatrix} s$.

Set $s = \dfrac{12}{19}$ to get $\mathbf{q} = \begin{bmatrix} 3/19 \\ 4/19 \\ 12/19 \end{bmatrix}$.

4. **(a)** Prove by induction that $p_{12}^{(n)} = 0$: Already true for $n = 1$. If true for $n - 1$, we have $P^n = P^{n-1}P$, so $p_{12}^{(n)} = p_{11}^{(n-1)}p_{12} + p_{12}^{(n-1)}p_{22}$. But $p_{12} = p_{12}^{(n-1)} = 0$ so $p_{12}^{(n)} = 0 + 0 = 0$. Thus, no power of P can have all positive entries, so P is not regular.

4. (b) If $\mathbf{x} = \begin{bmatrix} x_1 \\ x_2 \end{bmatrix}$, $P\mathbf{x} = \begin{bmatrix} \dfrac{1}{2}x_1 \\ \dfrac{1}{2}x_1 + x_2 \end{bmatrix}$, $P^2\mathbf{x} = \begin{bmatrix} \dfrac{1}{4}x_1 \\ \dfrac{1}{4}x_1 + \dfrac{1}{2}x_1 + x_2 \end{bmatrix}$

etc. We use induction to show $P^n\mathbf{x} = \begin{bmatrix} \left(\dfrac{1}{2}\right)^n x_1 \\ \left(1 - \left(\dfrac{1}{2}\right)^n\right)x_1 + x_2 \end{bmatrix}$.

Already true for $n = 1, 2$. If true for $n - 1$, then

$$P^n = P(P^{n-1}\mathbf{x}) = P\begin{bmatrix} \left(\dfrac{1}{2}\right)^{n-1} x_1 \\ \left(1 - \left(\dfrac{1}{2}\right)^{n-1}\right)x_1 + x_2 \end{bmatrix}$$

$$= \begin{bmatrix} \left(\dfrac{1}{2}\right)\left(\dfrac{1}{2}\right)^{n-1} x_1 \\ \left(1 - \left(\dfrac{1}{2}\right)^{n-1} + \left(\dfrac{1}{2}\right)^{n}\right)x_1 + x_2 \end{bmatrix} = \begin{bmatrix} \left(\dfrac{1}{2}\right)^{n} x_1 \\ \left(1 - \left(\dfrac{1}{2}\right)^{n}\right)x_1 + x_2 \end{bmatrix}.$$

Since $\lim\limits_{n\to\infty}\left[\dfrac{1}{2}\right]^n = 0$, we get $\lim\limits_{n\to\infty} P^n\mathbf{x} = \begin{bmatrix} 0 \\ x_1 + x_2 \end{bmatrix} = \begin{bmatrix} 0 \\ 1 \end{bmatrix}$ if $\mathbf{x}$ is a state vector.

(c) Theorem 3 says that the entries of the steady state vector should be positive; they are not for $\begin{bmatrix} 0 \\ 1 \end{bmatrix}$.

5. Let $\mathbf{q} = \begin{bmatrix} \dfrac{1}{k} & \dfrac{1}{k} & \cdots & \dfrac{1}{k} \end{bmatrix}^t$. Then $(P\mathbf{q})_i = \sum\limits_{j=1}^{k} p_{ij}q_j = \sum\limits_{j=1}^{k} \dfrac{1}{k}p_{ij} = \dfrac{1}{k}\sum\limits_{j=1}^{k} p_{ij} = \dfrac{1}{k}$, since the row sums of P are 1. Thus $(P\mathbf{q})_i = q_i$ for all i.

6. Since P has zero entries, consider $P^2 = \begin{bmatrix} \frac{1}{2} & \frac{1}{4} & \frac{1}{4} \\ \frac{1}{4} & \frac{1}{2} & \frac{1}{4} \\ \frac{1}{4} & \frac{1}{4} & \frac{1}{2} \end{bmatrix}$, so P is regular.

Note that all the rows of P sum to 1. Since P is 3×3, Exercise 5 implies $\mathbf{q} = \begin{bmatrix} \frac{1}{3} & \frac{1}{3} & \frac{1}{3} \end{bmatrix}^t$.

7. Let $\mathbf{x} = \begin{bmatrix} x_1 & x_2 \end{bmatrix}^t$ be the state vector, with $x_1 = $ probability that John is happy and $x_2 = $ probability that John is sad. The transition matrix P will be

$$P = \begin{bmatrix} 4/5 & 2/3 \\ 1/5 & 1/3 \end{bmatrix}$$

since the columns must sum to one. We find the steady state vector for P by solving

$$\begin{bmatrix} -1/5 & 2/3 \\ 1/5 & -2/3 \end{bmatrix} \begin{bmatrix} q_1 \\ q_2 \end{bmatrix} = \begin{bmatrix} 0 \\ 0 \end{bmatrix},$$

i.e., $\frac{1}{5}q_1 = \frac{2}{3}q_2$, so $\mathbf{q} = \begin{bmatrix} 10/3 \\ 1 \end{bmatrix} s$. Let $s = \frac{3}{13}$ and get $\mathbf{q} = \begin{bmatrix} 10/13 \\ 3/13 \end{bmatrix}$, so $10/13$ is the probability that John will be happy on a given day.

8. The state vector $\mathbf{x} = \begin{bmatrix} x_1 & x_2 & x_3 \end{bmatrix}^t$ will represent the proportion of the population living in regions *1*, *2* and *3*, respectively. In the transition matrix, p_{ij} will represent the proportion of the people in region j who move to region i, yielding

$$P = \begin{bmatrix} .90 & .15 & .10 \\ .05 & .75 & .05 \\ .05 & .10 & .85 \end{bmatrix}.$$

To find the steady state vector, solve

$$
\begin{bmatrix} -.10 & .15 & .10 \\ .05 & -.25 & .05 \\ .05 & .10 & -.15 \end{bmatrix} \begin{bmatrix} q_1 \\ q_2 \\ q_3 \end{bmatrix} = \begin{bmatrix} 0 \\ 0 \\ 0 \end{bmatrix} .
$$

First reduce to row echelon form $\begin{bmatrix} 1 & 0 & -13/7 \\ 0 & 1 & -4/7 \\ 0 & 0 & 0 \end{bmatrix}$, yielding $\mathbf{q} = \begin{bmatrix} 13/7 \\ 4/7 \\ 1 \end{bmatrix} s$.

Set $s = \dfrac{7}{24}$ and get $q = \begin{bmatrix} 13/24 \\ 4/24 \\ 7/24 \end{bmatrix}$, i.e., in the long run 13/24 $\left(\text{or } 54\frac{1}{6}\%\right)$ of the people

reside in region 1, 4/24 $\left(\text{or } 16\frac{2}{3}\%\right)$ in region 2 and 7/24 $\left(\text{or } 29\frac{1}{6}\%\right)$ in region 3.

EXERCISE SET 11.7

1. Note that the matrix has the same number of rows and columns as the graph has vertices, and that ones in the matrix correspond to arrows in the graph. We obtain

(a) $\begin{bmatrix} 0 & 0 & 0 & 1 \\ 1 & 0 & 1 & 1 \\ 1 & 1 & 0 & 1 \\ 0 & 0 & 0 & 0 \end{bmatrix}$

(b) $\begin{bmatrix} 0 & 1 & 1 & 0 & 0 \\ 0 & 0 & 0 & 0 & 1 \\ 1 & 0 & 0 & 1 & 0 \\ 0 & 0 & 1 & 0 & 0 \\ 0 & 0 & 1 & 0 & 0 \end{bmatrix}$

(c) $\begin{bmatrix} 0 & 1 & 0 & 1 & 0 & 0 \\ 1 & 0 & 0 & 0 & 0 & 0 \\ 0 & 1 & 0 & 1 & 1 & 1 \\ 0 & 0 & 0 & 0 & 0 & 1 \\ 0 & 0 & 0 & 0 & 0 & 1 \\ 0 & 0 & 1 & 0 & 1 & 0 \end{bmatrix}$

2. See the remark in problem 1; we obtain

(a)   (b)   (c)

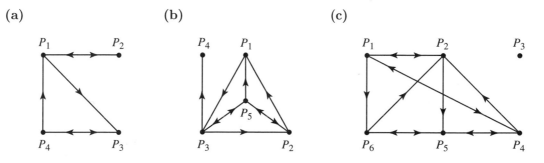

591

3. **(a)** As in problem 2, we obtain

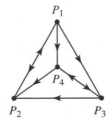

(b) $m_{12} = 1$, so there is one 1-step connection from P_1 to P_2.

$$M^2 = \begin{bmatrix} 1 & 2 & 1 & 1 \\ 0 & 1 & 1 & 1 \\ 1 & 1 & 1 & 0 \\ 1 & 1 & 0 & 1 \end{bmatrix} \quad \text{and} \quad M^3 = \begin{bmatrix} 2 & 3 & 2 & 2 \\ 1 & 2 & 1 & 1 \\ 1 & 2 & 1 & 2 \\ 1 & 2 & 2 & 1 \end{bmatrix}.$$

So $m_{12}^{(2)} = 2$ and $m_{12}^{(3)} = 3$ meaning there are two 2-step and three 3-step connections from P_1 to P_2 by Theorem 1. These are:

1-step: $P_1 \to P_2$

2-step: $P_1 \to P_4 \to P_2$ and $P_1 \to P_3 \to P_2$

3-step: $P_1 \to P_2 \to P_1 \to P_2$, $P_1 \to P_3 \to P_4 \to P_2$,

and $P_1 \to P_4 \to P_3 \to P_2$.

(c) Since $m_{14} = 1$, $m_{14}^{(2)} = 1$ and $m_{14}^{(3)} = 2$, there are one 1-step, one 2-step and two 3-step connections from P_1 to P_4. These are:

1-step: $P_1 \to P_4$

2-step: $P_1 \to P_3 \to P_4$

3-step: $P_1 \to P_2 \to P_1 \to P_4$ and $P_1 \to P_4 \to P_3 \to P_4$.

4. **(a)** Note that to be contained in a clique, a vertex must have "two-way" connections with at least two other vertices. Thus, P_4 could not be in a clique, so $\{P_1, P_2, P_3\}$ is the only possible clique. Inspection shows that this is indeed a clique.

(b) Not only must a clique vertex have two-way connections to at least two other vertices, but the vertices to which it is connected must share a two-way connection. This consideration eliminates P_1 and P_2, leaving $\{P_3, P_4, P_5\}$ as the only possible clique. Inspection shows that it is indeed a clique.

(c)   The above considerations eliminate P_1, P_3 and P_7 from being in a clique. Inspection shows that each of the sets

$\{P_2, P_4, P_6\}, \{P_4, P_6, P_8\}, \{P_2, P_6, P_8\}, \{P_2, P_4, P_8\}$ and $\{P_4, P_5, P_6\}$ satisfy conditions (i) and (ii) in the definition of a clique. But note that P_8 can be added to the first set and we still satisfy the conditions. P_5 may not be added, so $\{P_2, P_4, P_6, P_8\}$ is a clique, containing all the other possibilities except $\{P_4, P_5, P_6\}$, which is also a clique.

5.   (a)   With the given M we get, as in Example 5, that

$$S = \begin{bmatrix} 0 & 1 & 0 & 1 & 0 \\ 1 & 0 & 1 & 0 & 0 \\ 0 & 1 & 0 & 0 & 1 \\ 1 & 0 & 0 & 0 & 1 \\ 0 & 0 & 1 & 1 & 0 \end{bmatrix}, \quad S^3 = \begin{bmatrix} 0 & 3 & 1 & 3 & 1 \\ 3 & 0 & 3 & 1 & 1 \\ 1 & 3 & 0 & 1 & 3 \\ 3 & 1 & 1 & 0 & 3 \\ 1 & 1 & 3 & 3 & 0 \end{bmatrix}.$$

Since $s_{ii}^{(3)} = 0$ for all i, there are no cliques in the graph represented by M.

(b)   As in (a),

$$S = \begin{bmatrix} 0 & 1 & 0 & 1 & 0 & 0 \\ 1 & 0 & 1 & 0 & 1 & 0 \\ 0 & 1 & 0 & 1 & 0 & 1 \\ 1 & 0 & 1 & 0 & 1 & 1 \\ 0 & 1 & 0 & 1 & 0 & 0 \\ 0 & 0 & 1 & 1 & 0 & 0 \end{bmatrix}, \quad S^3 = \begin{bmatrix} 0 & 6 & 1 & 7 & 0 & 2 \\ 6 & 0 & 7 & 1 & 6 & 3 \\ 1 & 7 & 2 & 8 & 1 & 4 \\ 7 & 1 & 8 & 2 & 7 & 5 \\ 0 & 6 & 1 & 7 & 0 & 2 \\ 2 & 3 & 4 & 5 & 2 & 2 \end{bmatrix}.$$

The elements along the main diagonal tell us that only P_3, P_4 and P_6 are members of a clique. Since a clique contains at least three vertices, we must have $\{P_3, P_4, P_6\}$ as the only clique.

6.   As in problem 1,

$$M = \begin{bmatrix} 0 & 0 & 1 & 1 \\ 1 & 0 & 0 & 0 \\ 0 & 1 & 0 & 1 \\ 0 & 1 & 0 & 0 \end{bmatrix}.$$

Then

$$M^2 = \begin{bmatrix} 0 & 2 & 0 & 1 \\ 0 & 0 & 1 & 1 \\ 1 & 1 & 0 & 0 \\ 1 & 0 & 0 & 0 \end{bmatrix} \quad \text{and} \quad M + M^2 = \begin{bmatrix} 0 & 2 & 1 & 2 \\ 1 & 0 & 1 & 1 \\ 1 & 2 & 0 & 1 \\ 1 & 1 & 0 & 0 \end{bmatrix}.$$

By summing the rows of $M + M^2$, we get that the power of P_1 is $2 + 1 + 2 = 5$, the power of P_2 is 3, of P_3 is 4, and of P_4 is 2.

7. Associating vertex P_1 with team A, P_2 with $B, \ldots, P_5$ with E, the game results yield the following dominance-directed graph:

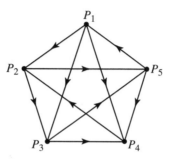

which has vertex matrix

$$M = \begin{bmatrix} 0 & 1 & 1 & 1 & 0 \\ 0 & 0 & 1 & 0 & 1 \\ 0 & 0 & 0 & 1 & 1 \\ 0 & 1 & 0 & 0 & 0 \\ 1 & 0 & 0 & 1 & 0 \end{bmatrix}.$$

Now,

$$M^2 = \begin{bmatrix} 0 & 1 & 1 & 1 & 2 \\ 1 & 0 & 0 & 2 & 1 \\ 1 & 1 & 0 & 1 & 0 \\ 0 & 0 & 1 & 0 & 1 \\ 0 & 2 & 1 & 1 & 0 \end{bmatrix}, \quad M + M^2 = \begin{bmatrix} 0 & 2 & 2 & 2 & 2 \\ 1 & 0 & 1 & 2 & 2 \\ 1 & 1 & 0 & 2 & 1 \\ 0 & 1 & 1 & 0 & 1 \\ 1 & 2 & 1 & 2 & 0 \end{bmatrix}.$$

Summing the rows, we get that the power of A is 8, of B is 6, of C is 5, of D is 3, and of E is 6. Thus ranking in decreasing order we get A in first place, B and E tie for second place, C in fourth place, and D last.

EXERCISE SET 11.8

1. **(a)** From Eq. (2), the expected payoff of the game is

$$\mathbf{p}A\mathbf{q} = \begin{bmatrix} \dfrac{1}{2} & 0 & \dfrac{1}{2} \end{bmatrix} \begin{bmatrix} -4 & 6 & -4 & 1 \\ 5 & -7 & 3 & 8 \\ -8 & 0 & 6 & -2 \end{bmatrix} \begin{bmatrix} \dfrac{1}{4} \\[4pt] \dfrac{1}{4} \\[4pt] \dfrac{1}{4} \\[4pt] \dfrac{1}{4} \end{bmatrix} = -\dfrac{5}{8}.$$

(b) If player R uses strategy $[p_1 \quad p_2 \quad p_3]$ against player C's strategy $\begin{bmatrix} \dfrac{1}{4} & \dfrac{1}{4} & \dfrac{1}{4} & \dfrac{1}{4} \end{bmatrix}^t$,
his payoff will be $\mathbf{p}A\mathbf{q} = (-1/4)\mathbf{p}_1 + (9/4)\mathbf{p}_2 - \mathbf{p}_3$. Since p_1, p_2 and p_3 are nonnegative and add up to 1, this is a weighted average of the numbers $-1/4$, $9/4$ and -1. Clearly this is the largest if $p_1 = p_3 = 0$ and $p_2 = 1$; that is, $\mathbf{p} = [0 \quad 1 \quad 0]$.

(c) As in (b), if player C uses $[q_1 \quad q_2 \quad q_3 \quad q_4]^t$ against $\begin{bmatrix} \dfrac{1}{2} & 0 & \dfrac{1}{2} \end{bmatrix}$, we get $\mathbf{p}A\mathbf{q} = -6q_1 +$
$3q_2 + q_3 - \frac{1}{2}q_4$. Clearly this is minimized over all strategies by setting $q_1 = 1$ and $q_2 = q_3 = q_4 = 0$. That is $\mathbf{q} = [1 \quad 0 \quad 0 \quad 0]^t$.

2. As per the hint, we will construct a 3×3 matrix with two saddle points, say $a_{11} = a_{33} = 1$. Such a matrix is

$$A = \begin{bmatrix} 1 & 2 & 1 \\ 0 & 7 & 0 \\ 1 & 2 & 1 \end{bmatrix}.$$

(The reader is invited to prove that if the matrix A has two saddle points a_{rs} and a_{tu}, then we must necessarily have $a_{rs} = a_{tu}$).

3. **(a)** Calling the matrix A, we see a_{22} is a saddle point, so the optimal strategies are pure, namely: $\mathbf{p} = [0 \quad 1]$, $\mathbf{q} = [0 \quad 1]^t$; the value of the game is $a_{22} = 3$.

    (b) As in (a), a_{21} is a saddle point, so optimal strategies are $\mathbf{p} = [0 \quad 1 \quad 0]$, $\mathbf{q} = [1 \quad 0]^t$; the value of the game is $a_{21} = 2$.

    (c) Here, a_{32} is a saddle point, so optimal strategies are $\mathbf{p} = [0 \quad 0 \quad 1]$, $\mathbf{q} = [0 \quad 1 \quad 0]^t$ and $v = a_{32} = 2$.

    (d) Here, a_{21} is a saddle point, so $\mathbf{p} = [0 \quad 1 \quad 0 \quad 0]$, $\mathbf{q} = [1 \quad 0 \quad 0]^t$ and $v = a_{21} = -2$.

4. **(a)** Calling the matrix A, the formulae of Theorem 2 yield $\mathbf{p} = [5/8 \quad 3/8]$, $\mathbf{q} = [1/8 \quad 7/8]^t$, $v = 27/8$ (A has no saddle points).

    (b) As in (a),

    $$\mathbf{p} = \begin{bmatrix} \dfrac{40}{60} & \dfrac{20}{60} \end{bmatrix} = \begin{bmatrix} \dfrac{2}{3} & \dfrac{1}{3} \end{bmatrix}, \mathbf{q} = \begin{bmatrix} \dfrac{10}{60} & \dfrac{50}{60} \end{bmatrix}^t = \begin{bmatrix} \dfrac{1}{6} & \dfrac{5}{6} \end{bmatrix}^t,$$

    $$v = \dfrac{1400}{60} = \dfrac{70}{3} \text{ (Again, } A \text{ has no saddle points).}$$

    (c) For this matrix, a_{11} is a saddle point, so $\mathbf{p} = [1 \quad 0]$, $\mathbf{q} = [1 \quad 0]^t$ and $v = a_{11} = 3$.

    (d) This matrix has no saddle points, so, as in (a),

    $$\mathbf{p} = \begin{bmatrix} \dfrac{-3}{-5} & \dfrac{-2}{-5} \end{bmatrix} = \begin{bmatrix} \dfrac{3}{5} & \dfrac{2}{5} \end{bmatrix}, \mathbf{q} = \begin{bmatrix} \dfrac{-3}{-5} & \dfrac{-2}{-5} \end{bmatrix}^t = \begin{bmatrix} \dfrac{3}{5} & \dfrac{2}{5} \end{bmatrix}^t \text{ and}$$

    $$v = \dfrac{-19}{-5} = \dfrac{19}{5}.$$

    (e) Again, A has no saddle points, so as in (a), $\mathbf{p} = \begin{bmatrix} \dfrac{3}{13} & \dfrac{10}{13} \end{bmatrix}$,

    $$\mathbf{q} = \begin{bmatrix} \dfrac{1}{13} & \dfrac{12}{13} \end{bmatrix}^t \text{ and } v = \dfrac{-29}{13}.$$

5.   Let a_{11} = payoff to R if the black ace and black two are played = 3.

$\qquad a_{12}$ = payoff to R if the black ace and red three are played = -4.

$\qquad a_{21}$ = payoff to R if the red four and black two are played = -6.

$\qquad a_{22}$ = payoff to R if the red four and red three are played = 7.

So, the payoff matrix for the game is $A = \begin{bmatrix} 3 & -4 \\ -6 & 7 \end{bmatrix}$.

A has no saddle points, so from Theorem 2, $\mathbf{p} = \begin{bmatrix} \dfrac{13}{20} & \dfrac{7}{20} \end{bmatrix}$,

$\mathbf{q} = \begin{bmatrix} \dfrac{11}{20} & \dfrac{9}{20} \end{bmatrix}^t$; that is, player R should play the black ace 65 percent of the time, and player C should play the black two 55 percent of the time. The value of the game is $-\dfrac{3}{20}$, that is, player C can expect to collect on the average 15 cents per game.

6.   If $\mathbf{p}^* = [0 \quad 0 \quad \cdots \quad 1 \quad \cdots \quad 0]$ (one in r-th place) and
$\mathbf{q}^* = [0 \quad 0 \quad \cdots \quad 1 \quad \cdots \quad 0]^t$ (one in s-th place) and
A is an $m \times n$ matrix (m = length of $\mathbf{p}$, n = length of $\mathbf{q}$), then

$$\mathbf{p}^* A \mathbf{q}^* = \mathbf{p}^*(A\mathbf{q}^*) = \sum_{i=1}^{m} p_i^* \left(\sum_{j=1}^{n} a_{ij} q_j^* \right) = \sum_{i=1}^{m} p_i^* a_{is} = a_{rs}$$

(since other entries of $\mathbf{p}^*$, $\mathbf{q}^*$ are zero), verifying Eq. (6). If a_{rs} is a saddle point then $a_{rs} \le a_{rj}$ for all j, since a_{rs} is the smallest in its row. So if $\mathbf{q} = [q_1 \quad \cdots \quad q_n]^t$ is any column strategy, we have

$$\mathbf{p}^* A \mathbf{q} = (\mathbf{p}^* A)\mathbf{q} = \sum_{j=1}^{n} \left(\sum_{i=1}^{m} p_i^* a_{ij} \right) q_j = \sum_{j=1}^{n} a_{rj} q_j \ge \sum_{j=1}^{n} a_{rs} q_j$$

$$= a_{rs} \sum_{j=1}^{n} q_j = a_{rs}(1) = a_{rs} \,,$$

demonstrating Eq. (7). For Eq. (8), note that a_{rs} is a saddle point implies $a_{rs} \geq a_{is}$ for all i, since a_{rs} is the largest in its column. So, if $\mathbf{p} = [p_1 \quad \cdots \quad p_n]$ is any row strategy, we have

$$\mathbf{p}A\mathbf{q}^* = \mathbf{p}(A\mathbf{q}^*) = \sum_{i=1}^{m} p_i \left(\sum_{j=1}^{n} a_{ij}q_j^* \right) = \sum_{i=1}^{m} p_i a_{is} \leq a_{rs} \sum_{i=1}^{m} p_i$$

$$= a_{rs}(1) = a_{rs}.$$

7.   Take $p_1^* = \dfrac{a_{22} - a_{21}}{a_{11} + a_{22} - a_{12} - a_{21}}$. Suppose $a_{22} - a_{21} > 0$. Since the game is not strictly determined, a_{21} is not a saddle point, so $a_{21} < a_{11}$. Similarly, a_{11} is not a saddle point, so $a_{11} > a_{12}$. Then $p_1^* = \dfrac{(a_{22} - a_{21})}{(a_{22} - a_{21}) + (a_{11} - a_{12})}$. All the parenthesized expressions being positive yields $0 < p_1^* < 1$. If $a_{22} - a_{21} < 0$, get $a_{22} < a_{12}$, then $a_{12} > a_{11}$, so all the parenthesized expressions are negative; again $0 < p_1^* < 1$. $p_2^* = \dfrac{a_{11} - a_{12}}{a_{11} + a_{22} - a_{12} - a_{21}} = \dfrac{(a_{11} - a_{12})}{(a_{11} - a_{12}) + (a_{22} - a_{21})}$. If $a_{11} > a_{12}$, then $a_{12} < a_{22}$, yielding $a_{22} > a_{21}$; i.e., all parenthesized expressions are positive, so $0 < p_2^* < 1$. If $a_{11} < a_{12}$, then $a_{11} < a_{21}$, so $a_{21} > a_{22}$, all parenthesized expressions are negative and $0 < p_2^* < 1$.

Expressing q_1^* and q_2^* as $\dfrac{(a_{22} - a_{12})}{(a_{22} - a_{12}) + (a_{11} - a_{21})}$ and $\dfrac{(a_{11} - a_{21})}{(a_{11} - a_{21}) + (a_{22} - a_{12})}$ respectively, and using the non-strictly determined property of A as above, we get $0 < q_1^* < 1$ and $0 < q_2^* < 1$ as well.

EXERCISE SET 11.9

1. **(a)** Calling the given matrix E, we need to solve

$$(I - E)\mathbf{p} = \begin{bmatrix} 1/2 & -1/3 \\ -1/2 & 1/3 \end{bmatrix} \begin{bmatrix} p_1 \\ p_2 \end{bmatrix} = \begin{bmatrix} 0 \\ 0 \end{bmatrix}.$$

This yields $\dfrac{1}{2}p_1 = \dfrac{1}{3}p_2$, that is, $\mathbf{p} = s[1 \quad 3/2]^t$. Set $s = 2$ and get $\mathbf{p} = [2 \quad 3]^t$.

(b) As in (a), solve

$$(I - E)\mathbf{p} = \begin{bmatrix} 1/2 & 0 & -1/2 \\ -1/3 & 1 & -1/2 \\ -1/6 & -1 & 1 \end{bmatrix} \begin{bmatrix} p_1 \\ p_2 \\ p_3 \end{bmatrix} = \begin{bmatrix} 0 \\ 0 \\ 0 \end{bmatrix}.$$

In row-echelon form, this reduces to

$$\begin{bmatrix} 1 & 0 & -1 \\ 0 & 1 & -5/6 \\ 0 & 0 & 0 \end{bmatrix} \begin{bmatrix} p_1 \\ p_2 \\ p_3 \end{bmatrix} = \begin{bmatrix} 0 \\ 0 \\ 0 \end{bmatrix}.$$

Solutions of this system have the form $\mathbf{p} = s[1 \quad 5/6 \quad 1]^t$. Set $s = 6$ and get $\mathbf{p} = [6 \quad 5 \quad 6]^t$.

(c) As in (a), solve

$$(I - E)\mathbf{p} = \begin{bmatrix} .65 & -.50 & -.30 \\ -.25 & .80 & -.30 \\ -.40 & -.30 & .60 \end{bmatrix} \begin{bmatrix} p_1 \\ p_2 \\ p_3 \end{bmatrix} = \begin{bmatrix} 0 \\ 0 \\ 0 \end{bmatrix},$$

which reduces to

$$
\begin{bmatrix} 1 & 3/4 & -3/2 \\ 0 & 1 & -54/79 \\ 0 & 0 & 0 \end{bmatrix} \begin{bmatrix} p_1 \\ p_2 \\ p_3 \end{bmatrix} = \begin{bmatrix} 0 \\ 0 \\ 0 \end{bmatrix}.
$$

Solutions are of the form $\mathbf{p} = s[78/79 \quad 54/79 \quad 1]^t$. Let $s = 79$ to obtain $\mathbf{p} = [78 \quad 54 \quad 79]^t$.

2. (a) By Corollary 1, this matrix is productive, since each of its row sums is .9.

   (b) By Corollary 2, this matrix is productive, since each of its column sums is less than one.

   (c) Try $\mathbf{x} = [2 \quad 1 \quad 1]^t$. Then $C\mathbf{x} = [1.9 \quad .9 \quad .9]^t$, i.e., $\mathbf{x} > C\mathbf{x}$, so this matrix is productive by Theorem 3.

3. Theorem 2 says there will be one linearly independent price vector for the matrix E if some positive power of E is positive. Since E is not positive, try E^2:

$$
E^2 = \begin{bmatrix} .2 & .34 & .1 \\ .2 & .54 & .6 \\ .6 & .12 & .3 \end{bmatrix} > 0.
$$

4. The exchange matrix for this arrangement (using A, B, and C in that order) is

$$
\begin{bmatrix} 1/2 & 1/3 & 1/4 \\ 1/3 & 1/3 & 1/4 \\ 1/6 & 1/3 & 1/2 \end{bmatrix}.
$$

For equilibrium, we must solve $(I - E)\mathbf{p} = 0$. That is

$$
\begin{bmatrix} 1/2 & -1/3 & -1/4 \\ -1/3 & 2/3 & -1/4 \\ -1/6 & -1/3 & 1/2 \end{bmatrix} \begin{bmatrix} p_1 \\ p_2 \\ p_3 \end{bmatrix} = \begin{bmatrix} 0 \\ 0 \\ 0 \end{bmatrix}.
$$

Row reduction yields solutions of the form $\mathbf{p} = [18/16 \quad 15/16 \quad 1 \ ]^t s$.

Set $s = 1600/15$ and obtain $\mathbf{p} = [120 \quad 100 \quad 106.67]^t$; i.e., the price of tomatoes was $120, corn was $100 and lettuce was $106.67.

5. Taking the CE, EE, and ME in that order, we form the consumption matrix C, where c_{ij} = the amount (per consulting dollar) of the i-th engineer's services purchased by the j-th engineer. Thus,

$$C = \begin{bmatrix} 0 & .2 & .3 \\ .1 & 0 & .4 \\ .3 & .4 & 0 \end{bmatrix} .$$

We want to solve $(I - C)\mathbf{x} = \mathbf{d}$, where $\mathbf{d}$ is the demand vector, i.e.

$$\begin{bmatrix} 1 & -.2 & -.3 \\ -.1 & 1 & -.4 \\ -.3 & -.4 & 1 \end{bmatrix} \begin{bmatrix} x_1 \\ x_2 \\ x_3 \end{bmatrix} = \begin{bmatrix} 500 \\ 700 \\ 600 \end{bmatrix} .$$

In row-echelon form this reduces to

$$\begin{bmatrix} 1 & -.2 & -.3 \\ 0 & 1 & -.43877 \\ 0 & 0 & 1 \end{bmatrix} \begin{bmatrix} x_1 \\ x_2 \\ x_3 \end{bmatrix} = \begin{bmatrix} 500 \\ 785.31 \\ 1556.19 \end{bmatrix} .$$

Back-substitution yields the solution $x = [1256.48 \quad 1448.12 \quad 1556.19]^t$.

6. The i-th column sum of E is $\sum_{j=1}^{n} e_{ji}$, and the elements of the i-th column of $I - E$ are the negatives of the elements of E, except for the ii-th, which is $1 - e_{ii}$. So, the i-th column sum of $I - E$ is $1 - \sum_{j=1}^{n} e_{ji} = 1 - 1 = 0$. Now, $(I - E)^t$ has zero row sums, so the vector $x = [1 \quad 1 \quad \cdots \quad 1]^t$ solves $(I - E)^t \mathbf{x} = 0$. This implies $\det(I - E)^t = 0$. But $\det(I - E)^t = \det(I - E)$, so $(I - E)\mathbf{p} = 0$ must have nontrivial (i.e., nonzero) solutions.

7. Let C be a consumption matrix whose column sums are less than one; then the row sums of C^t are less than one. By Corollary 1, C^t is productive so $(I - C^t)^{-1} \geq 0$. But

$$(I - C)^{-1} = \left(\left((I - C)^t\right)^{-1}\right)^t = \left((I - C^t)^{-1}\right)^t \geq 0.$$

Thus, C is productive.

8. **(I)** Let $\mathbf{y}$ be a strictly positive vector, and $\mathbf{x} = (I - C)^{-1}\mathbf{y}$. Since C is productive $(I - C)^{-1} \geq 0$, so $\mathbf{x} = (I - C)^{-1}\mathbf{y} \geq 0$. But then $(I - C)\mathbf{x} = \mathbf{y} > 0$, i.e., $\mathbf{x} - C\mathbf{x} > 0$, i.e., $\mathbf{x} > C\mathbf{x}$.

(II) Step 1: Since both $\mathbf{x}^*$ and C are ≥ 0, so is $C\mathbf{x}^*$. Thus $\mathbf{x}^* > C\mathbf{x}^* \geq 0$.

Step 2: Since $\mathbf{x}^* > C\mathbf{x}^*$, $\mathbf{x}^* - C\mathbf{x}^* > 0$. Let ϵ be the smallest element in $\mathbf{x}^* - C\mathbf{x}^*$, and M the largest element in $\mathbf{x}^*$. Then $\mathbf{x}^* - C\mathbf{x}^* > \dfrac{\epsilon}{2M}\mathbf{x}^* > 0$, i.e., $\mathbf{x}^* - \dfrac{\epsilon}{2M}\mathbf{x}^* > C\mathbf{x}^*$. Setting $\lambda = 1 - \dfrac{\epsilon}{2M} < 1$, we get $C\mathbf{x}^* < \lambda\mathbf{x}^*$.

Step 3: First, we show that if $\mathbf{x} > \mathbf{y}$, then $C\mathbf{x} > C\mathbf{y}$. But this is clear since $(Cx)_i = \displaystyle\sum_{j=1}^{n} c_{ij}x_j > \sum_{j=1}^{n} c_{ij}y_j = (Cy)_i$. Now we prove Step 3 by induction on n, the case $n = 1$ having been done in Step 2. Assuming the result for $n - 1$, then $C^{n-1}\mathbf{x}^* < \lambda^{n-1}\mathbf{x}^*$. But then $C^n\mathbf{x}^* = C(C^{n-1}\mathbf{x}^*) < C(\lambda^{n-1}\mathbf{x}^*) = \lambda^{n-1}(C\mathbf{x}^*) < \lambda^{n-1}(\lambda\mathbf{x}^*) = \lambda^n\mathbf{x}^*$, proving Step 3.

Step 4: Clearly, $C^n\mathbf{x}^* \geq 0$ for all n. So we have

$$0 \leq \lim_{n\to\infty} C^n\mathbf{x}^* \leq \lim_{n\to\infty} \lambda^n\mathbf{x}^* = 0, \text{ i.e., } \lim_{n\to\infty} C^n\mathbf{x}^* = 0.$$

Denote the elements of $\displaystyle\lim_{n\to\infty} C^n$ by $\tilde{c}_{ij}$. Then we have $0 = \displaystyle\sum_{j=1}^{n} \tilde{c}_{ij}x_j^*$ for all i. But $\tilde{c}_{ij} \geq 0$ and $\mathbf{x}^* > 0$ imply $\tilde{c}_{ij} = 0$ for all i and j, proving Step 4.

Step 5: By induction on n, the case $n = 1$ is trivial. Assume the result true for $n - 1$. Then

$$(I - C)(I + C + C^2 + \cdots + C^{n-1}) = (I - C)$$
$$(I + C + \cdots + C^{n-2}) + (I - C)C^{n-1} = (I - C^{n-1}) + (I - C)C^{n-1}$$
$$= I - C^{n-1} + C^{n-1} - C^n,$$
$$= I - C^n,$$

proving Step 5.

Step 6: First we show $(I - C)^{-1}$ exists. If not, then there would be a nonzero vector z such that $C\mathbf{z} = \mathbf{z}$. But then $C^n\mathbf{z} = \mathbf{z}$ for all n, so $\mathbf{z} = \lim_{n\to\infty} C^n\mathbf{z} = 0$, a contradiction, thus $I - C$ is invertible. Thus, $I + C + \cdots + C^{n-1} = (I - C)^{-1}(I - C^n)$, so $S = \lim_{n\to\infty}(I - C)^{-1}(I - C^n) = (I - C)^{-1}(I - \lim_{n\to\infty} C^n) = (I - C)^{-1}$, proving Step 6.

Step 7: Since S is the (infinite) sum of nonnegative matrices, S itself must be non-negative.

Step 8: We have shown in Steps 6 and 7 that $(I - C)^{-1}$ exists and is nonnegative, thus C is productive.

EXERCISE SET 11.10

1. Using Eq. (18), we calculate

$$Yld_2 = \frac{30s}{2} = 15s$$

$$Yld_3 = \frac{50s}{2 + \dfrac{3}{2}} = \frac{100s}{7}\,.$$

So all the trees in the second class should be harvested for an optimal yield (since $s = 1000$) of \$15,000.

2. From the solution to Example 1, we see that for the fifth class to be harvested in the optimal case we must have

$$p_5 s / \left(.28^{-1} + .31^{-1} + .25^{-1} + .23^{-1}\right) > 14.7s,$$

yielding $p_5 > \$222.63$.

3. Assume $p_2 = 1$, then $Yld_2 = \dfrac{s}{(.28)^{-1}} = .28s$. Thus, for all the yields to be the same we must have

$$p_3 s/(.28^{-1} + .31^{-1}) = .28s$$

$$p_4 s/(.28^{-1} + .31^{-1} + .25^{-1}) = .28s$$

$$p_5 s/(.28^{-1} + .31^{-1} + .25^{-1} + .23^{-1}) = .28s$$

$$p_6 s/(.28^{-1} + .31^{-1} + .25^{-1} + .23^{-1} + .27^{-1}) = 28s$$

Solving these successively yields $p_3 = 1.90$, $p_4 = 3.02$, $p_5 = 4.24$ and $p_6 = 5.00$. Thus the ratio

$$p_2 \ : \ p_3 \ : \ p_4 \ : \ p_5 \ : \ p_6 = 1 \ : \ 1.90 \ : \ 3.02 \ : \ 4.24 \ : \ 5.00 \ .$$

4. From Eq. (17) we have

$$x_1 = \frac{s}{1 + \dfrac{g_1}{g_2} + \dfrac{g_1}{g_3} + \cdots + \dfrac{g_1}{g_{k-1}}} = \frac{s}{g_1 \left(\dfrac{1}{g_1} + \dfrac{1}{g_2} + \cdots + \dfrac{1}{g_{k-1}} \right)} \ .$$

Then Eq. (16) yields

$$(*) \ x_i = \frac{g_1 x_1}{g_i} = \frac{g_1}{g_i} \frac{s}{g_1 \left(\dfrac{1}{g_1} + \cdots + \dfrac{1}{g_{k-1}} \right)} = \frac{1}{g_i} \frac{s}{\left(\dfrac{1}{g_1} + \cdots + \dfrac{1}{g_{k-1}} \right)}$$

for $1 \le i \le k - 1$. Then (*) and Eq. (13) yield Eq. (19).

5. Since $\mathbf{y}$ is the harvest vector, $N = \sum\limits_{i=1}^{n} y_i$ is the number of trees removed from the forest. Then Eq. (7) and the first of Eqs. (8) yield $N = g_1 x_1$, and from Eq. (17) we obtain

$$N = \frac{g_1 s}{1 + \dfrac{g_1}{g_2} + \cdots + \dfrac{g_1}{g_{k-1}}} = \frac{s}{\dfrac{1}{g_1} + \cdots + \dfrac{1}{g_{k-1}}} \ .$$

6. Set $g_1 = \cdots = g_{n-1} = g$, and $p_2 = 1$. Then from Eq. (18), $Yld_2 = \dfrac{p_2 s}{\dfrac{1}{y_1}} = gs$. Since we

want all of the Yld_k's to be the same, we need to solve $Yld_k = \dfrac{p_k s}{(k-1)\dfrac{1}{g}} = gs$ for p_k for

$3 \le k \le n$. Thus $p_k = k-1$. So the ratio

$$p_2 : p_3 : p_4 : \ldots : p_n = 1 : 2 : 3 : \ldots : (n-1).$$

EXERCISE SET 11.11

1. (a) Using the coordinates of the points as the columns of a matrix we obtain

$$\begin{bmatrix} 0 & 1 & 1 & 0 \\ 0 & 0 & 1 & 1 \\ 0 & 0 & 0 & 0 \end{bmatrix}.$$

(b) The scaling is accomplished by multiplication of the coordinate matrix on the left by

$$\begin{bmatrix} \dfrac{3}{2} & 0 & 0 \\ 0 & \dfrac{1}{2} & 0 \\ 0 & 0 & 1 \end{bmatrix},$$

resulting in the matrix

$$\begin{bmatrix} 0 & \dfrac{3}{2} & \dfrac{3}{2} & 0 \\ 0 & 0 & \dfrac{1}{2} & \dfrac{1}{2} \\ 0 & 0 & 0 & 0 \end{bmatrix},$$

which represents the vertices $(0,0,0)$, $\left(\dfrac{3}{2},\ 0,\ 0\right)$, $\left(\dfrac{3}{2},\ \dfrac{1}{2},\ 0\right)$ and $\left(0,\ \dfrac{1}{2},\ 0\right)$ as shown below.

1. **(c)** Adding the matrix

$$\begin{bmatrix} -2 & -2 & -2 & -2 \\ -1 & -1 & -1 & -1 \\ 3 & 3 & 3 & 3 \end{bmatrix}$$

to the original matrix yields

$$\begin{bmatrix} -2 & -1 & -1 & -2 \\ -1 & -1 & 0 & 0 \\ 3 & 3 & 3 & 3 \end{bmatrix},$$

which represents the vertices $(-2,-1,3)$, $(-1,-1,3)$, $(-1,0,3)$, and $(-2,0,3)$ as shown below.

(d) Multiplying by the matrix

$$\begin{bmatrix} \cos(-30°) & -\sin(-30°) & 0 \\ \sin(-30°) & \cos(-30°) & 0 \\ 0 & 0 & 1 \end{bmatrix},$$

we obtain

$$
\begin{bmatrix}
0 & \cos(-30°) & \cos(-30°) - \sin(-30°) & -\sin(-30°) \\
0 & \sin(-30°) & \cos(-30°) + \sin(-30°) & \cos(-30°) \\
0 & 0 & 0 & 0
\end{bmatrix} =
$$

$$
\begin{bmatrix}
0 & .866 & 1.366 & .500 \\
0 & -.500 & .366 & .866 \\
0 & 0 & 0 & 0
\end{bmatrix}.
$$

The vertices are then $(0,0,0)$, $(.866, -.500, 0)$, $(1.366, .366, 0)$, and $(.500, .866, 0)$ as shown:

2. **(a)** Simply perform the matrix multiplication

$$
\begin{bmatrix}
1 & \dfrac{1}{2} & 0 \\
0 & 1 & 0 \\
0 & 0 & 1
\end{bmatrix}
\begin{bmatrix}
x_i \\
y_i \\
z_i
\end{bmatrix} =
\begin{bmatrix}
x_i + \dfrac{1}{2}y_i \\
y_i \\
z_i
\end{bmatrix}.
$$

(b) We multiply

$$
\begin{bmatrix}
1 & \dfrac{1}{2} & 0 \\
0 & 1 & 0 \\
0 & 0 & 1
\end{bmatrix}
\begin{bmatrix}
0 & 1 & 1 & 0 \\
0 & 0 & 1 & 1 \\
0 & 0 & 0 & 0
\end{bmatrix} =
\begin{bmatrix}
0 & 1 & \dfrac{3}{2} & \dfrac{1}{2} \\
0 & 0 & 1 & 1 \\
0 & 0 & 0 & 0
\end{bmatrix}
$$

yielding the vertices $(0,0,0)$, $(1,0,0)$, $\left(\dfrac{3}{2},\ 1,\ 0\right)$, and $\left(\dfrac{1}{2},\ 1,\ 0\right)$.

2.   **(c)**   Obtain the vertices via

$$
\begin{bmatrix} 1 & 0 & 0 \\ .6 & 1 & 0 \\ 0 & 0 & 1 \end{bmatrix}
\begin{bmatrix} 0 & 1 & 1 & 0 \\ 0 & 0 & 1 & 1 \\ 0 & 0 & 0 & 0 \end{bmatrix}
=
\begin{bmatrix} 0 & 1 & 1 & 0 \\ 0 & .6 & 1.6 & 1 \\ 0 & 0 & 0 & 0 \end{bmatrix},
$$

yielding $(0,0,0)$, $(1,.6,0)$, $(1,1.6,0)$ and $(0,1,0)$, as shown:

3.   **(a)**   This transformation looks like scaling by the factors 1, -1, 1, respectively and indeed its matrix is

$$
\begin{bmatrix} 1 & 0 & 0 \\ 0 & -1 & 0 \\ 0 & 0 & 1 \end{bmatrix}.
$$

(b)   For this reflection we want to transform $(x_i,\, y_i,\, z_i)$ to $(-x_i, y_i, z_i)$ with the matrix

$$
\begin{bmatrix} -1 & 0 & 0 \\ 0 & 1 & 0 \\ 0 & 0 & 1 \end{bmatrix}.
$$

Negating the x-coordinates of the 12 points in view 1 yields the 12 points $(-1.000, -.800, .000)$, $(-.500, -.800, -.866)$, etc., as shown:

(c) Here we want to negate the z-coordinates, with the matrix

$$\begin{bmatrix} 1 & 0 & 0 \\ 0 & 1 & 0 \\ 0 & 0 & -1 \end{bmatrix}.$$

This does not change View 1.

4. (a) The formulas for scaling, translation and rotation yield the matrices

$$M_1 = \begin{bmatrix} 1/2 & 0 & 0 \\ 0 & 2 & 0 \\ 0 & 0 & 1/3 \end{bmatrix}, \qquad M_2 = \begin{bmatrix} \dfrac{1}{2} & \dfrac{1}{2} & \dfrac{1}{2} & \cdots & \dfrac{1}{2} \\ 0 & 0 & 0 & \cdots & 0 \\ 0 & 0 & 0 & \cdots & 0 \end{bmatrix},$$

$$M_3 = \begin{bmatrix} 1 & 1 & 0 \\ 0 & \cos 20^\circ & -\sin 20^\circ \\ 0 & \sin 20^\circ & \cos 20^\circ \end{bmatrix},$$

$$M_4 = \begin{bmatrix} \cos(-45^\circ) & 0 & \sin(-45^\circ) \\ 0 & 1 & 0 \\ -\sin(-45^\circ) & 0 & \cos(-45^\circ) \end{bmatrix},$$

and

$$M_5 = \begin{bmatrix} \cos 90^\circ & -\sin 90^\circ & 0 \\ \sin 90^\circ & \cos 90^\circ & 0 \\ 0 & 0 & 1 \end{bmatrix}.$$

(b) Clearly, $P' = M_5 M_4 M_3 (M_1 P + M_2)$.

5. (a) As in 4(a),

$$M_1 = \begin{bmatrix} .3 & 0 & 0 \\ 0 & .5 & 0 \\ 0 & 0 & 1 \end{bmatrix}, \qquad M_2 = \begin{bmatrix} 1 & 0 & 0 \\ 0 & \cos 45° & -\sin 45° \\ 0 & \sin 45° & \cos 45° \end{bmatrix},$$

$$M_3 = \begin{bmatrix} 1 & 1 & 1 & \cdots & 1 \\ 0 & 0 & 0 & \cdots & 0 \\ 0 & 0 & 0 & \cdots & 0 \end{bmatrix}, \qquad M_4 = \begin{bmatrix} \cos 35° & 0 & \sin 35° \\ 0 & 1 & 0 \\ -\sin 35° & 0 & \cos 35° \end{bmatrix},$$

$$M_5 = \begin{bmatrix} \cos(-45°) & -\sin(-45°) & 0 \\ \sin(-45°) & \cos(-45°) & 0 \\ 0 & 0 & 1 \end{bmatrix},$$

$$M_6 = \begin{bmatrix} 0 & 0 & 0 & \cdots & 0 \\ 0 & 0 & 0 & \cdots & 0 \\ 1 & 1 & 1 & \cdots & 1 \end{bmatrix}, \qquad \text{and} \qquad M_7 = \begin{bmatrix} 2 & 0 & 0 \\ 0 & 1 & 0 \\ 0 & 0 & 1 \end{bmatrix}.$$

(b) As in 4(b), $P' = M_7(M_6 + M_5M_4(M_3 + M_2M_1P))$.

6. Using the hint given, we have

$$R_1 = \begin{bmatrix} \cos\beta & 0 & \sin\beta \\ 0 & 1 & 0 \\ -\sin\beta & 0 & \cos\beta \end{bmatrix}, \qquad R_2 = \begin{bmatrix} \cos\alpha & -\sin\alpha & 0 \\ \sin\alpha & \cos\alpha & 0 \\ 0 & 0 & 1 \end{bmatrix},$$

$$R_3 = \begin{bmatrix} \cos\theta & 0 & \sin\theta \\ 0 & 1 & 0 \\ -\sin\theta & 0 & \cos\theta \end{bmatrix}, \qquad R_4 = \begin{bmatrix} \cos(-\alpha) & -\sin(-\alpha) & 0 \\ \sin(-\alpha) & \cos(-\alpha) & 0 \\ 0 & 0 & 1 \end{bmatrix},$$

$$\text{and } R_5 = \begin{bmatrix} \cos(-\beta) & 0 & \sin(-\beta) \\ 0 & 1 & 0 \\ -\sin(-\beta) & 0 & \cos(-\beta) \end{bmatrix}.$$

7. **(a)** We rewrite the formula for v_i' as

$$v_i' = \begin{bmatrix} 1 \cdot x_i + x_0 \cdot 1 \\ 1 \cdot y_i + y_0 \cdot 1 \\ 1 \cdot z_i + z_0 \cdot 1 \\ 1 \cdot 1 \end{bmatrix}.$$

So

$$v_i' = \begin{bmatrix} 1 & 0 & 0 & x_0 \\ 0 & 1 & 0 & y_0 \\ 0 & 0 & 1 & z_0 \\ 0 & 0 & 0 & 1 \end{bmatrix} \begin{bmatrix} x_i \\ y_i \\ z_i \\ 1 \end{bmatrix}.$$

(b) We want to translate x_i by -5, y_i by $+9$, z_i by -3, so $x_0 = -5$, $y_0 = +9$, $z_0 = -3$. The matrix is

$$\begin{bmatrix} 1 & 0 & 0 & -5 \\ 0 & 1 & 0 & 9 \\ 0 & 0 & 1 & -3 \\ 0 & 0 & 0 & 1 \end{bmatrix}.$$

8. This can be done most easily performing the multiplication RR^t and showing that this is I. For example, for the rotation matrix about the x-axis we obtain

$$RR^t = \begin{bmatrix} 1 & 0 & 0 \\ 0 & \cos\theta & -\sin\theta \\ 0 & \sin\theta & \cos\theta \end{bmatrix} \begin{bmatrix} 1 & 0 & 0 \\ 0 & \cos\theta & \sin\theta \\ 0 & -\sin\theta & \cos\theta \end{bmatrix} = \begin{bmatrix} 1 & 0 & 0 \\ 0 & 1 & 0 \\ 0 & 0 & 1 \end{bmatrix}.$$

EXERCISE SET 11.12

1. (a) The discrete mean value property yields the four equations

$$t_1 = \frac{1}{4}(t_2 + t_3)$$

$$t_2 = \frac{1}{4}(t_1 + t_4 + 1 + 1)$$

$$t_3 = \frac{1}{4}(t_1 + t_4)$$

$$t_4 = \frac{1}{4}(t_2 + t_3 + 1 + 1).$$

Translated into matrix notation, this becomes

$$
\begin{bmatrix} t_1 \\ t_2 \\ t_3 \\ t_4 \end{bmatrix} = \begin{bmatrix} 0 & \frac{1}{4} & \frac{1}{4} & 0 \\ \frac{1}{4} & 0 & 0 & \frac{1}{4} \\ \frac{1}{4} & 0 & 0 & \frac{1}{4} \\ 0 & \frac{1}{4} & \frac{1}{4} & 0 \end{bmatrix} \begin{bmatrix} t_1 \\ t_2 \\ t_3 \\ t_4 \end{bmatrix} + \begin{bmatrix} 0 \\ \frac{1}{2} \\ 0 \\ \frac{1}{2} \end{bmatrix}.
$$

1. (b) To solve the system in part (a), we solve $(I - M)\mathbf{t} = \mathbf{b}$ for $\mathbf{t}$:

$$
\begin{bmatrix}
1 & -\dfrac{1}{4} & -\dfrac{1}{4} & 0 \\[2mm]
-\dfrac{1}{4} & 1 & 0 & -\dfrac{1}{4} \\[2mm]
-\dfrac{1}{4} & 0 & 1 & -\dfrac{1}{4} \\[2mm]
0 & -\dfrac{1}{4} & -\dfrac{1}{4} & 1
\end{bmatrix}
\begin{bmatrix}
t_1 \\[2mm] t_2 \\[2mm] t_3 \\[2mm] t_4
\end{bmatrix}
=
\begin{bmatrix}
0 \\[2mm] \dfrac{1}{2} \\[2mm] 0 \\[2mm] \dfrac{1}{2}
\end{bmatrix}.
$$

In row-echelon form, this is

$$
\begin{bmatrix}
1 & -\dfrac{1}{4} & -\dfrac{1}{4} & 0 \\[2mm]
0 & 1 & -15 & 4 \\[2mm]
0 & 0 & 1 & -\dfrac{1}{2} \\[2mm]
0 & 0 & 0 & 1
\end{bmatrix}
\begin{bmatrix}
t_1 \\[2mm] t_2 \\[2mm] t_3 \\[2mm] t_4
\end{bmatrix}
=
\begin{bmatrix}
0 \\[2mm] 0 \\[2mm] -1/8 \\[2mm] 3/4
\end{bmatrix}.
$$

Back substitution yields the result $\mathbf{t} = [1/4 \quad 3/4 \quad 1/4 \quad 3/4]^t$.

(c)

$$
\mathbf{t}^{(1)} = M\mathbf{t}^{(0)} + \mathbf{b} =
\begin{bmatrix}
0 & \dfrac{1}{4} & \dfrac{1}{4} & 0 \\[2mm]
\dfrac{1}{4} & 0 & 0 & \dfrac{1}{4} \\[2mm]
\dfrac{1}{4} & 0 & 0 & \dfrac{1}{4} \\[2mm]
0 & \dfrac{1}{4} & \dfrac{1}{4} & 0
\end{bmatrix}
\begin{bmatrix}
0 \\[2mm] 0 \\[2mm] 0 \\[2mm] 0
\end{bmatrix}
+
\begin{bmatrix}
0 \\[2mm] \dfrac{1}{2} \\[2mm] 0 \\[2mm] \dfrac{1}{2}
\end{bmatrix}
=
\begin{bmatrix}
0 \\[2mm] \dfrac{1}{2} \\[2mm] 0 \\[2mm] \dfrac{1}{2}
\end{bmatrix}
$$

$$
\mathbf{t}^{(2)} = M\mathbf{t}^{(1)} + \mathbf{b} = [1/8 \quad 5/8 \quad 1/8 \quad 5/8]^t
$$

$$
\mathbf{t}^{(3)} = M\mathbf{t}^{(2)} + \mathbf{b} = [3/16 \quad 11/16 \quad 3/16 \quad 11/16]^t
$$

$$\mathbf{t}^{(4)} = M\mathbf{t}^{(3)} + \mathbf{b} = [7/32 \quad 23/32 \quad 7/32 \quad 23/32]^t$$

$$\mathbf{t}^{(5)} = M\mathbf{t}^{(4)} + \mathbf{b} = [15/64 \quad 47/64 \quad 15/64 \quad 47/64]^t$$

from Eq. (10).

(d) Using percentage error $= \dfrac{\text{computed value} - \text{actual value}}{\text{actual value}} \times 100\%$ we have that the

percentage error for t_1 and t_3 was $\dfrac{.0129}{.2371} \times 100\% = 5.4\%$, and for t_2 and t_4 was

$\dfrac{-.0129}{.7629} \times 100\% = -1.7\%$.

2. The average value of the temperature on the circle is

$$\frac{1}{2\pi r} \int_{-\pi}^{\pi} f(\theta) r d\theta,$$

where r is the radius of the circle and $f(\theta)$ is the temperature at the point of the circumference where the radius to that point makes the angle θ with the horizontal. Clearly $f(\theta) = 1$ for $-\pi/2 < \theta < \pi/2$ and is zero otherwise. Consequently, the value of the integral above (which equals the temperature at the center of the circle) is $1/2$.

3. As in $1(c)$, but using M and $\mathbf{b}$ as in the problem statement, we obtain

$$\mathbf{t}^{(1)} = M\mathbf{t}^{(0)} + \mathbf{b} = [3/4 \quad 5/4 \quad 1/2 \quad 5/4 \quad 1 \quad 1/2 \quad 5/4 \quad 1 \quad 3/4]^t$$

$$\mathbf{t}^{(2)} = M\mathbf{t}^{(1)} + \mathbf{b} = [13/16 \quad 9/8 \quad 9/16 \quad 11/8 \quad 13/16 \quad 7/16 \quad 21/16 \quad 1 \quad 5/8]^t.$$

4. Using the telephone number 555-1212, we start in row 1, column 2. The ten boundary values we obtain are 0, 0, 2, 2, 0, 0, 0, 2, 1, and 1. The last arrow used was in row 4, column 9, and the average value obtained was .8.

EXERCISE SET 11.13

1. **(a)** Using the lines in Eq. (5) we have

$$\mathbf{a}_1 = \begin{bmatrix} 1 \\ 1 \end{bmatrix} \qquad \mathbf{a}_2 = \begin{bmatrix} 1 \\ -2 \end{bmatrix} \qquad \mathbf{a}_3 = \begin{bmatrix} 3 \\ -1 \end{bmatrix}$$

$$\mathbf{a}_1^t \mathbf{a}_1 = 2 \qquad \mathbf{a}_2^t \mathbf{a}_2 = 5 \qquad \mathbf{a}_3^t \mathbf{a}_3 = 10$$

$$b_1 = 2 \qquad b_2 = -2 \qquad b_3 = 3$$

Setting $\mathbf{x}_k^{(p)} = \begin{bmatrix} x_{k1}^{(p)} \\ x_{k2}^{(p)} \end{bmatrix}$ we have

$$\mathbf{a}_1^t \mathbf{x}_0^{(p)} = x_{01}^{(p)} + x_{02}^{(p)}$$

$$\mathbf{a}_2^t \mathbf{x}_1^{(p)} = x_{11}^{(p)} - 2x_{12}^{(p)}$$

$$\mathbf{a}_3^t \mathbf{x}_2^{(p)} = 3x_{21}^{(p)} - x_{22}^{(p)}$$

$$\frac{b_1 - \mathbf{a}_1^t \mathbf{x}_0^{(p)}}{\mathbf{a}_1^t \mathbf{a}_1} = \frac{2 - x_{01}^{(p)} - x_{02}^{(p)}}{2}$$

$$\frac{b_2 - \mathbf{a}_2^t \mathbf{x}_1^{(p)}}{\mathbf{a}_2^t \mathbf{a}_2} = \frac{-2 - x_{11}^{(p)} + 2x_{12}^{(p)}}{5}$$

$$\frac{b_3 - \mathbf{a}_3^t \mathbf{x}_2^{(p)}}{\mathbf{a}_3^t \mathbf{a}_3} = \frac{3 - 3x_{21}^{(p)} - x_{22}^{(p)}}{10} .$$

Writing $\mathbf{x}_k^{(p)} = \mathbf{x}_{k-1}^{(p)} + \left(\dfrac{b_k - \mathbf{a}_k^t \mathbf{x}_{k-1}^{(p)}}{\mathbf{a}_k^t \mathbf{a}_k} \right) \mathbf{a}_k$

in the form:

$$
\begin{bmatrix} x_{k1}^{(p)} \\ \\ x_{k2}^{(p)} \end{bmatrix} = \begin{bmatrix} x_{k-1,1}^{(p)} \\ \\ x_{k-1,2}^{(p)} \end{bmatrix} + \left(\frac{b_k - \mathbf{a}_k^t \mathbf{x}_{k-1}^{(p)}}{\mathbf{a}_k^t \mathbf{a}_k} \right) \begin{bmatrix} a_{k1} \\ \\ a_{k2} \end{bmatrix}
$$

we have

$$
x_{k1}^{(p)} = x_{k-1,1}^{(p)} + \left(\frac{b_k - \mathbf{a}_k^t \mathbf{x}_{k-1}^{(p)}}{\mathbf{a}_k^t \mathbf{a}_k} \right) a_{k1}
$$

$$
x_{k2}^{(p)} = x_{k-1,2}^{(p)} + \left(\frac{b_k - \mathbf{a}_k^t \mathbf{x}_{k-1}^{(p)}}{\mathbf{a}_k^t \mathbf{a}_k} \right) a_{k2}.
$$

Substituting for the expressions $\left(\dfrac{b_k - \mathbf{a}_k^t \mathbf{x}_{k-1}^{(p)}}{\mathbf{a}_k^t \mathbf{a}_k} \right)$ gives for $k = 1$:

$$
x_{11}^{(p)} = x_{01}^{(p)} + \frac{2 - x_{01}^{(p)} - x_{02}^{(p)}}{2} \cdot 1 = \frac{1}{2} \left[2 + x_{01}^{(p)} - x_{02}^{(p)} \right]
$$

for $k = 2$:

$$
x_{12}^{(p)} = x_{02}^{(p)} + \frac{2 - x_{01}^{(p)} - x_{02}^{(p)}}{2} \cdot 1 = \frac{1}{2} \left[2 - x_{01}^{(p)} + x_{02}^{(p)} \right]
$$

$$
x_{21}^{(p)} = x_{11}^{(p)} + \frac{-2 - x_{11}^{(p)} + 2x_{12}^{(p)}}{5} \cdot 1 = \frac{1}{5} \left[-2 + 4x_{11}^{(p)} + 2x_{12}^{(p)} \right]
$$

$$
x_{22}^{(p)} = x_{12}^{(p)} + \frac{-2 - x_{11}^{(p)} + 2x_{12}^{(p)}}{5} \cdot (-2) = \frac{1}{5} \left[4 + 2x_{11}^{(p)} + 2x_{12}^{(p)} \right]
$$

for $k = 3$:

$$
x_{31}^{(p)} = x_{21}^{(p)} + \frac{3 - 3x_{21}^{(p)} + x_{22}^{(p)}}{10} \cdot 3 = \frac{1}{10} \left[9 + x_{21}^{(p)} + 3x_{22}^{(p)} \right]
$$

$$
x_{32}^{(p)} = x_{22}^{(p)} + \frac{3 - 3x_{21}^{(p)} + x_{22}^{(p)}}{10} \cdot (-1) = \frac{1}{10} \left[-3 + 3x_{21}^{(p)} + 9x_{22}^{(p)} \right]
$$

where $x_0^{(p+1)} = x_3^{(p)}$, which means

$$
\left(x_{01}^{(p+1)}, x_{02}^{(p+1)} \right) = \left(x_{31}^{(p)}, x_{32}^{(p)} \right).
$$

1. **(b)** From part (a) we have

$$x_{31}^{(p)} = \frac{1}{10}\left[9 + x_{21}^{(p)} + 3x_{22}^{(p)}\right] \ .$$

Substituting the expressions for $x_{21}^{(p)}$ and $x_{22}^{(p)}$ from part (a) gives

$$x_{31}^{(p)} = \frac{1}{10}\left[9 + \frac{1}{5}\left(-2 + 4x_{11}^{(p)} + 2x_{12}^{(p)}\right) + \frac{3}{5}\left(4 + 2x_{11}^{(p)} + x_{12}^{(p)}\right)\right]$$

$$= \frac{1}{50}\left[55 + 10x_{11}^{(p)} + 5x_{12}^{(p)}\right] \ .$$

Substituting the expression for $x_{11}^{(p)}$ and $x_{12}^{(p)}$ from part (a) gives

$$x_{31}^{(p)} = \frac{1}{50}\left[55 + \frac{10}{2}\left(2 + x_{01}^{(p)} - x_{02}^{(p)}\right) + \frac{5}{2}\left(2 - x_{01}^{(p)} + x_{02}^{(p)}\right)\right]$$

$$= \frac{1}{20}\left[28 + x_{01}^{(p)} - x_{02}^{(p)}\right] \ .$$

From part (a) we have

$$x_{32}^{(p)} = \frac{1}{10}\left[-3 + 3x_{21}^{(p)} + 9x_{22}^{(p)}\right] \ .$$

Substituting for $x_{21}^{(p)}$ and $x_{22}^{(p)}$ from part (a) gives

$$x_{32}^{(p)} = \frac{1}{10}\left[-3 + \frac{3}{5}\left(-2 + 4x_{11}^{(p)} + 2x_{12}^{(p)}\right) + \frac{9}{5}\left(4 + 2x_{11}^{(p)} + x_{12}^{(p)}\right)\right]$$

$$= \frac{3}{10}\left[1 + 2x_{11}^{(p)} + x_{12}^{(p)}\right] \ .$$

Substituting for $x_{11}^{(p)}$ and $x_{12}^{(p)}$ from part (a) gives

$$x_{32}^{(p)} = \frac{3}{10}\left[1 + \frac{2}{2}\left(2 + x_{01}^{(p)} - x_{02}^{(p)}\right) + \frac{1}{2}\left(2 - x_{01}^{(p)} + x_{02}^{(p)}\right)\right]$$

$$= \frac{1}{20}\left[24 + 3x_{01}^{(p)} - 3x_{02}^{(p)}\right] \ .$$

Now $x_0^{(p)} = x_3^{(p-1)}$ or $\left(x_{01}^{(p)}, x_{02}^{(p)}\right) = \left(x_{31}^{(p-1)}, x_{32}^{(p-1)}\right)$,

and so

$$x_{31}^{(p)} = \frac{1}{20}\left[28 + x_{31}^{(p-1)} - x_{32}^{(p-1)}\right]$$

$$x_{32}^{(p)} = \frac{1}{20}\left[24 + 3x_{31}^{(p-1)} - 3x_{02}^{(p-1)}\right].$$

1. **(c)** The linear system

$$x_{31}^* = \frac{1}{20}\left[28 + x_{31}^* - x_{32}^*\right]$$

$$x_{32}^* = \frac{1}{20}\left[24 + 3x_{31}^* - 3x_{32}^*\right]$$

can be rewritten as

$$19x_{31}^* + x_{32}^* = 28$$

$$-3x_{31}^* + 23x_{32}^* = 24,$$

which has the solution

$$x_{31}^* = 31/22$$

$$x_{32}^* = 27/22.$$

2. **(a)** Setting $\mathbf{x}_0^{(1)} = \left(x_{01}^{(1)}, x_{02}^{(1)}\right) = \left(x_{31}^{(0)}, x_{32}^{(0)}\right) = (0,0)$, and using part (b) of Exercise 1, we have

$$x_{31}^{(1)} = \frac{1}{20}[28] = 1.40000$$

$$x_{32}^{(1)} = \frac{1}{20}[24] = 1.20000$$

$$x_{31}^{(2)} = \frac{1}{20}[28 + 1.4 - 1.2] = 1.41000$$

$$x_{32}^{(2)} = \frac{1}{20}[24 + 3(1.4) - 3(1.2)] = 1.23000$$

$$x_{31}^{(3)} = \frac{1}{20}[28 + 1.41 - 1.23] = 1.40900$$

$$x_{32}^{(3)} = \frac{1}{20}[24 + 3(1.41) - 3(1.23)] = 1.22700$$

$$x_{31}^{(4)} = \frac{1}{20}[28 + 1.409 - 1.227] = 1.40910$$

$$x_{32}^{(4)} = \frac{1}{20}[24 + 3(1.409) - 3(1.227)] = 1.22730$$

$$x_{31}^{(5)} = \frac{1}{20}[28 + 1.4091 - 1.2273] = 1.40909$$

$$x_{32}^{(5)} = \frac{1}{20}[24 + 3(1.4091) - 3(1.2273)] = 1.22727$$

$$x_{31}^{(6)} = \frac{1}{20}[28 + 1.40909 - 1.22727] = 1.40909$$

$$x_{32}^{(6)} = \frac{1}{20}[24 + 3(1.40909) - 3(1.22727)] = 1.22727 \ .$$

(b)

$$\mathbf{x}_0^{(1)} = (1, 1) = \left(x_{31}^{(0)} , x_{32}^{(0)} \right)$$

$$x_{31}^{(1)} = \frac{1}{20}[28 + 1 - 1] = 1.4$$

$$x_{32}^{(2)} = \frac{1}{20}[24 + 3(1) - 3(1)] = 1.2$$

Since $\mathbf{x}_3^{(1)}$ in this part is the same as $\mathbf{x}_3^{(1)}$ in part (a), we will get $\mathbf{x}_3^{(2)}$ as in part (a) and therefore $\mathbf{x}_3^{(3)}, \ldots, \mathbf{x}_3^{(6)}$ will also be the same as in part (a).

(c)

$$\mathbf{x}_0^{(1)} = (148 , -15) = \left(x_{31}^{(0)} , x_{32}^{(0)} \right)$$

$$x_{31}^{(1)} = \frac{1}{20}[28 + 148 - (-15)] = 9.55000$$

$$x_{32}^{(1)} = \frac{1}{20}[24 + 3(148) - 3(-15)] = 25.65000$$

$$x_{31}^{(2)} = \frac{1}{20}[28 + 9.55 - 25.65] = 0.59500$$

$$x_{32}^{(2)} = \frac{1}{20}[24 + 3(9.55) - 3(25.65)] = -1.21500$$

$$x_{31}^{(3)} = \frac{1}{20}[28 + 0.595 + 1.215] = 1.49050$$

$$x_{32}^{(3)} = \frac{1}{20}[24 + 3(0.595) + 3(1.215)] = 1.47150$$

$$x_{31}^{(4)} = \frac{1}{20}[28 + 1.4905 - 1.4715] = 1.40095$$

$$x_{32}^{(4)} = \frac{1}{20}[24 + 3(1.4905) - 3(1.4715)] = 1.20285$$

$$x_{31}^{(5)} = \frac{1}{20}[28 + 1.40095 - 1.20285] = 1.40991$$

$$x_{32}^{(5)} = \frac{1}{20}[24 + 3(1.40095) - 3(1.20285)] = 1.22972$$

$$x_{31}^{(6)} = \frac{1}{20}[28 + 1.40991 - 1.22972] = 1.40901$$

$$x_{32}^{(6)} = \frac{1}{20}[24 + 3(1.40991) - 3(1.22972)] = 1.22703$$

3. (a) The three lines are

$$L_1: \quad x_1 + \quad x_2 = \quad 2$$
$$L_2: \quad x_1 - 2x_2 = -2$$
$$L_3: 3x_1 - \quad x_2 = \quad 3$$

x_1^* lies on L_1 since $\qquad \dfrac{12}{11} + \dfrac{10}{11} = \dfrac{22}{11} = 2$

x_2^* lies on L_2 since $\dfrac{46}{55} - 2\left(\dfrac{78}{55}\right) = \dfrac{-110}{55} = -2$

x_3^* lies on L_3 since $3\left(\dfrac{31}{22}\right) - \dfrac{27}{22} = \dfrac{66}{22} = 3$.

The slope of line L_1 is -1.

The slope of the line $\overline{x_1^* x_3^*}$ is $\dfrac{27/22 - 10/11}{31/22 - 12/11} = \dfrac{7}{7} = 1$.

Since $(-1) \cdot 1 = -1$, it follows $\overline{x_1^* x_2^*}$ is perpendicular to L_1.

The slope of line L_2 is $\dfrac{1}{2}$.

The slope of the line $\overline{x_1^* x_2^*}$ is $\dfrac{78/55 - 10/11}{46/55 - 12/11} = \dfrac{28}{-14} = -2$.

Since $\dfrac{1}{2} \cdot (-2) = -1$, it follows $\overline{x_1^* x_2^*}$ is perpendicular to L_2.

The slope of line L_3 is 3.

The slope of the line $\overline{x_2^* x_3^*}$ is $\dfrac{27/22 - 78/55}{31/22 - 46/55} = \dfrac{-21}{63} = \dfrac{-1}{3}$.

Since $3 \cdot \left(-\dfrac{1}{3}\right) = -1$, it follows $\overline{x_2^* x_3^*}$ is perpendicular to L_3.

(b) Substituting $\mathbf{x}_0^{(1)} = \left(\dfrac{31}{22}, \dfrac{27}{22}\right)$ in the equations for $\mathbf{x}_1^{(1)}$ from part (a) of Exercise 1 gives

$$x_{11}^{(1)} = \frac{1}{2}\left[2 + \frac{31}{22} - \frac{27}{22}\right] = \frac{48}{44} = \frac{12}{11}$$

$$x_{12}^{(1)} = \frac{1}{2}\left[2 - \frac{31}{22} + \frac{27}{22}\right] = \frac{40}{44} = \frac{10}{11}$$

Then $\mathbf{x}_1^{(1)} = \left(\dfrac{12}{11}, \dfrac{10}{11}\right)$ gives

$$x_{21}^{(1)} = \frac{1}{5}\left[-2 + \left(4 \cdot \frac{12}{11}\right) + \left(2 \cdot \frac{10}{11}\right)\right] = \frac{46}{55}$$

$$x_{22}^{(1)} = \frac{1}{5}\left[4 + \left(2 \cdot \frac{12}{11}\right) + \frac{10}{11}\right] = \frac{78}{55}$$

and $\mathbf{x}_2^{(1)} = \left(\dfrac{46}{55}, \dfrac{78}{55}\right)$ gives

$$x_{31}^{(1)} = \frac{1}{10}\left[9 + \frac{46}{55} + \left(3 \cdot \frac{78}{55}\right)\right] = \frac{775}{550} = \frac{31}{22}$$

$$x_{32}^{(1)} = \frac{1}{10}\left[-3 + \left(3 \cdot \frac{46}{55}\right) + \left(9 \cdot \frac{78}{55}\right)\right] = \frac{675}{550} = \frac{27}{22}$$

or $\mathbf{x}_3^{(1)} = \left(\dfrac{31}{22}, \dfrac{27}{22}\right).$

4. Referring to the figure below and starting with $\mathbf{x}_0^{(1)} = (0,0)$:

$\mathbf{x}_0^{(1)}$ is projected to $\mathbf{x}_1^{(1)}$ on L_1 ,

$\mathbf{x}_1^{(1)}$ is projected to $\mathbf{x}_2^{(1)}$ on L_2 ,

$\mathbf{x}_2^{(1)}$ is projected to $\mathbf{x}_3^{(1)}$ on L_3 , and so on.

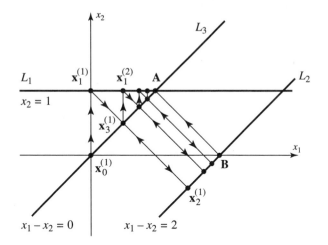

As seen from the graph the points of the limit cycle are

$$\mathbf{x}_1^* = \mathbf{A}, \qquad \mathbf{x}_2^* = \mathbf{B}, \qquad \mathbf{x}_3^* = \mathbf{A}.$$

Since $\mathbf{x}_1^*$ is the point of intersection of L_1 and L_3 it follows on solving the system

$$x_{12}^* = 1$$
$$x_{11}^* - x_{12}^* = 0$$

that $\mathbf{x}_1^* = (1,1)$. Since $\mathbf{x}_2^* = (x_{21}^*, x_{22}^*)$ is on L_2, it follows that $x_{21}^* - x_{22}^* = 2$. Now $\overline{\mathbf{x}_1^* \mathbf{x}_2^*}$ is perpendicular to L_2; therefore

$$\left(\frac{x_{22}^* - 1}{x_{21}^* - 1}\right)(1) = -1$$

so we have $x_{22}^* - 1 = 1 - x_{21}^*$ or $x_{21}^* + x_{22}^* = 2$.

Solving the system

$$x_{21}^* - x_{22}^* = 2$$

$$x_{21}^* + x_{22}^* = 2$$

gives $x_{21}^* = 2$ and $x_{22}^* = 0$. Thus the points on the limit cycle are

$$\mathbf{x}_1^* = (1,1), \qquad \mathbf{x}_2^* = (2,0), \qquad \mathbf{x}_3^* = (1,1).$$

5. **(a)** From Theorem 1:

$$\mathbf{x}_p = \mathbf{x}^* + \left(\frac{b - \mathbf{a}^t \mathbf{x}^*}{\mathbf{a}^t \mathbf{a}}\right)\mathbf{a}$$

and so

$$\mathbf{a}^t \mathbf{x}_p = \mathbf{a}^t \mathbf{x}^* + \left(\frac{b - \mathbf{a}^t \mathbf{x}^*}{\mathbf{a}^t \mathbf{a}}\right)\mathbf{a}^t \mathbf{a}$$

$$= \mathbf{a}^t \mathbf{x}^* + b - \mathbf{a}^t \mathbf{x}^* = b.$$

Therefore $\mathbf{x}_p$ lies on the line $\mathbf{a}^t \mathbf{x} = b$.

(b) Let $\mathbf{z}$ be a point on the line L, so that $\mathbf{a}^t \mathbf{z} = b$. We will prove that $\mathbf{x}_p - \mathbf{x}^*$ is perpendicular to $\mathbf{x}_p - \mathbf{z}$ by showing that $(\mathbf{x}_p - \mathbf{z})^t (\mathbf{x}_p - \mathbf{x}^*) = 0$ (i.e., their dot product is 0). Now

$$(\mathbf{x}_p - \mathbf{z})^t (\mathbf{x}_p - \mathbf{x}^*) = (\mathbf{x}_p - \mathbf{z})^t \left(\frac{b - \mathbf{a}^t \mathbf{x}^*}{\mathbf{a}^t \mathbf{a}}\right)\mathbf{a}$$

$$= \left(\frac{b - \mathbf{a}^t \mathbf{x}^*}{\mathbf{a}^t \mathbf{a}}\right)(\mathbf{x}_p - \mathbf{z})^t \mathbf{a}$$

Since $\mathbf{x}_p$ and $\mathbf{z}$ are on L, we have

$$\mathbf{a}^t\mathbf{x}_p = b$$
$$\mathbf{a}^t\mathbf{z} = b \ ,$$

and so $\mathbf{a}^t\left(\mathbf{x}_p - \mathbf{z}\right) = 0$ or $\left(\mathbf{x}_p - \mathbf{z}\right)^t \mathbf{a} = 0$.

Thus $\left(\mathbf{x}_p - \mathbf{z}\right)^t \left(\mathbf{x}_p - \mathbf{x}^*\right) = 0$, and so $\mathbf{x}_p - \mathbf{x}^*$ is perpendicular to L.

6. Suppose that the center of some pixel, say the K-th pixel, is not crossed by any of the M beams in the scan. Then we have $\mathbf{a}_{iK} = 0$ for $i = 1, 2, \ldots, M$. That is, the K-th component of each of the N-dimensional vectors $\mathbf{a}_1, \mathbf{a}_2, \ldots, \mathbf{a}_M$ is zero. But then they cannot span R^N since, in particular, no linear combination of them could equal $\mathbf{e}_K$, where $\mathbf{e}_K$ is the N-dimensional vector whose K-th component is one and whose other components are all zero.

7. Let us choose units so that each pixel is one unit wide. Then

$$a_{ij} = length \ of \ the \ center \ line \ of \ the \ i\text{-}th \ beam \ that \ lies \ in \ the \ j\text{-}th \ pixel$$

If the i-th beam crosses the j-th pixel squarely, it follows that $a_{ij} = 1$. From Fig. 11 in the text, it is then clear that

$$a_{17} = a_{18} = a_{19} = 1$$
$$a_{24} = a_{25} = a_{26} = 1$$
$$a_{31} = a_{32} = a_{33} = 1$$
$$a_{73} = a_{76} = a_{79} = 1$$
$$a_{82} = a_{85} = a_{88} = 1$$
$$a_{91} = a_{94} = a_{97} = 1$$

since beams 1, 2, 3, 7, 8, and 9 cross the pixels squarely.

Next, the centerlines of beams 5 and 11 lie along the diagonals of pixels 3, 5, 7 and 1, 5, 9, respectively. Since these diagonals have length $\sqrt{2}$, we have

$$a_{53} = a_{55} = a_{57} = \sqrt{2} = 1.41421$$

$$a_{11,1} = a_{11,5} = a_{11,9} = \sqrt{2} = 1.41421.$$

In the following diagram, the hypotenuse of triangle A is the portion of the centerline of the 10th beam that lies in the 2nd pixel. The length of this hypotenuse is twice the height of triangle A, which in turn is $\sqrt{2} - 1$. Thus,

$$a_{10,2} = 2(\sqrt{2} - 1) = .82843.$$

By symmetry we also have

$$a_{10,2} = a_{10,6} = a_{12,4} = a_{12,8}$$
$$= a_{62} = a_{64} = a_{46} = a_{48} = .82843.$$

Also from the diagram, we see that the hypotenuse of triangle B is the portion of the centerline of the 10th beam that lies in the 3rd pixel. Thus,

$$a_{10,3} = 2 - \sqrt{2} = .58579.$$

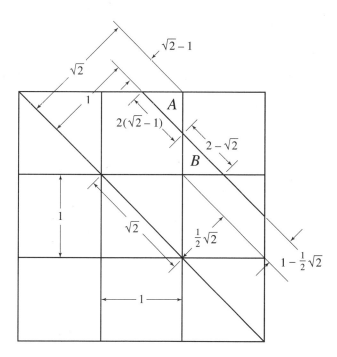

By symmetry we have

$$a_{10,3} = a_{12,7} = a_{61} = a_{49} = .58579.$$

The remaining a_{ij}'s are all zero, and so the 12 beam equations (4) are

$$x_7 + x_8 + x_9 \qquad\qquad = 13.00$$

$$x_4 + x_5 + x_6 \qquad\qquad = 15.00$$

$$x_1 + x_2 + x_3 \qquad\qquad = 8.00$$

$$.82843\,(x_6 + x_8) + .58579x_9 = 14.79$$

$$1.41421\,(x_3 + x_5 + x_7) \quad = 14.31$$

$$.82843\,(x_2 + x_4) + .58579x_1 = 3.81$$

$$x_3 + x_6 + x_9 \qquad\qquad = 18.00$$

$$x_2 + x_5 + x_8 \qquad\qquad = 12.00$$

$$x_1 + x_4 + x_7 \qquad\qquad = 6.00$$

$$.82843\,(x_2 + x_6) + .58579x_3 = 10.51$$

$$1.41421\,(x_1 + x_5 + x_9) \quad = 16.13$$

$$.82843\,(x_4 + x_8) + .58579x_7 = 7.04$$

8. Let us choose units so that each pixel is one unit wide. Then

$$a_{ij} = \textit{area of the i-th beam that lies in the j-th pixel}$$

Since the width of each beam is also one unit it follows that $a_{ij} = 1$ if the i-th beam crosses the j-th pixel squarely. From Fig. 11 in the text, it is then clear that

$$a_{17} = a_{18} = a_{19} = 1$$

$$a_{24} = a_{25} = a_{26} = 1$$

$$a_{31} = a_{32} = a_{33} = 1$$

$$a_{73} = a_{76} = a_{79} = 1$$

$$a_{82} = a_{85} = a_{88} = 1$$

$$a_{91} = a_{94} = a_{97} = 1$$

since beams 1, 2, 3, 7, 8, and 9 cross the pixels squarely.

For the remaining a_{ij}'s, first observe from the figure on the right that an isosceles right triangle of height h, as indicated, has area h^2. From the diagram of the nine pixels, we then have

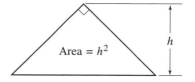

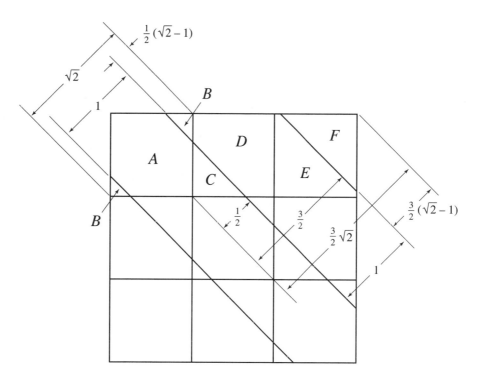

$$\text{Area of triangle } B = \left[\frac{1}{2}(\sqrt{2}-1)\right]^2 = \frac{1}{4}(3 - 2\sqrt{2}) = .04289$$

$$\text{Area of triangle } C = \left[\frac{1}{2}\right]^2 = \frac{1}{4} = .25000$$

$$\text{Area of triangle } F = \left[\frac{3}{2}(\sqrt{2}-1)\right]^2 = \frac{9}{4}(3 - 2\sqrt{2}) = .38604$$

We also have

$$Area\ of\ polygon\ A = 1 - 2 \times (Area\ of\ triangle\ B)$$
$$= \sqrt{2} - \frac{1}{2} = .91421$$
$$Area\ of\ polygon\ D = 1 - (Area\ of\ triangle\ C)$$
$$= 1 - \frac{1}{4} = \frac{3}{4} = .75000$$
$$Area\ of\ polygon\ E = 1 - (Area\ of\ triangle\ F)$$
$$= \frac{1}{4}(18\sqrt{2} - 23) = .61396$$

Referring back to Fig. 11, we see that

$$a_{11,1} = Area\ of\ polygon\ A = .91421$$
$$a_{10,1} = Area\ of\ triangle\ B = .04289$$
$$a_{11,2} = Area\ of\ triangle\ C = .25000$$
$$a_{10,2} = Area\ of\ polygon\ D = .75000$$
$$a_{10,3} = Area\ of\ polygon\ E = .61396$$

By symmetry we then have

$$a_{11,1} = a_{11,5} = a_{11,9} = a_{53} = a_{55} = a_{57} = .91421$$
$$a_{10,1} = a_{10,5} = a_{10,9} = a_{12,1} = a_{12,5} = a_{12,9}$$
$$= a_{63} = a_{65} = a_{67} = a_{43} = a_{45} = a_{47} = .04289$$
$$a_{11,2} = a_{11,4} = a_{11,6} = a_{11,8}$$
$$= a_{52} = a_{54} = a_{56} = a_{58} = .25000$$
$$a_{10,2} = a_{10,6} = a_{12,4} = a_{12,8}$$
$$= a_{62} = a_{64} = a_{46} = a_{48} = .75000$$
$$a_{10,3} = a_{12,7} = a_{61} = a_{49} = .61396.$$

The remaining a_{ij}'s are all zero, and so the 12 beam equations (4) are

$$x_7 + x_8 + x_9 \qquad\qquad\qquad = 13.00$$
$$x_4 + x_5 + x_6 \qquad\qquad\qquad = 15.00$$
$$x_1 + x_2 + x_3 \qquad\qquad\qquad = \ \ 8.00$$
$$0.04289\,(x_3 + x_5 + x_7) + 0.75\,(x_6 + x_8) + 0.61396 x_9 = 14.79$$

$$0.91421\left(x_3 + x_5 + x_7\right) + 0.25\left(x_2 + x_4 + x_6 + x_8\right) = 14.31$$

$$0.04289\left(x_3 + x_5 + x_7\right) + 0.75\left(x_2 + x_4\right) + 0.61396x_1 = 3.81$$

$$x_3 + x_6 + x_9 = 18.00$$

$$x_2 + x_5 + x_8 = 12.00$$

$$x_1 + x_4 + x_7 = 6.00$$

$$0.04289\left(x_1 + x_5 + x_7\right) + 0.75\left(x_2 + x_6\right) + 0.61396x_3 = 10.51$$

$$0.91421\left(x_1 + x_5 + x_7\right) + 0.25\left(x_2 + x_4 + x_6 + x_8\right) = 16.13$$

$$0.04289\left(x_1 + x_5 + x_7\right) + 0.75\left(x_4 + x_8\right) + 0.61396x_7 = 7.04$$

EXERCISE SET 11.14

1. Each of the subsets S_1, S_2, S_3, S_4 in the figure is congruent to the entire set scaled by a factor of $12/25$. Also, the rotation angles for the four subsets are all $0°$. The displacement distances can be determined from the figure to find the four similitudes that map the entire set onto the four subsets S_1, S_2, S_3, S_4. These are, respectively,

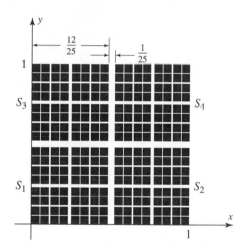

$$T_i\left(\begin{bmatrix} x \\ y \end{bmatrix}\right) = \frac{12}{25}\begin{bmatrix} 1 & 0 \\ 0 & 1 \end{bmatrix}\begin{bmatrix} x \\ y \end{bmatrix} + \begin{bmatrix} e_i \\ f_i \end{bmatrix}, \; i = 1, 2, 3, 4, \text{ where the four values of}$$

$$\begin{bmatrix} e_i \\ f_i \end{bmatrix} \text{ are } \begin{bmatrix} 0 \\ 0 \end{bmatrix}, \begin{bmatrix} 13/25 \\ 0 \end{bmatrix}, \begin{bmatrix} 0 \\ 13/25 \end{bmatrix}, \text{ and } \begin{bmatrix} 13/25 \\ 13/25 \end{bmatrix}.$$

Because $s = 12/25$ and $k = 4$ in the definition of a self-similar set, the Hausdorff dimension of the set is $d_H(S) = \ln(k)/\ln(1/s) = \ln(4)/\ln(25/12) = 1.889\ldots$. The set is a fractal because its Hausdorff dimension is not an integer.

2. The rough measurements indicated in the figure give an approximate scale factor of $s \approx (15/16)/(2) = .47$ to two decimal places. Since $k = 4$, the Hausdorff dimension of the set is approximately $d_H(S) = \ln(k)/\ln(1/s) \approx \ln(4)/\ln(1/.47) = 1.8$ to two significant digits. Examination of the figure reveals rotation angles of $180°$, $180°$, $0°$, and $-90°$ for the sets S_1, S_2, S_3, and S_4, respectively.

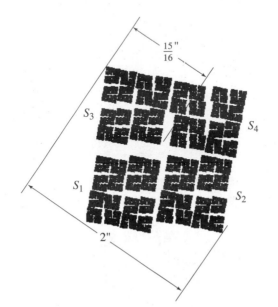

3. **(a)** The figure shows the original self-similar set and a decomposition of the set into seven nonoverlapping congruent subsets, each of which is congruent to the original set scaled by a factor $s = 1/3$. By inspection, the rotations angles are $0°$ for all seven subsets. The Hausdorff dimension of the set is $d_H(S) = \ln(k)/\ln(1/s) = \ln(7)/\ln(3) = 1.771\dots$. Because its Hausdorff dimension is not an integer, the set is a fractal.

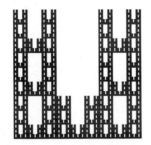 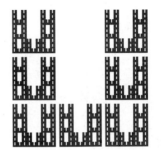

(b) The figure shows the original self-similar set and a decomposition of the set into three nonoverlapping congruent subsets, each of which is congruent to the original set scaled by a factor $s = 1/2$. By inspection, the rotation angles are $180°$ for all three subsets. The Hausdorff dimension of the set is $d_H(S) = \ln(k)/\ln(1/s) = \ln(3)/\ln(2) = 1.584\ldots$. Because its Hausdorff dimension is not an integer, the set is a fractal.

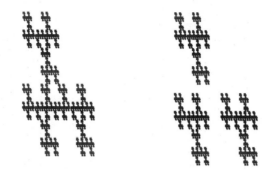

3. **(c)** The figure shows the original self-similar set and a decomposition of the set into three nonoverlapping congruent subsets, each of which is congruent to the original set scaled by a factor $s = 1/2$. By inspection, the rotation angles are $180°$, $180°$, and $-90°$ for S_1, S_2, and S_3, respectively. The Hausdorff dimension of the set is $d_H(S) = \ln(k)/\ln(1/s) = \ln(3)/\ln(2) = 1.584\ldots$. Because its Hausdorff dimension is not an integer, the set is a fractal.

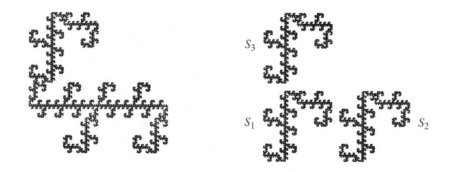

(d) The figure shows the original self-similar set and a decomposition of the set into three nonoverlapping congruent subsets, each of which is congruent to the original set scaled by a factor $s = 1/2$. By inspection, the rotation angles are $180°$, $180°$, and $-90°$ for S_1, S_2, and S_3, respectively. The Hausdorff dimension of the set is $d_H(S) = \ln(k)/\ln(1/s) = \ln(3)/\ln(2) = 1.584\ldots$. Because its Hausdorff dimension is not an integer, the set is a fractal.

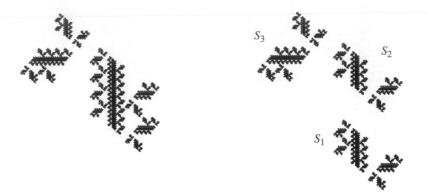

4. The matrix portion of the affine transformation in question is $\begin{bmatrix} .85 & .04 \\ -.04 & .85 \end{bmatrix}$. The matrix portion of a similitude is of the form $s \begin{bmatrix} \cos\theta & -\sin\theta \\ \sin\theta & \cos\theta \end{bmatrix}$. Consequently, we must have $s\cos\theta = .85$ and $s\sin\theta = -.04$. Solving this pair of equations gives $s = \sqrt{(.85)^2 + (-.04)^2} = .8509\ldots$ and $\theta = \tan^{-1}(-.04/.85) = -2.69\ldots°$.

5. Letting $\begin{bmatrix} x \\ y \end{bmatrix}$ be the vector to the tip of the fern and using the hint, we have $\begin{bmatrix} x \\ y \end{bmatrix} = T_2\left(\begin{bmatrix} x \\ y \end{bmatrix}\right)$ or $\begin{bmatrix} x \\ y \end{bmatrix} = \begin{bmatrix} .85 & .04 \\ -.04 & .85 \end{bmatrix}\begin{bmatrix} x \\ y \end{bmatrix} + \begin{bmatrix} .075 \\ .180 \end{bmatrix}$. Solving this matrix equation gives $\begin{bmatrix} x \\ y \end{bmatrix} = \begin{bmatrix} .15 & -.04 \\ .04 & .15 \end{bmatrix}^{-1}\begin{bmatrix} .075 \\ .180 \end{bmatrix} = \begin{bmatrix} .766 \\ .996 \end{bmatrix}$ rounded to three decimal places.

6. As the figure indicates, the unit square can be expressed as the union of 16 nonoverlapping congruent squares, each of side length 1/4. Consequently, the Hausdorff dimension of the unit square as given by Equation (2) of the text is $d_H(S) = \ln(k)/\ln(1/s) = \ln(16)/\ln(4) = 2$.

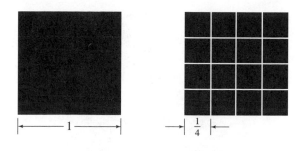

7. The similitude T_1 maps the unit square (whose vertices are $(0,0)$, $(1,0)$, $(1,1)$, and $(0,1)$) onto the square whose vertices are $(0,0)$, $(3/4,0)$, $(3/4,3/4)$, and $(0,3/4)$. The similitude T_2 maps the unit square onto the square whose vertices are $(1/4,0)$, $(1,0)$, $(1,3/4)$, and $(1/4,3/4)$. The similitude T_3 maps the unit square onto the square whose vertices are $(0,1/4)$, $(3/4,1/4)$, $(3/4,1)$, and $(0,1)$. Finally, the similitude T_4 maps the unit square onto the square whose vertices are $(1/4,1/4)$, $(1,1/4)$, $(1,1)$, and $(1/4,1)$. The union of these four smaller squares is the unit square, but the four smaller squares overlap. Each of the four smaller squares has side length of $3/4$, so that the common scale factor of the similitudes is $s = 3/4$. The right-hand side of Equation (2) of the text gives $\ln(k)/\ln(1/s) = \ln(4)/\ln(4/3) = 4.818\ldots$. This is not the correct Hausdorff dimension of the square (which is 2) because the four smaller squares overlap.

8. Because $s = 1/2$ and $k = 8$, Equation (2) of the text gives $d_H(S) = \ln(k)/\ln(1/s) = \ln(8)/\ln(2) = 3$ for the Hausdorff dimension of a unit cube. Because the Hausdorff dimension of the cube is the same as its topological dimension, the cube is not a fractal.

9. A careful examination of Figure 27 of the text shows that the Menger sponge can be expressed as the union of 20 smaller nonoverlapping congruent Menger sponges each of side length $1/3$. Consequently, $k = 20$ and $s = 1/3$, and so the Hausdorff dimension of the Menger sponge is $d_H(S) = \ln(k)/\ln(1/s) = \ln(20)/\ln(3) = 2.726\ldots$. Because its Hausdorff dimension is not an integer, the Menger sponge is a fractal.

10. The figure shows the first four iterates as determined by Algorithm 1 and starting with the unit square as the initial set. Because $k = 2$ and $s = 1/3$, the Hausdorff dimension of the Cantor set is $d_H(S) = \ln(k)/\ln(1/s) = \ln(2)/\ln(3) = 0.6309\ldots$. Notice that the Cantor set is a subset of the unit interval along the x-axis and that its topological dimension must be 0 (since the topological dimension of any set is a nonnegative integer less than or equal to its Hausdorff dimension).

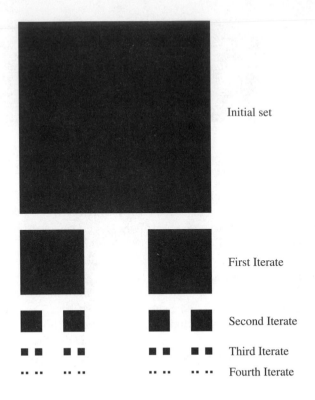

Initial set

First Iterate

Second Iterate

Third Iterate

Fourth Iterate

11. The area of the unit square S_0 is, of course, 1. Each of the eight similitudes $T_1, T_2, \ldots, T_8$ given in Equation 8 of the text has scale factor $s = 1/3$, and so each maps the unit square onto a smaller square of area $1/9$. Because these eight smaller squares are nonoverlapping, their total area is $8/9$, which is then the area of the set S_1. By a similar argument, the area of the set S_2 is $8/9$-th the area of the set S_1. Continuing the argument further, we find that the areas of $S_0, S_1, S_2, S_3, S_4, \ldots$ form the geometric sequence 1, $8/9$, $(8/9)^2$, $(8/9)^3$, $(8/9)^4, \ldots$. (Notice that this implies that the area of the Sierpinski carpet is 0, since the limit of $(8/9)^n$ as n tends to infinity is 0.)

EXERCISE SET 11.15

1. Because $250 = 2 \cdot 5^3$ it follows from (i) that $\prod(250) = 3 \cdot 250 = 750$.

   Because $25 = 5^2$ it follows from (ii) that $\prod(25) = 2 \cdot 25 = 50$.

   Because $125 = 5^3$ it follows from (ii) that $\prod(125) = 2 \cdot 125 = 250$.

   Because $30 = 6 \cdot 5$ it follows from (ii) that $\prod(30) = 2 \cdot 30 = 60$.

   Because $10 = 2 \cdot 5$ it follows from (i) that $\prod(10) = 3 \cdot 10 = 30$.

   Because $50 = 2 \cdot 5^2$ it follows from (i) that $\prod(50) = 3 \cdot 50 = 150$.

   Because $3750 = 6 \cdot 5^4$ it follows from (ii) that $\prod(3750) = 2 \cdot 3750 = 7500$.

   Because $6 = 6 \cdot 5^0$ it follows from (ii) that $\prod(6) = 2 \cdot 6 = 12$.

   Because $5 = 5^1$ it follows from (ii) that $\prod(5) = 2 \cdot 5 = 10$.

2. The point $(0,0)$ is obviously a 1-cycle. We now choose another of the 36 points of the form $(m/6, n/6)$, say $(0, 1/6)$. Its iterates produce the 12-cycle

$$\begin{bmatrix} 0 \\ 1/6 \end{bmatrix} \rightarrow \begin{bmatrix} 1/6 \\ 2/6 \end{bmatrix} \rightarrow \begin{bmatrix} 3/6 \\ 5/6 \end{bmatrix} \rightarrow \begin{bmatrix} 2/6 \\ 1/6 \end{bmatrix} \rightarrow \begin{bmatrix} 3/6 \\ 4/6 \end{bmatrix} \rightarrow \begin{bmatrix} 1/6 \\ 5/6 \end{bmatrix}$$

$$\rightarrow \begin{bmatrix} 0 \\ 5/6 \end{bmatrix} \rightarrow \begin{bmatrix} 5/6 \\ 4/6 \end{bmatrix} \rightarrow \begin{bmatrix} 3/6 \\ 1/6 \end{bmatrix} \rightarrow \begin{bmatrix} 4/6 \\ 5/6 \end{bmatrix} \rightarrow \begin{bmatrix} 3/6 \\ 6/6 \end{bmatrix} \rightarrow \begin{bmatrix} 5/6 \\ 1/6 \end{bmatrix}.$$

So far we have accounted for 13 of the 36 points. Taking one of the remaining points, say $(0, 2/6)$, we arrive at the 4-cycle $\begin{bmatrix} 0 \\ 2/6 \end{bmatrix} \rightarrow \begin{bmatrix} 2/6 \\ 4/6 \end{bmatrix} \rightarrow \begin{bmatrix} 0 \\ 4/6 \end{bmatrix} \rightarrow \begin{bmatrix} 4/6 \\ 2/6 \end{bmatrix}$. We

continue in this way, each time starting with some point of the form $(m/6, n/6)$ that has not yet appeared in a cycle, until we exhaust all such points. This yields a 3-cycle: $\begin{bmatrix} 3/6 \\ 0 \end{bmatrix} \rightarrow$

$\begin{bmatrix} 3/6 \\ 3/6 \end{bmatrix} \rightarrow \begin{bmatrix} 0 \\ 3/6 \end{bmatrix}$; another 4-cycle: $\begin{bmatrix} 4/6 \\ 0 \end{bmatrix} \rightarrow \begin{bmatrix} 4/6 \\ 4/6 \end{bmatrix} \rightarrow \begin{bmatrix} 2/6 \\ 0 \end{bmatrix} \rightarrow \begin{bmatrix} 2/6 \\ 2/6 \end{bmatrix}$;

and another 12-cycle: $\begin{bmatrix} 1/6 \\ 0 \end{bmatrix} \rightarrow \begin{bmatrix} 1/6 \\ 1/6 \end{bmatrix} \rightarrow \begin{bmatrix} 2/6 \\ 3/6 \end{bmatrix} \rightarrow \begin{bmatrix} 5/6 \\ 2/6 \end{bmatrix} \rightarrow \begin{bmatrix} 1/6 \\ 3/6 \end{bmatrix} \rightarrow$

$\begin{bmatrix} 4/6 \\ 1/6 \end{bmatrix} \rightarrow \begin{bmatrix} 5/6 \\ 0 \end{bmatrix} \rightarrow \begin{bmatrix} 5/6 \\ 5/6 \end{bmatrix} \rightarrow \begin{bmatrix} 4/6 \\ 3/6 \end{bmatrix} \rightarrow \begin{bmatrix} 1/6 \\ 4/6 \end{bmatrix} \rightarrow \begin{bmatrix} 55/6 \\ 3/6 \end{bmatrix} \rightarrow \begin{bmatrix} 2/6 \\ 5/6 \end{bmatrix}$.

The possible periods of points for the form $(m/6, n/6)$ are thus 1, 3, 4, and 12. The least common multiple of these four numbers is 12, and so $\prod(6) = 12$.

3. **(a)** We are given that $x_0 = 3$ and $x_1 = 7$. With $p = 15$ we have

$$x_3 = x_2 + x_1 \mod 15 = 7 + 3 \mod 15 = 10 \mod 15 = 10,$$
$$x_4 = x_3 + x_2 \mod 15 = 10 + 7 \mod 15 = 17 \mod 15 = 2,$$
$$x_5 = x_4 + x_3 \mod 15 = 2 + 10 \mod 15 = 12 \mod 15 = 12,$$
$$x_6 = x_5 + x_4 \mod 15 = 12 + 2 \mod 15 = 14 \mod 15 = 14,$$
$$x_7 = x_6 + x_5 \mod 15 = 14 + 12 \mod 15 = 26 \mod 15 = 11,$$
$$x_8 = x_7 + x_6 \mod 15 = 11 + 14 \mod 15 = 25 \mod 15 = 10,$$
$$x_9 = x_8 + x_7 \mod 15 = 10 + 11 \mod 15 = 21 \mod 15 = 6,$$
$$x_{10} = x_9 + x_8 \mod 15 = 6 + 10 \mod 15 = 16 \mod 15 = 1,$$
$$x_{11} = x_{10} + x_9 \mod 15 = 1 + 6 \mod 15 = 7 \mod 15 = 7,$$
$$x_{12} = x_{11} + x_{10} \mod 15 = 7 + 1 \mod 15 = 8 \mod 15 = 8,$$
$$x_{13} = x_{12} + x_{11} \mod 15 = 8 + 7 \mod 15 = 15 \mod 15 = 0,$$
$$x_{14} = x_{13} + x_{12} \mod 15 = 0 + 8 \mod 15 = 8 \mod 15 = 8,$$
$$x_{15} = x_{14} + x_{13} \mod 15 = 8 + 0 \mod 15 = 8 \mod 15 = 8,$$
$$x_{16} = x_{15} + x_{14} \mod 15 = 8 + 8 \mod 15 = 16 \mod 15 = 1,$$
$$x_{17} = x_{16} + x_{15} \mod 15 = 1 + 8 \mod 15 = 9 \mod 15 = 9,$$

$$x_{18} = x_{17} + x_{16} \quad \text{mod } 15 = \quad 9 + \ 1 \quad \text{mod } 15 = 10 \quad \text{mod } 15 = 10,$$

$$x_{19} = x_{18} + x_{17} \quad \text{mod } 15 = 10 + \ 9 \quad \text{mod } 15 = 19 \quad \text{mod } 15 = \ 4,$$

$$x_{20} = x_{19} + x_{18} \quad \text{mod } 15 = \quad 4 + 10 \quad \text{mod } 15 = 14 \quad \text{mod } 15 = 14,$$

$$x_{21} = x_{20} + x_{19} \quad \text{mod } 15 = 14 + \ 4 \quad \text{mod } 15 = 18 \quad \text{mod } 15 = \ 3,$$

$$x_{22} = x_{21} + x_{20} \quad \text{mod } 15 = \quad 3 + 14 \quad \text{mod } 15 = 17 \quad \text{mod } 15 = \ 2,$$

$$x_{23} = x_{22} + x_{21} \quad \text{mod } 15 = \quad 2 + \ 3 \quad \text{mod } 15 = \ 5 \quad \text{mod } 15 = \ 5,$$

$$x_{24} = x_{23} + x_{22} \quad \text{mod } 15 = \quad 5 + \ 2 \quad \text{mod } 15 = \ 7 \quad \text{mod } 15 = \ 7,$$

$$x_{25} = x_{24} + x_{23} \quad \text{mod } 15 = \quad 7 + \ 5 \quad \text{mod } 15 = 12 \quad \text{mod } 15 = 12,$$

$$x_{26} = x_{25} + x_{24} \quad \text{mod } 15 = 12 + \ 7 \quad \text{mod } 15 = 19 \quad \text{mod } 15 = \ 4,$$

$$x_{27} = x_{26} + x_{25} \quad \text{mod } 15 = \quad 4 + 12 \quad \text{mod } 15 = 16 \quad \text{mod } 15 = \ 1,$$

$$x_{28} = x_{27} + x_{26} \quad \text{mod } 15 = \quad 1 + \ 4 \quad \text{mod } 15 = \ 5 \quad \text{mod } 15 = \ 5,$$

$$x_{29} = x_{28} + x_{27} \quad \text{mod } 15 = \quad 5 + \ 1 \quad \text{mod } 15 = \ 6 \quad \text{mod } 15 = \ 6,$$

$$x_{30} = x_{29} + x_{28} \quad \text{mod } 15 = \quad 6 + \ 5 \quad \text{mod } 15 = 11 \quad \text{mod } 15 = 11,$$

$$x_{31} = x_{30} + x_{29} \quad \text{mod } 15 = 11 + \ 6 \quad \text{mod } 15 = 17 \quad \text{mod } 15 = \ 2,$$

$$x_{32} = x_{31} + x_{30} \quad \text{mod } 15 = \quad 2 + 11 \quad \text{mod } 15 = 13 \quad \text{mod } 15 = 13,$$

$$x_{33} = x_{32} + x_{31} \quad \text{mod } 15 = 13 + \ 2 \quad \text{mod } 15 = 15 \quad \text{mod } 15 = \ 0,$$

$$x_{34} = x_{33} + x_{32} \quad \text{mod } 15 = \quad 0 + 13 \quad \text{mod } 15 = 13 \quad \text{mod } 15 = 13,$$

$$x_{35} = x_{34} + x_{33} \quad \text{mod } 15 = 13 + \ 0 \quad \text{mod } 15 = 13 \quad \text{mod } 15 = 13,$$

$$x_{36} = x_{35} + x_{34} \quad \text{mod } 15 = 13 + 13 \quad \text{mod } 15 = 26 \quad \text{mod } 15 = 11,$$

$$x_{37} = x_{36} + x_{35} \quad \text{mod } 15 = 11 + 13 \quad \text{mod } 15 = 24 \quad \text{mod } 15 = \ 9,$$

$$x_{38} = x_{37} + x_{36} \quad \text{mod } 15 = \quad 9 + 11 \quad \text{mod } 15 = 20 \quad \text{mod } 15 = \ 5,$$

$$x_{39} = x_{38} + x_{37} \quad \text{mod } 15 = \quad 5 + \ 9 \quad \text{mod } 15 = 14 \quad \text{mod } 15 = 14,$$

$$x_{40} = x_{39} + x_{38} \quad \text{mod } 15 = 14 + \ 5 \quad \text{mod } 15 = 19 \quad \text{mod } 15 = \ 4,$$

$$x_{41} = x_{40} + x_{39} \quad \text{mod } 15 = \quad 4 + 14 \quad \text{mod } 15 = 18 \quad \text{mod } 15 = \ 3,$$

$$x_{42} = x_{41} + x_{40} \quad \text{mod } 15 = \quad 3 + \ 4 \quad \text{mod } 15 = \ 7 \quad \text{mod } 15 = \ 7,$$

and finally: $x_{41} = x_0$ and $x_{42} = x_1$. Thus this sequence is periodic with period 41.

(b) Step (ii) of the algorithm is

$$x_{n+1} = x_n + x_{n-1} \bmod p. \tag{A}$$

Replacing n in this formula by $n+1$ gives

$$x_{n+2} = x_{n+1} + x_n \mod p = (x_n + x_{n-1}) + x_n \mod p = 2x_n + x_{n-1} \mod p. \quad \text{(B)}$$

Equations (A) and (B) can be written as

$$x_{n+1} = x_{n-1} + x_n \mod p$$
$$x_{n+2} = x_{n-1} + 2x_n \mod p$$

which in matrix form are

$$\begin{bmatrix} x_{n+1} \\ x_{n+2} \end{bmatrix} = \begin{bmatrix} 1 & 1 \\ 1 & 2 \end{bmatrix} \begin{bmatrix} x_{n-1} \\ x_n \end{bmatrix} \mod p.$$

3. (c) Beginning with $\begin{bmatrix} x_0 \\ x_1 \end{bmatrix} = \begin{bmatrix} 5 \\ 5 \end{bmatrix}$, we obtain

$$\begin{bmatrix} x_2 \\ x_3 \end{bmatrix} = \begin{bmatrix} 1 & 1 \\ 1 & 2 \end{bmatrix} \begin{bmatrix} 5 \\ 5 \end{bmatrix} \mod 21 = \begin{bmatrix} 10 \\ 15 \end{bmatrix} \mod 21 = \begin{bmatrix} 10 \\ 15 \end{bmatrix}$$

$$\begin{bmatrix} x_4 \\ x_5 \end{bmatrix} = \begin{bmatrix} 1 & 1 \\ 1 & 2 \end{bmatrix} \begin{bmatrix} 10 \\ 15 \end{bmatrix} \mod 21 = \begin{bmatrix} 25 \\ 40 \end{bmatrix} \mod 21 = \begin{bmatrix} 4 \\ 19 \end{bmatrix}$$

$$\begin{bmatrix} x_6 \\ x_7 \end{bmatrix} = \begin{bmatrix} 1 & 1 \\ 1 & 2 \end{bmatrix} \begin{bmatrix} 4 \\ 19 \end{bmatrix} \mod 21 = \begin{bmatrix} 23 \\ 42 \end{bmatrix} \mod 21 = \begin{bmatrix} 2 \\ 0 \end{bmatrix}$$

$$\begin{bmatrix} x_8 \\ x_9 \end{bmatrix} = \begin{bmatrix} 1 & 1 \\ 1 & 2 \end{bmatrix} \begin{bmatrix} 2 \\ 0 \end{bmatrix} \mod 21 = \begin{bmatrix} 2 \\ 2 \end{bmatrix} \mod 21 = \begin{bmatrix} 2 \\ 2 \end{bmatrix}$$

$$\begin{bmatrix} x_{10} \\ x_{11} \end{bmatrix} = \begin{bmatrix} 1 & 1 \\ 1 & 2 \end{bmatrix} \begin{bmatrix} 2 \\ 2 \end{bmatrix} \mod 21 = \begin{bmatrix} 4 \\ 6 \end{bmatrix} \mod 21 = \begin{bmatrix} 4 \\ 6 \end{bmatrix}$$

$$\begin{bmatrix} x_{12} \\ x_{13} \end{bmatrix} = \begin{bmatrix} 1 & 1 \\ 1 & 2 \end{bmatrix} \begin{bmatrix} 4 \\ 6 \end{bmatrix} \bmod 21 = \begin{bmatrix} 10 \\ 16 \end{bmatrix} \bmod 21 = \begin{bmatrix} 10 \\ 16 \end{bmatrix}$$

$$\begin{bmatrix} x_{14} \\ x_{15} \end{bmatrix} = \begin{bmatrix} 1 & 1 \\ 1 & 2 \end{bmatrix} \begin{bmatrix} 10 \\ 16 \end{bmatrix} \bmod 21 = \begin{bmatrix} 26 \\ 42 \end{bmatrix} \bmod 21 = \begin{bmatrix} 5 \\ 0 \end{bmatrix}$$

$$\begin{bmatrix} x_{16} \\ x_{17} \end{bmatrix} = \begin{bmatrix} 1 & 1 \\ 1 & 2 \end{bmatrix} \begin{bmatrix} 5 \\ 0 \end{bmatrix} \bmod 21 = \begin{bmatrix} 5 \\ 5 \end{bmatrix} \bmod 21 = \begin{bmatrix} 5 \\ 5 \end{bmatrix}.$$

and we see that $\begin{bmatrix} x_{16} \\ x_{17} \end{bmatrix} = \begin{bmatrix} x_0 \\ x_1 \end{bmatrix}$.

4. (a) We have that

$$\Gamma^{25}\left(\begin{bmatrix} \dfrac{m}{101} \\ \dfrac{n}{101} \end{bmatrix}\right) = \begin{bmatrix} 7,778,742,049 & 12,586,269,025 \\ 12,586,269,025 & 20,365,011,074 \end{bmatrix} \begin{bmatrix} \dfrac{m}{101} \\ \dfrac{n}{101} \end{bmatrix} \bmod 1$$

$$= \begin{bmatrix} \dfrac{7,778,742,049}{101}m + \dfrac{12,586,269,025}{101}n \\ \dfrac{12,586,269,025}{101}m + \dfrac{20,365,011,074}{101}n \end{bmatrix} \bmod 1$$

$$= \begin{bmatrix} \left(r + \dfrac{1}{101}\right)m + sn \\ sm + \left(t + \dfrac{1}{101}\right)n \end{bmatrix} \bmod 1 \text{ (where } m, n, r, s, \text{ and } t \text{ are integers)}$$

$$= \begin{bmatrix} (rm + sn) + \dfrac{m}{101} \\ (sm + tn) + \dfrac{n}{101} \end{bmatrix} \bmod 1$$

$$= \begin{bmatrix} \dfrac{m}{101} \\ \dfrac{n}{101} \end{bmatrix} \text{ (because } rm + sn \text{ and } sm + tn \text{ are integers).}$$

4. **(b)** From (a) we have that every point of the form $(m/101, n/101)$ returns to its starting position after 25 iterations, and so every such point is a periodic point with a period that must divide 25. The only integers that divide 25 are 1, 5, and 25, and so every such point must have period 1, 5, or 25.

(c) We have that

$$\Gamma\left(\begin{bmatrix} \dfrac{1}{101} \\ 0 \end{bmatrix}\right) = \begin{bmatrix} 1 & 1 \\ 1 & 2 \end{bmatrix}\begin{bmatrix} \dfrac{1}{101} \\ 0 \end{bmatrix} \text{ mod } 1 = \begin{bmatrix} \dfrac{1}{101} \\ \dfrac{1}{101} \end{bmatrix},$$

$$\Gamma^2\left(\begin{bmatrix} \dfrac{1}{101} \\ 0 \end{bmatrix}\right) = \begin{bmatrix} 1 & 1 \\ 1 & 2 \end{bmatrix}\begin{bmatrix} \dfrac{1}{101} \\ \dfrac{1}{101} \end{bmatrix} \text{ mod } 1 = \begin{bmatrix} \dfrac{2}{101} \\ \dfrac{3}{101} \end{bmatrix},$$

$$\Gamma^3\left(\begin{bmatrix} \dfrac{1}{101} \\ 0 \end{bmatrix}\right) = \begin{bmatrix} 1 & 1 \\ 1 & 2 \end{bmatrix}\begin{bmatrix} \dfrac{2}{101} \\ \dfrac{3}{101} \end{bmatrix} \text{ mod } 1 = \begin{bmatrix} \dfrac{5}{101} \\ \dfrac{8}{101} \end{bmatrix},$$

$$\Gamma^4\left(\begin{bmatrix} \dfrac{1}{101} \\ 0 \end{bmatrix}\right) = \begin{bmatrix} 1 & 1 \\ 1 & 2 \end{bmatrix}\begin{bmatrix} \dfrac{5}{101} \\ \dfrac{8}{101} \end{bmatrix} \text{ mod } 1 = \begin{bmatrix} \dfrac{13}{101} \\ \dfrac{21}{101} \end{bmatrix},$$

$$\Gamma^5\left(\begin{bmatrix} \dfrac{1}{101} \\ 0 \end{bmatrix}\right) = \begin{bmatrix} 1 & 1 \\ 1 & 2 \end{bmatrix}\begin{bmatrix} \dfrac{13}{101} \\ \dfrac{21}{101} \end{bmatrix} \text{ mod } 1 = \begin{bmatrix} \dfrac{34}{101} \\ \dfrac{55}{101} \end{bmatrix}.$$

Because all five iterates are different, the period of the periodic point $(1/101, 0)$ must be greater than 5.

(d) From part (b) all points of the form $(m/101, n/101)$ must have period 1, 5, or 25. From part (c) the point $(1/101, 0)$ has period greater that 5, and so must have period 25. Because the least common multiple of 1, 5, and 25 is 25, it follows that $\prod(101) = 25$.

5. If $0 \le x < 1$ and $0 \le y < 1$, then $T(x, y) = (x + 5/12, y)$ mod 1, and so $T^2(x, y) = (x + 10/12, y)$ mod 1, $T^3(x, y) = (x + 15/12, y)$ mod 1, $\ldots, T^{12}(x, y) = (x + 60/12, y)$ mod 1 $= (x + 5, y)$ mod 1 $= (x, y)$. Thus every point in S returns to its original position after 12

iterations and so every point in S is a periodic point with period at most 12. Because every point is a periodic point, no point can have a dense set of iterates, and so the mapping cannot be chaotic.

6. **(a)** The matrix of Arnold's cat map, $\begin{bmatrix} 1 & 1 \\ 1 & 2 \end{bmatrix}$, is one in which (i) the entries are all integers, (ii) the determinant is 1, and (iii) the eigenvalues, $(3 + \sqrt{5})/2 = 2.6180\ldots$ and $(3 - \sqrt{5})/2 = 0.3819\ldots$, do not have magnitude 1. The three conditions of an Anosov automorphism are thus satisfied.

(b) The eigenvalues of the matrix $\begin{bmatrix} 0 & 1 \\ 1 & 0 \end{bmatrix}$ are ± 1, both of which have magnitude 1. By part (iii) of the definition of an Anosov automorphism, this matrix is not the matrix of an Anosov automorphism.

The entries of the matrix $\begin{bmatrix} 3 & 2 \\ 1 & 1 \end{bmatrix}$ are integers; its determinant is 1; and neither of its eigenvalues, $2 \pm \sqrt{3}$, has magnitude 1. Consequently, this is the matrix of an Anosov automorphism.

The eigenvalues of the matrix $\begin{bmatrix} 1 & 0 \\ 0 & 1 \end{bmatrix}$ are both equal to 1, and so both have magnitude 1. By part (iii) of the definition, this is not the matrix of an Anosov automorphism.

The entries of the matrix $\begin{bmatrix} 5 & 7 \\ 2 & 3 \end{bmatrix}$ are integers; its determinant is 1; and neither of its eigenvalues, $4 \pm \sqrt{15}$, has magnitude 1. Consequently, this is the matrix of an Anosov automorphism.

The determinant of the matrix $\begin{bmatrix} 6 & 2 \\ 5 & 2 \end{bmatrix}$ is 2, and so by part (ii) of the definition, this is not the matrix of an Anosov automorphism.

(c) The eigenvalues of the matrix $\begin{bmatrix} 0 & 1 \\ -1 & 0 \end{bmatrix}$ are $\pm i$; both of which have magnitude 1. By part (iii) of the definition, this cannot be the matrix of an Anosov automorphism.

Starting with an arbitrary point $\begin{bmatrix} x \\ y \end{bmatrix}$ in the interior of S, (that is, with $0 < x < 1$ and $0 < y < 1$) we obtain

$$\begin{bmatrix} 0 & 1 \\ -1 & 0 \end{bmatrix} = \begin{bmatrix} x \\ y \end{bmatrix} \bmod 1 = \begin{bmatrix} y \\ -x \end{bmatrix} \bmod 1 = \begin{bmatrix} y \\ 1-x \end{bmatrix},$$

$$\begin{bmatrix} 0 & 1 \\ -1 & 0 \end{bmatrix} = \begin{bmatrix} y \\ 1-x \end{bmatrix} \bmod 1 = \begin{bmatrix} 1-x \\ -y \end{bmatrix} \bmod 1 = \begin{bmatrix} 1-x \\ 1-y \end{bmatrix},$$

$$\begin{bmatrix} 0 & 1 \\ -1 & 0 \end{bmatrix} = \begin{bmatrix} 1-x \\ 1-y \end{bmatrix} \bmod 1 = \begin{bmatrix} 1-y \\ -1+x \end{bmatrix} \bmod 1 = \begin{bmatrix} 1-y \\ x \end{bmatrix},$$

$$\begin{bmatrix} 0 & 1 \\ -1 & 0 \end{bmatrix} = \begin{bmatrix} 1-y \\ x \end{bmatrix} \bmod 1 = \begin{bmatrix} x \\ -1+y \end{bmatrix} \bmod 1 = \begin{bmatrix} x \\ y \end{bmatrix}.$$

Thus every point in the interior of S is a periodic point with period at most 4. The geometric effect of this transformation, as seen by the iterates, is to rotate each point in the interior of S clockwise by $90°$ about the center point $\begin{bmatrix} 1/2 \\ 1/2 \end{bmatrix}$ of S.

Consequently, each point in the interior of S has period 4 with the exception of the center point $\begin{bmatrix} 1/2 \\ 1/2 \end{bmatrix}$, which is a fixed point.

For points not in the interior of S, we first observe that the origin $\begin{bmatrix} 0 \\ 0 \end{bmatrix}$ is a fixed point, which can easily be verified. Starting with a point of the form $\begin{bmatrix} x \\ 0 \end{bmatrix}$ with $0 < x < 1$, we obtain a 4-cycle $\begin{bmatrix} x \\ 0 \end{bmatrix} \rightarrow \begin{bmatrix} 0 \\ 1-x \end{bmatrix} \rightarrow \begin{bmatrix} 1-x \\ 0 \end{bmatrix} \rightarrow \begin{bmatrix} 0 \\ x \end{bmatrix} \rightarrow$

$\begin{bmatrix} x \\ 0 \end{bmatrix}$ if $x \neq 1/2$, otherwise we obtain the 2-cycle $\begin{bmatrix} 1/2 \\ 0 \end{bmatrix} \rightarrow \begin{bmatrix} 0 \\ 1/2 \end{bmatrix} \rightarrow \begin{bmatrix} 1/2 \\ 0 \end{bmatrix}.$

Similarly, starting with a point of the form $\begin{bmatrix} 0 \\ y \end{bmatrix}$ with $0 < y < 1$, we obtain a 4-cycle

$$\begin{bmatrix} 0 \\ y \end{bmatrix} \rightarrow \begin{bmatrix} y \\ 0 \end{bmatrix} \rightarrow \begin{bmatrix} 0 \\ 1-y \end{bmatrix} \rightarrow \begin{bmatrix} 1-y \\ 0 \end{bmatrix} \rightarrow \begin{bmatrix} 0 \\ y \end{bmatrix}$$ if $y \neq 1/2$, otherwise we

obtain the 2-cycle $\begin{bmatrix} 0 \\ 1/2 \end{bmatrix} \rightarrow \begin{bmatrix} 1/2 \\ 0 \end{bmatrix} \rightarrow \begin{bmatrix} 0 \\ 1/2 \end{bmatrix}$. Thus every point not in the

interior of S is a periodic point with period 1, 2, or 4. Finally, because all points in S are periodic points, no point in S can have a dense set of iterates and so the mapping cannot be chaotic.

7. That Arnold's cat map is a one-to-one mapping over the unit square S and that its range is S is fairly obvious geometrically. An analytical proof would proceed as follows: To show that it is one-to-one, suppose that (x_1, y_1) and (x_2, y_2), with $0 \leq x_i < 1$ and $0 \leq y_i < 1$, are two points in S such that $\Gamma(x_1, y_1) = \Gamma(x_2, y_2)$. From the definition of Γ it follows that

$$(x_1 + y_1) \bmod 1 = (x_2 + y_2) \bmod 1$$

$$(x_1 + 2y_1) \bmod 1 = (x_2 + 2y_2) \bmod 1.$$

From the definition of "mod 1", there are integers $r_1, s_1, r_2,$ and s_2 such that

$$x_1 + y_1 - r_1 = x_2 + y_2 - r_2 \tag{A}$$

$$x_1 + 2y_1 - s_1 = x_2 + 2y_2 - s_2. \tag{B}$$

Subtracting the first equation from the second gives

$$y_1 - s_1 + r_1 = y_2 - s_2 + r_2$$

or

$$y_1 - y_2 = s_1 - s_2 + r_2 - r_1. \tag{C}$$

The right-hand side of Eq. (C) is an integer and the left-hand side satisfies $-1 < y_1 - y_2 < 1$. Consequently, the integer on the right-hand side must be zero, and so $y_1 = y_2$. Putting $y_1 = y_2$ into Eq. (A) gives $x_1 - r_1 = x_2 - r_2$ and a similar argument shows that $x_1 = x_2$. Thus $(x_1, y_1) = (x_2, y_2)$, which shows that the mapping is one-to-one.

To show that the range of Γ is S, let (u, v) be any point in S, so that $0 \le u < 1$ and $0 \le v < 1$. We want to show that there exists a point (x, y) in S such that $\Gamma(x, y) = (u, v)$; that is, such that

$$(x + y) \bmod 1 = u$$

$$(x + 2y) \bmod 1 = v.$$

From the definition of "mod 1", this means that we must find integers r and s that

$$x + y - r = u$$

$$x + 2y - s = v$$

and for which x and y lie in $[0, 1)$. Solving for x and y in this linear system gives

$$x = (2u - v) + (2r - s)$$

$$y = (-u + 2v) + (-r + s).$$

Let a and b be integers such that $(2u - v) + a$ and $(-u + 2v) + b$ lie in the interval $[0, 1)$. Solving the system

$$2r - s = a$$

$$-r + s = b$$

gives $r = a + b$ and $s = a + 2b$, both of which are integers. This choice of r and s then provides the desired values of x and y.

8. Let (x, y) lie in S, so that $0 \le x < 1$ and $0 \le y < 1$. We must show that $\Gamma^{-1}(\Gamma(x, y)) = (x, y)$. From the definition of Γ^{-1} and Γ we obtain

$$\Gamma^{-1}(\Gamma(x, y)) = \Gamma^{-1}(x + y \bmod 1, x + 2y \bmod 1)$$

$$= \Gamma^{-1}(x + y - r, x + 2y - s), \text{ for certain integers } r \text{ and } s,$$

$$= (\, 2(x + y - r) - (x + 2y - s), \; -(x + y - r) + (x + 2y - s)) \bmod 1$$

$$= (x - 2r + s, y + r - s) \bmod 1$$

$$= (x, y),$$

where the last step follows because $-2r + s$ and $r - s$ are integers and $0 \le x, \, y < 1$.

9. As per the hint, we wish to find the regions in S that map onto the four indicated regions in the figure below.

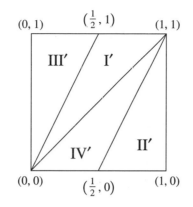

We first consider region $\mathbf{I}'$ with vertices $(0,0)$, $(1/2, 1)$, and $(1, 1)$. We seek points (x_1, y_1), (x_2, y_2), and (x_3, y_3), with entries that lie in $[0, 1]$, that map onto these three points under

the mapping $\begin{bmatrix} x \\ y \end{bmatrix} \rightarrow \begin{bmatrix} 1 & 2 \\ 1 & 2 \end{bmatrix} \begin{bmatrix} x \\ y \end{bmatrix} + \begin{bmatrix} a \\ b \end{bmatrix}$ for certain integer values of a and b to

be determined. This leads to the three equations

$$\begin{bmatrix} 1 & 1 \\ 1 & 2 \end{bmatrix} \begin{bmatrix} x_1 \\ y_1 \end{bmatrix} + \begin{bmatrix} a \\ b \end{bmatrix} = \begin{bmatrix} 0 \\ 0 \end{bmatrix},$$

$$\begin{bmatrix} 1 & 1 \\ 1 & 2 \end{bmatrix} \begin{bmatrix} x_2 \\ y_2 \end{bmatrix} + \begin{bmatrix} a \\ b \end{bmatrix} = \begin{bmatrix} 1/2 \\ 1 \end{bmatrix},$$

$$\begin{bmatrix} 1 & 1 \\ 1 & 2 \end{bmatrix} \begin{bmatrix} x_3 \\ y_3 \end{bmatrix} + \begin{bmatrix} a \\ b \end{bmatrix} = \begin{bmatrix} 1 \\ 1 \end{bmatrix}.$$

The inverse of the matrix $\begin{bmatrix} 1 & 1 \\ 1 & 2 \end{bmatrix}$ is $\begin{bmatrix} 2 & -1 \\ -1 & 1 \end{bmatrix}$. We multiply the above three matrix

equations by this inverse and set $\begin{bmatrix} c \\ d \end{bmatrix} = \begin{bmatrix} 2 & -1 \\ -1 & 1 \end{bmatrix} \begin{bmatrix} a \\ b \end{bmatrix}$. Notice that c and d must be integers. This leads to

$$\begin{bmatrix} x_1 \\ y_1 \end{bmatrix} = \begin{bmatrix} 2 & -1 \\ -1 & 1 \end{bmatrix} \begin{bmatrix} 0 \\ 0 \end{bmatrix} - \begin{bmatrix} c \\ d \end{bmatrix} = -\begin{bmatrix} c \\ d \end{bmatrix},$$

$$\begin{bmatrix} x_2 \\ y_2 \end{bmatrix} = \begin{bmatrix} 2 & -1 \\ -1 & 1 \end{bmatrix} \begin{bmatrix} 1/2 \\ 1 \end{bmatrix} - \begin{bmatrix} c \\ d \end{bmatrix} = \begin{bmatrix} 0 \\ 1/2 \end{bmatrix} - \begin{bmatrix} c \\ d \end{bmatrix},$$

$$\begin{bmatrix} x_3 \\ y_3 \end{bmatrix} = \begin{bmatrix} 2 & -1 \\ -1 & 1 \end{bmatrix} \begin{bmatrix} 1 \\ 1 \end{bmatrix} - \begin{bmatrix} c \\ d \end{bmatrix} = \begin{bmatrix} 1 \\ 0 \end{bmatrix} - \begin{bmatrix} c \\ d \end{bmatrix}.$$

The only integer values of c and d that will give values of x_i and y_i in the interval $[0,1]$ are $c = d = 0$. This then gives $a = b = 0$ and the mapping $\begin{bmatrix} x \\ y \end{bmatrix} \rightarrow \begin{bmatrix} 1 & 1 \\ 1 & 2 \end{bmatrix} \begin{bmatrix} x \\ y \end{bmatrix}$ that maps the three points $(0,0)$, $(0,1/2)$, and $(1,0)$ to the three points $(0,0)$, $(0,1/2)$, and $(1,1)$, respectively. The three points $(0,0)$, $(0,1/2)$, and $(1,0)$ define the triangular region labelled **I** in the diagram below, which then maps onto the region **I′** above.

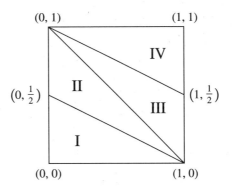

For region **II′**, the calculations are as follows:

$$\begin{bmatrix} 1 & 1 \\ 1 & 2 \end{bmatrix} \begin{bmatrix} x_1 \\ y_1 \end{bmatrix} + \begin{bmatrix} a \\ b \end{bmatrix} = \begin{bmatrix} 1/2 \\ 0 \end{bmatrix},$$

$$\begin{bmatrix} 1 & 1 \\ 1 & 2 \end{bmatrix} \begin{bmatrix} x_2 \\ y_2 \end{bmatrix} + \begin{bmatrix} a \\ b \end{bmatrix} = \begin{bmatrix} 1 \\ 1 \end{bmatrix},$$

$$\begin{bmatrix} 1 & 1 \\ 1 & 2 \end{bmatrix} \begin{bmatrix} x_3 \\ y_3 \end{bmatrix} + \begin{bmatrix} a \\ b \end{bmatrix} = \begin{bmatrix} 1 \\ 0 \end{bmatrix};$$

$$\begin{bmatrix} x_1 \\ y_1 \end{bmatrix} = \begin{bmatrix} 2 & -1 \\ -1 & 1 \end{bmatrix} \begin{bmatrix} 1/2 \\ 0 \end{bmatrix} - \begin{bmatrix} c \\ d \end{bmatrix} = \begin{bmatrix} 1 \\ -1/2 \end{bmatrix} - \begin{bmatrix} c \\ d \end{bmatrix},$$

$$\begin{bmatrix} x_2 \\ y_2 \end{bmatrix} = \begin{bmatrix} 2 & -1 \\ -1 & 1 \end{bmatrix} \begin{bmatrix} 1 \\ 1 \end{bmatrix} - \begin{bmatrix} c \\ d \end{bmatrix} = \begin{bmatrix} 0 \\ 1 \end{bmatrix} - \begin{bmatrix} c \\ d \end{bmatrix},$$

$$\begin{bmatrix} x_3 \\ y_3 \end{bmatrix} = \begin{bmatrix} 2 & -1 \\ -1 & 1 \end{bmatrix} \begin{bmatrix} 1 \\ 0 \end{bmatrix} - \begin{bmatrix} c \\ d \end{bmatrix} = \begin{bmatrix} 2 \\ -1 \end{bmatrix} - \begin{bmatrix} c \\ d \end{bmatrix}.$$

Only $c = 1$ and $d = -1$ will work. This leads to $a = 0$, $b = -1$ and the mapping

$$\begin{bmatrix} x \\ y \end{bmatrix} \rightarrow \begin{bmatrix} 1 & 1 \\ 1 & 2 \end{bmatrix} \begin{bmatrix} x \\ y \end{bmatrix} + \begin{bmatrix} 0 \\ -1 \end{bmatrix}$$ that maps region **II** with vertices $(0, 1/2)$, $(0, 1)$,

and $(1, 0)$ onto region **II**$'$.

For region **III**$'$, the calculations are as follows:

$$\begin{bmatrix} 1 & 1 \\ 1 & 2 \end{bmatrix} \begin{bmatrix} x_1 \\ y_1 \end{bmatrix} + \begin{bmatrix} a \\ b \end{bmatrix} = \begin{bmatrix} 0 \\ 0 \end{bmatrix},$$

$$\begin{bmatrix} 1 & 1 \\ 1 & 2 \end{bmatrix} \begin{bmatrix} x_2 \\ y_2 \end{bmatrix} + \begin{bmatrix} a \\ b \end{bmatrix} = \begin{bmatrix} 1/2 \\ 1 \end{bmatrix},$$

$$\begin{bmatrix} 1 & 1 \\ 1 & 2 \end{bmatrix} \begin{bmatrix} x_3 \\ y_3 \end{bmatrix} + \begin{bmatrix} a \\ b \end{bmatrix} = \begin{bmatrix} 0 \\ 1 \end{bmatrix};$$

$$\begin{bmatrix} x_1 \\ y_1 \end{bmatrix} = \begin{bmatrix} 2 & -1 \\ -1 & 1 \end{bmatrix} \begin{bmatrix} 0 \\ 0 \end{bmatrix} - \begin{bmatrix} c \\ d \end{bmatrix} = - \begin{bmatrix} c \\ d \end{bmatrix}$$

$$\begin{bmatrix} x_2 \\ y_2 \end{bmatrix} = \begin{bmatrix} 2 & -1 \\ -1 & 1 \end{bmatrix} \begin{bmatrix} 1/2 \\ 1 \end{bmatrix} - \begin{bmatrix} c \\ d \end{bmatrix} = \begin{bmatrix} 0 \\ 1/2 \end{bmatrix} - \begin{bmatrix} c \\ d \end{bmatrix},$$

$$\begin{bmatrix} x_3 \\ y_3 \end{bmatrix} = \begin{bmatrix} 2 & -1 \\ -1 & 1 \end{bmatrix} \begin{bmatrix} 0 \\ 1 \end{bmatrix} - \begin{bmatrix} c \\ d \end{bmatrix} = \begin{bmatrix} -1 \\ 1 \end{bmatrix} - \begin{bmatrix} c \\ d \end{bmatrix}.$$

Only $c = -1$ and $d = 0$ will work. This leads to $a = -1$, $b = -1$ and the mapping

$$\begin{bmatrix} x \\ y \end{bmatrix} \rightarrow \begin{bmatrix} 1 & 1 \\ 1 & 2 \end{bmatrix} \begin{bmatrix} x \\ y \end{bmatrix} + \begin{bmatrix} -1 \\ -1 \end{bmatrix}$$ that maps region **III** with vertices $(1,0)$, $(1, 1/2)$,

and $(0, 1)$ onto region **III**$'$.

For region **IV**$'$, the calculations are as follows:

$$\begin{bmatrix} 1 & 1 \\ 1 & 2 \end{bmatrix} \begin{bmatrix} x_1 \\ y_1 \end{bmatrix} + \begin{bmatrix} a \\ b \end{bmatrix} = \begin{bmatrix} 0 \\ 0 \end{bmatrix},$$

$$\begin{bmatrix} 1 & 1 \\ 1 & 2 \end{bmatrix} \begin{bmatrix} x_2 \\ y_2 \end{bmatrix} + \begin{bmatrix} a \\ b \end{bmatrix} = \begin{bmatrix} 1 \\ 1 \end{bmatrix},$$

$$\begin{bmatrix} 1 & 1 \\ 1 & 2 \end{bmatrix} \begin{bmatrix} x_3 \\ y_3 \end{bmatrix} + \begin{bmatrix} a \\ b \end{bmatrix} = \begin{bmatrix} 1/2 \\ 0 \end{bmatrix};$$

$$\begin{bmatrix} x_1 \\ y_1 \end{bmatrix} = \begin{bmatrix} 2 & -1 \\ -1 & 1 \end{bmatrix} \begin{bmatrix} 0 \\ 0 \end{bmatrix} - \begin{bmatrix} c \\ d \end{bmatrix} = -\begin{bmatrix} c \\ d \end{bmatrix}$$

$$\begin{bmatrix} x_2 \\ y_2 \end{bmatrix} = \begin{bmatrix} 2 & -1 \\ -1 & 1 \end{bmatrix} \begin{bmatrix} 1 \\ 1 \end{bmatrix} - \begin{bmatrix} c \\ d \end{bmatrix} = \begin{bmatrix} 1 \\ 0 \end{bmatrix} - \begin{bmatrix} c \\ d \end{bmatrix},$$

$$\begin{bmatrix} x_3 \\ y_3 \end{bmatrix} = \begin{bmatrix} 2 & -1 \\ -1 & 1 \end{bmatrix} \begin{bmatrix} 1/2 \\ 0 \end{bmatrix} - \begin{bmatrix} c \\ d \end{bmatrix} = \begin{bmatrix} 1 \\ -1/2 \end{bmatrix} - \begin{bmatrix} c \\ d \end{bmatrix}.$$

Only $c = 0$ and $d = -1$ will work. This leads to $a = -1$, $b = -2$ and the mapping

$$\begin{bmatrix} x \\ y \end{bmatrix} \rightarrow \begin{bmatrix} 1 & 1 \\ 1 & 2 \end{bmatrix} \begin{bmatrix} x \\ y \end{bmatrix} + \begin{bmatrix} -1 \\ -2 \end{bmatrix}$$ that maps region **IV** with vertices $(0,0)$, $(1,1)$, and

$(1/2, 0)$ onto region **IV**$'$.

10.   We first establish the result for $n = 2$. Given (x_0, y_0) we have

$$\begin{bmatrix} x_1 \\ y_1 \end{bmatrix} = \begin{bmatrix} 1 & 1 \\ 1 & 2 \end{bmatrix} \begin{bmatrix} x_0 \\ y_0 \end{bmatrix} \text{ mod } 1 = \begin{bmatrix} 1 & 1 \\ 1 & 2 \end{bmatrix} \begin{bmatrix} x_0 \\ y_0 \end{bmatrix} - \begin{bmatrix} r_0 \\ s_0 \end{bmatrix}$$

and

$$\begin{bmatrix} x_2 \\ y_2 \end{bmatrix} = \begin{bmatrix} 1 & 1 \\ 1 & 2 \end{bmatrix} \begin{bmatrix} x_1 \\ y_1 \end{bmatrix} \text{ mod } 1 = \begin{bmatrix} 1 & 1 \\ 1 & 2 \end{bmatrix} \begin{bmatrix} x_1 \\ y_1 \end{bmatrix} - \begin{bmatrix} r_1 \\ s_1 \end{bmatrix},$$

$$= \begin{bmatrix} 1 & 1 \\ 1 & 2 \end{bmatrix} \left(\begin{bmatrix} 1 & 1 \\ 1 & 2 \end{bmatrix} \begin{bmatrix} x_0 \\ y_0 \end{bmatrix} - \begin{bmatrix} r_0 \\ s_0 \end{bmatrix} \right) - \begin{bmatrix} r_1 \\ s_1 \end{bmatrix}$$

$$= \begin{bmatrix} 1 & 1 \\ 1 & 2 \end{bmatrix}^2 \begin{bmatrix} x_0 \\ y_0 \end{bmatrix} - \begin{bmatrix} r_0 + s_0 + r_1 \\ r_0 + 2s_0 + s_1 \end{bmatrix}$$

where r_0, s_0, r_1 and s_1 are integers. This last expression is the same as $\begin{bmatrix} 1 & 1 \\ 1 & 2 \end{bmatrix}^2 \begin{bmatrix} x_0 \\ y_0 \end{bmatrix}$ mod 1, which is what we wanted to show.

A similar argument establishes the result for any integer n.

11.   As per the hint, we want to show that the only solution of the matrix equation

$$\begin{bmatrix} x_0 \\ y_0 \end{bmatrix} = \begin{bmatrix} 1 & 1 \\ 1 & 2 \end{bmatrix} \begin{bmatrix} x_0 \\ y_0 \end{bmatrix} - \begin{bmatrix} r \\ s \end{bmatrix}$$

with $0 \leq x_0 < 1$, $0 \leq y_0 < 1$, and integer r and s is $x_0 = y_0 = 0$. This matrix equation is equivalent to the system

$$x_0 = x_0 + y_0 - r$$

$$y_0 = x_0 + 2y_0 - s.$$

The first equation is $y_0 = r$, which can only have the solution $y_0 = r = 0$ with the given constraints on y_0 and r. Putting $y_0 = 0$ into the second equation yields $x_0 = s$, which likewise has only the solution $x_0 = s = 0$.

12. As per the hint, we want to find all solutions of the matrix equation

$$\begin{bmatrix} x_0 \\ y_0 \end{bmatrix} = \begin{bmatrix} 2 & 3 \\ 3 & 5 \end{bmatrix} \begin{bmatrix} x_0 \\ y_0 \end{bmatrix} - \begin{bmatrix} r \\ s \end{bmatrix}$$

where $0 \le x_0 < 1$, $0 \le y_0 < 1$, and r and s are nonnegative integers. This equation can be rewritten as

$$\begin{bmatrix} 1 & 3 \\ 3 & 4 \end{bmatrix} \begin{bmatrix} x_0 \\ y_0 \end{bmatrix} = \begin{bmatrix} r \\ s \end{bmatrix},$$

which has the solution $x_0 = \dfrac{-4r + 3s}{5}$ and $y_0 = \dfrac{3r - s}{5}$. First trying $r = 0$ and $s = 0, 1, 2, \ldots$, then $r = 1$ and $s = 0, 1, 2, \ldots$, etc., we find that the only values of r and s that yield values of x_0 and y_0 lying in $[0, 1)$ are:

$r = 1$ and $s = 2$, which give $x_0 = 2/5$ and $y_0 = 1/5$;

$r = 2$ and $s = 3$, which give $x_0 = 1/5$ and $y_0 = 3/5$;

$r = 2$ and $s = 4$, which give $x_0 = 4/5$ and $y_0 = 2/5$;

$r = 3$ and $s = 5$, which give $x_0 = 3/5$ and $y_0 = 4/5$.

We can then check that $(2/5, 1/5)$ and $(3/5, 4/5)$ form one 2-cycle and $(1/5, 3/5)$ and $(4/5, 2/5)$ form another 2-cycle.

13. In the matrix equation $\begin{bmatrix} x_0 \\ y_0 \end{bmatrix} = \begin{bmatrix} 1 & 1 \\ 1 & 2 \end{bmatrix}^n \begin{bmatrix} x_0 \\ y_0 \end{bmatrix}$ mod 1, let the matrix $\begin{bmatrix} 1 & 1 \\ 1 & 2 \end{bmatrix}^n$

be denoted by $\begin{bmatrix} a_n & b_n \\ c_n & d_n \end{bmatrix}$, all of whose entries are positive integers. The matrix equation can then be written as

$$\begin{bmatrix} x_0 \\ y_0 \end{bmatrix} = \begin{bmatrix} a_n & b_n \\ c_n & d_n \end{bmatrix} \begin{bmatrix} x_0 \\ y_0 \end{bmatrix} - \begin{bmatrix} r \\ s \end{bmatrix} \quad \text{or} \quad \begin{bmatrix} a_n - 1 & b_n \\ c_n & d_n - 1 \end{bmatrix} \begin{bmatrix} x_0 \\ y_0 \end{bmatrix} = \begin{bmatrix} r \\ s \end{bmatrix}$$

where r and s are integers. By Cramer's rule, this last equation has the solution

$$x_0 = \frac{(d_n - 1)r - b_n s}{(a_n - 1)(d_n - 1) - b_n c_n} \quad \text{and} \quad y_0 = \frac{(a_n - 1)r - c_n r}{(a_n - 1)(d_n - 1) - b_n c_n}.$$

These two expressions identify x_0 and y_0 as the quotients of integers, and hence as rational numbers. One point remains: to verify that the determinant of the matrix $\begin{bmatrix} a_n - 1 & b_n \\ c_n & d_n - 1 \end{bmatrix}$ is not equal to zero, so that Cramer's rule is valid. This is equivalent to showing that the matrix $\begin{bmatrix} a_n & b_n \\ c_n & d_n \end{bmatrix}$ does not have the eigenvalue 1. Now, the eigenvalues of this matrix are λ_1^n and λ_2^n, where λ_1 and λ_2 are the eigenvalues of $\begin{bmatrix} 1 & 1 \\ 1 & 2 \end{bmatrix}$. Because these eigenvalues are $\lambda_1 = (3 + \sqrt{5})/2 = 2.6180\ldots$ and $\lambda_2 = (3 - \sqrt{5})/2 = 0.3819\ldots$, it is not possible that their nth powers can be 1 for $n = 1, 2, \ldots$.

1. First we group the plaintext into pairs and add the dummy letter T:

$$DA \qquad RK \qquad NI \qquad GH \qquad TT$$

or equivalently from Table 1:

$$4 \quad 1 \qquad\qquad 18 \quad 11 \qquad\qquad 14 \quad 9 \qquad\qquad 7 \quad 8 \qquad\qquad 20 \quad 20$$

(a) For the enciphering matrix $A = \begin{bmatrix} 1 & 3 \\ 2 & 1 \end{bmatrix}$ we have

$$\begin{bmatrix} 1 & 3 \\ 2 & 1 \end{bmatrix} \begin{bmatrix} 4 \\ 1 \end{bmatrix} = \begin{bmatrix} 7 \\ 9 \end{bmatrix} \qquad\qquad \begin{matrix} G \\ I \end{matrix}$$

$$\begin{bmatrix} 1 & 3 \\ 2 & 1 \end{bmatrix} \begin{bmatrix} 18 \\ 11 \end{bmatrix} = \begin{bmatrix} 51 \\ 47 \end{bmatrix} = \begin{bmatrix} 25 \\ 21 \end{bmatrix} \qquad \begin{matrix} Y \\ U \end{matrix}$$

$$\begin{bmatrix} 1 & 3 \\ 2 & 1 \end{bmatrix} \begin{bmatrix} 14 \\ 9 \end{bmatrix} = \begin{bmatrix} 41 \\ 37 \end{bmatrix} = \begin{bmatrix} 15 \\ 11 \end{bmatrix} \qquad \begin{matrix} O \\ K \end{matrix} \qquad (\text{mod } 26)$$

$$\begin{bmatrix} 1 & 3 \\ 2 & 1 \end{bmatrix} \begin{bmatrix} 7 \\ 8 \end{bmatrix} = \begin{bmatrix} 31 \\ 22 \end{bmatrix} = \begin{bmatrix} 5 \\ 22 \end{bmatrix} \qquad \begin{matrix} E \\ V \end{matrix}$$

$$\begin{bmatrix} 1 & 3 \\ 2 & 1 \end{bmatrix} \begin{bmatrix} 20 \\ 20 \end{bmatrix} = \begin{bmatrix} 80 \\ 60 \end{bmatrix} = \begin{bmatrix} 2 \\ 8 \end{bmatrix} \qquad \begin{matrix} B \\ H \end{matrix}$$

The Hill cipher is

$$GIYUOKEVBH$$

1. **(b)** For the enciphering matrix $A = \begin{bmatrix} 4 & 3 \\ 1 & 2 \end{bmatrix}$ we have

$$\begin{bmatrix} 4 & 3 \\ 1 & 2 \end{bmatrix} \begin{bmatrix} 4 \\ 1 \end{bmatrix} = \begin{bmatrix} 19 \\ 6 \end{bmatrix} \qquad \begin{matrix} S \\ F \end{matrix}$$

$$\begin{bmatrix} 4 & 3 \\ 1 & 2 \end{bmatrix} \begin{bmatrix} 18 \\ 11 \end{bmatrix} = \begin{bmatrix} 105 \\ 40 \end{bmatrix} = \begin{bmatrix} 1 \\ 14 \end{bmatrix} \qquad \begin{matrix} A \\ N \end{matrix}$$

$$\begin{bmatrix} 4 & 3 \\ 1 & 2 \end{bmatrix} \begin{bmatrix} 14 \\ 9 \end{bmatrix} = \begin{bmatrix} 83 \\ 32 \end{bmatrix} = \begin{bmatrix} 5 \\ 6 \end{bmatrix} \qquad \begin{matrix} E \\ F \end{matrix} \qquad (\text{mod } 26)$$

$$\begin{bmatrix} 4 & 3 \\ 1 & 2 \end{bmatrix} \begin{bmatrix} 7 \\ 8 \end{bmatrix} = \begin{bmatrix} 52 \\ 23 \end{bmatrix} = \begin{bmatrix} 0 \\ 23 \end{bmatrix} \qquad \begin{matrix} Z \\ W \end{matrix}$$

$$\begin{bmatrix} 4 & 3 \\ 1 & 2 \end{bmatrix} \begin{bmatrix} 20 \\ 20 \end{bmatrix} = \begin{bmatrix} 140 \\ 60 \end{bmatrix} = \begin{bmatrix} 10 \\ 8 \end{bmatrix} \qquad \begin{matrix} J \\ H \end{matrix}$$

The Hill cipher is

$$SFANEFZWJH$$

2. **(a)** For $A = \begin{bmatrix} 9 & 1 \\ 7 & 2 \end{bmatrix}$, we have $\det(A) = 18 - 7 = 11$, which is not divisible by 2 or 13.

Therefore by Corollary 2, A is invertible. From Eq. (2):

$$A^{-1} = (11)^{-1} \begin{bmatrix} 2 & -1 \\ -7 & 9 \end{bmatrix} = 19 \begin{bmatrix} 2 & -1 \\ -7 & 9 \end{bmatrix}$$

$$= \begin{bmatrix} 38 & -19 \\ -133 & 171 \end{bmatrix} = \begin{bmatrix} 12 & 7 \\ 23 & 15 \end{bmatrix} \pmod{26}.$$

Checking:

$$AA^{-1} = \begin{bmatrix} 9 & 1 \\ 7 & 2 \end{bmatrix} \begin{bmatrix} 12 & 7 \\ 23 & 15 \end{bmatrix} = \begin{bmatrix} 131 & 78 \\ 130 & 79 \end{bmatrix} = \begin{bmatrix} 1 & 0 \\ 0 & 1 \end{bmatrix} \pmod{26}$$

$$A^{-1}A = \begin{bmatrix} 12 & 7 \\ 23 & 15 \end{bmatrix} \begin{bmatrix} 9 & 1 \\ 7 & 2 \end{bmatrix} = \begin{bmatrix} 157 & 26 \\ 312 & 53 \end{bmatrix} = \begin{bmatrix} 1 & 0 \\ 0 & 1 \end{bmatrix} \pmod{26}.$$

(b) For $A = \begin{bmatrix} 3 & 1 \\ 5 & 3 \end{bmatrix}$, we have $\det(A) = 9 - 5 = 4$, which is divisible by 2. Therefore by Corollary 2, A is not invertible.

(c) For $A = \begin{bmatrix} 8 & 11 \\ 1 & 9 \end{bmatrix}$, we have $\det(A) = 72 - 11 = 61 = 9 \pmod{26}$, which is not divisible by 2 or 13. Therefore by Corollary 2, A is invertible. From (2):

$$A^{-1} = (9)^{-1} \begin{bmatrix} 9 & -11 \\ -1 & 8 \end{bmatrix} = 3 \begin{bmatrix} 9 & -11 \\ -1 & 8 \end{bmatrix}$$

$$= \begin{bmatrix} 27 & -33 \\ -3 & 24 \end{bmatrix} = \begin{bmatrix} 1 & 19 \\ 23 & 24 \end{bmatrix} \pmod{26}.$$

Checking:

$$AA^{-1} = \begin{bmatrix} 8 & 11 \\ 1 & 9 \end{bmatrix} \begin{bmatrix} 1 & 19 \\ 23 & 24 \end{bmatrix} = \begin{bmatrix} 261 & 416 \\ 208 & 235 \end{bmatrix} = \begin{bmatrix} 1 & 0 \\ 0 & 1 \end{bmatrix} \quad (\text{mod } 26)$$

$$A^{-1}A = \begin{bmatrix} 1 & 19 \\ 23 & 24 \end{bmatrix} \begin{bmatrix} 8 & 11 \\ 1 & 9 \end{bmatrix} = \begin{bmatrix} 27 & 182 \\ 208 & 469 \end{bmatrix} = \begin{bmatrix} 1 & 0 \\ 0 & 1 \end{bmatrix} \quad (\text{mod } 26) .$$

2. **(d)** For $A = \begin{bmatrix} 2 & 1 \\ 1 & 7 \end{bmatrix}$, we have $\det(A) = 14 - 1 = 13$, which is divisible by 13. Therefore by Corollary 4, A is not invertible.

(e) For $A = \begin{bmatrix} 3 & 1 \\ 6 & 2 \end{bmatrix}$, we have $\det(A) = 6 - 6 = 0$, so that A is not invertible by Corollary 2.

(f) For $A = \begin{bmatrix} 1 & 8 \\ 1 & 3 \end{bmatrix}$, we have $\det(A) = 3 - 8 = -5 = 21 \ (\text{mod } 26)$, which is not divisible by 2 or 13. Therefore by Corollary 2, A is invertible. From (2):

$$A^{-1} = (21)^{-1} \begin{bmatrix} 3 & -8 \\ -1 & 1 \end{bmatrix} = 5 \begin{bmatrix} 3 & -8 \\ -1 & 1 \end{bmatrix}$$

$$= \begin{bmatrix} 15 & -40 \\ -5 & 5 \end{bmatrix} = \begin{bmatrix} 15 & 12 \\ 21 & 5 \end{bmatrix} \quad (\text{mod } 26) .$$

Checking:

$$AA^{-1} = \begin{bmatrix} 1 & 8 \\ 1 & 3 \end{bmatrix} \begin{bmatrix} 15 & 12 \\ 21 & 5 \end{bmatrix} = \begin{bmatrix} 183 & 52 \\ 78 & 27 \end{bmatrix} = \begin{bmatrix} 1 & 0 \\ 0 & 1 \end{bmatrix} \quad (\text{mod } 26)$$

$$A^{-1}A = \begin{bmatrix} 15 & 12 \\ 21 & 5 \end{bmatrix} \begin{bmatrix} 1 & 8 \\ 1 & 3 \end{bmatrix} = \begin{bmatrix} 27 & 156 \\ 26 & 183 \end{bmatrix} = \begin{bmatrix} 1 & 0 \\ 0 & 1 \end{bmatrix} \quad (\text{mod } 26) .$$

3. From Table 1 the numerical equivalent of this ciphertext is

$$19 \quad 1 \qquad 11 \quad 14 \qquad 15 \quad 24 \qquad 1 \quad 15 \qquad 10 \quad 24$$

Now we have to find the inverse of $A = \begin{bmatrix} 4 & 1 \\ 3 & 2 \end{bmatrix}$.

Since $\det(A) = 8 - 3 = 5$, we have from Corollary 2:

$$A^{-1} = (5)^{-1} \begin{bmatrix} 2 & -1 \\ -3 & 4 \end{bmatrix} = 21 \begin{bmatrix} 2 & -1 \\ -3 & 4 \end{bmatrix} = \begin{bmatrix} 42 & -21 \\ -63 & 84 \end{bmatrix} = \begin{bmatrix} 16 & 5 \\ 15 & 6 \end{bmatrix} \pmod{26}.$$

To obtain the plaintext, we multiply each ciphertext vector by A^{-1}:

$$\begin{bmatrix} 16 & 5 \\ 15 & 6 \end{bmatrix} \begin{bmatrix} 19 \\ 1 \end{bmatrix} = \begin{bmatrix} 309 \\ 291 \end{bmatrix} = \begin{bmatrix} 23 \\ 5 \end{bmatrix} \begin{matrix} W \\ E \end{matrix}$$

$$\begin{bmatrix} 16 & 5 \\ 15 & 6 \end{bmatrix} \begin{bmatrix} 11 \\ 14 \end{bmatrix} = \begin{bmatrix} 246 \\ 249 \end{bmatrix} = \begin{bmatrix} 12 \\ 15 \end{bmatrix} \begin{matrix} L \\ O \end{matrix}$$

$$\begin{bmatrix} 16 & 5 \\ 15 & 6 \end{bmatrix} \begin{bmatrix} 15 \\ 24 \end{bmatrix} = \begin{bmatrix} 360 \\ 369 \end{bmatrix} = \begin{bmatrix} 22 \\ 5 \end{bmatrix} \begin{matrix} V \\ E \end{matrix} \qquad \pmod{26}$$

$$\begin{bmatrix} 16 & 5 \\ 15 & 6 \end{bmatrix} \begin{bmatrix} 1 \\ 15 \end{bmatrix} = \begin{bmatrix} 91 \\ 105 \end{bmatrix} = \begin{bmatrix} 13 \\ 1 \end{bmatrix} \begin{matrix} M \\ A \end{matrix}$$

$$\begin{bmatrix} 16 & 5 \\ 15 & 6 \end{bmatrix} \begin{bmatrix} 10 \\ 24 \end{bmatrix} = \begin{bmatrix} 280 \\ 294 \end{bmatrix} = \begin{bmatrix} 20 \\ 8 \end{bmatrix} \begin{matrix} T \\ H \end{matrix}$$

The plaintext is thus

$$WE\ LOVE\ MATH$$

4. From Table 1 the numerical equivalent of the known plaintext is

$$AR \qquad\qquad MY$$
$$1 \qquad 18 \qquad 13 \qquad 25$$

and the numerical equivalent of the corresponding ciphertext is

$$SL \qquad\qquad HK$$
$$19 \qquad 12 \qquad 8 \qquad 11$$

so the corresponding plaintext and ciphertext vectors are

$$\mathbf{p}_1 = \begin{bmatrix} 1 \\ 18 \end{bmatrix} \longleftrightarrow \mathbf{c}_1 = \begin{bmatrix} 19 \\ 12 \end{bmatrix}$$

$$\mathbf{p}_2 = \begin{bmatrix} 13 \\ 25 \end{bmatrix} \longleftrightarrow \mathbf{c}_2 = \begin{bmatrix} 8 \\ 11 \end{bmatrix}$$

We want to reduce

$$C = \begin{bmatrix} \mathbf{c}_1^t \\ \mathbf{c}_2^t \end{bmatrix} = \begin{bmatrix} 19 & 12 \\ 8 & 11 \end{bmatrix}$$

to I by elementary row operations and simultaneously apply these operations to

$$P = \begin{bmatrix} \mathbf{p}_1^t \\ \mathbf{p}_2^t \end{bmatrix} = \begin{bmatrix} 1 & 18 \\ 13 & 25 \end{bmatrix}.$$

The calculations are as follows:

$$\left[\begin{array}{cc|cc} 19 & 12 & 1 & 18 \\ 8 & 11 & 13 & 25 \end{array}\right] \qquad \boxed{\text{We formed the matrix } \left[C \mid P\right].}$$

$$\left[\begin{array}{cc|cc} 1 & 132 & 11 & 198 \\ 8 & 11 & 13 & 25 \end{array}\right] \qquad \boxed{\begin{array}{l} \text{We multiplied the first row by} \\ 19^{-1} = 11 \;(\text{mod } 26)\,. \end{array}}$$

$$\left[\begin{array}{cc|cc} 1 & 2 & 11 & 16 \\ 8 & 11 & 13 & 25 \end{array}\right]$$

We replaced 132 and 198 by their residues modulo 26 .

$$\left[\begin{array}{cc|cc} 1 & 2 & 11 & 16 \\ 0 & -5 & -75 & -103 \end{array}\right]$$

We added -8 times the first row to the second.

$$\left[\begin{array}{cc|cc} 1 & 2 & 11 & 16 \\ 0 & 21 & 3 & 1 \end{array}\right]$$

We replaced the entries in the second row by their residues modulo 26.

$$\left[\begin{array}{cc|cc} 1 & 2 & 11 & 16 \\ 0 & 1 & 15 & 5 \end{array}\right]$$

We multiplied the second row by $21^{-1} = 5 \pmod{26}$.

$$\left[\begin{array}{cc|cc} 1 & 0 & -19 & 6 \\ 0 & 1 & 15 & 5 \end{array}\right]$$

We added -2 times the second row to the first.

$$\left[\begin{array}{cc|cc} 1 & 0 & 7 & 6 \\ 0 & 1 & 15 & 5 \end{array}\right]$$

We replaced -19 by its residue modulo 26.

Thus $\left(A^{-1}\right)^{t} = \begin{bmatrix} 7 & 6 \\ 15 & 5 \end{bmatrix}$ so the deciphering matrix is

$$A^{-1} = \begin{bmatrix} 7 & 15 \\ 6 & 5 \end{bmatrix} \pmod{26} .$$

Since $\det\left(A^{-1}\right) = 35 - 90 = -55 = 23 \pmod{26}$, we have

$$A = \left(A^{-1}\right)^{-1} = (23)^{-1} \begin{bmatrix} 5 & -15 \\ -6 & 7 \end{bmatrix} = 17 \begin{bmatrix} 5 & -15 \\ -6 & 7 \end{bmatrix}$$

$$= \begin{bmatrix} 85 & -255 \\ -102 & 119 \end{bmatrix} = \begin{bmatrix} 7 & 5 \\ 2 & 15 \end{bmatrix} \pmod{26}$$

for the enciphering matrix.

5. From Table 1 the numerical equivalent of the known plaintext is

$$\begin{array}{cccc} AT & & OM & \\ 1 & 20 & 15 & 13 \end{array}$$

and the numerical equivalent of the corresponding ciphertext is

$$\begin{array}{cccc} JY & & QO & \\ 10 & 25 & 17 & 15 \end{array}$$

The corresponding plaintext and ciphertext vectors are:

$$\mathbf{p}_1 = \begin{bmatrix} 1 \\ 20 \end{bmatrix} \qquad \longleftrightarrow \qquad \mathbf{c}_1 = \begin{bmatrix} 10 \\ 25 \end{bmatrix}$$

$$\mathbf{p}_2 = \begin{bmatrix} 15 \\ 13 \end{bmatrix} \qquad \longleftrightarrow \qquad \mathbf{c}_2 = \begin{bmatrix} 17 \\ 15 \end{bmatrix}$$

We want to reduce

$$C = \begin{bmatrix} 10 & 25 \\ 17 & 15 \end{bmatrix}$$

to I by elementary row operations and simultaneously apply these operations to

$$P = \begin{bmatrix} 1 & 20 \\ 15 & 13 \end{bmatrix}.$$

The calculations are as follows:

$$\left[\begin{array}{cc|cc} 10 & 25 & 1 & 20 \\ 17 & 15 & 15 & 13 \end{array} \right]$$
     We formed the matrix $\left[C \mid P \right]$.

$$\left[\begin{array}{cc|cc} 27 & 40 & 16 & 33 \\ 17 & 15 & 15 & 13 \end{array} \right]$$
     We added the second row to the first (since 10^{-1} does not exist mod 26).

$$\left[\begin{array}{cc|cc} 1 & 14 & 16 & 7 \\ 17 & 15 & 15 & 13 \end{array}\right]$$
We replaced the entries in the first row by their residues modulo 26.

$$\left[\begin{array}{cc|cc} 1 & 14 & 16 & 7 \\ 0 & -223 & -257 & -106 \end{array}\right]$$
We added -17 times the first row to the second.

$$\left[\begin{array}{cc|cc} 1 & 14 & 16 & 7 \\ 0 & 11 & 3 & 24 \end{array}\right]$$
We replaced the entries in the second row by their residues modulo 26.

$$\left[\begin{array}{cc|cc} 1 & 14 & 16 & 7 \\ 0 & 1 & 57 & 456 \end{array}\right]$$
We multiplied the second row by $11^{-1} = 19 \pmod{26}$.

$$\left[\begin{array}{cc|cc} 1 & 14 & 16 & 7 \\ 0 & 1 & 5 & 14 \end{array}\right]$$
We replaced the entries in the second row by their residues modulo 26 .

$$\left[\begin{array}{cc|cc} 1 & 0 & -54 & -189 \\ 0 & 1 & 5 & 14 \end{array}\right]$$
We added -14 times the second row to the first.

$$\left[\begin{array}{cc|cc} 1 & 0 & 24 & 19 \\ 0 & 1 & 5 & 14 \end{array}\right]$$
We replaced -54 and -189 by their residues modulo 26.

Thus $\left(A^{-1}\right)^{t} = \left[\begin{array}{cc} 24 & 19 \\ 5 & 14 \end{array}\right]$, and so the deciphering matrix is

$$A^{-1} = \left[\begin{array}{cc} 24 & 5 \\ 19 & 14 \end{array}\right].$$

From Table 1 the numerical equivalent of the given ciphertext is

LN		GI		HG		YB		VR		EN		JY		QO	
12	14	7	9	8	7	25	2	22	18	5	14	10	25	17	15

To obtain the plaintext pairs, we multiply each ciphertext vector by A^{-1}:

$$\begin{bmatrix} 24 & 5 \\ 19 & 14 \end{bmatrix} \begin{bmatrix} 12 \\ 14 \end{bmatrix} = \begin{bmatrix} 358 \\ 424 \end{bmatrix} = \begin{bmatrix} 20 \\ 8 \end{bmatrix} \begin{matrix} T \\ H \end{matrix}$$

$$\begin{bmatrix} 24 & 5 \\ 19 & 14 \end{bmatrix} \begin{bmatrix} 7 \\ 9 \end{bmatrix} = \begin{bmatrix} 213 \\ 259 \end{bmatrix} = \begin{bmatrix} 5 \\ 25 \end{bmatrix} \begin{matrix} E \\ Y \end{matrix}$$

$$\begin{bmatrix} 24 & 5 \\ 19 & 14 \end{bmatrix} \begin{bmatrix} 8 \\ 7 \end{bmatrix} = \begin{bmatrix} 227 \\ 250 \end{bmatrix} = \begin{bmatrix} 19 \\ 16 \end{bmatrix} \begin{matrix} S \\ P \end{matrix}$$

$$\begin{bmatrix} 24 & 5 \\ 19 & 14 \end{bmatrix} \begin{bmatrix} 25 \\ 2 \end{bmatrix} = \begin{bmatrix} 610 \\ 503 \end{bmatrix} = \begin{bmatrix} 12 \\ 9 \end{bmatrix} \begin{matrix} L \\ I \end{matrix} \quad (\bmod\ 26)$$

$$\begin{bmatrix} 24 & 5 \\ 19 & 14 \end{bmatrix} \begin{bmatrix} 22 \\ 18 \end{bmatrix} = \begin{bmatrix} 618 \\ 670 \end{bmatrix} = \begin{bmatrix} 20 \\ 20 \end{bmatrix} \begin{matrix} T \\ T \end{matrix}$$

$$\begin{bmatrix} 24 & 5 \\ 19 & 14 \end{bmatrix} \begin{bmatrix} 5 \\ 14 \end{bmatrix} = \begin{bmatrix} 190 \\ 291 \end{bmatrix} = \begin{bmatrix} 8 \\ 5 \end{bmatrix} \begin{matrix} H \\ E \end{matrix}$$

$$\begin{bmatrix} 24 & 5 \\ 19 & 14 \end{bmatrix} \begin{bmatrix} 10 \\ 25 \end{bmatrix} = \begin{bmatrix} 365 \\ 540 \end{bmatrix} = \begin{bmatrix} 1 \\ 20 \end{bmatrix} \begin{matrix} A \\ T \end{matrix}$$

$$\begin{bmatrix} 24 & 5 \\ 19 & 14 \end{bmatrix} \begin{bmatrix} 17 \\ 15 \end{bmatrix} = \begin{bmatrix} 483 \\ 533 \end{bmatrix} = \begin{bmatrix} 15 \\ 13 \end{bmatrix} \begin{matrix} O \\ M \end{matrix}$$

which yields the message

THEY SPLIT THE ATOM

6. Since we want a Hill 3-cipher, we will group the letters in triples. From Table 1 the numerical equivalent of the known plaintext are

I	H	A	V	E	C	O	M	E
9	8	1	22	5	3	15	13	5

and the numerical eqiuvalents of the corresponding ciphertext are

$$
\begin{array}{ccccccccc}
H & P & A & F & Q & G & G & D & U \\
8 & 16 & 1 & 6 & 17 & 7 & 7 & 4 & 21
\end{array}
$$

The corresponding plaintext and ciphertext are

$$
\mathbf{p}_1 = \begin{bmatrix} 9 \\ 8 \\ 1 \end{bmatrix} \qquad \longleftrightarrow \qquad \mathbf{c}_1 = \begin{bmatrix} 8 \\ 16 \\ 1 \end{bmatrix}
$$

$$
\mathbf{p}_2 = \begin{bmatrix} 22 \\ 5 \\ 3 \end{bmatrix} \qquad \longleftrightarrow \qquad \mathbf{c}_2 = \begin{bmatrix} 6 \\ 17 \\ 7 \end{bmatrix}
$$

$$
\mathbf{p}_3 = \begin{bmatrix} 15 \\ 13 \\ 5 \end{bmatrix} \qquad \longleftrightarrow \qquad \mathbf{c}_3 = \begin{bmatrix} 7 \\ 4 \\ 21 \end{bmatrix}
$$

We want to reduce

$$
C = \begin{bmatrix} 8 & 16 & 1 \\ 6 & 17 & 7 \\ 7 & 4 & 21 \end{bmatrix}
$$

to I by elementary row operations and simultaneously apply these operations to

$$
P = \begin{bmatrix} 9 & 8 & 1 \\ 22 & 5 & 3 \\ 15 & 13 & 5 \end{bmatrix}.
$$

The calculations are as follows:

$$
\left[\begin{array}{ccc|ccc}
8 & 16 & 1 & 9 & 8 & 1 \\
6 & 17 & 7 & 22 & 5 & 3 \\
7 & 4 & 21 & 15 & 13 & 5
\end{array} \right]
$$

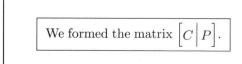

We formed the matrix $\begin{bmatrix} C & | & P \end{bmatrix}$.

$$\begin{bmatrix} 15 & 20 & 22 & \bigm| & 24 & 21 & 6 \\ 6 & 17 & 7 & \bigm| & 22 & 5 & 3 \\ 7 & 4 & 21 & \bigm| & 15 & 13 & 5 \end{bmatrix}$$

We added the third row to the first since 8^{-1} does not exist modulo 26.

$$\begin{bmatrix} 1 & 140 & 154 & \bigm| & 168 & 147 & 42 \\ 6 & 17 & 7 & \bigm| & 22 & 5 & 3 \\ 7 & 4 & 21 & \bigm| & 15 & 13 & 5 \end{bmatrix}$$

We multiplied the first row by $15^{-1} = 7 \pmod{26}$.

$$\begin{bmatrix} 1 & 10 & 24 & \bigm| & 12 & 17 & 16 \\ 6 & 17 & 7 & \bigm| & 22 & 5 & 3 \\ 7 & 4 & 21 & \bigm| & 15 & 13 & 5 \end{bmatrix}$$

We replaced the entries in the first row by their residues modulo 26.

$$\begin{bmatrix} 1 & 10 & 24 & \bigm| & 12 & 17 & 16 \\ 0 & -43 & -137 & \bigm| & -50 & -97 & -93 \\ 0 & -66 & -147 & \bigm| & -69 & -106 & -107 \end{bmatrix}$$

We added -6 times the first row to the second and -7 times the first row to the third.

$$\begin{bmatrix} 1 & 10 & 24 & \bigm| & 12 & 17 & 16 \\ 0 & 9 & 19 & \bigm| & 2 & 7 & 11 \\ 0 & 12 & 9 & \bigm| & 9 & 24 & 23 \end{bmatrix}$$

We replaced the entries in the second and third rows by their residues modulo 26.

$$\begin{bmatrix} 1 & 10 & 24 & \bigm| & 12 & 17 & 16 \\ 0 & 1 & 57 & \bigm| & 6 & 21 & 33 \\ 0 & 12 & 9 & \bigm| & 9 & 24 & 23 \end{bmatrix}$$

We multiplied the second row by $9^{-1} = 3$ modulo 26.

$$\begin{bmatrix} 1 & 10 & 24 & \bigm| & 12 & 17 & 16 \\ 0 & 1 & 5 & \bigm| & 6 & 21 & 7 \\ 0 & 12 & 9 & \bigm| & 9 & 24 & 23 \end{bmatrix}$$

We replaced the entries in the second row by their residues modulo 26.

$$\begin{bmatrix} 1 & 0 & -26 & \bigm| & -48 & -193 & -54 \\ 0 & 1 & 5 & \bigm| & 6 & 21 & 7 \\ 0 & 0 & -51 & \bigm| & -63 & -228 & -61 \end{bmatrix}$$

We added -10 times the second row to the first and -12 times the second row to the third.

$$
\left[\begin{array}{ccc|ccc}
1 & 0 & 0 & 4 & 15 & 24 \\
0 & 1 & 5 & 6 & 21 & 7 \\
0 & 0 & 1 & 15 & 6 & 17
\end{array}\right]
$$

We replaced the entries in the first and second row by their residues modulo 26.

$$
\left[\begin{array}{ccc|ccc}
1 & 0 & 0 & 4 & 15 & 24 \\
0 & 1 & 0 & -69 & -9 & -78 \\
0 & 0 & 1 & 15 & 6 & 17
\end{array}\right]
$$

We added -5 times the third row to the second.

$$
\left[\begin{array}{ccc|ccc}
1 & 0 & 0 & 4 & 15 & 24 \\
0 & 1 & 0 & 9 & 17 & 0 \\
0 & 0 & 1 & 15 & 6 & 17
\end{array}\right]
$$

We replaced the entries in the second row by their residues modulo 26.

Thus,

$$
\left(A^{-1}\right)^{t} = \left[\begin{array}{ccc}
4 & 15 & 24 \\
9 & 17 & 0 \\
15 & 6 & 17
\end{array}\right]
$$

and so the deciphering matrix is

$$
A^{-1} = \left[\begin{array}{ccc}
4 & 9 & 15 \\
15 & 17 & 6 \\
24 & 0 & 17
\end{array}\right].
$$

From Table 1 the numerical equivalent of the given ciphertext is

H	P	A	F	Q	G	G	D	U	G	D	D
8	16	1	6	17	7	7	4	21	7	4	4

H	P	G	O	D	Y	N	O	R
8	16	7	15	4	25	14	15	18

To obtain the plaintext triples, we multiply each ciphertext vector by A^{-1}:

$$\begin{bmatrix} 4 & 9 & 15 \\ 15 & 17 & 6 \\ 24 & 0 & 17 \end{bmatrix} \begin{bmatrix} 8 \\ 16 \\ 1 \end{bmatrix} = \begin{bmatrix} 191 \\ 398 \\ 209 \end{bmatrix} = \begin{bmatrix} 9 \\ 8 \\ 1 \end{bmatrix} \begin{matrix} I \\ H \\ A \end{matrix}$$

$$\begin{bmatrix} 4 & 9 & 15 \\ 15 & 17 & 6 \\ 24 & 0 & 17 \end{bmatrix} \begin{bmatrix} 6 \\ 17 \\ 7 \end{bmatrix} = \begin{bmatrix} 282 \\ 421 \\ 263 \end{bmatrix} = \begin{bmatrix} 22 \\ 5 \\ 3 \end{bmatrix} \begin{matrix} V \\ E \\ C \end{matrix}$$

$$\begin{bmatrix} 4 & 9 & 15 \\ 15 & 17 & 6 \\ 24 & 0 & 17 \end{bmatrix} \begin{bmatrix} 7 \\ 4 \\ 21 \end{bmatrix} = \begin{bmatrix} 379 \\ 299 \\ 525 \end{bmatrix} = \begin{bmatrix} 15 \\ 13 \\ 5 \end{bmatrix} \begin{matrix} O \\ M \\ E \end{matrix}$$

$$\begin{bmatrix} 4 & 9 & 15 \\ 15 & 17 & 6 \\ 24 & 0 & 17 \end{bmatrix} \begin{bmatrix} 7 \\ 4 \\ 4 \end{bmatrix} = \begin{bmatrix} 124 \\ 197 \\ 236 \end{bmatrix} = \begin{bmatrix} 20 \\ 15 \\ 2 \end{bmatrix} \begin{matrix} T \\ O \\ B \end{matrix}$$

$$(\text{mod } 26)$$

$$\begin{bmatrix} 4 & 9 & 15 \\ 15 & 17 & 6 \\ 24 & 0 & 17 \end{bmatrix} \begin{bmatrix} 8 \\ 16 \\ 7 \end{bmatrix} = \begin{bmatrix} 281 \\ 434 \\ 311 \end{bmatrix} = \begin{bmatrix} 21 \\ 18 \\ 25 \end{bmatrix} \begin{matrix} U \\ R \\ Y \end{matrix}$$

$$\begin{bmatrix} 4 & 9 & 15 \\ 15 & 17 & 6 \\ 24 & 0 & 17 \end{bmatrix} \begin{bmatrix} 15 \\ 4 \\ 25 \end{bmatrix} = \begin{bmatrix} 471 \\ 443 \\ 785 \end{bmatrix} = \begin{bmatrix} 3 \\ 1 \\ 5 \end{bmatrix} \begin{matrix} C \\ A \\ E \end{matrix}$$

$$\begin{bmatrix} 4 & 9 & 15 \\ 15 & 17 & 6 \\ 24 & 0 & 17 \end{bmatrix} \begin{bmatrix} 14 \\ 15 \\ 18 \end{bmatrix} = \begin{bmatrix} 461 \\ 573 \\ 642 \end{bmatrix} = \begin{bmatrix} 19 \\ 1 \\ 18 \end{bmatrix} \begin{matrix} S \\ A \\ R \end{matrix}$$

Finally, the message is

I HAVE COME TO BURY CAESAR

7. Since 29 is a prime number, by Corollary 1 a matrix A with entries in Z_{29} is invertible if and only if $\det(A) \neq 0 \pmod{29}$.

8. Testing $x = 0, 1, 2, \ldots, 25$ in $4x = 1 \pmod{26}$ gives

$$
\begin{aligned}
4(\ 0) &= \quad 0 \qquad\ \neq 1 \\
4(\ 1) &= \quad 4 \qquad\ \neq 1 \\
4(\ 2) &= \quad 8 \qquad\ \neq 1 \\
4(\ 3) &= \ 12 \qquad\ \neq 1 \\
4(\ 4) &= \ 16 \qquad\ \neq 1 \\
4(\ 5) &= \ 20 \qquad\ \neq 1 \\
4(\ 6) &= \ 24 \qquad\ \neq 1 \\
4(\ 7) &= \ 28 = \ \ 2 \neq 1 \\
4(\ 8) &= \ 32 = \ \ 6 \neq 1 \\
4(\ 9) &= \ 36 = 10 \neq 1 \\
4(10) &= \ 40 = 14 \neq 1 \qquad (\text{mod } 26) \\
4(11) &= \ 44 = 18 \neq 1 \\
4(12) &= \ 48 = 22 \neq 1 \\
4(13) &= \ 52 = \ \ 0 \neq 1 \\
4(14) &= \ 56 = \ \ 4 \neq 1 \\
4(15) &= \ 60 = \ \ 8 \neq 1 \\
4(16) &= \ 64 = 12 \neq 1 \\
4(17) &= \ 68 = 16 \neq 1 \\
4(18) &= \ 72 = 20 \neq 1 \\
4(19) &= \ 76 = 24 \neq 1 \\
4(20) &= \ 80 = \ \ 2 \neq 1 \\
4(21) &= \ 84 = \ \ 6 \neq 1 \\
4(22) &= \ 88 = 10 \neq 1 \\
4(23) &= \ 92 = 14 \neq 1 \\
4(24) &= \ 96 = 18 \neq 1 \\
4(25) &= 100 = 22 \neq 1
\end{aligned}
$$

9. **(a)** We have

$$P = \begin{bmatrix} \mathbf{p}_1^t \\ \mathbf{p}_2^t \\ \vdots \\ \mathbf{p}_n^t \end{bmatrix} \qquad C = \begin{bmatrix} \mathbf{c}_1^t \\ \mathbf{c}_2^t \\ \vdots \\ \mathbf{c}_n^t \end{bmatrix}$$

where $A\mathbf{p}_i = \mathbf{c}_i$ $\qquad i = 1, 2, \ldots, n.$

Assuming A^{-1} exists, we can write

$$\mathbf{p}_i = A^{-1}\mathbf{c}_i \qquad\qquad i = 1, 2, \ldots, n$$

or $\qquad\qquad \mathbf{p}_i^t = \mathbf{c}_i^t \left(A^{-1}\right)^t \qquad\qquad i = 1, 2, \ldots, n$

which in matrix form is

$$P = C \left(A^{-1}\right)^t.$$

(b) From part (a), $P = C \left(A^{-1}\right)^t$. Multiply both sides by $E_n \cdots E_2 E_1$ from the left to obtain

$$E_n \cdots E_2 E_1 P = E_n \cdots E_2 E_1 C \left(A^{-1}\right)^t.$$

But $E_n \cdots E_2 E_1 C = I$, so that

$$E_n \cdots E_2 E_1 P = \left(A^{-1}\right)^t.$$

10. **(a)** We have $P = \begin{bmatrix} \mathbf{p}_1^t \\ \mathbf{p}_2^t \\ \vdots \\ \mathbf{p}_n^t \end{bmatrix}$ $\qquad C = \begin{bmatrix} \mathbf{c}_1^t \\ \mathbf{c}_2^t \\ \vdots \\ \mathbf{c}_n^t \end{bmatrix}$

where $A\mathbf{p}_i = \mathbf{c}_i$ $\qquad (\bmod\ 26)$

or $\qquad \mathbf{p}_i = A^{-1}\mathbf{c}_i \qquad (\bmod\ 26)$

or $\qquad \mathbf{p}_i^t = \mathbf{c}_i^t \left(A^{-1}\right)^t \qquad (\bmod\ 26), \qquad i = 1, 2, \ldots, n$

which in matrix form is

$$P = C \left(A^{-1} \right)^t \qquad (\mathrm{mod}\ 26) \ .$$

Multiplying both sides by C^{-1} gives

$$C^{-1}P = \left(A^{-1} \right)^t$$

or

$$A^{-1} = \left(C^{-1}P \right)^t \qquad (\mathrm{mod}\ 26) \ .$$

(b) From Example 8, we have

$$P = \begin{bmatrix} 4 & 5 \\ 1 & 18 \end{bmatrix}, \qquad C = \begin{bmatrix} 9 & 15 \\ 19 & 2 \end{bmatrix} .$$

Using Eq. (2) to find C^{-1} we have

$$C^{-1} = (18 - 285)^{-1} \begin{bmatrix} 2 & -15 \\ -19 & 9 \end{bmatrix} = (-267)^{-1} \begin{bmatrix} 2 & -15 \\ -19 & 9 \end{bmatrix}$$

$$= 19^{-1} \begin{bmatrix} 2 & -15 \\ -19 & 9 \end{bmatrix} = 11 \begin{bmatrix} 2 & -15 \\ -19 & 9 \end{bmatrix}$$

$$= \begin{bmatrix} 22 & -165 \\ -209 & 99 \end{bmatrix} = \begin{bmatrix} 22 & 17 \\ 25 & 21 \end{bmatrix} \qquad (\mathrm{mod}\ 26),$$

and

$$C^{-1}P = \begin{bmatrix} 22 & 17 \\ 25 & 21 \end{bmatrix} \begin{bmatrix} 4 & 5 \\ 1 & 18 \end{bmatrix}$$

$$= \begin{bmatrix} 105 & 416 \\ 121 & 503 \end{bmatrix} = \begin{bmatrix} 1 & 0 \\ 17 & 9 \end{bmatrix} \qquad (\mathrm{mod}\ 26) \ .$$

Thus from part (a):

$$A^{-1} = \left(C^{-1}P\right)^t = \begin{bmatrix} 1 & 17 \\ 0 & 9 \end{bmatrix}.$$

EXERCISE SET 11.17

1. Use induction on n, the case $n = 1$ being already given. If the result is true for $n - 1$, then $M^n = M^{n-1}M = (PD^{n-1}P^{-1})(PDP^{-1}) = PD^{n-1}(P^{-1}P)DP^{-1} = PD^{n-1}DP^{-1} = PD^nP^{-1}$, proving the result.

2. Using Table 1 and the notations of Example 1, we derive the following equations:

$$a_n = \frac{1}{2}a_{n-1} + \frac{1}{4}b_{n-1}$$

$$b_n = \frac{1}{2}a_{n-1} + \frac{1}{2}b_{n-1} + \frac{1}{2}c_{n-1}$$

$$c_n = \frac{1}{4}b_{n-1} + \frac{1}{2}c_{n-1}\ .$$

The transition matrix is thus

$$M = \begin{bmatrix} \dfrac{1}{2} & \dfrac{1}{4} & 0 \\[2mm] \dfrac{1}{2} & \dfrac{1}{2} & \dfrac{1}{2} \\[2mm] 0 & \dfrac{1}{4} & \dfrac{1}{2} \end{bmatrix}.$$

The characteristic polynomial of M is $\det(\lambda I - M) = \lambda^3 - (3/2)\lambda^2 + (1/2)\lambda = \lambda(\lambda - 1)\left(\lambda - \dfrac{1}{2}\right)$, so the eigenvalues of M are $\lambda = 1$, $\lambda_2 = \dfrac{1}{2}$ and $\lambda_3 = 0$. Corresponding

eigenvectors (found by solving $(\lambda I - M)\mathbf{x} = 0$) are $\mathbf{e}_1 = \begin{bmatrix} 1 & 2 & 1 \end{bmatrix}^t$, $\mathbf{e}_2 = \begin{bmatrix} 1 & 0 & -1 \end{bmatrix}^t$ and $\mathbf{e}_3 = \begin{bmatrix} 1 & -2 & 1 \end{bmatrix}^t$. Thus

$$
M^n = PD^nP^{-1} = \begin{bmatrix} 1 & 1 & 1 \\ 2 & 0 & -2 \\ 1 & -1 & 1 \end{bmatrix} \begin{bmatrix} 1 & 0 & 0 \\ 0 & \left(\dfrac{1}{2}\right)^n & 0 \\ 0 & 0 & 0 \end{bmatrix} \begin{bmatrix} \dfrac{1}{4} & \dfrac{1}{4} & \dfrac{1}{4} \\ \dfrac{1}{2} & 0 & -\dfrac{1}{2} \\ \dfrac{1}{4} & -\dfrac{1}{4} & \dfrac{1}{4} \end{bmatrix}.
$$

This yields

$$
\mathbf{x}^{(n)} = \begin{bmatrix} a_n \\ b_n \\ c_n \end{bmatrix} = \begin{bmatrix} \dfrac{1}{4} + \left(\dfrac{1}{2}\right)^{n+1} & \dfrac{1}{4} & \dfrac{1}{4} - \left(\dfrac{1}{2}\right)^{n+1} \\ \dfrac{1}{2} & \dfrac{1}{2} & \dfrac{1}{2} \\ \dfrac{1}{4} - \left(\dfrac{1}{2}\right)^{n+1} & \dfrac{1}{4} & \dfrac{1}{4} + \left(\dfrac{1}{2}\right)^{n+1} \end{bmatrix} \begin{bmatrix} a_0 \\ b_0 \\ c_0 \end{bmatrix}.
$$

Remembering $a_0 + b_0 + c_0 = 1$, we obtain

$$
a_n = \frac{1}{4}a_0 + \frac{1}{4}b_0 + \frac{1}{4}c_0 + \left(\frac{1}{2}\right)^{n+1}a_0 - \left(\frac{1}{2}\right)^{n+1}c_0
$$

$$
= \frac{1}{4} + \left(\frac{1}{2}\right)^{n+1}(a_0 - c_0)
$$

$$
b_n = \frac{1}{2}a_0 + \frac{1}{2}b_0 + \frac{1}{2}c_0 = \frac{1}{2}
$$

$$
c_n = \frac{1}{4}a_0 + \frac{1}{4}b_0 + \frac{1}{2}c_0 - \left(\frac{1}{2}\right)^{n+1}a_0 + \left(\frac{1}{2}\right)^{n+1}c_0
$$

$$
= \frac{1}{4} - \left(\frac{1}{2}\right)^{n+1}(a_0 - c_0).
$$

Since $\left(\dfrac{1}{2}\right)^{n+1}$ approaches zero as $n \to \infty$, we obtain $a_n \to \dfrac{1}{4}$, $b_n \to \dfrac{1}{2}$, and $c_n \to \dfrac{1}{4}$ as $n \to \infty$.

3. Call M_1 the matrix of Example 1, and M_2 the matrix of Exercise 2. Then $\mathbf{x}^{(2n)} = (M_2 M_1)^n \mathbf{x}^{(0)}$ and $\mathbf{x}^{(2n+1)} = M_1 (M_2 M_1)^n \mathbf{x}^{(0)}$. We have

$$M_2 M_1 = \begin{bmatrix} \dfrac{1}{2} & \dfrac{1}{4} & 0 \\[2mm] \dfrac{1}{2} & \dfrac{1}{2} & \dfrac{1}{2} \\[2mm] 0 & \dfrac{1}{4} & \dfrac{1}{2} \end{bmatrix} \begin{bmatrix} 1 & \dfrac{1}{2} & 0 \\[2mm] 0 & \dfrac{1}{2} & 1 \\[2mm] 0 & 0 & 0 \end{bmatrix} = \begin{bmatrix} \dfrac{1}{2} & \dfrac{3}{8} & \dfrac{1}{4} \\[2mm] \dfrac{1}{2} & \dfrac{1}{2} & \dfrac{1}{2} \\[2mm] 0 & \dfrac{1}{8} & \dfrac{1}{4} \end{bmatrix}.$$

The characteristic polynomial of this matrix is $\lambda^3 - \dfrac{5}{4}\lambda^2 + \dfrac{1}{4}\lambda$, so the eigenvalues are $\lambda_1 = 1$, $\lambda_2 = \dfrac{1}{4}$, $\lambda_3 = 0$. Corresponding eigenvectors are $\mathbf{e}_1 = \begin{bmatrix} 5 & 6 & 1 \end{bmatrix}^t$, $\mathbf{e}_2 = \begin{bmatrix} -1 & 0 & 1 \end{bmatrix}^t$, and $\mathbf{e}_3 = \begin{bmatrix} 1 & -2 & 1 \end{bmatrix}^t$. Thus,

$$(M_2 M_1)^n = P D^n P^{-1} = \begin{bmatrix} 5 & -1 & 1 \\ 6 & 0 & -2 \\ 1 & 1 & 1 \end{bmatrix} \begin{bmatrix} 1 & 0 & 0 \\ 0 & \left(\dfrac{1}{4}\right)^n & 0 \\ 0 & 0 & 0 \end{bmatrix} \begin{bmatrix} \dfrac{1}{12} & \dfrac{1}{12} & \dfrac{1}{12} \\[2mm] -\dfrac{1}{3} & \dfrac{1}{6} & \dfrac{2}{3} \\[2mm] \dfrac{1}{4} & -\dfrac{1}{4} & \dfrac{1}{4} \end{bmatrix}$$

Using the notation of Example 1 (recall $a_0 + b_0 + c_0 = 1$), we obtain

$$a_{2n} = \frac{5}{12} + \frac{1}{6 \cdot 4^n}(2a_0 - b_0 + 4c_0)$$

$$b_{2n} = \frac{1}{2}$$

$$c_{2n} = \frac{1}{12} - \frac{1}{6 \cdot 4^n}(2a_0 - b_0 + 4c_0).$$

and

$$a_{2n+1} = \frac{2}{3} + \frac{1}{6 \cdot 4^n}(2a_0 - b_0 - 4c_0)$$

$$b_{2n+1} = \frac{1}{3} - \frac{1}{6 \cdot 4^n}(2a_0 - b_0 - 4c_0)$$

$$c_{2n+1} = 0.$$

4. The characteristic polynomial of M is $(\lambda - 1)\left(\lambda - \dfrac{1}{2}\right)$, so the eigenvalues are $\lambda_1 = 1$ and $\lambda_2 = \dfrac{1}{2}$. Corresponding eigenvectors are easily found to be $\mathbf{e}_1 = \begin{bmatrix} 1 & 0 \end{bmatrix}^t$ and $\mathbf{e}_2 = \begin{bmatrix} 1 & -1 \end{bmatrix}^t$. From this point, the verification of Eq. (7) is in the text.

5. From Eq. (9), if $b_0 = .25 = \dfrac{1}{4}$, we get $b_1 = \dfrac{1/4}{9/8} = \dfrac{2}{9}$, then $b_2 = \dfrac{2/9}{10/9} = \dfrac{1}{5}$, $b_3 = \dfrac{1/5}{11/10} = \dfrac{2}{11}$, and in general $b_n = \dfrac{2}{8 + n}$. We will reach $\dfrac{2}{20} = .10$ in 12 generations. According to Eq. (8), under the controlled program the percentage would be $\dfrac{1}{2^{14}}$ in 12 generations, or $\dfrac{1}{16384} = .00006 = .006\%$.

6.

$$P^{-1}\mathbf{x}^{(0)} = \begin{bmatrix} \dfrac{1}{2} & \dfrac{1}{2} & 0 & 0 & \dfrac{1}{2} + \dfrac{\sqrt{5}}{2} & \dfrac{1}{2} - \dfrac{\sqrt{5}}{2} \end{bmatrix}^t$$

$$D^n P^{-1}\mathbf{x}^{(0)} = \begin{bmatrix} \dfrac{1}{2} & \dfrac{1}{2} & 0 & 0 & 2\left(\dfrac{1 + \sqrt{5}}{4}\right)^{n+1} & 2\left(\dfrac{1 - \sqrt{5}}{4}\right)^{n+1} \end{bmatrix}^t$$

$$PD^n P^{-1}\mathbf{x}^{(0)} = \begin{bmatrix} \dfrac{1}{2} - \dfrac{1}{2}(3 + \sqrt{5})\left(\dfrac{1 + \sqrt{5}}{4}\right)^{n+1} - \dfrac{1}{2}(3 - \sqrt{5})\left(\dfrac{1 - \sqrt{5}}{4}\right)^{n+1} \\[2ex] 2\left(\dfrac{1 + \sqrt{5}}{4}\right)^{n+1} + 2\left(\dfrac{1 - \sqrt{5}}{4}\right)^{n+1} \\[2ex] \dfrac{1}{2}\left(\dfrac{1 + \sqrt{5}}{4}\right)^{n} + \dfrac{1}{2}\left(\dfrac{1 - \sqrt{5}}{4}\right)^{n} \\[2ex] \dfrac{1}{2}\left(\dfrac{1 + \sqrt{5}}{4}\right)^{n} + \dfrac{1}{2}\left(\dfrac{1 - \sqrt{5}}{4}\right)^{n} \\[2ex] 2\left(\dfrac{1 + \sqrt{5}}{4}\right)^{n+1} + 2\left(\dfrac{1 - \sqrt{5}}{4}\right)^{n+1} \\[2ex] \dfrac{1}{2} - \dfrac{1}{2}(3 + \sqrt{5})\left(\dfrac{1 + \sqrt{5}}{4}\right)^{n+1} - \dfrac{1}{2}(3 - \sqrt{5})\left(\dfrac{1 - \sqrt{5}}{4}\right)^{n+1} \end{bmatrix}$$

Since all exponentiated terms above are less than one in absolute value, we have $\lim\limits_{n\to\infty} \mathbf{x}^{(n)} =$
$\begin{bmatrix} \dfrac{1}{2} & 0 & 0 & 0 & 0 & \dfrac{1}{2} \end{bmatrix}^t.$

7. From (13) we have that the probability that the limiting sibling-pairs will be type (A, AA) is

$$a_0 + \frac{2}{3}b_0 + \frac{1}{3}c_0 + \frac{2}{3}d_0 + e_0 \ .$$

The proportion of A genes in the population at the outset is as follows: all the type (A, AA) genes, 2/3 of the type (A, Aa) genes, 1/3 the type (A, aa) genes, etc. ... yielding

$$a_0 + \frac{2}{3}b_0 + \frac{1}{3}c_0 + \frac{2}{3}d_0 + e_0 \ .$$

8. From an (A, AA) pair we get only (A, AA) pairs and similarly for (a, aa). From either (A, aa) or (a, AA) pairs we must get an Aa female, who will not mature. Thus no offspring will come from such pairs. The transition matrix is then

$$\begin{bmatrix} 1 & 0 & 0 & 0 \\ 0 & 0 & 0 & 0 \\ 0 & 0 & 0 & 0 \\ 0 & 0 & 0 & 1 \end{bmatrix}.$$

9. For the first column of M we realize that parents of type (A, AA) can produce offspring only of that type, and similarly for the last column. The fifth column is like the second column, and follows the analysis in the text. For the middle two columns, say the third, note that male offspring from (A, aa) must be type a, and females are of type Aa, because of the way the genes are inherited.

EXERCISE SET 11.18

1. **(a)** The characteristic polynomial of L is $\lambda^2 - \lambda - 3/4$, so the eigenvalues are $\lambda_1 = 3/2$, $\lambda_2 = -1/2$. The eigenvector corresponding to λ_1 is $\begin{bmatrix} 3 & 1 \end{bmatrix}^t$.

   (b) $\mathbf{x}^{(1)} = L\mathbf{x}^{(0)} = \begin{bmatrix} 100 \\ 50 \end{bmatrix}, \mathbf{x}^{(2)} = \begin{bmatrix} 175 \\ 50 \end{bmatrix}, \mathbf{x}^{(3)} = \begin{bmatrix} 250 \\ 88 \end{bmatrix},$

   $\mathbf{x}^{(4)} = \begin{bmatrix} 382 \\ 125 \end{bmatrix}, \mathbf{x}^{(5)} = \begin{bmatrix} 570 \\ 191 \end{bmatrix}.$

   (c) $\mathbf{x}^{(6)} = L\mathbf{x}^{(5)} = \begin{bmatrix} 857 \\ 285 \end{bmatrix}. \quad \lambda_1\mathbf{x}^{(5)} = \begin{bmatrix} 855 \\ 287 \end{bmatrix}.$

2. For L in Eq. (4), we obtain

$$\lambda I - L = \begin{bmatrix} \lambda - a_1 & -a_2 & -a_3 \cdots -a_{n-1} & -a_n \\ -b_1 & \lambda & 0 \cdots 0 & 0 \\ 0 & -b_2 & \lambda \cdots 0 & 0 \\ \vdots & \vdots & \vdots & \vdots & \vdots \\ 0 & 0 & 0 \cdots -b_{n-1} & \lambda \end{bmatrix}.$$

The determinant of this matrix is obtained most easily using cofactors along the first row. Each minor is a lower-triangular matrix. Thus, the characteristic polynomial of L is

$$(\lambda - a_1)\lambda^{n-1} - a_2 b_1 \lambda^{n-2} - a_3 b_1 b_2 \lambda^{n-3} - \cdots - a_n b_1 b_2 \cdots b_{n-1}.$$

3. From the solution in Exercise 2, we see that the derivative of the characteristic polynomial evaluated at λ_1 is

$$n\lambda_1^{n-1} - (n-1)a_1\lambda_1^{n-2}$$

$$- (n-2)a_2b_1\lambda_1^{n-3} - \cdots - a_{n-1}b_1b_2\cdots b_{n-2}.$$

Recalling that $a_i \geq 0$ and at least one $a_i > 0$, and $b_i > 0$, we see that the above is strictly greater than

$$n\left(\lambda_1^{n-1} - a_1\lambda_1^{n-2} - a_2b_1\lambda_1^{n-3} - \cdots - a_{n-1}b_1\cdots b_{n-2}\right)$$

$$\geq \frac{n}{\lambda_1}a_nb_1\cdots b_{n-1} \geq 0.$$

So the derivative is positive and λ_1 is a simple root.

4. The product

$$\begin{bmatrix} 1 & 0 & 0 & \cdots & 0 \\ 0 & 0 & 0 & \cdots & 0 \\ \vdots & \vdots & \vdots & & \vdots \\ 0 & 0 & 0 & \cdots & 0 \end{bmatrix} P^{-1}\mathbf{x}^{(0)}$$

is $\begin{bmatrix} c & 0 & 0 & \cdots & 0 \end{bmatrix}^t$, where c is the first element of $P^{-1}\mathbf{x}^{(0)}$. But then $P\begin{bmatrix} c & 0 & 0 & \cdots & 0 \end{bmatrix}^t$ returns c times the first column of P which is $c\mathbf{x}_1$, $\mathbf{x}_1$ being the first eigenvector, i.e., thus the first column of P.

5. a_1 is the average number of offspring produced in the first age period. a_2b_1 is the number of offspring produced in the second period times the probability that the female will live into the second period, i.e., it is the expected number of offspring per female during the second period, and so on for all the periods. Thus, the sum of these, which is the net reproduction rate, is the expected number of offspring produced by a given female during her expected lifetime.

6. If the population is eventually decreasing, then $\lambda_1 < 1$ and the characteristic polynomial of L is positive for all $\lambda > \lambda_1$, in particular for $\lambda = 1$. But the polynomial evaluated at 1 (which is 1 minus the net reproduction rate) is positive. Thus the net reproduction rate is less than one. Conversely, if the $R < 1$ we get that the characteristic polynomial at 1 is positive; thus $\lambda_1 < 1$ and the population is eventually decreasing. For $R > 1$, just switch all inequality symbols above.

7. $R = 0 + 4\left(\dfrac{1}{2}\right) + 3\left(\dfrac{1}{2}\right)\left(\dfrac{1}{4}\right) = 19/8.$

8. $R = 0 + (.00024)(.99651) + \cdots + (.00240)(.99651)\cdots(.987)$

$\quad = 1.49611.$

9.   Let λ be any eigenvalue of L, and write $\lambda = re^{i\theta} = r(\cos\theta + i\sin\theta)$. We know $\lambda^n = r^n e^{in\theta}$,

so $1 = q(\lambda) = \dfrac{a_1}{r}(\cos\theta - i\sin\theta) + \cdots + \dfrac{a_n b_1 b_2 \cdots b_{n-1}}{r^n}(\cos n\theta - i\sin n\theta).$

Since $q(\lambda)$ is real, we can ignore all the sine terms in the formula.

Thus

$$1 = q(\lambda) = \frac{a_1}{r}\cos\theta + \frac{a_2 b_1}{r^2}\cos 2\theta + \cdots + \frac{a_n b_1 b_2 \cdots b_{n-1}}{r^2}\cos n\theta$$

$$\leq \frac{a_1}{r} + \frac{a_2 b_1}{r} + \cdots + \frac{a_n b_1 b_2 \cdots b_{n-1}}{r^n} = q(r)$$

i.e., $q(r) \geq 1$. Since q is a decreasing function, we have $r \leq \lambda_1$.

EXERCISE SET 11.19

1. (a) The characteristic polynomial of L is $\lambda^3 - 2\lambda - 3/8 = (\lambda - 3/2)[\lambda^2 + (3/2)\lambda + 1/4]$, so $\lambda_1 = 3/2$. Thus h, the fraction harvested of each age group, is $1 - \dfrac{2}{3} = \dfrac{1}{3}$ from Eq. (6), so the yield is $33\frac{1}{3}\%$ of the population. The eigenvector corresponding to $\lambda_1 = 3/2$ is $\begin{bmatrix} 1 & 1/3 & 1/18 \end{bmatrix}^t$; this is the age distribution vector after each harvest.

   (b) From Eq. (10), the age distribution vector $\mathbf{x}_1$ is $\begin{bmatrix} 1 & 1/2 & 1/8 \end{bmatrix}^t$. Eq. (9) tells us that $h_1 = 1 - 1/(19/8) = 11/19$, so we harvest $11/19$ or 57.9% of the youngest age class. Since $L\mathbf{x}_1 = \begin{bmatrix} 19/8 & 1/2 & 1/8 \end{bmatrix}^t$, the youngest class contains 79.2% of the population. Thus the yield is 57.9% of 79.2%, or 45.8% of the population.

2. The Leslie matrix of Example 1 has $b_1 = .845$, $b_2 = .975$, $b_3 = .965$, etc. This, together with the harvesting data from Eq. (13) and the formula of Eq. (5) yields

$$\mathbf{x}_1 = \begin{bmatrix} 1 & .845 & .824 & .795 & .755 & .699 & .626 & .532 & 0 & 0 & 0 & 0 \end{bmatrix}^t.$$

$$L\mathbf{x}_1 = \begin{bmatrix} 2.090 & .845 & .824 & .795 & .755 & .699 & .626 & .532 & .418 & 0 & 0 & 0 \end{bmatrix}^t.$$

The total of the entries of $L\mathbf{x}_1$ is 7.584. The proportion of sheep harvested is $h_1(L\mathbf{x}_1)_2 + h_9(L\mathbf{x}_1)_9 = 1.51$, or 19.9% of the population.

3. Using the L of Eq. (3) and the $\mathbf{x}_1$ of Eq. (10) we obtain for the first coordinate of $L\mathbf{x}_1$,

$$a_1 + a_2 b_1 + a_3 b_1 b_2 + \cdots + a_n b_1 b_2 b_3 \cdots b_{n-1}.$$

The other coordinates of $L\mathbf{x}_1$ are, respectively, b_1, $b_1 b_2$, $b_1 b_2 b_3, \ldots, b_1 b_2 \cdots b_{n-1}$. Thus,

$$
L\mathbf{x}_1 - \mathbf{x}_1 = \begin{bmatrix} a_1 + a_2 b_1 + a_3 b_1 b_2 + \cdots + a_n b_1 b_2 \cdots b_{n-1} - 1 \\ 0 \\ 0 \\ \vdots \\ 0 \end{bmatrix}
$$

$$
= \begin{bmatrix} R - 1 \\ 0 \\ 0 \\ \vdots \\ 0 \end{bmatrix}.
$$

4. In this situation we have $h_I \neq 0$, and $h_1 = h_2 = \cdots = h_{I-1} = h_{I+1} = \cdots = h_n = 0$. Eq. (4) then takes the form

$$
a_1 + a_2 b_1 + a_3 b_1 b_2 + \cdots a_I b_1 b_2 \cdots b_{I-1}(1 - h_I) + a_{I+1} b_1 b_2 \cdots b_I (1 - h_I)
$$

$$
+ \cdots + a_n b_1 b_2 \cdots b_{n-1}(1 - h_I) = 1.
$$

Factoring out $(1 - h_I)$ we get:

$$
(1 - h_I)[a_I b_1 b_2 \cdots b_{I-1} + a_{I+1} b_1 b_2 \cdots b_I + \cdots + a_n b_1 b_2 \cdots b_{n-1}]
$$

$$
= 1 - a_1 - a_2 b_1 - \cdots - a_{I-1} b_1 b_2 \cdots b_{I-2}.
$$

So

$$h_I = \frac{a_1 + a_2 b_1 + \cdots + a_{I-1} b_1 b_2 \cdots b_{I-2} - 1}{a_I b_1 b_2 \cdots b_{I-1} + \cdots + a_n b_1 b_2 \cdots b_{n-1}} + 1$$

$$= \frac{(a_1 + a_2 b_1 + \cdots + a_{I-1} b_1 b_2 \cdots b_{I-2} - 1 \; + a_I b_1 b_2 \cdots b_{I-1} + \cdots + a_n b_1 b_2 \cdots b_{n-1})}{a_I b_1 b_2 \cdots b_{I-1} + \cdots + a_n b_1 b_2 \cdots b_{n-1}}$$

$$= \frac{R - 1}{a_I b_1 b_2 \cdots b_{I-1} + \cdots + a_n b_1 b_2 \cdots b_{n-1}}$$

5. Here $h_J = 1$, $h_I \neq 0$, and all the other h_k are zero. Then Eq. (4) becomes

$$a_1 + a_2 b_1 + \cdots + a_{I-1} b_1 b_2 \cdots b_{I-2} + (1 - h_I)[a_I b_1 b_2 \cdots b_{I-1} + \cdots + a_{J-1} b_1 b_2 \cdots b_{J-2}] = 1.$$

We solve for h_I to obtain

$$h_I = \frac{a_1 + a_2 b_1 + \cdots + a_{I-1} b_1 b_2 \cdots b_{I-2} - 1}{a_I b_1 b_2 \cdots b_{I-1} + \cdots + a_{J-1} b_1 b_2 \cdots b_{J-2}} + 1$$

$$= \frac{a_1 + a_2 b_1 + \cdots + a_{J-1} b_1 b_2 \cdots b_{J-2} - 1}{a_I b_1 b_2 \cdots b_{I-1} + \cdots + a_{J-1} b_1 b_2 \cdots b_{J-2}}.$$

EXERCISE SET 11.20

1. From Theorem 2, we compute $a_0 = \dfrac{1}{\pi} \displaystyle\int_0^{2\pi} (t-\pi)^2 \, dt = \dfrac{2}{3}\pi^2$, $a_k = \dfrac{1}{\pi} \displaystyle\int_0^{2\pi} (t-\pi)^2 \cos\ kt \, dt = \dfrac{4}{k^2}$, and $b_k = \dfrac{1}{\pi} \displaystyle\int_0^{2\pi} (t-\pi)^2 \sin\ kt \, dt = 0$. So the least-squares trigonometric polynomial is

$$\frac{\pi^2}{3} + 4\ \cos\ t + \cos\ 2t + \frac{4}{9}\cos\ 3t.$$

2. From Theorem 3, we compute $a_0 = \dfrac{2}{T} \displaystyle\int_0^T t^2 \, dt = \dfrac{2}{3}T^2$,

$$a_k = \frac{2}{T} \int_0^T t^2 \cos \frac{2k\pi t}{T} \, dt = \frac{T^2}{k^2 \pi^2}$$

and

$$b_k = \frac{2}{T} \int_0^T t^2 \sin \frac{2k\pi t}{T} \, dt = -\frac{T^2}{k\pi}\ .$$

So the least-squares trigonometric polynomial is

$$\frac{T^2}{3} + \frac{T^2}{\pi^2}\left(\cos \frac{2\pi t}{T} + \frac{1}{4}\cos \frac{4\pi t}{T} + \frac{1}{9}\cos \frac{6\pi t}{T} + \frac{1}{16}\cos \frac{8\pi t}{T} \right)$$

$$-\frac{T^2}{\pi}\left(\sin \frac{2\pi t}{T} + \frac{1}{2}\sin \frac{4\pi t}{T} + \frac{1}{3}\sin \frac{6\pi t}{T} + \frac{1}{4}\sin \frac{8\pi t}{T} \right)\ .$$

3. As in Exercise 1, $a_0 = \dfrac{1}{\pi} \displaystyle\int_0^{2\pi} f(t)\, dt = \dfrac{1}{\pi} \displaystyle\int_0^{\pi} \sin t\, dt = \dfrac{2}{\pi}$

(Note the upper limit on second integral),

$$a_k = \frac{1}{\pi} \int_0^{\pi} \sin t\, \cos kt\, dt$$

$$= \frac{1}{\pi} \left(\frac{1}{k^2 - 1} \left[k\, \sin kt\, \sin t + \cos kt\, \cos t \right] \right) \Big|_0^{\pi}$$

$$= \frac{1}{\pi} \left(\frac{1}{k^2 - 1} \left[0 + (-1)^{k+1} - 1 \right] \right)$$

$$= \begin{cases} 0 & \text{if } k \text{ is odd} \\ -\dfrac{2}{\pi(k^2 - 1)} & \text{if } k \text{ is even.} \end{cases}$$

$$b_k = \frac{1}{\pi} \int_0^{\pi} \sin kt\, \sin t\, dt = \begin{cases} \dfrac{1}{2} & \text{if } k = 1 \\ 0 & \text{if } k > 1 \end{cases}.$$

So the least-squares trigonometric polynomial is

$$\frac{1}{\pi} + \frac{1}{2} \sin t - \frac{2}{3\pi} \cos 2t - \frac{2}{15\pi} \cos 4t.$$

4. As in Exercise 1, $a_0 = \dfrac{1}{\pi} \displaystyle\int_0^{2\pi} \sin \frac{1}{2} t\, dt = \dfrac{4}{\pi}$,

$$a_k = \frac{1}{\pi} \int_0^{2\pi} \sin \frac{1}{2} t\, \cos kt\, dt = -\frac{4}{\pi(4k^2 - 1)}$$

$$= -\frac{4}{\pi(2k - 1)(2k + 1)},$$

$$b_k = \frac{1}{\pi} \int_0^{2\pi} \sin \frac{1}{2} t\, \sin kt\, dt = 0.$$

So the least-squares trigonometric polynomial is

$$\frac{2}{\pi} - \frac{4}{\pi} \left(\frac{\cos t}{1 \cdot 3} + \frac{\cos 2t}{3 \cdot 5} + \frac{\cos 3t}{5 \cdot 7} + \cdots + \frac{\cos nt}{(2n - 1)(2n + 1)} \right).$$

5. As in Exercise 2,

$$a_0 = \frac{2}{T} \int_0^T f(t) \, dt = \frac{2}{T} \int_0^{\frac{1}{2}T} t \, dt + \frac{2}{T} \int_{\frac{1}{2}T}^T (T - t) \, dt = \frac{T}{2}.$$

$$a_k = \frac{2}{T} \int_0^{\frac{1}{2}T} t \, \cos \frac{2k\pi t}{T} \, dt + \frac{2}{T} \int_{\frac{1}{2}T}^T (T - t) \cos \frac{2k\pi t}{T} \, dt$$

$$= \frac{4T}{4k^2\pi^2}((-1)^k - 1) = \begin{cases} 0 & \text{if } k \text{ is even} \\ \dfrac{8T}{(2k)^2\pi^2} & \text{if } k \text{ is odd} \end{cases}.$$

$$b_k = \frac{2}{T} \int_0^{\frac{1}{2}T} t \, \cos \frac{2k\pi t}{T} \, dt + \frac{2}{T} \int_{\frac{1}{2}T}^T (T - t) \cos \frac{2k\pi t}{T} \, dt = 0.$$

So the least-squares trigonometric polynomial is:

$$\frac{T}{4} - \frac{8T}{\pi^2} \left(\frac{1}{2^2} \cos \frac{2\pi t}{T} + \frac{1}{6^2} \cos \frac{6\pi t}{T} + \frac{1}{10^2} \cos \frac{10\pi t}{T} + \cdots + \frac{1}{(2n)^2} \cos \frac{2\pi n t}{T} \right)$$

if n is even; the last term involves $n - 1$ if n is odd.

6. (a) $\|1\| = \sqrt{\int_0^{2\pi} dt} = \sqrt{2\pi}.$

(b) $\| \cos kt \| = \sqrt{\int_0^{2\pi} \cos^2 kt \, dt} = \sqrt{\int_0^{2\pi} \left(\frac{1 + \cos 2t}{2} \right) dt} = \sqrt{\pi}.$

(c) $\| \sin kt \| = \sqrt{\int_0^{2\pi} \sin^2 kt \, dt} = \sqrt{\int_0^{2\pi} \left(\frac{1 - \cos 2t}{2} \right) dt} = \sqrt{\pi}.$

7. There are several cases to consider. First, the function 1 is orthogonal to all the others because

$$\int_0^{2\pi} \cos kt \, dt = \int_0^{2\pi} \sin kt \, dt = 0.$$

Then consider $\int_0^{2\pi} \cos mt \cos nt \, dt = 0$ if $m \neq n$. Similarly $\int_0^{2\pi} \sin mt \sin nt \, dt = 0$ if $m \neq n$. Finally $\int_0^{2\pi} \cos mt \sin nt \, dt = 0$ for all values of m and n.

8. $f\left(\dfrac{T\tau}{2\pi}\right)$ is defined for all τ in the interval $[0, 2\pi]$ because $0 \leq \tau \leq 2\pi$ immediately implies $0 \leq \dfrac{T\tau}{2\pi} \leq T$, and f is defined in the interval $[0, T]$. $f\left(\dfrac{T\tau}{2\pi}\right)$ is a continuous function of τ for τ in the interval $[0, 2\pi]$, since there it is the composition of two continuous functions, namely f and $g(\tau) = \dfrac{T\tau}{2\pi}$. Theorem 2 tells us that the trigonometric polynomial $g(\tau)$ that minimizes

$$\int_0^{2\pi} \left(f\left(\frac{T\tau}{2\pi}\right) - g(\tau) \right)^2 d\tau \text{ has coefficients}$$

$$a_k = \frac{1}{\pi} \int_0^{2\pi} f\left(\frac{T\tau}{2\pi}\right) \cos k\tau \, d\tau \qquad \text{and} \qquad b_k = \frac{1}{\pi} \int_0^{2\pi} f\left(\frac{T\tau}{2\pi}\right) \sin k\tau \, d\tau \ .$$

But then by making the change of variable $t = \dfrac{T\tau}{2\pi}$ we obtain Theorem 3 immediately.

EXERCISE SET 11.21

1. **(a)** Equation (2) is

$$c_1 \begin{bmatrix} 1 \\ 1 \\ 1 \end{bmatrix} + c_2 \begin{bmatrix} 3 \\ 5 \end{bmatrix} + c_3 \begin{bmatrix} 4 \\ 2 \end{bmatrix} = \begin{bmatrix} 3 \\ 3 \end{bmatrix}$$

and Equation (3) is $c_1 + c_2 + c_3 = 1$. These equations can be written in combined matrix form as

$$\begin{bmatrix} 1 & 3 & 4 \\ 1 & 5 & 2 \\ 1 & 1 & 1 \end{bmatrix} \begin{bmatrix} c_1 \\ c_2 \\ c_3 \end{bmatrix} = \begin{bmatrix} 3 \\ 3 \\ 1 \end{bmatrix}.$$

This system has the unique solution $c_1 = 1/5$, $c_2 = 2/5$, and $c_3 = 2/5$. Because these coefficients are all nonnegative, it follows that $\mathbf{v}$ is a convex combination of the vectors $\mathbf{v}_1$, $\mathbf{v}_2$, and $\mathbf{v}_3$.

(b) As in part (a) the system for c_1, c_2 and c_3 is

$$\begin{bmatrix} 1 & 3 & 4 \\ 1 & 5 & 2 \\ 1 & 1 & 1 \end{bmatrix} \begin{bmatrix} c_1 \\ c_2 \\ c_3 \end{bmatrix} = \begin{bmatrix} 2 \\ 4 \\ 1 \end{bmatrix}$$

which has the unique solution $c_1 = 2/5$, $c_2 = 4/5$, and $c_3 = -1/5$. Because one of these coefficients is negative, it follows that $\mathbf{v}$ is not a convex combination of the vectors $\mathbf{v}_1$, $\mathbf{v}_2$, and $\mathbf{v}_3$.

1. **(c)** As in part (a) the system for c_1, c_2 and c_3 is

$$
\begin{bmatrix}
3 & -2 & 3 \\
3 & -2 & 0 \\
1 & 1 & 1
\end{bmatrix}
\begin{bmatrix}
c_1 \\
c_2 \\
c_3
\end{bmatrix}
=
\begin{bmatrix}
0 \\
0 \\
1
\end{bmatrix}
$$

which has the unique solution $c_1 = 2/5$, $c_2 = 3/5$, and $c_3 = 0$. Because these coefficients are all nonnegative, it follows that $\mathbf{v}$ is a convex combination of the vectors $\mathbf{v}_1$, $\mathbf{v}_2$, and $\mathbf{v}_3$.

(d) As in part (a) the system for c_1, c_2 and c_3 is

$$
\begin{bmatrix}
3 & -2 & 3 \\
3 & -2 & 0 \\
1 & 1 & 1
\end{bmatrix}
\begin{bmatrix}
c_1 \\
c_2 \\
c_3
\end{bmatrix}
=
\begin{bmatrix}
1 \\
0 \\
1
\end{bmatrix}
$$

which has the unique solution $c_1 = 4/15$, $c_2 = 6/15$, and $c_3 = 5/15$. Because these coefficients are all nonnegative, it follows that $\mathbf{v}$ is a convex combination of the vectors $\mathbf{v}_1$, $\mathbf{v}_2$, and $\mathbf{v}_3$.

2. For both triangulations the number of triangles, m, is equal to 7; the number of vertex points, n, is equal to 7; and the number of boundary vertex points, k, is equal to 5. Equation (7), $m = 2n - 2 - k$, becomes $7 = 2(7) - 2 - 5$, or $7 = 7$.

3. Combining everything that is given in the statement of the problem, we obtain:

$$
\mathbf{w} = M\mathbf{v} + \mathbf{b} = M(c_1\mathbf{v}_1 + c_2\mathbf{v}_2 + c_3\mathbf{v}_3) + (c_1 + c_2 + c_3)\mathbf{b}
$$
$$
= c_1(M\mathbf{v}_1 + \mathbf{b}) + c_2(M\mathbf{v}_2 + \mathbf{b}) + c_3(M\mathbf{v}_3 + \mathbf{b}) = c_1\mathbf{w}_1 + c_2\mathbf{w}_2 + c_3\mathbf{w}_3.
$$

4. A triangulation of the points in Figure 11.21.3 in which the points $\mathbf{v}_3$, $\mathbf{v}_5$, and $\mathbf{v}_6$ form the vertices of a single triangle is given in diagram (a), and a triangulation in which the points $\mathbf{v}_2$, $\mathbf{v}_5$, and $\mathbf{v}_7$ do not form the vertices of a single triangle is given in diagram (b):

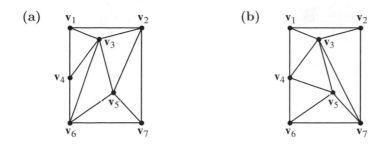

5.   (a) Let $M = \begin{bmatrix} m_{11} & m_{12} \\ m_{21} & m_{22} \end{bmatrix}$ and $\mathbf{b} = \begin{bmatrix} b_1 \\ b_2 \end{bmatrix}$. Then the three matrix equations $M\mathbf{v}_i + \mathbf{b} = \mathbf{w}_i$, $i = 1, 2, 3$, can be written as the six scalar equations

$$m_{11} + m_{12} + b_1 = 4$$
$$m_{21} + m_{22} + b_2 = 3$$

$$2m_{11} + 3m_{12} + b_1 = 9$$
$$2m_{21} + 3m_{22} + b_2 = 5$$

$$2m_{11} + m_{12} + b_1 = 5$$
$$2m_{21} + m_{22} + b_2 = 3$$

The first, third, and fifth equations can be written in matrix form as

$$\begin{bmatrix} 1 & 1 & 1 \\ 2 & 3 & 1 \\ 2 & 1 & 1 \end{bmatrix} \begin{bmatrix} m_{11} \\ m_{12} \\ b_1 \end{bmatrix} = \begin{bmatrix} 4 \\ 9 \\ 5 \end{bmatrix}$$

and the second, fourth, and sixth equations as

$$\begin{bmatrix} 1 & 1 & 1 \\ 2 & 3 & 1 \\ 2 & 1 & 1 \end{bmatrix} \begin{bmatrix} m_{21} \\ m_{22} \\ b_2 \end{bmatrix} = \begin{bmatrix} 3 \\ 5 \\ 3 \end{bmatrix}.$$

The first system has the solution $m_{11} = 1$, $m_{12} = 2$, $b_1 = 1$ and the second system has the solution $m_{21} = 0$, $m_{22} = 1$, $b_2 = 2$. Thus we obtain $M = \begin{bmatrix} 1 & 2 \\ 0 & 1 \end{bmatrix}$ and $\mathbf{b} = \begin{bmatrix} 1 \\ 2 \end{bmatrix}$.

(b) As in part (a), we are led to the following two linear systems:

$$\begin{bmatrix} -2 & 2 & 1 \\ 0 & 0 & 1 \\ 2 & 1 & 1 \end{bmatrix} \begin{bmatrix} m_{11} \\ m_{12} \\ b_1 \end{bmatrix} = \begin{bmatrix} -8 \\ 0 \\ 5 \end{bmatrix} \quad \text{and} \quad \begin{bmatrix} -2 & 2 & 1 \\ 0 & 0 & 1 \\ 2 & 1 & 1 \end{bmatrix} \begin{bmatrix} m_{21} \\ m_{22} \\ b_2 \end{bmatrix} = \begin{bmatrix} 1 \\ 1 \\ 4 \end{bmatrix}.$$

Solving these two linear systems leads to $M = \begin{bmatrix} 3 & -1 \\ 1 & 1 \end{bmatrix}$ and $\mathbf{b} = \begin{bmatrix} 0 \\ 1 \end{bmatrix}$.

(c) As in part (a), we are led to the following two linear systems:

$$\begin{bmatrix} -2 & 1 & 1 \\ 3 & 5 & 1 \\ 1 & 0 & 1 \end{bmatrix} \begin{bmatrix} m_{11} \\ m_{12} \\ b_1 \end{bmatrix} = \begin{bmatrix} 0 \\ 5 \\ 3 \end{bmatrix}$$

and

$$\begin{bmatrix} -2 & 1 & 1 \\ 3 & 5 & 1 \\ 1 & 0 & 1 \end{bmatrix} \begin{bmatrix} m_{21} \\ m_{22} \\ b_2 \end{bmatrix} = \begin{bmatrix} -2 \\ 2 \\ -3 \end{bmatrix}.$$

Solving these two linear systems leads to $M = \begin{bmatrix} 1 & 0 \\ 0 & 1 \end{bmatrix}$ and $\mathbf{b} = \begin{bmatrix} 2 \\ -3 \end{bmatrix}$.

(d) As in part (a), we are led to the following two linear systems:

$$\begin{bmatrix} 0 & 2 & 1 \\ 2 & 2 & 1 \\ -4 & -2 & 1 \end{bmatrix} \begin{bmatrix} m_{11} \\ m_{12} \\ b_1 \end{bmatrix} = \begin{bmatrix} 5/2 \\ 7/2 \\ -7/2 \end{bmatrix}$$

and

$$
\begin{bmatrix} 0 & 2 & 1 \\ 2 & 2 & 1 \\ -4 & -2 & 1 \end{bmatrix} \begin{bmatrix} m_{21} \\ m_{22} \\ b_2 \end{bmatrix} = \begin{bmatrix} -1 \\ 3 \\ -9 \end{bmatrix}.
$$

Solving these two linear systems leads to $M = \begin{bmatrix} 1/2 & 1 \\ 2 & 0 \end{bmatrix}$ and $\mathbf{b} = \begin{bmatrix} 1/2 \\ -1 \end{bmatrix}$.

6. **(a)** From Section 3.1, a point $\mathbf{v}$ lies on the line segment connecting $\mathbf{a}$ and $\mathbf{b}$ if $\mathbf{v} = \mathbf{a} + t(\mathbf{b} - \mathbf{a})$ for some t with $0 \le t \le 1$. This can be written as $\mathbf{v} = (1 - t)\mathbf{a} + t\mathbf{b}$ or $\mathbf{v} = c_1\mathbf{a} + c_2\mathbf{b}$ where $c_1 = 1 - t$ and $c_2 = t$. We note that c_1 and c_2 are nonnegative numbers such that $c_1 + c_2 = 1$.

   (b) If $c_1 + c_2 = 0$, then $c_1 = c_2 = 0$ and so $c_1\mathbf{a} + c_2\mathbf{b} = \mathbf{0}$ and we are finished. If $c_1 + c_2 > 0$, set $d_1 = c_1/(c_1 + c_2)$ and $d_2 = c_2/(c_1 + c_2)$. Then d_1 and d_2 both lie in the interval $[0, 1]$ and $d_1 + d_2 = 1$. Consequently, the vector $\mathbf{w} = d_1\mathbf{a} + d_2\mathbf{b}$ lies on the line segment connecting the vectors $\mathbf{a}$ and $\mathbf{b}$. The vector $(c_1 + c_2)\mathbf{w}$ then lies on the line segment connecting the origin and the vector $\mathbf{w}$ since $0 < c_1 + c_2 \le 1$. Because this line segment is contained in the triangle connecting the origin and the tips of the vectors $\mathbf{a}$ and $\mathbf{b}$, it follows that the vector $(c_1 + c_2)\mathbf{w}$, or $c_1\mathbf{a} + c_2\mathbf{b}$, is also contained in that triangle.

   (c) Because c_1, c_2, and c_3 are nonnegative numbers such that $c_1 + c_2 + c_3 = 1$, it follows that $0 \le c_1 + c_2 \le 1$. As per the hint, set $\mathbf{a} = \mathbf{v}_1 - \mathbf{v}_3$ and $\mathbf{b} = \mathbf{v}_2 - \mathbf{v}_3$. Then from part (b) the vector $\mathbf{w} = c_1\mathbf{a} + c_2\mathbf{b}$ lies in the triangle connecting the origin and the tips of the vectors $\mathbf{a}$ and $\mathbf{b}$. The vector $\mathbf{w} + \mathbf{v}_3$ then lies on the displaced triangle connecting the vectors $\mathbf{v}_3$, $\mathbf{a} + \mathbf{v}_3 (= \mathbf{v}_1)$, and $\mathbf{b} + \mathbf{v}_3 (= \mathbf{v}_2)$. Finally,

$$
\mathbf{w} + \mathbf{v}_3 = c_1\mathbf{a} + c_2\mathbf{b} + \mathbf{v}_3 = c_1(\mathbf{v}_1 - \mathbf{v}_3) + c_2(\mathbf{v}_2 - \mathbf{v}_3) + \mathbf{v}_3
$$

$$
= c_1\mathbf{v}_1 + c_2\mathbf{v}_2 + (1 - c_1 - c_2)\mathbf{v}_3 = c_1\mathbf{v}_1 + c_2\mathbf{v}_2 + c_3\mathbf{v}_3.
$$

7. **(a)** The vertices $\mathbf{v}_1$, $\mathbf{v}_2$, and $\mathbf{v}_3$ of a triangle can be written as the convex combinations $\mathbf{v}_1 = (1)\mathbf{v}_1 + (0)\mathbf{v}_2 + (0)\mathbf{v}_3$, $\mathbf{v}_2 = (0)\mathbf{v}_1 + (1)\mathbf{v}_2 + (0)\mathbf{v}_3$, and $\mathbf{v}_3 = (0)\mathbf{v}_1 + (0)\mathbf{v}_2 + (1)\mathbf{v}_3$. In each of these cases, precisely two of the coefficients are zero and one coefficient is one.

   (b) If, for example, $\mathbf{v}$ lies on the side of the triangle determined by the vectors $\mathbf{v}_1$ and $\mathbf{v}_2$ then from Exercise 6(a) we must have that $\mathbf{v} = c_1\mathbf{v}_1 + c_2\mathbf{v}_2 + (0)\mathbf{v}_3$ where $c_1 + c_2 = 1$. Thus at least one of the coefficients, in this example c_3, must equal zero.

(c) From part (b), if at least one of the coefficients in the convex combination is zero, then the vector must lie on one of the sides of the triangle. Consequently, none of the coefficients can be zero if the vector lies in the interior of the triangle.

8. Consider the vertex $\mathbf{v}_1$ of the triangle and its opposite side determined by the vectors $\mathbf{v}_2$ and $\mathbf{v}_3$. The midpoint $\mathbf{v}_m$ of this opposite side is $(\mathbf{v}_2 + \mathbf{v}_3)/2$ and the point on the line segment from $\mathbf{v}_1$ to $\mathbf{v}_m$ that is two-thirds of the distance to $\mathbf{v}_m$ is given by

$$\frac{1}{3}\mathbf{v}_1 + \frac{2}{3}\mathbf{v}_m = \frac{1}{3}\mathbf{v}_1 + \frac{2}{3}\left(\frac{\mathbf{v}_2 + \mathbf{v}_3}{2}\right) = \frac{1}{3}\mathbf{v}_1 + \frac{1}{3}\mathbf{v}_2 + \frac{1}{3}\mathbf{v}_3.$$